TOXIC IMMANENCE

TOXIC IMMANENCE

Decolonizing Nuclear Legacies and Futures

Edited by **LIVIA MONNET**

McGILL-QUEEN'S UNIVERSITY PRESS
Montreal & Kingston • London • Chicago

ISBN 978-0-2280-1136-1 (cloth)
ISBN 978-0-2280-1326-6 (ePDF)

Legal deposit third quarter 2022
Bibliothèque nationale du Québec

Printed in Canada on acid-free paper that is 100% ancient forest free
(100% post-consumer recycled), processed chlorine free

Funded by the Government of Canada | Financé par le gouvernement du Canada | Canada

Canada Council for the Arts | Conseil des arts du Canada

We acknowledge the support of the Canada Council for the Arts.
Nous remercions le Conseil des arts du Canada de son soutien.

Library and Archives Canada Cataloguing in Publication

Title: Toxic immanence : decolonizing nuclear legacies and futures / edited by Livia Monnet.
Names: Monnet, Livia, editor.
Description: Includes bibliographical references and index.
Identifiers: Canadiana (print) 20220211825 | Canadiana (ebook) 2022021185X | ISBN 9780228011361 (cloth) | ISBN 9780228013266 (ePDF)
Subjects: LCSH: Nuclear weapons. | LCSH: Nuclear accidents—Social aspects. | LCSH: Nuclear energy—Social aspects. | LCSH: Geopolitics. | LCSH: Radioactive wastes. | LCSH: Art and nuclear warfare.
Classification: LCC U263 .T69 2022 | DDC 355.02/17—dc23

Contents

TWO

The Cold War and Post–Cold War Nuclear State and Its Geopolitics: Imaginaries and Contestations

THREE

Archaeologies and Heritages

FOUR

Nuclear Aesthetics: Contemporary Art, Nuclear Colonialism, and the Transformation of Life and the Environment

FIVE

Artists' Contributions

Table and Figures

Table

Figures

Foreword

The Atomic Now

MAGDALENA E. STAWKOWSKI

Many people may remember where they were when news reached them of the 2011 nuclear disaster at the Fukushima Daiichi Nuclear Power Plant in Japan. I was in Kazakhstan when it occurred, deep into my own ethnographic fieldwork about radiation and environmental change. The village where I lived sits on the unmarked border of the Semipalatinsk Nuclear Test Site. It was March and crisp snow still covered the ground. This is where the Soviets tested more than 450 nuclear weapons, 113 above ground that appeared as fiery mushroom clouds in the sky. The sum of the energy released by Soviet testing in Kazakhstan was equivalent to 2,500 Hiroshima bombs. For forty years, the land in and around the Kazakh village of Koyan had been poisoned with layers of radioactive dust dispersed by the winds with each successive blast. People got sick; many died from cancer or other ailments caused by radioactive fallout. From my window, I understood, as many there did, that radiation knows no borders. It is a global problem and a local one, both a universal experience and a confined form of knowledge bound up with heterogenous parts that make up what we have come to call the universalizing epoch of the Anthropocene. Fukushima reminded us that we don't live in a post-nuclear world in which there is a neat division between then and now – we are all, to one degree or another, nuclear beings, and each and every one of us is carrying trace amounts of radioactive elements from nuclear fallout.

This collection, *Toxic Immanence*, brings together something that is utterly human in its diversity of texts, media, poetry, and prose. Collectively, these contributions symbolize and strive defiantly to give voice to and create a space for the places and people that have come to exist in our nuclear age. Its task is rooted in an uphill struggle to lay the groundwork for a "decolonizing pedagogy of the nuclear," to claim a place for alternative and decolonized understandings of atomic histories and atomic presents, in order to bring about the categorical abolition of nuclear technologies in all their forms. It is a call to action that promotes social and environmental justice, rooted in a critical-theoretical and methodological heteroglossia to advocate for the mutual care of all life and non-life on Earth.

This edited volume takes into account how life has been lived since humans went nuclear; how people have *become with* and are *becoming with* nuclear technologies and energy; and how, by decoupling from the hegemonic Eurocentric models of industrial modernity, we can critically account for its lethality.

In the fall of 2020 I debuted a class at the University of South Carolina, one that I had long yearned to teach, titled "Humans Going Nuclear: Atomic Bombs, Cold War, and the Fallout." Most of my students knew very little about nuclear weapons, their lives having begun after the Cold War ended. Most were too young to even remember Fukushima. As many of them noted during the semester, there is a palpable lack of opportunity for them to learn about nuclear weapons, why they have been used, why they are tested so frequently, to learn about the materials in them and where the materials have come from, or to learn about the links between nuclear power and nuclear weapons. Most were upset that they knew so little about the nuclear age. I was upset in solidarity with them and wondered what erasures and cooptions must have happened to silence the voices of those most impacted by the nuclear colonization of their lands and bodies. After all, the university is only fifty miles from the Department of Energy Savannah River Site that produced tritium and plutonium for nuclear bombs, and ten miles from the Westinghouse factory that manufactures fuel for nuclear power plants. The two facilities have left a deadly radioactive legacy in their wake that has impacted the environment, the ecology, and human health in disproportionally minority and low-income areas of the state. The students' excitement made it all too clear to me that it is vital to keep conversations about nuclear legacies going, lest young students, and the rest of us for that matter, think the atomic age has long passed.

Toxic Immanence shows how, since the dawn of the Cold War, nuclear technologies have been legitimated and normalized for reasons of national security, and later as a cleaner and greener alternative to fossil fuels. This has meant that, to no small degree, measures have been enacted to safeguard against alternative perceptions of the ingredients of nuclear energy. That is to say – as the Fukushima disaster has shown most recently – tremendous effort has been invested in making people in the world trust the science and technology and governance to provide answers and solutions to what are framed as exceptions to the rule in an otherwise very well-working field of science and energy production. Of course, there is much more to shaping and controlling the meanings of "the nuclear" when it comes to sensibly and openly speaking about environmental and human health legacies. From my own research in Koyan, I have learned that with the Western science that we have on hand it is nearly impossible to suggest, let alone conclusively prove, for example, that exposure to radiation at low doses, such as those found near nuclear power plants, poses any determinable effects on people.

As of 27 January 2021, the *Bulletin of the Atomic Scientists Doomsday Clock* is 100 seconds to midnight, closer than ever to a global catastrophe that could permanently extinguish all life on Earth. We are living at a time of an unprecedented pandemic that has, as I write this, killed more than 5 million people worldwide and infected many more. We are also facing a climate emergency with the extraordinary destruction of the environment and loss of biodiversity. Catastrophic wildfires and scorching heat in the American West, Siberia, and Australia, and record floods in Europe, are our new normal. At the same time, we are confronted with a renewed nuclear arms race between Russia and the United States, which have terminated the Intermediate-Range Nuclear Forces Treaty. Meanwhile, China, North Korea, India, and Pakistan are expanding their nuclear arsenals, and Iran is seeking to develop its own nuclear arsenal. The actions we take now will determine what our future will be.

For the reader, *Toxic Immanence* poses a difficult question: how can we talk effectively and inclusively about the relationships between people, their environments, and the changes to the fabric of their lives as a result of nuclear energy? How has nuclear colonialism invaded our lives, our languages, our landscapes, our institutions, our imaginations? What and how have we learned? This edited volume is an excellent place to start exploring these questions.

Acknowledgments

This book has been a long time in the making. I would like to thank all authors in the volume for their inspiring contributions as well as for their patience, support, and understanding. I am indebted to Peter C. van Wyck for his constructive feedback in the initial phase of conception of this volume. My thanks go also to the University of Montreal's Vice-rectorat à la recherche (Vice-Rectorate of Research), which provided funding for the 2015 conference "Rethinking Radiation Ecologies" – the launchpad for some of the ideas that went into the conception and production of this volume. Many thanks also to Magda, Jessica, Dan, Karena, Ruby, and Kyveli, who provided thoughtful and important reflections on this volume's contributions. I owe a debt of gratitude to the graduate students and research assistants who offered much-needed help for part of the research related to this book, and whose know-how made the thankless labour of formatting and editing *Toxic Immanence* less onerous: Flora Roussel, Gabriel Rémy-Handfield, Adina Blanariu, Alexandra Kakon, and Mathieu Li-Goyette. I am also deeply indebted to Jonathan Crago and Elli Stylianou at McGill-Queen's University Press, whose patience, support, and professional advice during the production of this volume have been exemplary. Many thanks also to Matthew Kudelka for his superlative editing. Finally, a debt of thanks is due to the Social Sciences and Humanities Research Council of Canada (SSHRC) for an Insight Grant that supported part of the production of this edited collection.

TOXIC
IMMANENCE

Introduction

Toxic Immanence: Toward Decolonizing Pedagogies of the Nuclear

LIVIA MONNET

The Nuclear in the Age of Climate Change and COVID-19

Let me first provide some context for the present edited collection. In March 2015 I organized a conference titled "Rethinking Radiation Ecologies" at the University of Montreal.[1] The impetus for the conference was the Fukushima Daiichi nuclear disaster in Tohoku, Japan. Caused by the Great East Japan Earthquake (Higashi Nihon daishinsai) of 11 March 2011, the meltdowns at the Fukushima Daiichi Nuclear Power Plant represent the most severe nuclear accident in Japan's postwar history and twenty-first century's most serious nuclear disaster so far.[2] The intense discussions that took place during and after the conference led to a call for papers for an edited collection, which in turn laid the foundation for the present volume. Animating both the conference and the call for an edited collection was a desire to test the conceptual framework for an emerging field that could be called nuclear (environmental) humanities or nuclear energy humanities.[3]

At the time the discussions that led to this volume took place, the call for an interdisciplinary field focusing on nuclear technologies and nuclear energy from a humanistic and social sciences perspective seemed both timely and necessary: the Fukushima nuclear disaster had once again revealed with searing clarity the enormous risks to the environment and entire populations entailed in nuclear power production, yet nuclear issues occupied a relatively minor position in the environmental and energy humanities. The field of nuclear studies seemed to be operating in relative independence from environmentally oriented research domains, and only rarely did scholars working with the tools and resources of nuclear criticism have recourse to the insights and perspectives of researchers focusing on ecocriticism, and vice versa.

Today the situation is different. Notwithstanding a series of commemorative events and a significant number of new publications occasioned by the tenth anniversary of the Fukushima nuclear disaster, the trauma and ongoing legacies of that disaster have been effectively erased from public consciousness; only rarely do they generate headlines in the media.[4] At the same time, nuclear energy in all its forms and expressions is now a visible feature in edited collections on environmental humanities and energy humanities. The recent work of scholars such as Joseph Masco, Molly Wallace, Jessica Hurley, Rebecca Hogue, Anais Maurer, Hsinya Huang, and Daniel Cordle – work that brings nuclear criticism, risk criticism, ecocriticism, environmental humanities, and energy humanities in conversation with anti- and post-colonial and decolonial theory as well as with critical Indigenous studies, critical science studies, and critical military studies, among other interdisciplinary fields – has shown not only that nuclear critique and nuclear environmental justice and activism are as essential and vital as ever, but also that nuclear theory and analytics – indeed, approaches to and studies of the nuclear age in general – have to be decolonized.[5]

While the *Toxic Immanence* collection was taking shape, and while scientists continued to debate the starting point and the stratigraphic markers of the current epoch of the Anthropocene, the disasters and catastrophes brought about by climate change and global neoliberal capitalism's relentless despoiling of the planet cascaded in frightening succession: ever more powerful cyclones and hurricanes; increasingly destructive and widespread bushfires in Australia – at the beginning of 2020 the fires burned through 10 million hectares and killed an estimated billion native animals; severe droughts in the Horn of Africa that wiped out crops and livestock and forced millions of people into migration as they faced acute water and food shortages; devastating floods and landslides in India, Nepal, and Bangladesh that have destroyed the homes and livelihoods of a vast population, with 12 million people forced to leave their homes in 2019 alone; Central America's Dry Corridor entering its sixth year of drought caused by the climate emergency's impact on El Niño; 1.6 million hectares burned by wildfires in California in 2020, with a fivefold increase in the total burned area between 1978 and 2018; severe, record-breaking heat waves in western Canada, the western United States, and the Pacific Northwest in June and July 2021. This list of climate-change–related disasters and hazards could be substantially expanded.[6]

As I finish revising this introduction, the SARS-COV-2 (COVID-19) pandemic continues to wreak havoc around the world. The pandemic has compounded the already dire effects of anthropogenic/capitalogenic climate change, industrial toxicity, food insecurity, and global social and economic inequality. It has also provoked an economic crisis of a severity not seen since Great Depression of the 1930s. At the time of this writing, more than 380 million confirmed cases of COVID-19 and more than 5.6 million deaths have been reported globally. The United States leads the global case count and death toll, followed closely by India and Brazil. The pandemic has stretched the capacity of health care systems around the world and severely disrupted the informal economy that employs a large portion of the world's people. It has also exposed the rich industrialized nations' reliance on natural resources and cheap labour from the Global South, as well as global capitalism's unpreparedness for the pandemic and indifference to the suffering of millions of people caused by the lockdowns and other restrictions governments have declared to stop the spread of the virus. The coronavirus pandemic has also shed new light on the links between the anthropogenic/capitalogenic climate emergency, biodiversity loss, increased military spending, and the development of new weapons technologies both in nuclear states and in emerging economies, the growing assertiveness and power of nationalist right-wing populisms and fascisms, and the rising violence and political instability around the world.[7]

The SARS-COV-2 pandemic and the recent disasters unleashed by the climate emergency have unfolded against the backdrop of yet another looming global crisis: the threat of a renewed nuclear arms race and of an increasingly unstable and dangerous global nuclear order. Under the Trump administration the United States withdrew from the US–Russia Intermediate-Range Nuclear Forces Treaty (INF). The Trump administration's ambivalence, even hostility, toward arms control and nuclear non-proliferation treaties went in tandem with a rolling back of about one hundred environmental-protection- and climate-change-related laws and rules, including those introduced by the preceding Obama administration, and with the increasing policing and criminalization of environmental justice and anti-racism movements. Other populist authoritarian governments, such as those of Brazil and the Philippines, are adopting similar policies. While the current Biden administration has restored many of the environmental protection regulations that had been lifted by the preceding Trump administration, it has reiterated its commitment to the nuclear modernization plans proposed by previous administrations. The Japanese government is pushing ahead with its plan to dump the highly radioactive water (officially designated as "treated water") used for cooling the damaged reactors at the Fukushima Daiichi nuclear power plant into the Pacific Ocean – a plan that has provoked strong criticism both in Japan and internationally. Meanwhile Russia, China, France, the US, Japan, and South Korea have been competing to sell their nuclear technologies and reactors to countries in the Middle East, South, Central, and Southeast Asia, Africa, and Latin America.[8]

Considering these recent developments, and in view of the entanglement of anthropogenic/capitalogenic climate change and the changing landscape of nuclear geopolitics, I argue that an urgent task for scholars working on the history, technologies, culture, and global impact of the nuclear age is to articulate *decolonizing pedagogies of the nuclear.* Such pedagogies would ensure that nuclear energy is a visible and constant concern within the environmental and energy humanities. These decolonizing practices could also contribute to the re-envisioning and repurposing of nuclear criticism and theory in the contemporary period.[9]

However, readers of this volume should not expect to find a full-blown decolonial theory of the nuclear or a fully elaborated decolonizing heuristics of the nuclear age. Rather, what they will encounter within and across the chapters of *Toxic Immanence* is an emergent thinking about and engagement with what I call decolonizing pedagogies of the nuclear. In some contributions this engagement translates into an explicit articulation of decolonizing approaches, methodologies, and praxes.

What is a decolonizing pedagogy of the nuclear, and what does its theory and practice involve? Before proposing a working definition of such a pedagogy and identifying some of its goals as I envision them, let me clarify three points or ideas that may be taken as premises for any theory and praxis of decolonizing pedagogies of the nuclear. They are as follows:

First, nuclear science and technology, like many other modern sciences and technologies, was originally developed by European imperial and colonial powers (and subsequently in the US). The development and expansion of nuclear technologies depended on the appropriation and exploitation of resources in outlying or internal colonies. The development of the atomic bomb by the Manhattan Project is a textbook case of such colonial relations, in that much of the uranium the project required was initially delivered by colonies such as the Belgian Congo. The uranium used to produce nuclear weapons and nuclear energy has been mined on Navajo Nation land in the US and on First Nations lands in Canada, as well as in Kazakhstan and in several African countries – territories that were colonized either by settler nations such as the US, Canada, and Australia or by the Soviet Union and European colonial powers. Nuclear technoscience has in many ways maintained a top-down, anti-democratic, (settler-)colonial power structure; it continued to promote a colonial discourse and practice even after the achievement of independence by former colonies in the twentieth century. *Nuclear science and technologies were always and continue largely to be colonial and colonizing practices.*[10]

Second, while the discourse and practices of nuclear colonialism and nuclear imperialism(s) – defined respectively as "a system of domination through which governments and corporations target Indigenous peoples and their lands [as well as ethnic minority, immigrant, downwinder, and poor communities and populations – my addition] to maintain the nuclear production process," and as "the state-sponsored, systemic mode of oppression in current or former sites of empire through any use of the nuclear complex"[11] – have often been debated, investigated, and vigorously opposed,[12] there has been little discussion about the *coloniality of the nuclear.* As theorized by Anibal Quijano and further developed by decolonial thinkers in Latin America and the United States, the concept of the *coloniality of power* designates the racialized, hierarchized, gendered systems and practices of governance, knowledge, and cultural production by means of which European colonial empires maintained their colonial domination first in the Americas and later in Asia, Africa, and Oceania. This concept also refers to the institutionalized, internalized, and normalized legacies of this colonial matrix of power in the former colonies even after they became independent states.[13] The *coloniality of the nuclear* may be defined as a complex system of entangled power relations, epistemologies, technoscientific practices, regulatory or advisory agencies and organizations, psychosocial and affective conditioning, and cultural representations, through which all nuclear technologies have been legitimated, normalized, and institutionalized in modern nuclear states and around the world. As a system of socio-environmental governance and geopolitical, (neo) colonial hegemony, the coloniality of the nuclear has sustained national nuclear regimes and the global nuclear order and perpetuated the colonization of life, the biosphere, and the planet. These regimes of nuclear governmentality have been top-down, militarized, racialized or racist, and colonially inflected ever since they emerged in the wake of the atomic bombing of Hiroshima and Nagasaki.

Third, closely related to the concept of nuclear regimes and nuclear technoscience as colonizing offshoots of a Eurocentric industrial-colonial modernity, and to the concept of the coloniality of the nuclear, is the notion of *decolonizing the nuclear.* As I envision it, this notion refers to any practice, theory, aesthetic, or cultural production that contests, resists, subverts, or deconstructs the hegemony, legitimacy, violence, and coloniality of the nuclear episteme and its biopolitics (or, as some critics have described it, necropolitics[14]). *Decolonizing the nuclear* is not shorthand for anti-nuclear thought or activism; rather, it refers to the difficult, risky task of uncovering, mapping, and critically documenting the histories, archives, and imaginaries both of nuclear technoscience and trans/national nuclear complexes and of the political and economic regimes that have built and sustained these assemblages. At the same time, it designates a critical mapping of and accounting for the way nuclear technologies and complexes have colonized and transformed not only bodies and territories but also the very materials and entanglements that constitute the living *and* the non-living. And it refers to the paradoxical, impossible task of disentangling the dizzying half-lives of radioactive matter and waste as well as the (im)materiality of ionizing radiation from contaminated non-human and geologic material existences and temporalities. It further designates a multitask practice of immanent critique: a critical highlighting of the *practical impossibility (an impossibility that may be a possibility or an alternative in the realm of the creative-speculative arts) of nuclear decolonization.* The sobering fact is that a thoroughly denuclearized and decontaminated sovereignty, or a complete existential independence or separation from the nuclear, is unattainable. This, clearly and relentlessly, exposes our complicity with, and willing amnesia and complacency about, the colonial nuclear episteme's instituting of local/national sacrifice zones as well as of a planetary ecology of radioactive contamination. *Decolonizing the nuclear, then, describes the immanent, reflexive critical activity of documenting and providing evidence of the irrationality, coloniality, (slow) violence, and lethality of the nuclear episteme and of nuclear complexes and infrastructures with the aim of bringing about their dismantling and final abolition.*[15]

Toward Decolonizing Pedagogies of the Nuclear

Expanding on these observations, and building on recent developments in the growing field of decolonizing pedagogies/education, I define a *decolonizing pedagogy of the nuclear as a critical, situated dual activity – that is, both a pedagogy or educational theory and praxis and a more general ecology of practices – that aims to achieve two main goals: first, the decolonizing of the nuclear; and second, total and unconditional nuclear abolition.*[16] Following Isabelle Stengers, an ecology of practices may be defined as a set of tools for thinking that "address[es] and actualize[s] … the power of a situation to make us think" (where thinking does not provide an ideal horizon, such as Truth or Rationality, and cannot be disentangled from the environment that produces its ethos).[17] A decolonizing pedagogy of the nuclear may similarly be envisioned as an immanent ecology of practices or a set of tools for critically and reflexively (re)thinking, stimulating practices of learning and unlearning about, and actualizing or redirecting and transforming, the toxic immanence of the nuclear in the contaminated presents and futures of the (post-)Anthropocene.

To decolonize effectively and ethically, and to pave the way for the refusal, prohibition, and abolition of transnational nuclear complexes and for the profound transformation of the socio-political regimes that have supported them, decolonizing pedagogies of the nuclear must insist that "colonial domination and its ideological frameworks operate and are reproduced in and through" education and the social organization of learning practices. That is, an anti- and decolonial and decolonizing educational theory and praxis of the nuclear needs to confront and work to transform the mutually constitutive forms of violence characterizing the internal (neo)colonialism of settler-colonial states such as the United States, as well as the neocolonial condition imposed by (neoliberal) global capitalism.[18] Such pedagogies aim to develop a critical decolonizing consciousness (of the nuclear, but also of the global capitalist system and its networks and apparatuses of domination, extraction, and biopolitical/necropolitical and geopolitical governance). Carlos Tejeda, Manuel Espinoza, and Kris Gutierrez have argued that the development of a decolonizing consciousness in educational, learning, and awareness-raising practices must be based on anti-colonial, post/anti-capitalist, anti-racist, anti-sexist, and anti-homophobic values. Decolonizing pedagogies must also be informed "by a theoretical heteroglossia that strategically utilizes theorizations and understandings from various fields and conceptual frameworks to unmask the logics, workings and effects of internal colonial domination, oppression and exploitation" in settler-colonial contexts such as that of the United States. They also insist that such a theoretical heteroglossia will enable scholars, educators, and activists to work toward establishing new social relations and systems of activity that will lay the groundwork for an instantiation of social justice defined by the perspectives of Indigenous and other colonized peoples.[19]

Expanding on the reflections of Tejeda and Espinoza, as well as on the work of Linda Tuhiwai Smith, Cash Ahenakew, Vanessa de Oliveira Andreotti, and other theorists of decolonizing methodologies and pedagogies in research and education, I argue that decolonizing pedagogies of the nuclear should centre, listen respectfully to, and work with the voices, storytelling, "tools for thinking," and traditional knowledges, cosmovisions, and epistemologies of Indigenous and Aboriginal peoples. Such pedagogies should also rely on the living experience and citizens' science of poor, black, brown, downwinder, and other populations that have been living with the harmful and disruptive legacies of uranium mining and nuclear weapons and energy production for decades. These pedagogies should also acknowledge that even a decolonizing ecology of practices sometimes cannot avoid articulating its questions and organizing its activism and sense of justice from within the grammar of (neo)colonial modernity. This latter conundrum may result, as Andreotti and colleagues have suggested, in a re-enacting, however unwitting and partial, of the onto-epistemological presuppositions that have produced modernity's injustices, violence, and unsustainability.[20] At the same time, a decolonizing pedagogy of the nuclear has the ethico-political and educational obligation to "construct the conditions for a different kind of encounter, an encounter that both opposes ongoing colonization and that seeks to heal the social, cultural, and spiritual ravages of (nuclear) colonial history."[21]

Decolonizing pedagogies of the nuclear should also strive to show unequivocally that military and "civilian" nuclear technologies and complexes have always been inseparable from and mutually supportive of each other and that they are incompatible with the flourishing of most living species and their habitats. This task is particularly important for building a decolonizing consciousness of the atomic age, as nuclear complexes around the world have always sought to strategically separate nuclear weapons from the production of "peaceful" nuclear power. Ever since Dwight Eisenhower's 1953 UN "Atoms for Peace" speech, nuclear power has been represented as a safe, cheap technology for producing electricity. Indeed, developing nuclear cooperation among states, it was argued, would mitigate against the threat of nuclear proliferation and nuclear war. Yet the plutonium that is produced in nuclear power plants can relatively easily be converted to weapons-grade plutonium, and research and test reactors can produce plutonium that can be used for making nuclear weapons. Scholars such as Matthew Fuhrmann have found that peaceful nuclear assistance and cooperation increases the risk of nuclear proliferation and nuclear insecurity.[22]

Decolonizing pedagogies of the nuclear should also marshal evidence that nuclear energy, far from being carbon-free, clean, "green energy" with the potential to mitigate climate change and save the planet – as nuclear industries and their advocates have contended in recent years – is not only a high-cost, hazardous form of energy that produces irreversible environmental and biological harm but also a major contributing factor to the disasters being caused by global warming. Wildfires, rising sea levels, and earthquakes increase the risk of nuclear disasters and environmental radioactive contamination. Forest fires in the Chernobyl exclusion zone and surrounding areas, for instance, raised radiation releases to ten times normal levels in 2015. Meanwhile, the radioactive runoff from melting glaciers (which contain large quantities of radioactive fallout from Cold War nuclear tests and various nuclear accidents) flows into the oceans, is absorbed into the atmosphere, and falls again as radioactive acid rain.[23]

Allying itself with Indigenous, black, queer, and feminist critiques of Anthropocene discourses,[24] as well as with Indigenous environmental justice movements defending their land, water, and natural resources from corporate pipeline construction, deforestation, mining, fracking, and nuclear waste storage projects, decolonizing pedagogies of the nuclear should also expose and debunk all forms of nuclear nationalism, nuclear imperialism, and nuclear exceptionalism (such as the nuclear deterrence doxa promoted by the governments of the United States, the Soviet Union, and the NATO states during the Cold War and its contemporary offshoots in various nuclear weapons "modernization" programs). Delinking from Eurocentric/white-privilege humanisms and environmental ethics, decolonizing pedagogies of the nuclear could instead promote a holistic cosmovision based on the foundational principle of the co-constitution and the mutual nurturing of living and non-living, human and non-human existents or natures.

Anchored in a perspective that advocates for denuclearization, demilitarization, and socio-environmental, racial, and gender justice as well as for Indigenous sovereignty and a nuclear-free future, decolonizing pedagogies of the nuclear should also show that the politics of national and supranational nuclear regulatory bodies and agencies such as NRC (Nuclear Regulatory Commission) in the US and the ASN (Autorité de sûreté nucléaire / Nuclear Safety Authority) in France, as well as the politics and policies of IAEA and UNSCEAR (UN Scientific Committee on the Effects of Atomic Radiation), have been less concerned about the protection and safety of populations and the environment and more implicated in serving the interests of nuclear industries and corporations. In their educational strategies, publications, activism, and public statements, such pedagogies could expand on and complement the research of scholars such as Kate Brown, Joseph Masco, Olga Kuchinskaya, Magdalena Stawkowski, Paul Jobin, Robert (Bo) Jacobs, Jeff Kingston, and many others who have compellingly demonstrated how the IAEA and national nuclear industries and corporations, as well as the government agencies mandated to manage, regulate, and oversee nuclear complexes and infrastructures, have consistently co-opted science, used disinformation and deceit, suppressed compromising documentation, and harassed or silenced whistle-blowers and critical scientists while at the same time expanding the sacrifice zones of nuclear colonization.[25] Kate Brown, for instance, has documented how the IAEA and the UNSCEAR deliberately ignored or dismissed evidence of the causal relation between thousands of cases of children's thyroid cancer in Belarus and Ukraine and the fallout and radiation of the Chernobyl nuclear accident.[26] Even though the Japanese government has tightened nuclear safety regulations and created a new, independent regulatory body, the NRA (Nuclear Regulation Authority), in the wake of the Fukushima nuclear accident, nuclear policy in Japan has continued to be dominated by the power of the "nuclear village" – the structure of vested interests comprised of electric companies, nuclear reactor manufacturers, LDP politicians, various business groups and organizations, nuclear advocacy groups, and the media. Critical reporting on and protest activities related to compensation and the health, economic, and environmental effects of the Fukushima disaster have dwindled considerably due to the impact of the State Secrecy Law (Tokutei himitsu no hogo ni kansuru hôritsu), government propaganda on and the media promotion of the huge reconstruction and recovery projects in the disaster area in Fukushima prefecture, and the ongoing lack of transparency and accountability of the government's nuclear policy in general.[27]

A final note on the "theoretical heteroglossia" informing decolonizing pedagogies of the nuclear that must be invested at once in public education, scholarship, and the building of a decolonizing consciousness, and in the struggle for denuclearization, demilitarization, Indigenous sovereignty, environmental justice, and climate change action: A cursory survey of recent nuclear-themed publications in the humanistic and social sciences indicates not only that they strategically mobilize, reconfigure, or adapt theoretical constructs and frameworks from various interdisciplinary fields (e.g., environmental humanities, feminist, postcolonial and Indigenous approaches to science and technology studies, critical Anthropocene studies), but also that decolonizing perspectives are becoming increasingly important for nuclear critics.[28] The essays in this collection attest to a "theoretical heteroglossia" that evokes, and sometimes explicitly articulates, a decolonizing positionality.

Important to any conceptualization of decolonizing pedagogies of the nuclear (which I would argue is of primary importance not only in nuclear humanities but also in critical Anthropocene studies and all subfields of the environmental humanities) is the theorizing of *toxic immanence* – the embodied experience of living with, (re)thinking and contesting, and offering resilience under the contaminated, increasingly unlivable conditions created by extractive neocolonial capitalism. Michelle Murphy, Joseph Masco, and Jessica Hurley have proposed notable theorizations of toxic immanence, which they articulate respectively as *alterlife*, *fallout*, and *the nuclear mundane*. Murphy argues that the ubiquitous, pervasive reality or immanence of industrial chemical (and radioactive) toxicity as well as the flagrant inequality, exploitative labour relations, and differential distribution of socio-ecological harm under late neoliberal capitalism compel us to re-envision life as *alterlife*. Alterlife designates "life already altered, which is also life open to alteration." While this notion may also refer to symbiotic, mutually nurturing relations between individuals or collectives of different species, it is crucial that we understand contemporary manifestations of alterlife as *toxic entanglements*. Murphy is eloquent, indeed almost visionary, in her conceptualization of these entanglements, which she insists are the result not merely of global (white) capitalist relations, but also of the latter's racialized, (neo) colonial and colonizing structure. Alterlife may thus be envisioned as a

> ubiquitous condition of chemically [and radioactively; my addition] altered living-being, a condition which divides us as much as it binds us ... and that enacts and extends colonialism and racism into the intergenerational future ... [Alterlife is a figure of life that also designates a condition of] varied enmeshment and enfleshment in infrastructures ... an entrapment in and a response to each other's life supports and conditions.

Murphy goes on to contend that while the conditions of contemporary alterlife do not allow for easy escape or for disentanglement through utopic schemes of geoengineering, a possibility for envisioning and laying the groundwork for "feminist decolonial futures" might emerge through "a non-deferral of the decolonial, seized now, despite its impossibility."[29]

Joseph Masco's writing has focused on the *fallout* – that is, the harmful, disruptive socio-ecological and psycho-affective impacts and legacies – of the United States' status as a nuclear-armed military superpower and its government's nuclear nationalism, as well as on the fallout of its interminable, unbounded war on terror. In his recent work Masco expands on the broader semantic field inherent in the notion of fallout, arguing that the Anthropocene may alternatively be called the Age of Fallout and that this notion can be extended to several interrelated phenomena: the geoengineering or transformation of the earth's systems through the operations and cumulative effects of petrochemical, nuclear, and biochemical synthetic industries, industrial agriculture, and capitalist resource extraction;

and the post-1945 production of a biological "stratum" or post-nuclear formation whose manifestations and governance have become – from the perspective of neoliberal states – both unmanageable and expendable. Fallout also refers to the development of new scientific fields and theories and to the revamping of several existing ones, as well as to the emergence of new technologies as offshoots of the Cold War nuclear arms race and geopolitics: the earth and environmental sciences, digital information technologies, and satellite surveillance systems on the one hand, and on the other hand, a vision of the earth as a unified system (or system of interrelated systems) that has been produced by these advanced technologies and scientific endeavours.[30]

Jessica Hurley has recently proposed the concept of the *nuclear mundane* as a decolonizing, ecocritical and nuclear-criticism-based, intersectional heuristic that examines the material realities of the nuclear age, "focusing on the environmental, infrastructural, bodily and social impacts of nuclear technologies and the politics that prioritize them." Hurley argues that the nuclear mundane makes visible the reach of the nuclear into all aspects of everyday life as well as its contestability, "as something that can be named and challenged." She further contends that this heuristic enables us to perceive the nuclear as enviro-techno-social systems or infrastructures that have been structured by, and continue to structure, "existing distributions of power along axes of race, class, gender, sexuality, ability and indigeneity."[31]

Among the theorizations and (re)formulations of toxic immanence in this volume (articulations that mostly do not name this notion as such but propose similar conceptual configurations), I would cite Deckard's conception of the enduring nuclear in Chapter 1; Topçu's contestation of the TINANT (There Is No Alternative to Nuclear Toxicity) ideology in Chapter 2; and the new subjectivity formation emerging from the mutual exploration of the capitalist disaster complex/nuclear military-industrial complex and the *otaku* complex in Lamarre's reading of Anno Hideaki's film *Shin Godzilla* in Chapter 4.

The final section of this introduction outlines this volume's contributions. Taken as a whole, these texts ask troubling questions about violent colonial pasts and presents, persistent toxic immanence and material-embodied legacies, and the unending futures of the nuclear referent. In this author's view, these unsettling questions may be regarded as contributions to a growing body of decolonizing pedagogies of the nuclear.

Structure of the Volume and Chapter Outline

Toxic Immanence is divided into five parts titled respectively "Aftereffects of Chernobyl and Fukushima"; "The Cold War and Post–Cold War Nuclear State and Its Geopolitics: Imaginaries and Contestations"; "Archaeologies and

Heritages"; "Nuclear Aesthetics: Contemporary Art, Nuclear Colonialism, and the Transformation of Life and the Environment"; and "Artists' Contributions." The chapters in Part 1 examine the short- and long-term consequences of and social movements and imaginaries engendered by the Chernobyl and Fukushima nuclear accidents. Part 2 gathers texts that focus largely on the Cold War nuclear arms race and its lasting legacies as well as on the discourses, anti-nuclear movements, and art practices it has produced. Both chapters in Part 3 rely on archeological methods of investigation but differ considerably in terms of their approaches to the cultural heritage of the nuclear age.

The chapters in Part 4 examine various conceptions of nuclear aesthetics, which art historian and cultural critic Sven Lütticken has described as a theory and practice of crisis trying to extract itself from numbing abstractions: "The post-war nuclear regime confirms that the aesthetic is a practice and theory of crisis, in which petrifying abstractions instill a need to work with whatever form or degree of concretion can be attained."[32] The texts in this section are less concerned with "petrifying abstractions" (Lütticken's term both for hyperobjects such as ionizing radiation and its effects on humans and non-human living beings, and for ideological constructs that have served various political positions and fuelled culture wars) than with the fraught relation between settler colonialism and nuclear colonialism. Finally, Part 5 offers a cluster of single-authored and collaborative artistic projects that highlight the violence and ravages inflicted by, and the potential for healing inherent in, the nuclear age. The poems and photography included in this section speak to the underlying decolonizing theme of the volume insofar as they address the coloniality and long-lasting legacies of both nuclear weapons and "peaceful" nuclear energy, and because they envision a nuclear-free future.

The short afterwords that follow Parts 1, 2, 3, and 4 were commissioned to reflect on the chapters assembled in each of those sections. They extend the readings and propositions found in the individual chapters, taking them in new directions that in turn generate new or different perspectives.

In Chapter 1, "'The Future Is Behind Them!': Post-Apocalypse and the Enduring Nuclear in Post-Soviet Russian Fiction," Sharae Deckard argues that Dmitry Glukhovsky's novel *Metro 2033* enacts the contradictions and paradoxes of post-Soviet post-catastrophist speculative fiction: the novel proposes a "Chernobyl model" of history in which nuclear wars and disasters occur repeatedly and nostalgia for the lost Soviet empire is often articulated in overtly xenophobic, racist, and misogynistic terms; it also demonstrates contemporary Russian post-apocalyptic fiction's incapacity to imagine alternative futures and modes of existence that transcend a Cold War–derived nuclear imaginary. At the same time, Deckard goes on to contend, the novel's articulation of the *enduring nuclear* – a concept that indexes at once the imperceptible nature and deep temporalities of radioactive decay, the lived slow violence and uncanny effect of post-disaster nuclear immanence, and the return of the Soviet past's repressed nuclear colonialism, as well as the impossible futurity of nuclear modernity's "missilic consciousness" – and its repurposing of video game conventions allow it to partly subvert the generic constraints of post-apocalyptic fiction.

Chapter 2 by Sezin Topçu, "From Toxic Lands to Toxic Rumours: Nuclear Accidents, Contaminated Territories, and the Production of (Radio)active Ignorance," is unequivocal in its critical, decolonizing approach to the problem of how to manage contaminated land and exposed populations in the wake of nuclear accidents. The text examines three related strategies of post-nuclear disaster (colonial) governance: geo-management or zoning, biological or body management, and, finally, social management. While the prevention or minimizing of the devaluation of contaminated land and the regulation of population evacuations were regarded as significant issues as early as the 1950s, the management of the bodies and psychology of populations exposed to ionizing radiation has become a major concern for nuclear states in the aftermath of the Chernobyl and Fukushima disasters. The production of ignorance, or the construction of powerful instruments and apparatuses of persuasion, coercion, and distraction with the aim of normalizing the exceptional condition of living in radioactively contaminated environments, has played a major role in post-Chernobyl and post-Fukushima policies. The process of ignorance production has involved strategies and approaches such as the European ETHOS program of "participatory rehabilitation" and the Japanese government's enforcement of a series of controversial policies, laws, and regulations in the wake of the 3.11 disaster. The ETHOS program aimed to educate communities in Belarus that were significantly impacted by the fallout from the Chernobyl accident to live responsibly as "empowered (radioactive) citizens" by closely monitoring and quantifying their daily exposure to and intake of radionuclides. Post-3.11 policies in Japan include the imposition of 20mSv per year as the new, acceptable level of radiation exposure, as well as the "war against rumours" (*fuhyô higai*), which involves the systematic denigration and censoring of reports and studies documenting the negative health and environmental effects of radioactive contamination. The governing of the biological body through the "TINANT (There Is No Alternative to Nuclear Toxicity) ideology," argues Topçu, represents a new form of authoritarian biopolitical capitalism that must be exposed and relentlessly deconstructed in order restore democracy and bring about social and environmental justice.

In Chapter 3, "The Fukushima Process," Sabu Kohso diagnoses the normalizing of the post-Fukushima state of exception by Abe Shinzo's government, This chapter resonates in many ways with Topçu's examination of the Abe administration's authoritarian biopolitical governance. Kohso argues that the state of emergency declared by the Japanese government in the wake of the triple disaster of 11 March 2011 set the stage for the increasingly authoritarian, neonationalist policies adopted by the Abe administration after 2013, as well as for the conflicted relationship between the Fukushima "event" and the Fukushima "process." The Fukushima event refers to the singular, epochal rupture caused by the Fukushima nuclear disaster. This event opened deep cracks in Japan's socio-political and economic system that seemed to announce the end of the postwar regime – a nuclear state whose prosperity and stability was to a great extent due to its servicing of American wars and military hegemony in the Asia-Pacific region. These cracks also exposed the corrupt, dysfunctional structure of Japan's notorious "nuclear village" as well as the miscommunication and divergent policy orientation between the central government in Tokyo and the local authorities in the disaster area. In contrast to the event, the process that has been unfolding since the Abe administration's launching of its massive reconstruction program in the areas most severely affected by the 2011 earthquake and tsunami has been top-down, largely indifferent to the real needs, emotions, and experiences of the affected population, and accompanied by a series of controversial, anti-democratic laws and policies (such as the Specially Designated Secrets Act/SDS Act, Tokutei himitsu no hogo ni kansuru hôritsu). These laws and policies have curtailed citizens' civil liberties and democratic rights while giving the government more power to control the population. Kohso argues that even though the potential for radical socio-political change that developed in the wake of the 3.11 disasters was crushed, the re-emergence of a political ontology that could reconnect Japan's population to global people's struggles against militarized, nuclearized, extractive neocolonial capitalism is still possible.

In Chapter 4, "Fukushima and the Rebuild of Godzilla: Multiplying Media in an Era of Multiplying Disaster," Thomas Lamarre examines Anno Hideaki's award-winning film *Shin Gojira* (Shin Godzilla/Godzilla Resurgence, 2016). The text discusses three major aspects in Anno's film: the multiplication of disaster, the nationalizing of trauma, and the work's inciting viewers to propose multiple interpretations of its narrative and visuals. *Shin Gojira* shows clearly that disaster capitalism is bound to produce multiple accidents and calamities (especially in the case of a nuclear political economy such as Japan's). But the film also endorses and glorifies the nationalizing of nuclear trauma. As might be expected, this latter aspect is embodied in the "new" Godzilla, who has ingested radioactive waste and acts like a fission reactor on the brink of explosion. Lamarre contends that Anno's revisiting of the iconic national allegory of the radioactive monster is interesting especially because it proposes, in typical *otaku* fashion (i.e., from the perspective of devoted fans of anime, manga, video games, and their transmedia franchises), at least four possible readings of the nuclear crisis unleashed by the monster. The "happy" reading or pathway of interpretation projects a possible overcoming of radiation, which in turn implies a smooth, effortless continuation of the nuclear energy regime. A second possible reading of the nuclear crisis is a stoic one – an interpretation that enjoins the audience to accept radiation (and its nationalization) as a new way of life. A third, mocking-parodic way of dealing with the crisis that is suggested in the film boils down to a proposal to recapitalize on radioactivity by way of the global contents industry (of manga, anime, video games, and related franchises). Finally, a fourth, uncanny reading of the nuclear crisis provoked by the new Godzilla hints at an ominous condition – the incorporation of humans into nuclear fission. All four possible readings of the nuclear catastrophe envisioned in the film hinge on a central paradox that remains unnamed, namely the fact that nuclear energy is always already weaponry. *Shin Gojira* thus ultimately seems to suggest that a new subjectivity is emerging from the mutual exploration of the capitalist disaster complex (or the nuclear military-industrial complex) and the *otaku* complex. This emerging subjective formation harbours the potential to envision alternative futures or forms of worlding.

Joe Masco's Chapter 5, "Shaking, Trembling, Rattling, Shouting: Seismic Politics in the Nuclear Age," examines the seismic politics, or what he calls the politics of the shake, of the US nuclear weapons complex and the US government during the Cold War. He argues that the Manhattan Project scientists' early dream of tracking a nuclear chain reaction leading to a nuclear explosion at the subatomic level evolved through the above-ground and underground testing regimes until it was eventually realized in the computer-simulated detonations that have dominated nuclear weapons research since the 1990s. He shows that nuclear detonations were also planetary seismic events that radically impacted the geo- and bio-spheres, remaking both geological and human time. He goes on to contend that Cold War militarism's vision of nuclear weapons detonations as a seismic force that could be harnessed for geopolitical, strategic, environmental, and population management also produced at least two more vibrational or shaking systems: one system was a literally trembling, shaking, and bouncing infrastructure consisting of giant springs, hardened yet flexible building materials, and data-collecting computers that were intended to withstand the destructive shock of nuclear explosions. Another "shaking" assemblage consisted of interrelated apparatuses of surveillance, control, and affective management predicated on nuclear fear and apocalyptic doom, which in turn created nervous publics living in constant anticipation of a nuclear war and the annihilation of the planet. The US Cold War nuclear state's colonial seismic politics and politics of deterrence, writes Masco, have since the beginning of the twenty-first century been repurposed and amplified by the counterterror state in the form of a global war on terror.

The Cold War nuclear arms race also energized a wide range of anti-nuclear activisms and mobilized huge popular movements for social, racial, and environmental justice as well as decolonization and Indigenous rights movements, which created their own forms of vibrational, relational politics of solidarity. An inspiring form of the popular, creative anti-nuclear "shaking" movement is the Give Peace a Dance marathon, which emerged in Seattle in the mid-1980s and soon became global. Art Chantry's humorous, subversive posters for these joyful, all-night dance-a-thons can be viewed as one of the happiest forms of anti-nuclear art.

In Chapter 6, "What Is the Matter with Nuclear Weapons Communication?," Bryan Taylor argues for an expanded critical engagement with the complex entanglements between nuclear weapons complexes and communication. The first section of the chapter develops a comparative analysis of the intertwined material and symbolic dimensions of both nuclear weapons and communication. The second section proposes two contrasting narratives about the relative independence and relative interdependence of these phenomena. Arguing that the socio-political, technological, environmental, and discursive aspects of nuclear weapons systems have always been "interconnected, recursive and ... negotiated," Taylor highlights in the concluding part of his chapter the interdependence of nuclear weapons and nuclear power production infrastructures and operations on the one hand, and on the other hand the entanglement between these infrastructures and (nuclear) communication. Enhanced critical attention to these interconnections would empower communities and populations affected by nuclear complexes and by radioactive waste siting; it would also help us identify "critical handholds and pressure points [that would] allow us to disrupt the communicative politics [of Western nuclear states] that currently suspend us in ... the endlessly terrifying condition of nuclear insecurity."

Sarah Kanouse and Shiloh Krupar, in Chapter 7, "The National Toxic Land/Labor Conservation Service: 10-Year Final Report," offer one of several collaborations between artists and scholars featured in this volume (see also Chapters 8 and 15). Parodying the style of an official government agency's annual report, Kanouse and Krupar's text offers a half-serious, half-satirical survey of the history, legacies, and commemorative politics of the US nuclear weapons complex. Their "report" is interspersed with personal memories and reflections on the development and activities of the "National TLC Service" – the fictional, tongue-in-cheek agency they founded in 2011 to map a "people's geography" of the Cold War and the US nuclear weapons complex and to craft a dissenting narrative on the devastating material effects of the latter complex's reckless experiments. Through public activities such as design charrettes and workshops for building fictional monuments and museums and for sharing stories about the experience of living in contaminated militarized landscapes, the National TLC Service has sought to counter the government's and the atomic commemorative complex's official narrative on the Cold War nuclear arms race and to correct and dispel misinformation and ignorance regarding the latter's material effects. The main goal of their project, write Kanouse and Krupar, has been to produce a queer, feminist, transnatural, and transnational ecological politics that experiments with forms of alternative social organization and care for the contaminated and the damaged while also proposing creative pedagogies that will enable action and community organizing for demilitarization, decolonization, and environmental and social justice.

Much like Kanouse and Krupar, Juliet Palmer, Julie Salverson, and Peter C. van Wyck in Chapter 8, "Sounding Out the Nuclear: An Atomic Opera," have crafted an experimental, poetic text that weaves together elements of scholarly writing, memoir, and philosophical reflections. Consisting of a series of apparently loosely connected fragments, "Sounding Out the Nuclear" centres on two main themes: the creation and production of the atomic opera *Shelter* (2012), a collaboration between Palmerson (music) and Salverson (libretto); and a particularly fraught chapter in Canada's contribution to the Manhattan Project, the mining and transport of uranium ore by Great Bear Lake Dene community members (Sahtugot'ine Bear Lake People in the Northwest Territories). *Shelter* is an allegorical comic opera that pivots on Jewish Austrian physicist Lise Meitner's refusal to participate in the Manhattan Project's production and use of the atomic bomb on the one hand, and on the other hand the romance between the "radioactive daughter" Hope and the Pilot. The Canadian government's callous employment of the Dene as uranium miners and labourers without informing them of the risks of radioactive exposure has been documented in a series of scholarly and filmic works, including van Wyck's *Highway of the Atom* and Peter Blow's documentary *Village of Widows* (1999). The chapter's final section suggests that the Dene who travelled to Hiroshima in August 1998 to apologize to *hibakusha* survivors of the atomic bomb, as well as artists and scholars who have explored and continue to uncover the (often unknown) embodied histories, imaginaries, and legacies of nuclear weapons and nuclear energy production complexes, may be said to share an ethic of infinite responsibility for the absurdity and limitedness of the human condition. Such an ethic, the authors conclude, is best expressed through comedy and humour.

Jim Kraus's Chapter 9, "Poetry and Anti-Nuclearism: Τεχνη and the 'Fundamental Project,'" examines a sample of anti-nuclear poetry and short prose texts by poets such as Allen Ginsberg, Daniel Berrigan, William Heyen, Galway Kinnell, and Kathy Jetnil-Kijiner. Kraus highlights the various poetic genres and forms of expression used by these poets to oppose and resist nuclear weapons testing, nuclear energy production, and the (il)logic of nuclear deterrence or to expose the devastating legacies of nuclear complexes. Ranging from Whitmanesque incantations and spiritual visions informed by a Christian ethic of love and compassion to graphic descriptions of the horrific destruction, injuries, and suffering wrought by the atomic bombing of Hiroshima or by the American nuclear tests in the Marshall Islands, the discourse of anti-nuclear resistance deployed in the works examined by Kraus also includes expressions of abjection and irony as well as a critical engagement with the nuclear sublime. The chapter's concluding section asserts that poetry, like other creative arts, can provide readers and audiences with resources and imaginaries for confronting the forces "pushing humanity toward … war and nuclear annihilation."

Thomas Pringle's contribution, in Chapter 10, "Emergency/Salvage Archaeology: Excavating Media and Uranium in the Glen Canyon," proposes a synthesis of recent approaches in environmental media archeology and environmental studies. Building on Jussi Parikka's theorizing of "medianatures" and environmental studies' emphasis on the "conjoint constitution" of environmental materiality and social knowledge, as well as on recent scholarship on the role of photography and paleontological or prehistorical materials such as fossils in promoting US settler colonialism and naturalizing Indigenous dispossession, Pringle sheds light on the shifting valuations of the notions of "nature" and "the natural" in the media environmental history of the Glen Canyon in Utah between the 1940s and the 1960s. Deploying the category of the natural through the metaphor of the lens to highlight the term's work in "registering a … shifting heuristic characterized by techniques of focus, framing and serialization," Pringle uncovers a fraught material-discursive history and imaginary of Glen Canyon's nature across several archives or knowledge formations. These archives include government documents from the early years of the Cold War attesting to a conception of the area as a resource for uranium extraction; the infrastructural history of the Glen Canyon Dam and Lake Powell, the artificial reservoir that flooded and gradually filled the canyon over a period of twenty years; and the Emergency Salvage Archeology project undertaken by a team of scientists and researchers from the mid-1950s to the early 1960s to document as exhaustively as possible the biological, ecological, and geological formations of the Glen Canyon area as well as its Indigenous cultural heritage. Pringle's examination of these and other archives (including documentation of Lake Powell's function as repository for a 26,000-ton uranium tailings pile buried at its bottom) shows not only that the "medianatural" histories of the Glen Canyon area attest to the violence of postwar America's extractive economy but also that settler colonialism is "an epistemic project that engineers geographical space across time, giving rise to the nature experienced today." The combined methods of media archaeology and media environmental history, asserts Pringle his conclusion, enable us to view settler colonial economics, politics, and environmental governance as systems predicated on violence and dispossession and to trace this system's contribution to the (un)making and (re) signifying of toxic (nuclear) landscapes such as that of the Glen Canyon.

Based on their previous and ongoing research on the possibilities afforded by the encounter between the practices of archaeology, cultural heritage management, and nuclear waste management, Cornelius Holtorf and Anders Högberg in

Chapter 11, "Nuclear Waste as Critical Heritage," argue that nuclear waste can be seen as a form of cultural heritage of the atomic age. They also argue that the planning and design of deep geological repositories for the long-term storage of radioactive waste should be based not only on technoscientific and engineering criteria but also on what they call a *future consciousness*. This type of consciousness projects a vision of the future that embraces unimagined future presents while also allowing future generations to evaluate and make decisions for their presents and futures according to their own experience, politics, and socio-cultural imaginaries. The second part of Holtorf and Högberg's chapter contends that future approaches to nuclear waste repositories and their contents will be shaped by future heritage values that will vary across time and space. This implies on the one hand that these repositories are likely to be regarded by future generations as something different from containers of toxic radioactive materials, and on the other hand that nuclear waste management companies and nuclear regulatory bodies need to consider that the meaning of storage facilities will also change in ways we cannot imagine today. Such a conception fosters in turn the understanding that nuclear waste can act as a catalyst for radioactive waste management planners and engineers to opt for designs that respond to the challenges of an open-ended future and for cultural heritage professionals to better address the question of the long-term futurity implied in their work. This conception may also inspire environmental/nuclear humanities scholars to engage in a more sustained way with their long-term responsibilities.[33]

In Chapter 12, "Atomic Aborigines: Appropriation and Colonization of Indigenous Australia during British Nuclear Testing," Mick Broderick begins by charting a settler-colonial practice in Australia – that of designating place names by borrowing British and European geographic names and names of historical personages, as well as appropriating and/or distorting expressions and names from Aboriginal languages. The latter practice of "indigenizing" place names without acknowledging the local Aboriginal language and cultural traditions may be seen in the naming of Emu and Maralinga, two of the sites where British nuclear weapons tests were carried out between 1952 and 1963. This was, argues Broderick, a practice of colonial cultural dispossession of First Peoples' languages, histories, and spiritual-epistemic imaginaries. As documented by a 1985 Australian Royal Commission that investigated the operations and legacies of the British nuclear tests in Australia, these detonations were carried out with nearly total disregard and ignorance of Aboriginal nomadic practices of habitation and land cultivation, and as such should be regarded as violent colonial acts directed at Aboriginal nations and communities. Broderick's chapter concludes by examining recent artworks and creative projects that attest to the grievous legacies of radioactive contamination, sickness, and death bequeathed by the British nuclear weapons tests. An outstanding work among these projects is the VR film *Collisions* (2016).

Collisions is a collaboration between filmmaker and new media artist Lynette Walworth and Martu elder Nyarri Nyarri Morgan. It reimagines Morgan's first-hand experience of a British atomic test through an immersive, powerful aesthetic that enables viewers to witness his initial interpretation of the blast as an overwhelming manifestation of ancestral gods and spirits – an interpretation that was, as Broderick notes, "entirely consistent with his localized Dreamtime engagement." While this spiritual interpretation was later displaced by an understanding that the "supernatural" appearance of the nuclear detonation concealed an invisible radioactive poison, Morgan's testimony as presented in *Collisions*, writes Broderick, belongs to a growing body of artworks that articulate "unique Indigenous perspective(s) [on their] encounter with the nuclear age."

N.A.J. Taylor's Chapter 13, "The Antipodean Stance of Pam Debenham's 1980s Screenprints," examines a selection of anti-nuclear posters by Australian artist Pam Debenham. A prominent member of the Sydney-based feminist art collective Tin Sheds, Debenham produced a series of screen-printed posters in the 1980s whose anti-nuclear visual politics galvanized the peace, environmental, and anti-nuclear movements in Australia while also inciting opposition to nuclear testing internationally. Taylor contends that Debenham's posters evince an Antipodean stance – an oppositional perspective grounded in the historical, geophysical, and cultural specificity of the Oceanic Pacific region. The posters' powerful aesthetic also expresses a sense of solidarity and empathy with the populations and environments that have been affected by the American, British, and French nuclear tests in the South Pacific. They further evince a radical feminist critique both of militarized masculinity and the (ir)rationality of the nuclear arms race and nuclear deterrence, and of the authority of nuclear technoscience. Debenham's anti-nuclear art also represents in Taylor's view a significant contribution to an anti-nuclear visual global politics and a corrective to the hegemony of (male-centred) Anglo-American discourses and imaginaries in nuclear environmental humanities and social sciences.

The concluding chapter in Part4 of this volume, Chapter 14, "The Immanation-Image: Immanent Experience and Kazakhstan's Socialist and Postsocialist Modernity in Almagul Menlibayeva's Video Installation *Transformation*," features my own reading of Kazakhstani artist Almagul Menlibayeva's video installation *Transformation* (2016). This monumental installation reimagines the Soviet nuclear tests, and other major episodes in Kazakhstan's Soviet and post-Soviet capitalist modernity, from the perspective and immanent experience of residents of the former Semipalatinsk Test Site area and more broadly of the people of northeastern Kazakhstan. The artist's reimagining of this localized immanent experience is articulated through a specific technique I call the *immanation-image*. A performative image-event that attempts to capture living and non-living existents' immersion and perseverance in the radical ecology of immanence of global capitalism and industrial (neo)colonial modernity, the immanation-image is also characterized by the simultaneously intensive and extensive-distributive function of a strange attractor. Through a detailed analysis of *Transformation*'s feminist decolonial aesthetic I show that the installation articulates a powerful immanent critique of the Soviet nuclear tests and the Soviet state's vision of "atomic-powered communism" as well as of post/socialist nuclear technoscience and its colonization of life, the environment, and the future. I also contend that Menlibayeva's installation further proposes a decolonizing, speculative vision of post-extinction futures that may (or may not) enable the emergence of new forms of (female-centred) alterlife and alterknowledges.

The first contribution in Part 5 of this volume, "20 Poems by Bo Jacobs for 20 Photographs by elin O'Hara slavick," consists of a series of photographs of objects and sites in Hiroshima, Nagasaki, and Fukushima by artist elin O'Hara slavick and of poems penned by nuclear culture historian Robert (Bo) Jacobs that were inspired by these photographs. A further poetic contribution is CHamoru (Chamorro) poet Craig Perez's long poem "Nuclear Family." While powerfully attesting to and expressing anger and resistance with respect to the lasting radioactive legacies of the atomic bombing of Hiroshima and Nagasaki, Cold War nuclear tests in the Pacific, and nuclear disasters such as Fukushima, these artworks also suggest that the difficult task of laying the groundwork for and building toward a democratic and humane future involves the decommissioning and elimination of nuclear weapons and nuclear energy production complexes without delay.

Coda: The Non-Deferral of Nuclear Abolition Now Despite Its (Seeming) Impossibility

In its argument for decolonizing pedagogies of the nuclear, this chapter has added its voice to the present collection's critical examination of various aspects of the nuclear age. In their trenchant analysis of the *toxic immanence* produced by nuclear complexes and infrastructures, of the latter's lasting legacies of contamination and production of apocalyptic presents and futures, and finally of these processes' enmeshment in industrial-colonial modernity's relentless quest to master and despoil the planet's interconnected systems, the chapters and artworks in this volume may be said to contribute to an emerging, transnational, transcultural work of building decolonizing pedagogies of the nuclear. In more concrete terms, the contributors to this volume propose decolonizing pedagogical methods, thought experiments, analytics, and creative practices such as the following: the theorizing of the new political ontology that emerged in Japan in the wake of the Fukushima nuclear disaster, which may still be revitalized so as to enable people to reconnect with global struggles against neoliberal disaster capitalism

(Kohso, Chapter 3); the revisiting of Cold War–era anti-nuclear movements and anti-nuclear art as potential incentives for re-energizing contemporary activism and artistic projects oriented toward the contestation and abolition of nuclear complexes (Masco, Kraus, and N.A.J. Taylor, in Chapters 5, 9, and 13 respectively); the potential for Indigenous knowledges, living practices, and spiritualities to mitigate the legacies and future impacts of the nuclear age and climate change and to envision a nuclear-free (though not radiation-free) "world of many worlds"[34] (Broderick and Monnet, in Chapters 12 and 14 respectively); and the transnatural and transnational, queer feminist ecopolitics underwriting Kanouse and Krupar's National TLC Service, as well as the latter project itself as described in Chapter 7.

Paraphrasing Michelle Murphy's eloquent call to seize the decolonial now, this edited collection's underlying ethico-political and pedagogical vocation may be summed up as a commitment to the *non-deferral of nuclear abolition, seized now, despite its (seeming) impossibility.*

NOTES

1 See conference website, https://rethinkingradiationecologies.wordpress.com.

2 The Fukushima Daiichi nuclear disaster was a Level 7 nuclear accident, the second to receive that classification for most severe nuclear accident on the International Nuclear Event Scale after the Chernobyl disaster. It led to the evacuation of 154,000 local residents. The 11 March 2011 earthquake and tsunami killed 15,900 people; 2,525 more are still considered missing. In 2012, the National Diet of Japan Fukushima Nuclear Accident Independent Investigation Commission published its final report, which found that the accident was a man-made disaster that could have been prevented and that the government, regulatory bodies, and the Tokyo Electric Power Company (TEPCO, the Fukushima Daiichi Nuclear Plant's operator) lacked a sense of responsibility for the protection of society as well as people's lives. Some thirty collective civil actions and nearly four hundred individual civil actions have been filed against the government and TEPCO; the nearly 12,000 plaintiffs are mostly evacuees and other people affected by the disaster. For a comprehensive archive in English and Japanese on the 3.11 disasters, see "Japan's 3.11 Earthquake, Tsunami, Atomic Meltdown," https://apjjf.org/3-11.html. For recent studies of literary responses to the Fukushima Daiichi Nuclear Disaster see DiNitto, *Fukushima Fiction.* On ongoing civil actions and lawsuits against TEPCO and the government, see Jobin, "The Fukushima Nuclear Disaster and Civil Actions." On post-Fukushima citizens' science and citizens' radiation measurement efforts in Japan, see Kimura, *Radiation Brain Moms,* and Kimura, "The Potentials and Challenges of Citizen Science." For a recent edited collection assessing the Fukushima nuclear disaster from a critical disaster studies perspective, see Cleveland, Knowles, and Shineha, eds., *Legacies of Fukushima: 3.11 in Context.*

3 On nuclear humanities approaches to the nuclear bombing of Hiroshima and Nagasaki, see Taylor and Jacobs, eds., *Reimagining Hiroshima and Nagasaki*. The University of London–based Nuclear Culture Research Group has organized workshops, exhibitions, film screenings, field trips to various nuclear sites, and other activities. Curator and nuclear arts researcher Ele Carpenter is the research group's convenor. See their website, https://nuclear.artscatalyst.org.

4 Commemorative events and memorials were held throughout Japan on 11 March 2021, exactly a decade after the devastating 3.11 triple disaster. A moment of silence was observed by residents in the affected prefectures of Fukushima, Miyagi, and Iwate at 2:46 p.m., the time when the massive earthquake and the huge tsunami it unleashed struck eastern and northeastern Japan. Cultural commemorative events and shows included the public broadcaster NHK World-Japan's *3.11 – 10 Years On*, a collection of new and encore programs on the triple disaster, which were aired in March and April 2021; *A Future for Memory: Art and Life after the Great East Japan Earthquake* / 記憶のための未来—東日本大震災後のアートと暮らし(Kioku no tame no mirai: Higashi Nihon daishinsaigo no âto to kurashi), an exhibition featuring works by several Japanese artists, documentation by citizens' groups and museum collections from the disaster area, which was shown at the Museum of Anthropology (MOA) at the University of British Columbia from 11 February to 19 September 2021; and the exhibition titled *Artists and the (3.11) Disaster: Imagining in the 10th Year* (3.11とアーティスト : 10年目の想像 / 3.11 to Atisuto: Jûnenme no sôzô), featured at the Art Tower Mito Museum in Ibaraki Prefecture, 20 February–9 May 2021. Recent publications of note occasioned by, or anticipating, the 10th anniversary of the 3.11 triple disaster include Cleveland, Knowles, and Shineha, eds., *Legacies of Fukushima*; Kohso, *Radiation and Revolution*; Asanuma-Brice, *Fukushima: 10 ans après*; Funahashi Yôichi, *Fukushima senki*, vols. 1 and 2; and NHK Reporting Collective,Fukushima Daiichi.

5 See Masco, *The Future of Fallout*; Wallace, *Risk Criticism*; Hurley, *Infrastructures of Apocalypse*; Hogue and Maurer, "Special Forum Introduction"; Cordle, "Climate Criticism and Nuclear Criticism"; and Huang and Rapongan, "Radiation Ecologies." On the growing presence of nuclear issues in energy humanities and environmental humanities, see Misic and Kujundzic, eds., *Energy Humanities*, in particular the chapters by Misic and Kujundzic, Sukhenko, and Pannekoek. See also Brannigan et al., "The Languo of Flows."

6 On recent extreme weather events provoked by climate change and their devastating legacies, see Oxfam International, "5 Natural Disasters." On anthropogenic disasters and hazards as seen from the perspective of Earth's deep history and on the need to transform the modernist approaches and theoretical perspectives of hazard and disaster scholarship, see Dominey-Howes, "Hazards and Disasters in the Anthropocene." On the relationship between California's increasing summer forest fires and climate change see Williams et al., "Observed impacts"; and McGrath, "Global Warming Driving California Wildfire." On the record-breaking heat wave

and the "heat dome" in June and July 2021 in western Canada, especially in British Columbia, and the western United States, see Samenow and Cappucci, "Severe Heatwave"; and Watts, "Canadian Inferno."

7 On regularly updated SARS-COV-2 statistics, see the World Health Organization's Weekly Epidemiological Update, https://www.who.int/emergencies/diseases/novel-coronavirus-2019/situation-reports; and the WHO Covid19 Dashboard, https://covid19.who.int. See also the European Centre for Disease Prevention and Control (ECDC)'s website, https://www.ecdc.europa.eu/en/geographical-distribution-2019-ncov-cases. Regularly updated data on COVID-19 in the United States and globally can also be found on the website of the John Hopkins University Corona Virus Resource Center, https://coronavirus.jhu.edu. On the economic, social, and political impact of the COVID-19 pandemic, see Lenzen et al., "Global Socio-Economic Losses." On the relationship between the coronavirus pandemic and climate change, militarism, nationalism, and global violence see Garcia, "Redirect Military Budgets."

8 On the Trump administration's withdrawal from nuclear arms control treaties and plans for the US nuclear arsenal's modernization see Goodman, "Trump's War on Seven Decades of Arms Control"; and Reif, "US Nuclear Modernization Programs." On the Biden administration's nuclear modernization plans see Harper, "Biden to Stay Course on Nuclear Modernization." On the Japanese government's plans to dump a million tons of radioactive water used in the cooling of the reactor cores into the ocean see Ruff and Beavis, "Japan Plans to Dump a Million Tons of Radioactive Water into the Pacific"; and McCurry, "Fukushima: Japan Will have to Dump." On the competition between the United States, Russia, China, France, Japan, South Korea, and other nuclear states to sell reactors and other nuclear technologies to countries in the Middle East, South, Central, and Southeast Asia, and Africa and Latin America see Kane, "Nuclear Reactors to Saudi Arabia"; Calabrese, "China and the US in the Middle East"; and Platte, "Exporting Nuclear Norms." On the relationship between climate activism/environmental justice movements and anti-racism activism during the COVID-19 pandemic see Hewett, "'Racial Justice Is Climate Justice.'" On the state's criminalization of and violent actions against environmental protests and activism see White, "Environmental Victims and Resistance"; Birss, "Criminalizing Environmental Activism"; Cagle, "Protesters as Terrorists"; and Federman," Climate Activist Group as Extremists."

9 On the revival and legitimacy of nuclear criticism in the post–Cold War period see Blouin, Shipley, and Taylor, "Introduction: The Silence of Fallout"; Cordle, "Legacy Waste"; Cordle, "The Futures of Nuclear Criticism"; and Cordle, "Climate Criticism and Nuclear Criticism." On contemporary nuclear theory and the imperative of questioning Derrida's classic essay on nuclear criticism, "No Apocalypse, Not Now," through the lens of contemporary era's apocalyptic Anthropocene imaginings, nuclear amnesia, and the slow violence of nuclear contamination and toxicity see Milner and Kinsella, "Nuclear Theory Degree Zero."

10 On the colonial power relations within nuclear technoscience and the nuclear regimes it has sustained see Hecht, "Colonial Networks of Power"; Masco, *Nuclear Borderlands*; and Endres, "The Rhetoric of Nuclear Colonialism." On the colonial character of

modern (Western) science and technology and the politics of decolonizing science see Arnold, "Europe, Technology, and Colonialism"; and Roy, "Decolonise Science – Time to End Another Imperial Era."

11 See Endres, "The Rhetoric of Nuclear Colonialism," 40; and Hogue and Maurer, "Special Forum Introduction," 27.

12 Scholarship on nuclear colonialism includes Danielle Endres's articles, all listed on her website, https://www.danielleendres.com; Valerie Kuletz's now classic work, *The Tainted Desert*; Edwards, "Nuclear Colonialism and the Social Construction of Landscape"; Masco, "The Age of Fallout"; and Danielsson and Danielsson, *Mururoa, notre bombe coloniale.* On the decades-long fight by Aboriginal and First Peoples, downwinders, and other populations in the United States, Canada, Australia, the Marshall Islands, French Polynesia, and other regions for recognition of the devastating legacies of nuclear testing, as well as against the siting of nuclear waste storage repositories on their lands, see LaDuke, *All Our Relations*; Edwards, "Nuclear Colonialism"; Maclellan, "The Nuclear Age in the Pacific Islands;" Simmonds and Liboiron, "Nuclear State, Nuclear Waste"; and Barker, *Bravo for the Marshallese.* Edited by Anais Maurer and Rebecca Hogue, the Special Forum on "Transnational Nuclear Imperialisms" in the *Journal of Transnational American Studies* 11, No. 2 (2020), gathers essays focusing on the ongoing colonial-imperial nuclear politics of nuclear states such the United States, Britain, France, and Russia. See especially the essays of Bahng and Schwartz in this Special Forum.

13 On Quijano's elaboration of the concept of coloniality of power see his essays "Coloniality of Power, Eurocentrism, and Latin America," "Coloniality and Modernity/Rationality," and "Colonialidad, Poder, Cultura y Conocimiento." Quijano's work has inspired the scholarship of numerous thinkers including Arturo Escobar, Walter Mignolo, Ramon Grosvoguel, Sylvia Wynter, Maria Lugones, Catherine Walsh, and Nelson MaldonadoTorres. See also Mignolo and Walsh, *On Decoloniality.*

14 The notion of necropolitics was introduced by Achille Mbembe in a series of essays, in particular "Necropolitics" (2003), and expanded upon in *Politiques de l'inimitié* (2016), published in English translation as *Necropolitics* (2019). Since Foucault's concepts of biopower and biopolitics are insufficient to account for these forms of subjugation, argues Mbembe, they must be supplemented by the concepts of necropower and necropolitics. These notions describe "the various ways in which, in our contemporary world, weapons are deployed in the interest of maximum destruction of persons and the creation of death-worlds, new and unique forms of social existence in which vast populations are subjected to conditions of life conferring upon them the status of *living dead.*" See Mbembe, "Necropolitics," 39–40. On the necropolitics of the Anthropocene see Bubandt, "Haunted Geologies"; and Lykke, "Making Live and Letting Die." On the necropolitics and slow violence of petrochemical pollution in a contaminated geography see Davies, "Toxic Space and Time."

15 A recent workshop on "Decolonising the Nuclear" was organized on 22–23 October 2019 by Ele Carpenter and Warren Harper at Goldsmith University in London, UK. See http://m-a-r-s.online/sessions/decolonising-the-nuclear-public-lecture-and-research-workshop. The public lecture at the workshop was delivered by Gabrielle Hecht.

16 The Treaty on the Prohibition of Nuclear Weapons, which was adopted on 7 July 2017 by the United Nations conference "to negotiate a legally binding instrument to prohibit nuclear weapons, leading towards their total elimination," provides a salutary precedent for the difficult work of achieving the "total elimination" of nuclear arsenals. Signed by thirty-three states, ratified by fifty-three states, and acceded to by two states, the TPNW does not provide a legally binding instrument for abolishing nuclear energy production.

17 See Stengers, "Introductory Notes,"185–7.

18 Tejeda, Espinoza, and Gutierrez argue that contemporary American society and culture are characterized by an internal neocolonialism that "has its origins in the mutually reinforcing systems of colonial and capitalist domination and exploitation that enslaved Africans and dispossessed indigenous populations throughout the 17th, 18th, and 19th centuries." They also contend that justice among and for all peoples inhabiting the United States can only be achieved through the dismantling of this internal neocolonial condition and "the abolishing of its multiple forms of violence." See "Toward a Decolonizing Pedagogy,"1-5.

19 Tejeda, Espinoza, Gutierrez, "Toward a Decolonizing Pedagogy," 7, 9.

20 See Andreotti et al., "Mapping Interpretations of Decolonization," 34–7. See also Ahenakew et al., "Beyond Epistemic Provincialism"; and Smith, *Decolonizing Methodologies.*

21 See Gaztambide-Fernandez, "Decolonization and the Pedagogy of Solidarity," 42.

22 See Fuhrmann, *Atomic Assistance*, esp. Chapters 7 and 8.

23 On the problematic viability of nuclear energy as potential mitigating solution to climate change see Hutner and Cirino, "Nuclear Power Is Not the Answer"; Johnson, "Can Nuclear Power Help Save Us from Climate Change"; and Jacobs, "Nuclear Stockholm Syndrome." On forest fires in Chernobyl and the risk of increased radiation levels see Braxton Little, "Forest Fires." See also Chapter 3, section 3.3, "Why Nuclear Power Represents an Opportunity Cost," in Jakobson, *100% Clean, Renewable Energy and Storage for Everything*, on the inappropriateness and risks of nuclear power for addressing and mitigating the challenges of climate emergency, air pollution, and energy insecurity.

24 Recent critical and decolonizing approaches to scientific and popular discourses on the Anthropocene include Todd, "Indigenizing the Anthropocene"; Whyte, "Indigenous Climate Change Studies"; Simpson, "The Anthropocene as Colonial Discourse"; Yusoff, *A Billion Black Anthropocenes*; Karera, "Blackness and the Pitfalls"; Erickson, "Anthropocene Futures"; and Åsberg, "Feminist Posthumanities in the Anthropocene."

25 Studies that have exposed the anti-democratic, opaque structure and corrupt practices of nuclear military-industrial complexes and nuclear regimes in the United States, Britain, France, the former Soviet Union, Japan and other nuclear states include Brown, *Plutopia* and *Manual for Survival*; Kuchinskaya, *The Politics of Invisibility*; Masco, *The Nuclear Borderlands*; Topçu, *La France nucléaire (Nuclear France)*; Tanter, "After Fukushima"; Stawkowski, "Radiophobia"; and Walton, "Deadly Fallout of Disinformation."

26 See Brown, *Manual for Survival*, 211-84.

27 On nuclear policy in Japan after the Fukushima nuclear disaster see Vivoda and Graetz, "Nuclear Policy and Regulation"; and Andrews-Speed, "Governing Nuclear Safety in Japan."

28 On critical approaches to nuclear colonialism in the work of contemporary Indigenous artists see Titterington, "Nuclear Colonialism." For a decolonizing approach to art history, contemporary art, and visual culture that also critiques dominant discourses on the Anthropocene and anthropogenic/capitalogenic extraction, consumption, and energy regimes see T.J. Demos, *Beyond the World's End*; *Against the Anthropocene*; and *Decolonizing Nature*. The essays in *Issue 5: Reimagine Maralinga* of the online journal *Unlikely: Journal for Creative Arts* propose decolonial or decolonizing readings of the artworks that were created both by white and by Aboriginal artists in connection with the project *Nuclear Futures*, https://nuclearfutures.org/about/nuclear-pasts.

29 Murphy, "Alterlife," 497–8, 501. For a discussion of Murphy's theorizing of alterlife see also Monnet's chapter in this volume.

30 See Masco, *The Future of Fallout*, Preface and Chapter 1; Masco, "The Age of Fallout"; and "Terraforming Planet Earth."

31 See Hurley, *Infrastructures of Apocalypse*, 9–11.

32 See Lütticken, "Nuclear Aesthetics: Beyond Big Bangs," 14. On critical nuclear aesthetics see also Lütticken, "Shattered Matter, Transformed Forms: Notes on Nuclear Aesthetics," parts 1 and 2. See also Mavrokordopoulou and de Vos, eds. "Nuclear Aesthetics."

33 I should note here that while I agree with Holtorf and Högberg's ideas about the unimagined futures or future presents of any form of cultural heritage, as well as with their contention that future generations will have different understandings of energy production and conservation, I do not share their optimism about nuclear waste's potential to be regarded, at some point in its open-ended futurity, as a (non-hazardous or non-toxic) resource. I consider Rosemary A. Joyce's argument that we will need "ways to imagine new cohabitation in the future" (and in the present) with radioactivity and nuclear waste more attuned to the latter's toxic reality and materiality. Rather than thinking that we can contain and safely store nuclear waste, or banish it from human company, or attempting to imagine a future that may shield us from the toxic legacies of the nuclear age, it seems more productive to indulge in speculative visions such as that imagined by Joyce in the conclusion of her study, *The Future of Nuclear Waste*. Joyce imagines a distant future in which radioactivity is well-understood and lived with as a toxic yet survivable part of an animate environment, and where the markers built by past civilizations to call attention to toxic repositories are as opaque and mysterious as other archeological remains attesting to those vanished cultures. See Joyce, *The Future of Nuclear Waste*, 226–33.

34 The expression "world of many worlds" references a well-known Zapatista declaration: "In the world we want everybody fits. The world we want is a world in which many worlds fit" (El mundo que queremos es uno donde quepan muchos mundos). Mario Blaser and Marisol de la Cadena redefine such a world of many worlds as a pluriverse and as an "invitation to reworlding possibilities": "heterogeneous worldings coming

together as a political ecology of practices, negotiating their difficult being together in heterogeneity." See Blaser and de la Cadena, "Pluriverse: Proposal for a World of Many Worlds," 1, 4.

BIBLIOGRAPHY

Ahenakew, Cash, Vanessa de Oliveira Andreotti, Garrick Cooper, and Hemi Hireme. "Beyond Epistemic Provincialism: De-provincializing Indigenous Resistance." *AlterNative* 10, no. 3 (2014): 216–31.

Andreotti, Vanessa de Oliveira, Sharon Stein, Cash Ahenakew, and Dallas Hunt. "Mapping Interpretations of Decolonization in the Context of Higher Education." *Decolonization: Indigeneity, Education, and Society* 4, no. 1 (2015): 21–40.

Andrews-Speed, Philip. "Governing Nuclear Safety in Japan after the Fukushima Nuclear Accident: Incremental or Radical Change?" *Journal of Energy and Natural Resources Law* 38, no. 2 (2020): 161–81. https://doi.org/10.1080/02646811.2020.1741990.

Arnold, David. "Europe, Technology, and Colonialism in the 20th Century." *History and Technology* 21, no. 1 (2005): 85–106. https://doi.org/10.1080/07341510500037537.

Asanuma Brice, Cécile. *Fukushima, 10 ans après: Sociologie d'un désastre* (Fukushima, 10 Years Later: A Sociological Study of the Disaster). Paris: Maison des sciences de l'homme, 2021.

Åsberg, Cecilia. "Feminist Posthumanities in the Anthropocene: Forays into the Postnatural." *Journal of Posthuman Studies* 1, no. 2 (2018): 185–204. doi:10.5325/jpoststud.1.2.0185.

Barker, Holly M. *Bravo for the Marshallese: Regaining Control in a Post-Colonial, Post-Nuclear World* [2004]. Belmont: Wadsworth, Cengage Learning, 2013.

Birss, Moira. "Criminalizing Environmental Activism." *NACLA Report on the Americas* 49, no. 3 (2017): 315–22. https://doi.org/10.1080/10714839.2017.1373958.

Blaser, Mario, and Marisol de la Cadena. "Pluriverse: Proposal for a World of Many Worlds." In *A World of Many Worlds*, edited by Marisol de la Cadena and Mario Blaser, 1–22. Durham: Duke University Presss, 2018.

Blouin, Michael, Morgan Shipley, and Jack Taylor. "Introduction: The Silence of Fallout." In *The Silence of Fallout: Nuclear Criticism in a Post–Cold War World*, edited by Michael Blouin, Morgan Shipley, and Jack Taylor, 1–15. Newcastle upon Tyne: Cambridge Scholars, 2013.

Brannigan, John, Frances Ryfield, Tasman Crowe, and David Cabana. "'The Languo of Flows': Ecosystem Services, Cultural Value, and the Nuclear Legacy in the Irish Sea." *Environmental Humanities* 11, no. 2 (2019): 280–301. https://doi.org/10.1215/22011919-7754456.

Braxton Little, Jane. "Forest Fires Are Setting Chernobyl's Radiation Free." *The Atlantic*, 10 August 2020. https://www.theatlantic.com/science/archive/2020/08/chernobyl-fires/615067.

Brown, Kate. *Plutopia: Nuclear Families, Atomic Cities, and the Great Soviet and American Plutonium Disasters*. New York: Oxford University Press, 2013.

– *Manual for Survival: A Chernobyl Guide to the Future.* London: Penguin, 2019.

Bubandt, Nils. "Haunted Geologies: Spirits, Stones, and the Necropolitics of the Anthropocene." In *Arts of Living on a Damaged Planet: Ghosts and Monsters of the Anthropocene*, edited by Anna Tsing, Heather Swanson, Elaine Gan, and Nils Bubandt, 121–41. Minneapolis: University of Minnesota Press, 2017. http://www.jstor.org/stable/10.5749/j.ctt1qft070.

Cagle, Susie. "'Protesters as Terrorists': Growing Number of States Turn Anti-Pipeline Activism into a Crime." *The Guardian*, 8 July 2019. https://www.theguardian.com/environment/2019/jul/08/wave-of-new-laws-aim-to-stifle-anti-pipeline-protests-activists-say.

Calabrese, John. "Intersections: China and the US in the Middle East." *Publications of the Middle East Institute*, 18 June 2019. https://www.mei.edu/publications/intersections-china-and-us-middle-east#1.

Cleveland, Kyle, Scott Gabriel Knowles, and Ryuma Shineha, eds. *Legacies of Fukushima: 3.11 in Context.* Philadelphia: University of Pennsylvania Press, 2021.

Cordle, Daniel. "Climate Criticism and Nuclear Criticism." In *Climate and Literature*, edited by Adeline Johns-Putra, 281–97. Cambridge Critical Concepts. Cambridge: Cambridge University Press, 2019. doi:10.1017/9781108505321.018.

– "The Futures of Nuclear Criticism." *Alluvium* 5, no. 3 (2016): n.pag. DOI:http://dx.doi.org/10.7766/alluvium.v5.3.03.

– "Legacy Waste: Nuclear Culture after the Cold War." In *The Silence of Fallout: Nuclear Criticism in a Post–Cold War World*, edited by Michael Blouin, Morgan Shipley, and Jack Taylor, 230–49. Newcastle upon Tyne: Cambridge Scholars, 2013.

Danielsson, Bengt, and Marie-Thérèse Danielsson. *Mururoa, notre bombe coloniale: histoire de la colonisation nucléaire de la Polynésie française* (Mururoa, Our Colonial Bomb: History of the Nuclear Colonization of French Polynesia). Paris: L'Harmattan, 1990.

Davies, Thom. "Toxic Space and Time: Slow Violence, Necropolitics, and Petrochemical Pollution." *Annals of the American Association of Geographers* 108, no. 6 (2018): 1537–53. https://doi.org/10.1080/24694452.2018.1470924.

Demos, T.J. *Against the Anthropocene: Visual Culture and Environment Today.* London: Sternberg Press, 2017.

– *Beyond the World's End: Arts of Living at the Crossing.* Durham: Duke University Press, 2020.

– *Decolonizing Nature: Contemporary Art and the Politics of Ecology.* London: Sternberg Press, 2016.

DiNitto, Rachel. *Fukushima Fiction: The Literary Landscape of Japan's Triple Disaster.* Honolulu: University of Hawai'i Press, 2019. muse.jhu.edu/book/67068.

Dominey-Howes, Dale. "Hazards and Disasters in the Anthropocene: Some Critical Reflections for the Future." *Geoscience Letters* 5 (2018). https://doi.org/10.1186/s40562-018-0107-x.

Edwards, Nelta. "Nuclear Colonialism and the Social Construction of Landscape in Alaska." *Environmental Justice* 4, no. 2 (2011): 109–14 https://doi.org/10.1089/env.2010.0023.

Endres, Danielle. "The Rhetoric of Nuclear Colonialism: Rhetorical Exclusion of American Indian Arguments in the Yucca Mountain Nuclear Waste Siting Decision." *Communication and Critical/Cultural Studies* 6, no. 1 (2009): 39–60. DOI:https://doi.org/10.1080/14791420802632103.

Erickson, Bruce. "Anthropocene Futures: Linking Colonialism and Environmentalism in an Age of Crisis." *Environment and Planning D: Society and Space* 38, no. 1 (February 2020): 111–28. https://doi.org/10.1177/0263775818806514.

Federman, Adam. "Revealed: US Listed Climate Activist Group as 'Extremists' alongside Mass Killers." *The Guardian*, 13 January 2020. https://www.theguardian.com/environment/2020/jan/13/us-listed-climate-activist-group-extremists.

Fuhrman, Matthew. *Atomic Assistance: How "Atoms for Peace" Programs Cause Nuclear Insecurity*. Ithaca: Cornell University Press, 2012.

Funahashi Yôichi. *Fukushima senki: Jûnengo no countdown meltdown* (War Report on Fukushima: Ten Years after the Disaster, Countdown to Meltdown). 船橋洋一著『福島戦記—１０年後のカウントダウン　メルトダウン』上、下。文藝春秋、. Tokyo: Bungei shunjû, 2021.

Garcia, Denise. "Redirect Military Budgets to Tackle Climate Change and Pandemics." *Nature* 584 (27 August 2020): 521–3. https://doi.org/10.1038/d41586-020-02460-9.

Gaztambide-Fernandez, Ruben A. "Decolonization and the Pedagogy of Solidarity." *Decolonization: Indigeneity, Education, and Society* 1, no. 1 (2012): 41–67.

Goodman, Melvin. "Trump's War on Arms Control and Disarmament." *Counterpunch*, 12 August 2020. https://www.counterpunch.org/2020/08/12/trumps-war-on-arms-control-and- disarmament-2.

Harper, Jon. "Biden to Stay Course on Nuclear Modernization." *National Defense*, 15 June 2021. https://www.nationaldefensemagazine.org/articles/2021/6/15/biden-to-stay-course-on-nuclear-modernization.

Hecht, Gabriele. "Colonial Networks of Power: The Far Reaches of Systems." *Annales historiques de l'électricité* 1, no. 2 (2004): 147–57. https://www.cairn.info/revue-annales-historiques-de-l-electricite-2004-1-page-147.htm#.

Hewett, Frederick. "'Racial Justice Is Climate Justice': Why the Climate Movement Needs to Be Anti-Racist." *Cognoscenti. WBUR, Boston's NPR News Station*, 6 June 2020. https://www.wbur.org/cognoscenti/2020/06/09/the-links-between-racism-and-the-environment-frederick-hewett.

Hogue, Rebecca, and Anais Maurer. "Special Forum Introduction: Transnational Nuclear Imperialisms." In "Special Forum on Transnational Nuclear Imperialisms," edited by Hogue and Maurer. *Transnational American Studies* 11, no. 2 (2020): 25–43. https://doi.org/10.5070/T8112050495.

Huang, Hsinya, and Syaman Rapongan. "Radiation Ecologies, Resistance, and Survivance on the Pacific Islands." In *Mushroom Clouds: Ecocritical Approaches to Militarization and the Environment in East Asia*, edited by Simon C.Estok, Iping Liang, and Shinji Iwamasa. New York: Routledge, 2021.

Hurley, Jessica. *Infrastructures of Apocalypse: American Literature and the Nuclear Complex*. Minneapolis: University of Minnesota Press, 2020.

Hutner, Heidi, and Erica Cirino. "Nuclear Power Is Not the Answer in a Time of Climate Change." *Aeon*, 28 May 2019. https://aeon.co/ideas/nuclear-power-is-not-the-answer-in-a-time-of-climate-change.

Jacobs, Robert. "Nuclear Stockholm Syndrome." *Counterpunch*, 9 July 2021. https://www.counterpunch.org/2021/07/09/nuclear-stockholm-syndrome.

Jacobson, Mark Z. *100% Clean, Renewable Energy and Storage for Everything*. Cambridge: Cambridge University Press, 2020. doi:10.1017/9781108786713.

Jobin, Paul. "The Fukushima Nuclear Disaster and Civil Actions as a Social Movement." *Asia-Pacific Journal: Japan Focus* 18, no. 9 (2020). https://apjjf.org/2020/9/Jobin.html.

Johnson, Jeff. "Can Nuclear Power Help Save Us from Climate Change?" *C&EN/Chemical and Engineering News*, 23 September 2019. ttps://cen.acs.org/energy/nuclear-power/nuclear-power-help-save-us/97/i37.

Joyce, Rosemary A. *The Future of Nuclear Waste: What Art and Archeology Can Tell Us about Securing the World's Most Hazardous Material*. Oxford: Oxford University Press, 2020.

Kane, Chen. "Why Proposals to Sell Nuclear Reactors to Saudi Arabia Raise Red Flags." *The Conversation*, 22 February 2019. https://theconversation.com/why-proposals-to-sell-nuclear-reactors-to-saudi-arabia-raise- red-flags-112276.

Karera, Axelle. "Blackness and the Pitfalls of Anthropocene Ethics." *Critical Philosophy of Race* 7, no. 1 (2019): 32–56. https://doi.org/10.5325/critphilrace.7.1.0032.

Kimura, Aya Hirata. "The Potentials and Challenges of Citizen Science: 9 Years of Experience from Post-Fukushima Japan." *Somatosphere*, 16 April 2020.

– *Radiation Brain Moms and Citizen Scientists: The Gender Politics of Food Contamination after Fukushima*. Durham: Duke University Press, 2016.

Kohso, Sabu. *Radiation and Revolution*. Durham: Duke University Press, 2020.

Kuchinskaya, Olga. *The Politics of Invisibility: Public Knowledge about Radiation Health Effects after Chernobyl*. Cambridge, MA: MIT Press, 2014.

Kuletz, Valerie L. *The Tainted Desert: Environmental and Social Ruin in the American West*. New York: Routledge, 2016. https://doi.org/10.4324/9781315538839.

LaDuke, Winona. *All Our Relations: Native Struggles for Land and Life* [1999]. Chicago: Haymarket Books, 2016.

Lenzen, Manfred, Mengyu Li, Arunima Malik, Francesco Pomponi, Ya-Yen Sun, Thomas Wiedmann, Futu Faturay, Jacob Fry, Blanca Gallego, Arne Geschke, Jorge Gómez-Paredes, Keiichiro Kanemoto, Steven Kenway, Keisuke Nansai, Mikhail Prokopenko, Takako Wakiyama, Yafei Wang, Moslem Yousefzadeh. "Global Socio-Economic Losses and Environmental Gains from the Coronavirus Pandemic." *PLOS ONE* 15, no. 7 (2020): e0235654. DOI: 10.1371/journal.pone.0235654.

Lütticken, Sven. "Nuclear Aesthetics: Beyond Big Bangs." In "Nuclear Aesthetics," edited by Kyveli Mavrokordopoulou and Ruby de Vos. Special Issue. *Kunstlicht* 39, nos. 3–4 (2018): 13–19.

– "Shattered Matter, Transformed Forms: Notes on Nuclear Aesthetics, Part 1." *E-Flux Journal* 94 (October 2018). https://www.e-flux.com/journal/94/221035/shattered-matter-transformed-forms-notes-on-nuclear-aesthetics-part-1.

– "Shattered Matter, Transformed Forms: Notes on Nuclear Aesthetics, Part 2." *E-Flux Journal* 96 (January 2019) https://www.e-flux.com/journal/96/243057/shattered-matter-transformed-forms-notes-on-nuclear-aesthetics-part-2.

Lykke, Nina. "Making Live and Letting Die: Cancerous Bodies between Anthropocene Necropolitics and Chthulucene Kinship." *Environmental Humanities* 11, no. 1 (2019): 108–36. https://doi.org/10.1215/22011919-7349444.

Maclellan, Nic. "The Nuclear Age in the Pacific Islands." *Contemporary Pacific* 17, no. 2 (2005): 363–72. http://www.jstor.org/stable/23722064.

Masco, Joseph. "The Age of Fallout." *History of the Present* 5, no. 2 (Fall 2015): 137–68. https://www.jstor.org/stable/10.5406/historypresent.5.2.0137.

– *The Future of Fallout, and Other Episodes in Radioactive World-Making.* Durham: Duke University Press, 2021.

– *The Nuclear Borderlands: The Manhattan Project in Post–Cold War New Mexico* [2006]. New Edition. Princeton: Princeton University Press, 2020.

– "Terraforming Planet Earth: The Age of Fallout." In *The Politics of Globality since 1945: Assembling the Planet*, edited by Rens van Munster and Casper Sylvest, 44–70. Abingdon: Routledge, 2016.

Mavrokordopoulou, Kyveli, and Ruby de Vos, eds. "NuclearAesthetics." Special Issue. *Kunstlicht* 39, nos. 3–4 (2018): 5–95.

Mbembe, Achille. "Necropolitics." Translated by Libby Meintjes. *Public Culture* 15, no. 1 (2003): 11–40.

– *Necropolitics.* Translated by Steven Corcoran. Durham: Duke University Press, 2019.

– *Politiques de l'inimitié* (Politics of Enmity). Paris: La Découverte, 2016.

McCurry, Justin. "Fukushima: Japan Will Have to Dump Radioactive Water into Pacific, Minister Says." *The Guardian*, 10 September 2019. https://www.theguardian.com/environment/2019/sep/10/fukushima-japan-will-have-to-dump-radioactive-water-into-pacific-minister-says.

McGrath, Matt. "Global Warming Driving California Wild Fire: Study." *BBC News*, 25 September 2020. https://www.bbc.com/news/science-environment-54278988.

Mignolo, Walter D., and Catherine E. Walsh. *On Decoloniality: Concepts, Analytics, Praxis.* Durham: Duke University Press, 2018.

Milne, Drew, and John Kinsella. "Nuclear Theory Degree Zero, with Two Cheers for Derrida." In "Nuclear Theory Degree Zero: Essays against the Nuclear Android," edited by Drew Milne and John Kinsella. Special Issue. *Angelaki* 22, no. 3 (2017): 1–16. https://doi.org/10.1080/0969725X.2017.1387358.

Misic, Mathus, and Nada Kujundzic, eds. *Energy Humanities: Current State and Future Directions.* Berlin and Heidelberg: Springer, 2021.

Murphy, Michelle. "Alterlife and Decolonial Chemical Relations." *Cultural Anthropology* 32, no. .4 (2017): 494–503. https://doi.org/10.14506/ca32.4.02.

NHK Merutodaun shuzaihan. *Fukushima Daiichi Genpatsu Jiko no Shinjitsu* (NHK Reporting Team: The Fukushima Meltdown. The Truth about the Fukushima Nuclear Disaster). メルトダウン取材班『福島第一原発事故の真実』。講談社、Tokyo: Kodansha, 2021.

Oxfam International. "5 Natural Disasters That Beg for Climate Action." https://www.oxfam.org/en/5-natural-disasters-beg-climate-action.

Platte, James E. "Exporting Nuclear Norms: Japan and South Korea in the International Nuclear Market." *Journal of Indo-Pacific Affairs* 3, no. 2 (Summer 2020): 127–45. https://www.airuniversity.af.edu/JIPA/Display/Article/2210987/exporting-nuclear-norms-japan-and-south-korea-in-the-international-nuclear-mark.

Quijano, Anibal. "Colonialidad, Poder, Cultura y Conocimiento en America latina (Coloniality, Power, Culture, and Knowledge in Latin America)." *Dispositio* 24, no. 51 (1999): 137–48. https://www.jstor.org/stable/41491587.

– "Coloniality and Modernity/Rationality." Translated by Sonia Therborn. *Cultural Studies* 21, nos. 2–3 (2007): 168–78. https://doi.org/10.1080/09502380601164353.

– "Coloniality of Power, Eurocentrism, and Latin America." Translated by Michael Ennis. *Nepantla: Views from South* 1, no. 3 (2000): 533–80. muse.jhu.edu/article/23906.

Reif, Kingston. "US Nuclear Modernization Programs." "Fact Sheets and Briefs." *Arms Control Association*, August 2018. https://www.armscontrol.org/factsheets/USNuclearModernization.

Roy, Rohan Deb. "Decolonise Science – Time to End Another Imperial Era." *The Conversation*, 5 April 2018. https://theconversation.com/decolonise-science-time-to-end-another-imperial-era-89189.

Ruff, Tilman, and Margaret Beavis. "Japan Plans to Dump a Million Tonnes of Radioactive Water into the Pacific. But Australia Has Nuclear Waste Problems, Too." *The Conversation*, 22 October 2020. https://theconversation.com/japan-plans-to-dump-a-million-tonnes-of-radioactive-water-into-the-pacific-but-australia-has-nuclear-waste-problems-too-148337.

Samenow, Jason, and Matthew Cappucci. "Severe Heat Wave Builds across Western U.S. after Nation's Hottest June on Record." *Seattle Times*, 9 July 2021.

Simmonds, Emily, with Max Liboiron. "Nuclear State, Nuclear Waste." *Discard Studies*, 11 May 2018. https://discardstudies.com/2018/11/05/nuclear-state-nuclear-waste.

Simpson, Michael. "The Anthropocene as Colonial Discourse." *Environment and Planning D: Society and Space* 38, no. 1 (February 2020): 53–71. https://doi.org/10.1177/0263775818764679.

Smith, Linda Tuhiwai. *Decolonizing Methodologies: Research and Indigenous Peoples*, 2nd ed. London: Zed Books, 2012.

Stawkowski, Magdalena E. "Radiophobia Had to Be Reinvented." In "Invisible Harm: Science, Subjectivity, and the Things We Cannot See," edited by Donna Goldstein. Special Issue. *Culture, Theory and Critique* 58, no. 4 (2017): 357–74. https://doi.org/10.1080/14735784.2017.1356740.

Stengers, Isabelle. "Introductory Notes on an Ecology of Practices." *Cultural Studies Review* 11, no. 1 (2005): 183–96. https://doi.org/10.5130/csr.v11i1.3459.

Tanter, Richard. "After Fukushima: A Survey of Corruption in the Global Nuclear Power Industry." *Asian Perspective* 37, no. 4 (2013): 475–500. http://www.jstor.org/stable/42704842.

Taylor, N.A.J., Paul Brown, and Ellise Barkley, eds. "Reimagining Maralinga." Special Issue. *Unlikely: Journal for Creative Arts* 5 (2018).

Taylor, N.A.J., and Robert Jacobs, eds. *Reimagining Hiroshima and Nagasaki: Nuclear Humanities in the Post–Cold War.* New York: Routledge, 2018. https://unlikely.net.au/issues/issue-05.

Tejeda, Carlos, Manuel Espinoza, and Kris Gutierrez. "Toward a Decolonizing Pedagogy: Social Justice Reconsidered." In *Pedagogies of Difference: Rethinking Education for Social Change*, edited by Peter Pericles Trifonas, 9–38. New York and London: Routledge, 2003.

Titterington, David. "Nuclear Colonialism." 5 June 2019. https://davidtitterington.medium.com/nuclear-colonialism-98463eb1d2f7.

Todd, Zoe. "Indigenizing the Anthropocene." In *Art in the Anthropocene: Encounters among Aesthetics, Politics, Environment, and Epistemology*, edited by Heather Davis and Etienne Turpin, 241–54. London: Open Humanities Press, 2015. http://library.oapen.org/handle/20.500.12657/3319110.26530/OAPEN_560010.

Topçu, Sezin. *La France nucléaire: L'art de gouverner une technologie contestée* (Nuclear France: The Art of Governing a Contested Technology). Paris: Editions du Seuil, 2013.

Vivoda, Vlado, and Geordan Graetz. "Nuclear Policy and Regulation in Japan after Fukushima: Navigating the Crisis." *Journal of Contemporary Asia* 45, no. 3 (2015): 490–509. https://doi.org/10.1080/00472336.2014.981283.

Wallace, Molly. *Risk Criticism: Precautionary Reading in an Age of Environmental Uncertainty.* Ann Arbor: University of Michigan Press, 2016.

Walton, Calder. "The Deadly Fallout of Disinformation: How the History of Soviet Lies after Chernobyl Sheds Light on Our Fight against Covid-19." *Washington Post*, 8 July 2020. https://www.washingtonpost.com/outlook/2020/07/08/deadly-fallout-disinformation.

Watts, Jonathan. "Canadian Inferno: Northern Heat Exceeds Worst Case Climate Models." *The Guardian*, 2 July 2021. https://www.theguardian.com/environment/2021/jul/02/canadian-inferno-northern-heat-exceeds-worst-case-climate-models

White, Rob. "Environmental Victims and Resistance to State Crime through Transnational Activism." *Social Justice* 36, no. 3 (2009): 46–60. http://www.jstor.org/stable/29768548.

Whyte, Kyle. "Indigenous Climate Change Studies: Indigenizing Futures, Decolonizing the Anthropocene." *English Language Notes* 55, nos. 1–2 (Fall 2017): 153–62. https://doi.org/10.1215/00138282-55.1-2.153.

Williams, Park A., John T. Abatzoglou, Alexander Gershunov, Janin Guzman-Morales, Daniel A. Bishop, Hennifer K.Balch, and Dennis P. Lettenmaier. "Observed Impacts of Anthropogenic Climate Change on Wildfire in California." *Earth's Future* 7, no. 8 (2019): 892–910. https://doi.org/10.1029/2019EF001210.

Yusoff, Kathryn. *A Billion Black Anthropocenes or None.* Minneapolis: University of Minnesota Press, 2019.

ONE

Aftereffects of Chernobyl and Fukushima

1

"The Future Is Behind Them!": Post-Apocalypse and the Enduring Nuclear in Post-Soviet Russian Fiction

SHARAE DECKARD

A powerful monologue in Svetlana Alexievich's polyphonic oral history, *Voices from Chernobyl*, captures the crisis of representation that surrounds nuclear disaster. Journalist Anatoly Shimanskiy reflects on the impossibility of writing about Chernobyl when he can't fathom the long temporality of radioactive decay or imagine a future in which anthropogenic radiation will no longer affect people and environments: "Beyond that my consciousness couldn't go. I couldn't even understand anymore: what is time? Where am I? To write about that now, when only ten years have gone by. Write about it? I think it's senseless. You can't explain it, you can't understand it." He concludes: "Show me a fantasy novel about Chernobyl – there isn't one! Because reality is more fantastic."[1]

Paradoxically, what Eliot Borenstein has called the "Chernobyl model" of nuclear catastrophe looms large in post-Soviet Russian literary imaginations of the future, most prominently in twenty-first-century post-catastrophic speculative fiction.[2] The giant Russian publisher Eksmo features an entire series, titled "Russian Apocalypse," of novels set in worlds devastated by atomic war, including Viktor Glumov's *Death City* (2012), Viacheslav Khvatov's *Atomic Autumn* (2012), and Oleg Kulagin's *Russian Dawn* (2011).[3] In these texts, wounded heroes battle Manichean forces in a hostile environment, their survival guaranteed by the "will to kill."[4] However, the nuclear events that are their *raison d'*être occur offstage in the historical past. The nuclear is conceived as a zone of alienation or as a durable catastrophe only perceptible in its transgenerational effects over the *longue durée*, as in the case of the genetic "Consequences" and uncanny metamorphoses of human-animals persisting long after "the Blast" in Tatyana Tolstaya's *The Slynx*, or the mutated life forms and post-human species evolving after global atomic war in Dmitry Glukhovsky's *Metro 2033*. This chapter explores generic and ideological contradictions in the post-catastrophist imaginary of the nuclear, using *Metro 2033* as a paradigmatic example of post-Soviet Russian speculative fiction, and examines the transmedial content and refashioning of Glukhovsky's novel for international video game markets. I will focus on the concept of the "enduring nuclear," a term chosen to evoke the long, seemingly imperceptible temporalities of radioactive decay and the "slow violence" of nuclear disaster,[5] as well as the enduring social legacies of the militarized, racialized, and colonially inflected regimes of nuclear governance, but which also alludes to the aesthetic dimensions of the post-apocalyptic genre and how its formal conventions are bound up with a sense of impossible or catastrophic futurity, resulting in the contradiction that speculative post-Soviet fictions such as *Metro 2033*, while post-apocalyptic, seem incapable of imagining a transition to a post-nuclear modernity.

If a central task of energy humanities is to ask "Who gets to imagine energy futures?,"[6] comparative criticism should be essential to the formulation of a nuclear energy humanities, especially one that seeks to decolonize the imaginaries and ideologies associated with forms of nuclear colonialism and development. A world-literary approach to nuclear environmental criticism can uncover not only imaginaries corresponding to British and American nuclear culture and the dominance of the US imperium, but also the enduring nuclear futures imagined in post-Soviet Russia and other global contexts of nuclear development in the twenty-first century, such as China, India, and Iran. Our capitalist nuclear modernity is above all a global modernity – singular, planetary, and world-historical – in which all forms of earthly life have been altered by the atmospheric effects of nuclear development. As Joseph Masco observes,

> there have been thousands of nuclear explosions conducted in test sites around the world – deliver[ing] vast amounts of radioactive materials into the global biosphere. Taken up by global wind currents as well as plants and animals, these materials were delivered into each and every person on the planet and deposited in varying amounts within their genomes. As a result, all of us carry in our bodies traces of Strontium-90 and other human-made radioactive elements from the nuclear test program, making life on planet Earth quite literally a post-nuclear formation.[7]

We may live in a "post-nuclear formation" in the sense of occupying a planet on which the genomic and geophysical mutations unleashed by nuclear disasters and warfare will continue to unfold over temporalities that exceed individual human lifetimes, but we do not live in a post-nuclear age in the sense of having abandoned nuclear energy regimes or weapons systems, or having overcome the social inequities of nuclear colonialism. The "amazing political flexibility of nuclear exceptionalism" lingers into the twenty-first century, long after its emergence in the postwar discourses of Cold Warriors.[8] The "nuclear state" persists as a marker of modernity in which geopolitical status is gauged in proportion to the number of nuclear missiles owned and the capacity to process nuclear waste, while the lack of weapons of mass destruction is a sign of "backwardness."

In twenty-first-century Russia, energy regimes have been central to the symbolic and material constitution of the post-Soviet nation as an enduringly nuclear state. In the decade from 2001 to 2011, the term "nuclear renaissance" was used throughout the global nuclear industry to refer to a revival of nuclear power, which experts hoped would be driven by Euro-American countries building new reactors in an attempt to reduce greenhouse gas emissions and escape the rising prices of fossil fuels, as well as to encompass the expansion of nuclear programs in nations such as China and India. In Russia, Putin's administration removed barriers to the expansion of the nuclear industry by dismantling expert assessment and regulation, extending the licences of old reactor models after their planned operational lifetimes, and setting three major goals for Rosatom, the state-owned nuclear corporation: to increase the nuclear contribution to electricity generation by 30 percent as of 2030 by building forty new reactors; to continue "sustainable" development of the nuclear weapons complex; and to expand into the international market of nuclear technology by selling products and services such as the reprocessing of imported nuclear waste and the construction of reactors for China, India, and Belarus.[9] An inherent tension between conceptions of nuclear power as *energy* and as *weaponry* characterizes the historical development of nuclear technology, where the idea of the nuclear as "safe" or "destructive" energy depends on the rate of fission.[10] If a hallmark of the Soviet Union's hegemony as a "Great Power" was its "atomic communism,"[11] grounded in a monopoly of uranium reserves and the construction of weapons of mass destruction, then a nuclear energy renaissance in post-Soviet Russia would symbolically reaffirm its geopolitical dominance as a nuclear state even after the loss of its Soviet territories.

In 2011, however, the disaster at the Daiichi nuclear power plant in Fukushima, Japan, exposed the global nuclear industry to renewed critique, undermining its claims to be a "clean," "safe" alternative to fossil fuels and leading energy-consuming nations to reconsider their plans for expansion. Russia's nuclear aspirations faltered when the state encountered strong protests from environmental groups in cities where proposed plants were to be constructed, and it failed to raise sufficient capital investment or machinery capacity to build new reactors, with the result that the number of new plants was cut by two-thirds, leading the World Information Service on Energy (WISE) to conclude: "The Russian 'nuclear renaissance' may well be over, even before it started. And this is good news because reactors are expensive, inefficient and dangerous, just as they were 24 years ago when Chernobyl happened."[12] Indeed, the eco-nationalist movements prompted by the breach in the core reactor of the Chernobyl power plant in 1986 were instrumental in hastening the collapse of the Soviet Union, dislodging "the discourses of benevolent technology, scientism, and 'humane' progress that been central to Soviet identity" and demanding that the nuclear state account for its catastrophic legacy of "damage to ecosystems and public health."[13] However aspirational, Russia's resurgent nuclear ambition demonstrates the persistence of nuclear ontologies in post–Cold War geopolitics, as well as the ways in which public imaginaries of energy futures can be swiftly renegotiated in contexts of nuclear emergency and activism.

Contemporary Russian literature is haunted by nuclear disasters and catastrophes, which most critics have read as allegorizing the trauma of the geopolitical collapse of the Soviet Union and the violence of the transition from a sovereign socialist state to a capitalist economy. Apocalyptic frameworks have a peculiar charge in post-Soviet fiction, employing as they do a "cyclical model of history" that establishes the dystopian future as "a repetition of a familiar past"[14] and expressing the sense of a paradoxically doubled temporality where the category of future seems to have "reversed its interpretational vector" and "the Russian present is no longer interpreted in terms of the great future to come, but in terms of what happened (and failed)."[15] The post-apocalyptic carries a register of post-imperial melancholy, bound up with nostalgia for the 2 million square miles of lost Soviet territory. It also evokes a "post-shock subjectivity" produced by economic shock therapy: "the privatization of oil and gas resources, intertwined with massive violence and social processes that were aimed at reshaping subjectivity according to the demands of profitability and the market."[16] However, the environmental content of post-apocalyptic fictions has less often been interpreted not only as allegory of the trauma of ideological transition and the betrayal of the bright Soviet future, but also as mediating the legacies of the nuclear catastrophes of the Soviet past and the ecological crises that have been compounded by post-Soviet neoliberalization.

The energy ontologies of oil and nuclear power have a particular salience in contemporary Russian fiction, corresponding to the energy regimes shaping the formation of the post-Soviet state. J.R. McNeill defines energy regimes as "the collection of arrangements whereby energy is harvested from the sun (or uranium atoms), directed, stored, bought, sold, used for work or wasted, and ultimately dissipated"[17] and distinguishes between somatic energy regimes, which rely on muscle power, and exosomatic regimes, which use fossil fuels or uranium. However, Michael Niblett emphasizes that these regimes are also social, understanding energy sources not as static inputs "but as simultaneously structured and structuring relations of the patterned totalities of historical systems."[18] In this chapter I argue that the toxic imaginaries of post-Soviet post-apocalyptic fictions such as *Metro 2033* are "energy forms"[19] that mediate the affects and social relations corresponding to the exosomatic energy regimes structuring the nuclear modernity of post-Soviet Russia. The enduringly nuclear futures they imagine "tell us more about the present than they do about the future," mediating anxieties around energy transition and the socio-ecological asymmetries of nuclear regimes, as well as registering the difficulty of imagining the post-nuclear organization of human societies.[20]

1. The Nuclear Uncanny

I begin by examining the imagination of catastrophic futurity in Dmitry Glukhovsky's *Metro 2033*, which turns on ecogothic revelations of a nuclear uncanny. *Metro 2033* is set in post-apocalyptic Russia in the near future, twenty years after a global atomic war has turned the earth into a radioactive wasteland. Moscow, instead of New York or London, functions as the apocalyptic metonym for global annihilation. The only human survivors are those who happened to be underground in the Moscow metro during the missile strikes, and a subterranean society evolves among the 70,000 subway inhabitants. The geopolitical causes of the atomic war are never revealed; the event happens offstage, its causality receding from memory in the consciousness of the survivors, who are immersed in permanent war, fighting for water, food, ammunition, and fuel; conducting wars over ideological differences between factions of Stalinists, Trotskyists, free market fundamentalists, and religious cultists; and combating psychic disturbances as well as waves of mutated creatures.

This world is distinctly ecophobic, brimming with representations of weirded nature in the tunnels and abandoned ranges where "the radioactive earth, disfigured by thousands of explosions, ploughed with trenches and pitted with catacombs, put forth monstrous sprouts."[21] The novel's protagonist, Artyom, is orphaned after a "living torrent" of mutated rats overruns his station.[22] If in Andrei Tarkovsky's film *Stalker*, one of Glukhovsky's inspirations for the novel, the eerie Zone into which the protagonist ventures is a space of mystic immanence hallowing the radical contingency of the natural world, then in *Metro 2033*, irradiated environments are portrayed as zones of excrescent fecundity, nuclear nature that has exceeded human control. The symbolic opposition of Artyom's home station of VDNKh, site of the Soviet-era Exhibition for the Achievements of the National Economy, and the Botanical Gardens occupied by post-human creatures, sets up a Cartesian dualism between culture and nuclear nature. The scientistic modernity of human civilization is opposed to the post-apocalyptic nature of a forest gone feral, and Artyom perceives the organic structures of the Dark Ones' city as cancerous mutations: "It was a strange and repulsive sight: a large city like a gigantic life-giving organ, pulsing and quivering, that stretched out for several square kilometres. The sky gradually was being painted with morning colours, and this terrible tumour was becoming ever more visible: a living membrane entangled with veins, tiny black figures crawling out of cesspool exits."[23]

Throughout the novel, characters articulate Hobbesian visions of a catastrophic futurity in which human culture is failing to hold back the tide of a frenzied nature, sharply differentiating what is conceived as human and as non-human. They invest the swarming life of nuclear nature with a terrifying alterity, couched in neo-Darwinian rhetoric of a battle for species survival against "strange, freakish, and dangerous creatures, the likes of which might well have brought Darwin himself to despair with their obvious lack of conformity to the laws of evolutionary development."[24] The novel's nuclear uncanny invokes a return of the repressed in which evolution itself is imagined as a revenge that threatens to replace *Homo sapiens* with a new species in retribution for the atomic destruction of the earth.

As Sukhoi, the protagonist's adoptive father, bitterly exclaims, the rise of the Dark Ones, whom he calls *Homo Novus*, marks the decline of human mastery: "Surrender, Homo sapiens! You are no longer the king of nature! … Evolution, the laws of which you understood, has already created its new branch, and you are no longer the latest stage."[25] This challenge to anthropocentrism presents profound anxiety, rather than an inspiration to imagine forms of multispecies cohabitation. Sukhoi mocks hopes that "the levels of radiation would lower, and people could return to the surface again," as a postlapsarian fantasy of return to pre-atomic nature, and similarly dismisses the suggestion that "the underground is now our natural habitat" (40), arguing that humans have become as "pale and sick as Wells' Morlocks" from dwelling underground, subsisting on a meagre diet comprised of mushrooms cultivated with human dung and pork from pigs. However, he can propose no other organization of life beyond the conquest of nature; his declensionist warning mourns the loss of species dominance: "We aren't at home there on the surface anymore. The world doesn't belong to us anymore."[26]

This Spenglerian scenario of "world-dissolution" and devolution can be read allegorically as a structure of feeling encapsulating nostalgia for Soviet hegemony over nature, couched in the "exhausted imagination of a dying class" that compensates for "the cancelled future of a vanished colonial and imperial destiny" by "intoxicat[ing] itself with images of death."[27] Ulrich Schmid argues that "the homelessness of the post-apocalyptic hero corresponds to the dominating self-perception of [post-Soviet] Russian society."[28] However, Sukhoi's profound sense of unworlding can also be interpreted as the ecological unmooring of humanity-in-nature, the novel's prolepsis of "the end of cheap nature," Jason W. Moore's phrase for the ecological crisis produced by the tendency of capitalism to appropriate energy, raw materials, and the food produced through the work of both human and extra-human natures as a source of unpaid surpluses to increase the productivity of its accumulation regime, without acknowledging biophysical limits.[29]

The Metro as giant air-raid shelter seems to offer a fantasy of the "cheap nuclear": of nuclear endurance without ecological consequences or costs, where humans magically bypass famine induced by the mass death of crops and livestock and the poisoning of water, soil, and air. Yet their subterranean environments are organized around unsustainable monocultures, wholly lacking biodiversity. Their late modernity is truly a "toxic modernity" in which "toxicity and alienation are the perverse conditions for modern subject formation."[30] If humans "are creatures constituted by radiation, solar and otherwise,"[31] then the station dwellers are cut off from the sustaining vitality of solar energy, grown "thin and pale" and fearful of the "searing rays of the sun."[32] They are both physically and psychically dispossessed, prone to etiolation, depression, and paranoia. Their ecological regime teeters on the edge of exhaustion, dependent on dwindling relics of exosomatic regimes: canisters of petrol scavenged from the surface, plastic bottles of synthesized multivitamins. Human culture, clinging tenuously to a tattered web of life, is subject to the contraction of the plenitude once promised by capitalist modernity: stations are lost, parts of the grid gutter into darkness, tools break, resources dwindle, and even language disappears as the sensuous totality of nature diminishes.

The Metro society focuses on replicating the technics of the capitalist production of nature, but it does not imagine how to *live differently* and is unable to think of how to increase the diversity of life, reverse ecological degradation, or adapt to the altered ecologies that surround them. It remains trapped in Malthusian conceptions of scarcity and perennial competition, unable to conceive an equality that "facilitates the coexistence of manifold forms of being – human, non-human, and post-human – and their various attachments to ecosystems in the world without placing these in hierarchies."[33] The novel concludes with a nuclear missile detonation, which follows the original atomic war's planetary ecocide with a second holocaust of the new forms of irradiated nature that have emerged on the surface: "A pitch-black cloud shot upwards, new explosions circled him from all sides and the city crashed down, emitting a tired, dying moan. It was clouded by the thick smoke of the burning forest. From the sky more missiles fell, and each death reverberated with a melancholy pain in Artyom's soul."[34] This tragic repetition not only fails to imagine a post-nuclear energy future, but worse, conceives that futurity in terms of doubled, compounded nuclear catastrophe.

The catastrophic future of *Metro 2033* is thus enduringly nuclear in two senses: as weaponry and as energy. The Metro is illuminated by scarlet emergency lights powered by secret reactors, suggesting a zombified endurance of energy infrastructure long after the collapse of the Russian nuclear state. The diurnal distinction between day and night collapses in the perpetually illumined stations, and dwellers depend on giant electric clocks – "considered to be as important as strategic objects like the arms store, the water filters and the electric generator" – to tell them when to sleep, their circadian rhythms having been unmoored from solar time.[35] The total dependency of the Metro dwellers on electricity-powered illumination – to be cut off from power is literally to become powerless, plunged into darkness and horror – can be read as an ecogothic assertion of the energy unconscious of a post-Soviet society "addicted" to electricity, unable to relinquish the assumption that "a standing reserve of energy" will always exist.[36]

This is "gridlife dependency" in its starkest form, exaggerated by the novel's setting in a wholly constructed environment that is irrevocably cut off from solar energy, paradoxically dark even while perpetually electrified.[37] The solipsism of the Moscow setting, in which the finitude of the energy regime is experienced as tragic loss, and where the irradiated natures produced by past atomic disaster are cast as externalized enemies, can also be seen as a return of the repressed reality of nuclear colonialism throughout the rest of the former Soviet Union, come back to haunt the post-imperial metropole. What is portrayed as a near-future apocalypse for the Muscovites is the energy present of real-life populations from the Urals to the Ukraine, left to inhabit "disabling environments" toxified by uranium ore extraction, nuclear fuel recycling, waste processing and disposal, and power plant disasters such as Kyshtym and Chernobyl.[38] For these people, radiation and toxicity are not perverse externalities but rather immanent to their daily lives, embedded in their very bodies as cellular mutations. Sukhoi's admonition that "the future is behind them!" can thus be read not only as registering fear that Russia's greatest achievements lay in the pre-catastrophic past, but also as an unconscious reflection on how the overriding sense of loss associated with the imagined future says something about what the novel does not portray: the hidden costs of Russia's nuclear past and present.

Significantly, the novel's setting draws on historical fact. Glukhovsky was inspired by revelations during *perestroika* that the Moscow subway was the "world's biggest nuclear shelter," equipped with doors that could be hermetically sealed and supplied with artesian wells and air filters to cleanse nuclear contamination.[39] A second subway, the Metro-2 system, had been constructed alongside the first, a labyrinth of secret stations beneath the Kremlin, the ministries, the Russian State Library, and the army headquarters, with the intention of preserving not only the *nomenklatura* and security services but also the intellectual and scientific elites: "The metro actually has two separate lines – one for common people and one for the rulers and the elite. And this was quite a shock to me … You're riding through a subway tunnel, you see a little door, and behind it is a small tunnel, which leads to this huge [bunker]."[40]

A large part of the novel's gothic effect derives from this *unheimlich* revelation of a concealed reality, the nuclear uncanny of the Soviet past expressed in an urban infrastructure that routinizes the prospect of nuclear annihilation and instant mass death. It is the potential deathworld beneath the surface of the everyday, the physical manifestation of the "new social orientation toward death" that underlies "atomic health" in nuclear modernity, "both normalized and rendered invisible as a new form of nature."[41] Just as the Metro-2 architecture lingers into the post-Soviet era like a phantom skeleton undergirding the political, cultural, scientific, and military apparatuses of the city, so too do the "nuclear effects" of energy regimes persist into the twenty-first century, when the threat of nuclear destruction seems to have waned in political consciousness, but weapons systems, ore refineries, waste processing plants, and aging reactors remain undismantled, disclosing an "inability to think beyond these atomic potentialities – let alone to reduce the technological, political, and environmental conditions that continue to support them."[42] The underground's "nuclear nature" is rooted in class and race asymmetries and an "instrumental calculus of death" that divides human life into those who are valued and whose health is worth "securing" in the specialized Metro-2 system, and the surplus population, whose lives are mediated by catastrophic risk. [43]

Artyom's odyssey through the Metro underworld performs a dark tourism of the necropolis and hecatomb, animated by spectres of the nuclear dead. Artyom possesses a special "supersensitivity" that enables him to read the changing aura of the subway by "listening to the tunnel."[44] In his epiphanic hallucinations, inanimate objects take on life and his consciousness expands to take in the totality of the underground environment:

> The flowing sounds from that pipe seemed to him the same as ether, slowly extending along the tunnel ... Suddenly it seemed to Artyom that he was standing on the threshold of an understanding of something important as though the last hour he had spent wandering in the pitch-black darkness of the tunnels ... had pulled the curtain of this great mystery slightly to the side, separating all rational beings from a knowledge of the true nature of this new world.[45]

Other explorers experience this psychic noise as "tunnel horror,"[46] a madness that breaks their minds, like the siren songs lashing the sailors of the *Odyssey*, which Horkheimer and Adorno read as nature's immanence confronting the "identical, purpose-directed, masculine character" of the rational ego.[47] By contrast, Artyom abandons himself to prophetic intuition and allows his subjectivity to dissolve into a wider sensorium, thus opening himself to the auditory manifestations of the repressed reality of nuclear nature. His extra-sensory telesthesia can be interpreted as the perception of catastrophic legacies and temporalities beyond the normal range of empirical senses, recalling Sarah D. Phillips's description of "Chernobyl's sixth sense" as an "ever-present awareness" that haunts those who share the experience of nuclear disaster, rupturing history into life *before* Chernobyl and life living *through* Chernobyl's still unfolding catastrophe.[48]

In Artyom's hallucinations, time and space are subject to curious expansion and contraction as he passes through particularly toxic or irradiated tunnels where mass deaths have occurred. This temporal relativity evokes the relatively short *durée* of individual human life against the *longue durées* of radioactive half-lives, invisible and seemingly impossible to perceive, yet everywhere shaping the Metro environment. The mystical Khan attributes the tunnel noise Artyom hears to the "voices of the dead," emanating from all those killed since the original atomic event: "How many megatons and bevatons does it take to disperse the noosphere? ... In our world the soul stays in the metro after death ... It will rush around under the arches of these underground tunnels until the end of time because there isn't anywhere for it to go ... We live amidst the souls of the dead."[49]

Here again is the nuclear uncanny, where the souls of the nuclear dead circulate through the infrastructure of the underground like the invisible particles of radioactive decay, impossibly animating the environment with reminders of catastrophic violence long after their physical deaths. Artyom's ability to bear witness posits him as a Virgil of the underworld with the potential for *anamnesis*; however, this testimonial capacity is held in tension with the text's use of gaming conventions, as I will examine in the next section.

2. Nuclear Transmediality

The transmediality of nuclear culture, with its explicit origins in imperialist and capitalist institutions, offers a rich terrain for nuclear humanities criticism. The emergence of video games as a unique medium is inseparable from nuclear modernity, for they are generated from the same military research matrix that resulted in the internet and the personal computer. The very ability to play such games relies on the material infrastructures of military–industrial capitalism and on the energy regimes that power electronic devices. Dyer-Witheford and De Peuter argue that "video games are a paradigmatic medi[um] of Empire – planetary, militarized hypercapitalism."[50] Digital games originated in Cold War nuclear simulations, invented in the 1950s and 1960s by the programmers tending Pentagon military computers used to war-game missile trajectories. These hackers transformed oscilloscope screens into fantastic playgrounds on which to fire space torpedoes at imagined virtual enemies.[51] In the gaming industry, nuclear scenarios were central to the plots of many games and were influenced by the recruitment of Japanese manga and anime artists in the 1980s, who cross-pollinated transnational subcultures shaped by nuclear events and regimes.

Manga is a cultural form that gained prominence among Japanese youth born in the era of post-nuclear reconstruction, a generation haunted by the sense of being "a new humankind" indelibly inflected by the trauma of Hiroshima.[52] Shaped by a post-traumatic, anti-authoritarian sensibility, manga was catapulted to a nationally significant genre with Keiji Nakazawa's iconic testimonial series *Barefoot Gen* (1973 to 1985), which narrated the story a six-year-old boy who survives the bombing of Hiroshima, loosely based on Nakazawa's own experiences as an "explosion-affected person." Similarly, the anime film *Akira* (1988), directed by Katsuhiro Otomo and based on Otomo's manga of the same name, wrestles with the horror and trauma of nuclear holocaust, depicting a tragic hero whose mutant psychic abilities trigger a big bang in an alternate universe. Twenty-first-century popular culture continues to be saturated with transmedial imaginations of nuclear apocalypse, often in contradictory ways that demonstrate the testimonial power of forms such as manga and anime to grapple with the socio-ecological fallout of nuclear disaster, but that also revel in fantasies of nuclear empowerment and mutant creativity, as in the ideological apparatuses of video games that reinforce the subjectivities of "worker-consumer and soldier-citizen."[53]

Metro 2033 is a paradigmatic example of the transmediality of nuclear culture in its serialized online composition and distribution across multiple media platforms, and of intermediality in its transnational allusions to other video games, films, and fiction. After the first version of the novel was rejected by publishers, Glukhovsky turned to the Internet, posting each chapter on LiveJournal and conducting interactive "literary beta-testing" to gauge reception and recalibrate plot elements and world-building according to reader feedback.[54] Fans protested tragic or mystical endings in which Artyom died or was dissolved into the collective consciousness of Homo Novus; thus Glukhovsky revived the "hero" and allowed him to consummate his ecocidal quest to fire nuclear missiles. Much of the online feedback was oriented toward the fine details of weapon fetishism and techno-military idolatry corresponding to first-person shooters: what calibre of ammo is used by a sniper gun, why a flamethrower can't be used in a tunnel, and so forth, despite the fact that little shooting actually takes place in the novel, in contrast to its game adaptation.[55] The resulting text was a more mainstream product calibrated for a market of readers influenced by gaming conventions. The finalized form of the novel, published by Eksmo, was a bestseller, selling more than 500,000 copies in Russia alone, and was developed into a video game by 4A Games in 2010.[56]

In interviews, Glukhovsky has attested that the novel was heavily influenced by his "passion" for playing the first *Fallout: A Post Nuclear Role Playing Game* (1997) from American video game developer Interplay Entertainment. The post-apocalyptic open-world RPG is set in an alternate-history version of the mid-twenty-second century, where the world has been destroyed by a nuclear war sparked by inter-state energy wars over the last of the non-renewable commodities, oil and uranium. The player's avatar, a Vault Dweller miraculously preserved from radiation damage, is tasked with exploring the Wasteland, which is littered with technological remnants from the cataclysmic war and teeming with mutated creatures, in search of a Water Chip, in order to save the inhabitants of the Vault.[57] The *Fallout* series' retrofuturist aesthetics create "an ironically safe space" to explore contemporary anxieties about nuclear technology through the simulation of US Cold War environments, inscribing the tension between critique of the glorification of atomic weapons and the ludic pleasures of nuclear power into the gameplay, while reinforcing ideologies of Malthusian competition over scarce resources.[58]

The novel's serialized online composition invited feedback that encouraged readers to treat the world of *Metro* as an RPG that they could explore and adjust to their liking. Structured around an epic quest akin to a game's main quest arc, the text's narrative translates the formal structure of the RPG into novel form; it is not strongly plot-driven, but rather episodic, reflecting the linear, bead-on-a-string structure of many video games as it passes from station to station, encountering different environments. Its world-building privileges atmosphere over content, recalling the emphasis on immersion typical of open-world games. However, it offers a bleaker, more ecogothic vision than that of the US video game, replacing the "Raygun Gothic" pop art and 1950s nostalgia of *Fallout* with brooding tunnels and survival horror. Its simplistic character development revolves around two archetypes, the rugged survivalist and the mystic seer, which conjoin *Fallout*'s Wanderer with the figure of the "stalker" appropriated from Arkady and Boris Strugatsky's 1971 science fiction novel *Roadside Picnic* and reimagined in Andrey Tarkovsky's 1979 film *Stalker.*

The Strugatskys' novel is set in a post-visitation world in which aliens have left behind six Zones full of mysterious objects. The "stalkers" are hardy adventurers who scavenge for artifacts in the dangerous Zones; their name, an Anglicism borrowed from Rudyard Kipling's "Stalky" stories (1899), evokes imperialist connotations. Tarkovsky reinvented his Stalker as a mystical guide on an anguished search for spiritual knowledge in a sentient Zone. Famously shot in the toxic ruins of power plants and chemical factories in Tallin, the film created an indelible image of a contaminated ecology littered by decaying human debris, which would haunt the Soviet nuclear imaginary like a spectre of the future, an uncanny prolepsis of Chernobyl's zone of restriction. As Žižek comments, for ex-citizens of the Soviet Union the notion of a "forbidden Zone" gives rise to multiple associations: as prison gulag or secluded domain of the *nomenklatura,* but also as "a territory poisoned or otherwise rendered uninhabitable by some technological (biochemical, nuclear …) catastrophe, like Chernobyl."[59]

In the video game *S.T.A.L.K.E.R.: Shadows of Chernobyl*, released by Ukraine developer GSC Game World (2010), the Zone is explicitly reconfigured as the site of a second Chernobyl disaster, into which players venture in search of their identity while combating mutated creatures and animate objects. While the game features more than one ending, the supposedly "happy" conclusion requires the player not only to kill his adversaries but also to destroy all the elements that contain the zone's consciousness, violently imposing his ego on the external environment in another version of the doubled catastrophe. If the Chernobyl exclusion zone has inspired real-life forms of "dark and toxic tourism"[60] that commodify sites of industrial death and disaster, video games like *S.T.A.L.K.E.R.* enable a virtual tourism, exploiting the libidinal pleasures of nuclear horror and imagining nuclear simulations not as traumatic environments but rather as ludonarrative sandboxes where catastrophe is "reincorporated into the ordinary and thus available for comic play."[61]

The *Metro 2033* novel borrows from Tarkovsky's metaphysical vision of toxic modernity but transplants the zone to the post-Soviet metropole and inverts the spiritual awe that Tarkovsky's Stalker feels for the zone into an affect of survivalist horror more akin to the video game adaptation. Artyom romanticizes stalkers as conquerors of the weird nature of the surface:

> Every stalker became a living legend, a demigod, whom everyone, young and old, regarded with rapt amazement. In a world in which there was nowhere left to sail or fly ... children dream of becoming stalkers. To strike out, clothed in shining armour ... climbing to the surface, to the realm of the gods, to do battle with monsters and, returning underground, to bring the people fuel, military supplies, light and fire.[62]

This reactivates an imperial imaginary for the post-Soviet era, casting the nuclear wasteland as a new frontier to be conquered by explorers in protective suits and gas masks; it borrows heavily from the fantasies of atomic health in *Fallout*, where the toxic effects of radiation exposure become a "element of ludo" that can be effortlessly overcome by wearing "power armour" and consuming "Rad-Away."[63]

These power fantasies are explicitly gendered, centred in worlds of exclusively homosocial relations and literally toxic masculinity. In the novel and game versions of *Metro 2033*, there is a glaring absence of speaking female characters. The only referenced woman is Artyom's dead mother, whose memory functions as an Oedipal consolation when his existential homelessness becomes overwhelming. The nuclear power fantasy is also overtly racialized. A defining feature of the Dark Ones is their black skin and alleged animality. Artyom is haunted by dreams of "blacks" marching on the barricade at VDNKh, "naked, covered in black glossy skins, with huge eyes and mouths like gashes."[64] Characters throughout the novel describe Homo Novus as dehumanized monsters, speculating that they are mutants, cannibals, vampires, carrion-eaters, vermin, or undead; one soldier calls them "black bitches," combining misogyny and racism.[65] The invasion of the Other has been integral to video games since their inception, naturalizing the mutant-as-enemy in order to provide a ludonarrative justification for "murder simulation"; consequently, many first-person shooters reinforce ideologies of popular nationalism. At first glance, the xenophobic, racialized imagination of the Dark Ones in *Metro 2033* seems to accommodate such conventions without friction, particularly when invoked alongside allusions to Chechens and Georgians, inviting their interpretation as foreign "enemies" threatening post-Soviet Russia. However, the novel's conclusion denaturalizes the enemy-figure, subverting the pleasures of demonizing alterity.

Glukhovsky describes himself as a humanist, arguing that *Metro 2033* criticizes the resurgence of popular nationalism and white supremacy in post-Soviet Russia, particularly during the wars in Chechnya and Georgia for strategic dominance of oil-rich central Asia: "The biggest discovery I have made … in observing what is happening in Russia, was the sudden need for an enemy. Both an internal enemy and an external one as well." He attributes this invention of enemies to a compensatory mechanism of Russia's post-imperial "complex": "Russians used to be a part of a huge empire … We are facing the after effects of a war that we lost … And so there is a great need to restore this subjective feeling of superiority."[66] In this context, Artyom's post-apocalyptic quest can be reinterpreted not as a heroic adventure to eliminate ethnic Others who are disturbing the purity of the Russian underground civilization, but as a picaresque "tale of xenophobia" that turns on the decision of a naive young man, indoctrinated by ideologies of masculinist, racialized violence, "to fight or not fight the Dark Ones."[67]

Daria Kabanova argues that post-apocalyptic post-Soviet texts are riven by a crisis of subjectivity stemming "from the loss of the identifiable enemy" after the end of the Cold War, and "are actively searching for a new other (or others), against which a renewed national identity can be constructed."[68] She sees the hybrid "manimals" of novels such as Tatyana Tolstaya's *The Slynx* as attempting to negotiate a way around the dichotomies of negative identity formation by "turning to the ultimate other available to the human species, the animal."[69] In *Metro 2033*, a similar function is performed by the imagination of the Dark Ones: the negative dichotomies in which they are initially portrayed (of race, colour, animality, inhumanity, natural immanence, etc.) are blurred by the conclusion, collapsing the protagonist's epistemic basis for his subjective differentiation from the environment. The novel's erasure of the causality of the original atomic war seems emblematic of the loss of "political-historical agency" that characterizes many twenty-first-century post-apocalyptic novels, in contrast to "the nuclear apocalypses that dominated the Cold War imagination of the future, [in which] agency is retained always in the spirit of an urgent but still-timely warning: living in the present, rather than the scorched and radioactive future, we can choose not to build the last bomb and choose not to push the button that will launch it."[70] However, the spectre of agency is resurrected by the novel's final plot twist, which ironizes Arytom's decision to launch the remaining missiles of the Moscow arsenal against the Dark Ones' hive city, razing the teeming forests of the Botanical Gardens.

Only after he has triggered the countdown does he discover that the "Commoners," as they name themselves collectively, are a post-human species whose telepathic attempts to establish communication with humanity were intended to establish multispecies solidarity. When Artyom finally sees through the Commoners' eyes, he is consumed with a sense of wonder at their beauty and adaptability to nuclear nature: "Having disappeared into the gaze of the dark one, he suddenly saw the universe with its eyes. New life was being reborn and hundreds and thousands of individual minds were being joined together into a single whole … The resilient black skin allowed the dark one to endure both the scorching sun and the January frosts, the soft telepathic tentacles enabled it to caress any creation."[71]

The idea of the evolution of a post-human species that can effortlessly tolerate radiation is a form of wish-fulfilment, which imagines evolution as "fixing" the problem of nuclear radiation. However, Artyom's perceptual insight can be understood as the registration of nuclear immanence in his own body and subjectivity, fundamentally altered by habitation of a post-nuclear environment; in this sense, he is as much a post-nuclear creature as the Commoners he had previously imagined as fundamentally Other. Radiation, after all, "knows no boundaries, either political or biological."[72] His final epiphany is that he is not a "chosen one" in the sense that he thought he was: not a messianic soldier-citizen destined to eliminate an enemy, but rather a medium for voices, of the human dead and of the soon-to-be-dead post-human. The Commoners offer an alternative vision of a nature-society founded in symbiotic cohabitation instead of individual dominance, within a shared web of life in which everything is connected: "He understood that there was nothing dividing people and the dark ones. He understood they were not competing for survival but were two organisms intended by nature to work together."[73] Tragically, however, Artyom is unable to think beyond what Kamau Brathwaite has called the "missilic" consciousness of mass destruction until it is already too late.[74]

Metro 2033 thus subverts many of the imperial gaming conventions and tropes incorporated in its form: the epic hero's quest to destroy one true evil, the missilic imagination and white, male imperial gaze, the power fantasies and "strong man" myths, the Manichean enemy as racialized Other, the nature-versus-culture dualism, the privileging of action over feeling, the survivalist motivated by a "hypertrophic instinct for self-preservation."[75] However, in concluding with atomic destruction, the novel invokes a nuclear fatalism that deterministically assumes a doomed world, raising the utopian prospect of a symbiotic ecological commons, only to destroy it. The novel's imagination of an enduringly nuclear future represents a failure to "demilitarize the mind,"[76] a blockage of the capacity to think beyond the nuclear imaginary to pose "an entirely new kind of nature."[77] It suggests that post-Soviet Russian fiction remains in a "state of toxic suspension,"[78] employing categories of terror and post-apocalypse to imagine the future, rather than conceiving new forms of political action and social organization aimed at nuclear abolition or energy transition.

However, this failure can be partly recuperated through a dialectical reading of *Metro 2033*'s transmediality. Sven Cavalcanti argues that the post-apocalyptic video game "is about a lost world, not about an averted world. It displays nightmare fantasies that already exist"[79] and, as such, demands that the player anticipate the destructive tendencies of their existing societies: "Like no other cultural domain, computer games insist on the world being a hostile place that is by no means a favourable location to live in – nowhere else is the future definitively labelled as abortive."[80] *Metro 2033* is thus exemplary of the representational limits of post-catastrophic futurity bound up with the "Chernobyl model" of post-Soviet fiction that employs fantastic aesthetics to depict an enduring nuclear disaster. The novel is simultaneously constrained by and subversive of genre conventions, in danger of fostering an apocalyptic fatality that promotes political paralysis, yet also prompting the reader/player to recognize that ecological catastrophe and contaminated nature are not merely end-time scenarios of virtual worlds, but are the daily routine of their own present, where the materiality of the nuclear is already immanent.

NOTES

1 Alexievich, *Voices from Chernobyl*, 124.
2 Borenstein, "Dystopias," 88.
3 English translations of titles from Schmid, "Post-Apocalypse," 4.
4 Schmid, "Post-Apocalypse," 4.
5 Nixon, *Slow Violence*, 6.
6 Petrocultures Research Group, *After Oil*, 55.
7 Masco, "Atomic," 340.
8 Hecht, "Nuclear Ontologies," 258.
9 Pomper, "Russian Nuclear," 2.
10 Flisfeder, "Nuclear 1," 243.
11 Josephson, *Red Atom*, 5.
12 WISE Russia, "The End of Russian," 2.
13 Costlow and Nelson, *Other Animals*, 195.
14 Borenstein, "Dystopias," 88.
15 Schmid, "Post-Apocalypse," 5.

16 Penzin, "Russia," 305.
17 McNeill, *Something New,* 297.
18 Niblett, "Energy Regimes," 137.
19 Clarke, *Energy Forms,* 1.
20 Petrocultures Research Group, *After Oil*, 63.
21 Glukhovsky, *Metro 2033*, 30.
22 Ibid., 8.
23 Ibid., 454.
24 Ibid., 9.
25 Ibid., 37.
26 Ibid., 36.
27 Jameson, "Progress," 152.
28 Schmid, "Post-Apocalypse," 4.
29 Moore, *Capitalism,* 291.
30 Masco, "Fallout," 158.
31 Deloughrey, "Radiation Ecologies," 468.
32 Glukhovsky, *Metro 2033*, 5.
33 Petrocultures Research Group, *After Oil*, 64.
34 Glukhovsky, *Metro 2033*, 458.
35 Ibid., 45.
36 Petrocultures Research Group, *After Oil*, 56.
37 Ibid.
38 Carrigan, "Postcolonial Disaster," 255.
39 Glukhovsky, "Interview," n.pag.
40 Ibid.
41 Masco, "Atomic," 338.
42 Ibid., 339.
43 Ibid.
44 Glukhovsky, *Metro 2033*, 89.
45 Ibid., 90.
46 Ibid., 111.
47 Horkheimer and Adorno, *Dialectic,* 26.
48 Phillips, "Chernobyl's Sixth Sense," 159.
49 Glukhovsky, *Metro 2033*, 111.
50 Dyer-Witheford and De Peuter, *Games of Empire*, xxix.
51 Ibid., xv.
52 Yoshimi, "Consuming 'America,'" 210.
53 Dyer-Witheford and De Peuter, *Games of Empire*, xxix.
54 Garratt, "Interview," n.pag.
55 Glukhovsky, "Interview," n.pag.
56 Glukhovsky now sponsors a mass personal book publishing program in which independent authors apply to create *Metro* sequels set in a cross-translatable universe around the world.
57 Glukhovsky, "Interview," n.pag.

58 Rowan, "Destroying Yesterday's World," 4.
59 Žižek, "The Thing," 227.
60 Yankovska and Hannam, "Dark and Toxic," 929.
61 Palmer, "Ordinary Catastrophes," 16.
62 Glukhovsky, *Metro 2033*, 30.
63 Podzerac-Chenevey, "A Direct Link," n.pag.
64 Glukhovsky, *Metro 2033*, 24.
65 Ibid., 455.
66 Glukhovsky, cited in Ieshchenko, "Russians," n.pag.
67 Glukhovsky, cited in Garratt, "Interview," n.pag.
68 Kabanova, "The Animal," 220.
69 Ibid.
70 Canavan, "Introduction," 4.
71 Glukhovsky, *Metro 2033*, 456.
72 Costlow and Nelson, *Other Animals*, 14.
73 Ibid.
74 Brathwaite, "Caribbean Culture," 9-54.
75 Glukhovsky, *Metro 2033*, 37.
76 Masco, "Atomic Health," 351.
77 Ibid.
78 Masco, "Fallout," 161.
79 Cavalcanti, "Preconscious Capitalism," 139.
80 Ibid.

BIBLIOGRAPHY

Alexievich, Svetlana. *Voices from Chernobyl.* Translated by Keith Gessen. New York: Picador, 2006.

Borenstein, Eliot. "Dystopias and Catastrophe Tales after Chernobyl." In *Russian Literature since 1991*, edited by Evgeny Dobrenko and Mark Lipovetsky, 86–103. Cambridge: Cambridge University Press, 2015.

Brathwaite, Kamau. "Caribbean Culture: Two Paradigms." In *Missile and Capsule*, edited by Jürgen Martini, 9–54. Bremen: Universität Bremen, 1983.

Canavan, Gerry. "Introduction: If This Goes On." In *Green Planets: Ecology and Science Fiction*, edited by Gerry Canavan and Kim Stanley Robinson, 1–24. Middletown: Wesleyan University Press, 2014.

Carrigan, Anthony. "Postcolonial Disaster, Pacific Nuclearization, and Disabling Environments." *Journal of Literary and Cultural Disability Studies* 4, no. 3 (2010): 255–72.

Cavalcanti, Sven. "Preconscious Capitalism: The Failure of Capitalism in Computer Games." In *Computer Games as a Sociocultural Phenomenon: Games without Frontiers, War without Tears*, edited by Andreas Jahn-Sudmann and Ralf Stockmann, 131–9. London: Palgrave Macmillan, 2008.

Clarke, Bruce. *Energy Forms: Allegory and Science in the Era of Classical Thermodynamics.* Ann Arbor: University of Michigan Press, 2001.

Costlow, Jane, and Amy Nelson, eds. *Other Animals: Beyond the Human in Russian Culture and History.* Pittsburgh: University of Pittsburgh Press, 2010.

DeLoughrey, Elizabeth. "Radiation Ecologies and the Wars of Light." *Modern Fiction Studies* 55, no. 3 (Fall 2009): 468–98.

Dyer-Witheford, Nick, and Greig De Peuter. *Games of Empire: Global Capitalism and Video Games.* Ann Arbor: University of Michigan Press, 2009.

Flisfeder, Matthew. "Nuclear 1." In *Fuelling Culture: 101 Words for Energy and Environment,* edited by Imre Szeman, Jennifer Wenzel, and Patricia Yaeger, 242–4. New York: Fordham University Press, 2017.

Garratt, Patrick. "Interview: Metro 2033's Dmitry Glukhovsky and Huw Beynon." *VG 24/7,* 25 February 2010. https://www.vg247.com/2010/02/25/interview-metro-2033s-dmitry-glukhovsky-and-huw-beynon.

Glukhovsky, Dmitry. "Interview with Dmitry Glukhovsky." *Kawerna,* 12 December 2011. http://www.kawerna.pl/biblioteka/english-corner/item/3240-interview-with-dmitry-glukhovsky-part-i.html.

– *Metro 2033.* Translated by Natasha Randall. London: Orion Books, 2009.

Hecht, Gabrielle. "Nuclear Ontologies." In *Energy Humanities,* edited by Imre Szeman and Dominic Boyer, 249–60. Baltimore: Johns Hopkins Press, 2017.

Horkheimer, Max, and Theodor W. Adorno. *Dialectic of the Enlightenment.* Translated by Edmund Jephcott. Stanford: Stanford University Press, 2002.

Ieshchenko, Maria. "Russians 'Were Offered an Enemy' Says Metro 2033 Author Dmitry Glukhovsky." *Euronews,* 9 October 2015. http://www.euronews.com/2015/10/09/russians-were-offered-an-enemy-says-metro-2033-author-dmitry-glukhovsky.

Jameson, Fredric. "Progress versus Utopia; or, Can We Imagine the Future?" *Science Fiction Studies* 9, no. 2 (1982): 147–58.

Josephson, Paul. *Red Atom: Russia's Nuclear Power Program from Stalin to Today.* Pittsburgh: University of Pittsburgh Press, 2005.

Kabanova, Daria. "The Animal Watches You: Identity 'After' History in Tatyana Tolstaya's *The Slynx.*" In *Other Animals: Beyond the Human in Russian Culture and History,* edited by *Jane* Costlow and Amy Nelson, 219–33. Pittsburgh: University of Pittsburgh Press, 2010.

Masco, Joseph. "Atomic Health, or How the Bomb Altered American Notions of Death." In *Energy Humanities,* edited by Imre Szeman and Dominic Boyer, 339–52. Baltimore: Johns Hopkins University Press, 2017.

– "Fallout." In *Fuelling Culture: 101 Words for Energy and Environment,* edited by Imre Szeman, Jennifer Wenzel, and Patricia Yaeger, 158–61. New York: Fordham University Press, 2017.

McNeill, J.R. *Something New under the Sun: An Environmental History of the Twentieth Century.* London: Allen Lane, 2000.

Moore, Jason W. *Capitalism in the Web of Life.* London: Verso, 2015.

Nakazawa, Keiji. *Barefoot Gen,* vol. 1. Translated by Project Gen. San Francisco: Last Gasp of San Francisco, 2004.

Niblett, Michael. "Energy Regimes." In *Fuelling Culture: 101 Words for Energy and Environment*, edited by Imre Szeman, Jennifer Wenzel, and Patricia Yaeger, 136–9. New York: Fordham University Press, 2017.

Nixon, Rob. *Slow Violence and the Environmentalism of the Poor.* Cambridge, MA: Harvard University Press, 2011.

Palmer, Christopher. "Ordinary Catastrophes: Paradoxes and Problems in Some Recent Post-Apocalypse Fictions." In *Green Planets: Ecology and Science Fiction*, edited by Gerry Canavan and Kim Stanley Robinson, 158–78. Middletown: Wesleyan University Press, 2014.

Penzin, Alexei. "Russia." In *Fuelling Culture: 101 Words for Energy and Environment*, edited by Imre Szeman, Jennifer Wenzel, and Patricia Yaeger, 303–6. New York: Fordham University Press, 2017.

Petrocultures Research Group. *After Oil.* Edmonton: Petrocultures Research Group, 2016.

Phillips, Sarah D. "Chernobyl's Sixth Sense: The Symbolism of an Ever-Present Awareness." *Anthropology and Humanism* 29, no. 2 (2004): 159–85.

Pomper, Miles. 2009. "The Russian Nuclear Industry: Status and Prospects." *Nuclear Energy Futures Papers* no. 3. Centre for International Governance Innovation. www.cigionline.org.

Podzerac-Chenevey, Sarah. "A Direct Link to the Past: Nostalgia and Semiotics in Video Game Music." *Divergence Press* 2 (2014): n.pag. http://divergencepress.net/articles/2016/11/3/a-direct-link-to-the-past-nostalgia-and-semiotics-in-video-game-music.

Rowan, Derrick. "Destroying Yesterday's World of Tomorrow Playing in the Wasteland." *Reconstruction: Studies in Contemporary Culture* 12, no. 2 (2012): 1–4.

Schmid, Ulrich. "Post-Apocalypse, Intermediality, and Social Distrust." *Russian Analytical Digest* 126 (2013): 2–5.

Strugatsky, Boris, and Arkady Strugatsky. *Roadside Picnic* [1972]. Translated by Olena Bormashenko. London: Gollancz, 2012.

Stalker. Directed by Andrey Tarkovsky. 1979.

Tolstaya, Tatyana. *The Slynx.* Translated by Jamey Gambrell. New York: New York Review Books, 2003.

WISE Russia. "The End of Russian 'Nuclear Renaissance'?" *Third World Resurgence* 236 (2010): 2.

Yankovska, Ganna, and Kevin Hannam. "Dark and Toxic Tourism in the Chernobyl Exclusion Zone." *Current Issues in Tourism* 17, no. 10 (2013): 929–39.

Yoshimi, Shunya. "Consuming 'America': from Symbol to System." In *Consumption in Asia: Lifestyles and Identities*, edited by Chua Beng-Huat, 202–24. New York: Routledge, 2000.

Žižek, Slavoj. "The Thing from Inner Space." *Angelaki: Journal of the Theoretical Humanities* 4, no. 3 (1999): 221–31.

2

From Toxic Lands to Toxic Rumours: Nuclear Accidents, Contaminated Territories, and the Production of (Radio)active Ignorance

SEZIN TOPÇU

In June 1986, Günter Anders wrote: "Today, after Chernobyl, now that no one can feign ignorance, their defenders have deliberately committed a crime. This crime is not only called 'genocide' – what a use of the adverb 'only'! – but 'globicide,' the destruction of the globe."[1] Three decades later, his thesis, according to which no one can feign ignorance any longer (about the risks and lasting toxic effects of nuclear energy) after Chernobyl, appears to have been too optimistic about the process of establishing scientific consensus on and public awareness of nuclear harm. The "anthropological shock" that Chernobyl represented, in Ulrich Beck's terms,[2] did not render idle the business of the "merchants of doubt and of ignorance,"[3] nor did it put an end to the ever-lasting controversy over the gravity of nuclear disasters. The Chernobyl accident demonstrated that the latter controversy was, to a significant extent, artificially created and maintained to produce ambivalence.[4] This did not prevent it from becoming more and more polarized between, on the one hand, *anti-nuclearists* and populations affected by nuclear accidents who denounce these accidents as crime, and on the other hand, *nuclearists*[5] or pro-nuclear advocates who instrumentalize issues of uncertainty and lack of scientific evidence to minimize all types of risks and harms engendered by nuclear system failures.

Clearly, the possibility that vast tracts of land could be sacrificed after a nuclear disaster played a crucial role in the historical lobbying of nuclearists to produce a specific form of ignorance in this field – ignorance that was paradoxically redefined as a form of enlightenment and empowerment. Indeed, the fact that a nuclear accident would sacrifice entire villages, towns, regions, even countries, continents, and oceans, for thousands of years, is almost never discussed openly in public debates about nuclear energy. Nuclear states, industries and agencies do their best to block such collective reflection, often by imposing more performative, less controversial framings. Without a doubt, the reason for this is that public negotiations over the amount of territorial sacrifice that is collectively acknowledged in the event of an accident represent an unacceptable, even explosive debate in sociological terms. It is also too risky, for the promoters of the atom, to admit publicly that entire territories could be contaminated and utterly abandoned for such a long time, since it would mean admitting that, far from being green (carbon-free, renewable, etc.), nuclear energy is a major destructive force, amounting to a "globocide" within the Anthropocene.

The promoters of nuclear energy thus frame the debate in a different way, so as to cast doubt on whether the evacuation of victims is justified and to what extent. They claim that evacuations should be minimized as much as possible, justifying this by emphasizing the trauma it would inflict on the evacuees, which is of course real but is in no way the only trauma experienced by populations in post-nuclear disaster zones. Various forms of post-traumatic stress and psychological distress and anxiety affect different categories of people, including those who were never able to leave the areas contaminated by the Chernobyl disaster. Yet nuclear industries and health agencies rarely care about the trauma suffered by these populations. They claim that people can continue to live happily in radioactively contaminated areas if they learn to be responsible, informed, and empowered – that is, if they learn to optimize their (radioactive) food intake, control their movements, and discipline their bodies.[6] In short, according to nuclearists, victims should adopt a *radiological culture* (i.e., learn to measure and control contamination with a Geiger counter or a dosimeter on a daily basis) and be cultivated enough to prevent the overconsumption of becquerels. Such a *culture* is supposed to help irradiated populations better protect themselves against the risks of radioactivity, but also, paradoxically, to teach them to live as if radioactive contamination has always been around, as if it were now a natural, unquestionable condition of being modern. Anti-nuclear activists retort that discourses and practices aiming to produce "radioactive empowerment" and "responsible radioactive citizenship" are unacceptable – that such discourses aim, above all, to naturalize and banalize radioactive harm and to pretend that it is possible to cope with the consequences of major nuclear accidents. This is the way debates on the risks of nuclear energy have been unfolding since the Chernobyl disaster.

In this chapter, I depart from the debates over the radioactive colonization of land, with the aim of highlighting the strategies, theories, and tools that nuclear states and agencies have developed to deal with large-scale fallouts from nuclear accidents. I explore how the problem of contaminated land was planned to be managed in the past at the international level, as well as how it was or is being managed in the regions affected by the Chernobyl and the Fukushima accidents. I analyze the contexts, policies, and discourses that led to the formulation of the problem of contaminated zones in terms of technical modalities and the legitimacy of evacuation, despite the urgent need to launch a public debate on whether one should continue to make use of an energy regime that has such huge destructive potential, exposing land, the environment, and living beings to lasting contamination and the slow violence of disease.

The significant historical, sociological, and anthropological work on nuclear accidents that has been published in recent years deals with the organizational factors leading to nuclear reactor failures as complex technological systems,[7] as well as with crisis management,[8] media coverage and political impacts,[9] environmental and health fallouts,[10] the construction of the public memory of catastrophes,[11] and, to a lesser extent, the compensation mechanisms and long-term management of victimhood.[12] However, in the social sciences, little attention has been paid to zone/land management, which is often regarded as a technical issue rather than a political one.[13] This chapter aims to fill that historiographic gap.

A key issue I raise in this chapter is the prevention of long-term devaluation of land. That issue has become a major concern for nuclearized countries since Chernobyl. This concern about minimizing land loss has gone hand in hand with a bifurcated regulatory framework for public health protection. I will deal with the fundamental question of how such a compromise between land and body contamination has been made socially acceptable. I will shed light on three principal managerial strategies, which are intertwined:

- Geo-management or zoning, whose ultimate form today appears to be cancellation of the very idea of "contaminated zones."
- Biological or body management, where, increasingly, *states of exception* serve as the norm for managing contaminated lands.
- Social management, which implies a shift from material management of disasters (in the short term) to the administration of social psychology in a longer term.

I contend that the production of ignorance is central to each of these strategies. Drawing on the growing field of agnotology,[14] and expanding on the seminal work of Olga Kuchinskaya on the production of invisibility regarding the health effects of Chernobyl in Belarus,[15] I will show that the production of the ignorance I have in mind is less about secrecy[16] than about generating oblivion or forgetfulness about that which cannot be kept entirely secret. In the case of nuclear accidents, such difficult-to-conceal evidence involves mutated species, large numbers of deaths and illnesses, overcrowded hospitals, and peer-reviewed publications and reports that demonstrate a relation of causality between these abnormal phenomena and the post-disaster radioactive environment. As the main part of the following discussion reveals, the work of ignorance production consists in establishing conditions that encourage individuals to stop informing themselves and doing research on or being concerned about their contaminated environment. Such dynamics, far from being specific to the nuclear industry, can be observed in many other techno-industrial sectors that generate large-scale pollution, such as the pesticide sector[17] and the asbestos industry.[18]

The empirical materials on which I rely include official reports dealing with the management of the consequences of major nuclear accidents, as well as expert scenarios and estimations that have been influential in evaluations of the environmental, health, and economic costs of possible accidents, in the frame of relevant plans for preparedness.[19]

1. From Sacrificed Lands to "Zones": A Historical Survey

1.1. EXCEPTIONAL ARRANGEMENTS FOR EXCEPTIONAL RISKS

In the nuclear energy sector, post-disaster management (management of territories and of affected populations) relies on a series of technical, juridical, economic, and insurance-related tools that were developed in the 1950s, first in the United States and later at the international level. Even before there was a nuclear industry, an expert consensus was established that in the event of a nuclear accident, vast territories would be highly contaminated. By the late 1940s, it was already being admitted that a significant number of people would need to be evacuated. This constraint played a central role in the choice of nuclear sites and the corresponding urban politics. In 1950, an expert committee struck by the US Atomic Energy Commission proposed an exclusion zone for nuclear reactor sites, namely, a 30 km empty zone for reactors producing 1000 MWe of power (with the size of the exclusion zone being proportional to the power generation of the reactor being built). Industrial firms were reluctant to accept this proposal, arguing that the money would be better spent on stronger reactor security. In 1956, the US AEC concurred, reasoning that evacuation zones (which would be put into practice in case of an accident) were preferable to exclusion zones (which were to be instituted *de facto*), since the latter could provoke "unnecessary" fear and anxiety among the public.[20] In 1957, at the request of the AEC, a research team at the Brookhaven National Laboratory developed the first scenarios for the evacuation of populations in the event of a nuclear accident. The findings of the Brookhaven report were extremely alarming: in a major accident at a 500 MWe reactor, 3,400 people would be killed and more than 40,000 would be irradiated; moreover, it would be necessary to evacuate about 240,000 km2 and the evacuation, land, and health costs would exceed $7 billion.[21]

The political fallout of this report was radical. To protect the industry against such gigantic financial risks, the US government, followed by the principal nuclear states, drastically limited the financial liability of nuclear operators in the event of an accident, first of all via the Price Anderson Act adopted in 1957, and then via the Paris and Vienna Conventions signed in 1960 and 1966 respectively. The juridical and insurance arrangements made by the major industrial states to shield the nuclear industries were exceptional – a dynamic I analyzed elsewhere as a process that erased the very notion of responsibility by transforming it into a purely discursive tool of pro-nuclear governments.[22] Much the same framework exists today. The French electricity supplier EDF (Electricité de France), for instance, is anticipated to confront a liability cost of only €94 million in the event of a major accident at a French nuclear park. This sum is insignificant relative to the "real" costs of such an accident. For Fukushima, official estimates (which have doubled since 2013) correspond currently to €175 billion.[23] As to cost estimations made by the French regulatory bodies based on various accident scenarios, they range from €430 billion for an average scenario (based on a meltdown of only one-third of a reactor's core, generating radioactive release for only a few hours), to €760 billion for a more devastating one (cf. IRSN 2007).[24]

1.2. AUTHORITARIAN ZONES AND THE HIERARCHIZING OF HARM

To persuade industrial firms to invest in the nuclear sector in 1960s, the liabilities for accidents had to be reduced to "symbolic" levels. This inevitably generated the need to categorize harms, to prioritize remedies and compensation, and also to control public perceptions of risk and damage. By the 1960s, a major assumption had been consolidated within this frame: the complete evacuation of territories contaminated by a nuclear accident was judged to be impossible (in technical, economic, and social terms), just as it was considered impossible for the nuclear industry to entirely compensate for the damages such an accident would cause (an assumption confirmed by the accidents at Three Mile Island, Chernobyl, and Fukushima). In other words, by the 1960s, nuclear experts had admitted, and were even suggesting, that evacuations and compensations should be optimized. This led to the emergence of zoning as a tool for managing major accidents.

Zone management in the event of a nuclear accident (i.e., zones whose level of contamination would determine the level of state intervention as well as the type of compensation) was already envisaged by the authors of the Wash 740 report. Five-level zoning was considered plausible,[25] and that method inspired the zoning and evacuation criteria that were established after the Chernobyl and Fukushima accidents. In 1991 (after the fall of the Soviet Union), yet another significant innovation took place in the field: so-called "optional evacuation zones."

Zone management gradually embraced a threefold discursive function. The first function, put into effect during the emergency period, allows governments and experts to affirm, thanks to a technical-rational tool they have developed, that the problem is localized and that the threat is under control. The second function, which relates to zones, signifies that there are different levels of being affected by radioactivity; one's experience of injury should be placed in relation to that of other affected groups, the implication being that compensation (also managed by the state) will be calculated in relation to the degree of harm experienced by others. The third performative function concerns the implication that recovery is possible and that the status of zones can and even should evolve. This problematic conception of zoning is especially strong in post-2011 Japan: according to current state discourses and policies related to post-accident management, far from being static or permanent, zones should quickly disappear.

Zones have thus been conceived, especially in recent years, as dynamic socio-technical tools for the authoritative – indeed, authoritarian – management of populations, land, and public health. Frontiers between these are not rigid. They are subject to permanent recategorization or redefinition. In Japan, from the beginning, the state adopted the strategy of "reconquest" of contaminated territories, which the IAEA described as post-accident recovery, aiming "to re-establish an acceptable basis for a fully functioning society in the affected areas."[26] In the Soviet Union, return to the evacuated zones was initially not a priority; after 1991, evacuations were even expanded by an initiative from the Ukrainian and Belarus governments. By the late 1990s, however, the return of evacuated populations had become a priority, just as in post-2011 Japan.[27]

These developments suggest that "zone recovery" has acquired great importance in nuclear states. In terms of politics, reclaiming a radioactive zone boosts a state's credibility, especially if it has maintained a nuclear industry after a catastrophe. In economic terms, a state will want to minimize the compensation it owes. The zoning apparatus is also crucial, more generally, for the future of the nuclear industry at the international level.

2. From Chernobyl to Fukushima: Zoning, Biopolitical Governance, and the Production of Ignorance

2.1. FROM NO MAN'S LAND TO "NO-ZONE" LAND

Based on a definition of contaminated land as land with radioactivity levels higher than 1 mSv per year, the accident at the Chernobyl no. 4 reactor contaminated around 200,000 km2 of land, 71 percent of it in Belarus, Ukraine, and Russia, and the rest in other parts of Europe. Regarding Fukushima, around 80 percent of the atmospheric release after the chain reactions at the three units was directed to the Pacific Ocean; the zones with radiation levels superior to 1 mSv corresponded to around 13,000 km2. However, the Japanese government has defined contaminated land as areas where levels exceed 20 mSv. The state has assumed responsibility only for the decontamination work in these zones (excluding an area of about 1000 km2 – mainly forests and mountains). Decontamination of zones with levels between 1 mSv and 20 mSv has been delegated to municipalities.

After both Chernobyl and Fukushima, the exclusion zone was initially decreed in a somewhat arbitrary manner: a circle was drawn around each destroyed reactor, and quick evacuation was ordered for those within that circle. This corresponded to a zone within a 20 km radius for Fukushima and a 30 km radius for Chernobyl. The exclusion circles were updated significantly during the first days, weeks, even months after the accidents, thanks to more sophisticated contamination maps. This took nine months in the case of Fukushima, several years in the case of Chernobyl. The initial exclusion affected around 70,000 people in Japan and 90,000 in the Soviet Union. Over the following months and years, the number of evacuees reached 200,000 in Japan, 350,000 in the Soviet Union (today's Ukraine, Belarus, and Russian Federation).

In large parts of the exclusion zone around Chernobyl,[28] no resettlement has been envisaged. By contrast, in Japan, in early 2012, state officials announced that it was crucial to proceed to the recovery (reconquest) of the evacuated zones (including towns within the 20 km circle) as much and as soon as possible. Thus, tremendous decontamination work was undertaken with nearly 15,000 people mobilized daily; however, the effectiveness of these operations is highly uncertain.[29] Around Chernobyl, the exclusion zone included one city (Pripyat) and several villages. At Fukushima, it initially included five towns (Namie, Futaba, Okuma, Tomioka, and Naraha) as well as portions of Tamura city and Kawauchi village. Nevertheless, by 2015, Naraha, Tamura, and Kawauchi were (re)categorized as habitable; they were followed by Minami-Sôma and Katsurao in 2016. By the end of March 2017, plans were to cancel the evacuation order for most of the officially evacuated zone (with financial compensation to be stopped shortly thereafter), except for very limited "difficult-to-return zones."[30] This latter category refers to zones where contamination exceeds 50 mSv (the authorized level is 20 mSv in Japan, 1 mSv elsewhere). Surprisingly, even these highly contaminated zones are in no way considered dead or definitively sacrificed by the Japanese government. That is why the euphemistic term "difficult-to-return" replaced "forbidden" only two years after Fukushima. And the Japanese government has announced recently that even in these zones, evacuation orders will be lifted by 2022. Furthermore, the same doctrine of "zero zone, zero sacrifice" holds even for the 16 km2 area around the Fukushima plant itself. The Japanese government has decided to utilize that plant as a temporary storage site for the tremendous amount of radioactive residues generated by the decontamination process (i.e., around 22 million m3). A new law that foresees the reclaiming and recovery of this waste zone within thirty years has recently been tabled.[31]

In Chernobyl, the revaluation of the restricted/abandoned zone has been managed by other means than a gradual lifting of its special designation. Efforts to recover some of the economic and historical value of Pripyat and its surroundings have focused mainly on tourism. A "dark" tourism of memorialization (with guided visits to the sarcophagus that now encases the reactor) has developed around Pripyat, as has a "scientific" tourism. In the 2000s, experts from the IAEA launched a conservationist marketing strategy, claiming that Chernobyl's exclusion zone, referred to as the "dead zone" for many years, *was not dead at all,* that it had become a "unique sanctuary for biodiversity."[32] This well-established myth ignores independent scientific evidence that the region's biodiversity has been seriously degraded.[33]

Beyond the strict exclusion zones are other zones. In particular there are "optional evacuation zones," which also evince a strategic importance for nuclearists. These zones are subject to negotiations, controversies, and ever-changing categorizations and regulations. The category of optional evacuation zones was elaborated during the early 1990s in new laws adopted in Ukraine, Belarus, and the Russian Federation. These zones are not among the most contaminated; in many hotspots, however, the contamination levels exceed what is authorized. In a neoliberal disaster management frame, individuals who were living there before the accident are said to be free to abandon their lands, homes, and villages, which implies that the responsibility for risk management has been transferred from the state to the victims, who are expected to shoulder the costs either way – that is, the economic costs if they leave, and the biological/health costs if they stay.[34] Especially in these areas, P&R campaigns were launched the second half of the 1990s with the aim of convincing people not to leave their homes, or to return to them if they had left. In Japan, most of those who left (voluntarily or otherwise) have been subjected to similar campaigns of persuasion. These campaigns appear to be aimed more at those who left on their own initiative (around 26,600 people, according to official records). Recent surveys report that they are often stigmatized directly or indirectly by state officials and by non-victims as cowards or troublemakers, jeopardizing the national effort to reconstruct Fukushima.[35]

2.2. BIOPOLITICAL ZONES: STATES OF EXCEPTION AND BODY MANAGEMENT

"Exceptional" sanitary norms were established after Chernobyl and Fukushima in order to limit evacuations. As a result, all those who needed to be evacuated were not even prepared for evacuation. In France, state bodies have developed a similar action plan in the event of an accident. A study by the French Institute for Radioprotection and Nuclear Safety (IRSN) conducted a few years before Fukushima (and made public in 2013, thanks to the efforts of anti-nuclear groups) argues that given the scale of the contamination provoked by a major accident, the measures to be undertaken in the event of a nuclear catastrophe in France will *necessarily* be "non-optimal" in order not to sacrifice too much agricultural and habitable land. The additional cancer cases brought about by such non-optimal choices have been folded into the accident cost estimations. According to experts from the IRSN, a major accident at the Dampierre nuclear plant (the case investigated), in the Loire region 50 kilometres from the city of Orléans, would theoretically require the evacuation of 2.5 million people, but the French authorities envisage evacuating only 25,000 people. The same experts add that such non-optimal evacuation measures would generate 17,500 cancer cases. The cost of these cancer cases for the national health system has been carefully calculated and integrated into the accident budget.[36]

The Japanese government adopted a non-optimal land and health management policy immediately after Fukushima by multiplying the acceptable dose level by twenty, that is, by raising it from 1 mSv (the international norm fixed by the ICRP) to 20 mSv, which is the maximum level for nuclear workers in most nuclear countries (the authorized exposure level for Fukushima workers was upgraded to 250 mSv). This ignorance production work created the conditions for a general memory loss about the good old days of radioprotection organized around the 1 mSv threshold level, by redefining as normal the 20 mSv level and by thereby transforming everyone (including children and the general public) into nuclear workers. Since 2011, the Japanese regulatory bodies have further decided that below the dose limit of 100 mSv, the probability of developing radiation-related cancers is negligible and that smoking or obesity are riskier than low-dose radioactivity. In this way they have end-run the international scientific consensus. Clearly, ignorance production work is also about shifting the line between truth and not-truth. At Chernobyl, the radiation exposure level in the post-emergency (i.e., recovery) period was raised to 5 mSv under the Soviet regime, then lowered back to 1 mSv after 1991 (in Russia, Ukraine, and Belarus).[37] In practice, though, many people continue to live in regions where radioactivity levels far exceed 1 mSv, for states have failed to proceed with their resettlement. According to data from the IAEA, twenty years after the Chernobyl, 5 million people were still living in contaminated areas. The total number of official victims was around 7 million (in the three countries).[38]

For evacuees who are returning home, management systems have been established based on guidelines that exceed sanitary norms. Land that was initially evacuated is being normalized not only depending on progress in decontamination work (which, in most cases, was found to be inefficient or temporary, because of the unavoidable migration of radioelements from one place to another), but also thanks to people's willingness to *obey* the state. Very often, people are asked to, and decide to, resettle back in their homelands despite insoluble radioactive contamination. The so-called guidelines for "participatory rehabilitation" play a key performative role here. They aim to teach victims how to optimize their daily food intake and other practices (use of water, walks in the forest) so that they absorb as few radionuclides as possible. The experts' assumption is that if the inhabitants who resettle in these *renormalized* areas continue to live as before (i.e., if they pay no attention to the new, radioactive condition of their environment), their land will gradually lose its normality (it will become "contaminated," "exceptional" again) and they will probably fall ill one day. If, on the contrary, they take some simple measures, if they act as responsible and informed citizens, they will not (i.e., not necessarily, or not immediately) be victims of radiation-related illnesses. Thus, the land on which they choose to (re)settle will preserve its normality and will not be stigmatized.

The participatory rehabilitation guidelines were elaborated by a group of French experts (the Ethos team) financed by the European Commission in the late 1990s. They were tested in the Brest region of Belarus, where one-quarter of the territory was contaminated.[39] Paying attention to one's own health, and managing radioactivity risks on an individual level, were praised as ways to be empowered, to get rid of the mentality of loser or victim, and to valorize oneself and one's land and property. Individual efforts to adopt new and exceptional life standards or habits were viewed as the means to remediate the toxic or contaminated land and make it functional, hospitable, fertile, and thus (economically) valuable again. Thanks to the collaborative efforts of the victims, land would no longer be sacrificed. But sacrifice was to silently shift from the territorial to the biological arena (i.e., victims' bodies).

The Ethos project further explored ways to revalorize land in Brest (Belarus) and to gradually reopen it to agriculture – a resilience strategy that had already been explored by the Soviet nuclear authorities during the Cold War, especially after the Mayak nuclear disaster, which was kept secret for thirty years.[40] After Chernobyl, the IAEA and the FAO, for instance, launched a large-scale project in the Gomel and Mogilov regions of Belarus (more contaminated than Brest) with the aim of developing clean rapeseed oil production.[41] Since 2011, the same strategies and guidelines have been in effect, under the leadership of Ethos and other organizations in the Fukushima region.

3. Toxic Rumours: From Management of Material Harm to Social Management

By the early 1990s, WHO and the IAEA experts, but also those from the OECD and the World Bank, began to critique the official state's management of the consequences of the Chernobyl accident in Ukraine, Belarus and the Russian Federation. The critique was not so much directed at these governments' failure to ensure that all resettlement operations they had promised to the affected populations be carried out, but at the fact that state institutions in the three countries had promised "too much." Indeed, almost univocally, all reports and studies from international agencies judged that the post-Chernobyl evacuation and compensation policies were too cautious, economically unsustainable, and politically counterproductive. They claimed that the governments' hasty examination of the situation (of radioactive contamination) reinforced the devaluation of too much land and resources, which in their opinion were unfairly labeled as contaminated as a whole. They argued that large-scale evacuations provoked anxiety, fear, trauma, depression, and even mental illnesses among local populations. The concept of radiophobia was developed based on such assumptions. The international bodies further considered that the excessive will of the Soviet and the post-Soviet governments to compensate and assist the victims favoured the consolidation of a mentality of permanent victimhood, and of heavy dependence of citizens on the state. For example, a group of international experts commissioned by the United Nations Development Programme suggested, in 2002, that

> bringing condemned land back into economic use would be a powerful marker of the process of recovery for potential investors and in terms of the *psychology* of the communities concerned. The issue needs to be carefully considered by these communities, working with appropriate specialists and local and national government agencies. Wherever possible, the assumption should be that local people should have the choice of where to live and work, provided the interests of vulnerable individuals, including children, can be properly protected.[42]

The (biological) health effects were thus minimized, the psychological issues were maximized. In short, it can safely be argued that, according to the international bodies, chiefly the IAEA and WHO, the most serious harm from the Chernobyl accident was psychological.[43]

In the case of Fukushima too, the psychologization of harm, and official strategies for disciplining public fear and concern have proven crucial. At an international level, the UNSCEAR and the IAEA have argued, since 2013, that "the most important health effect [from the accident] is on mental and social well-being, related to the enormous impact of the earthquake, tsunami and nuclear accident, and the fear and stigma related to the perceived risk of exposure to ionizing radiation."[44] At a national level, since the beginning, state experts have assumed the mission of teaching people "to fear correctly," by popularizing two distinct categories, *anzen* and *anshin*, which mean objective, officially defined safety and the feeling of safety (i.e., subjective, emotional safety) respectively.[45] They put forth scientists' and experts' way of defining and measuring risks as the unique manner of evaluating the fallout from the catastrophe – a phenomenon that Hirakawa and

Shirabe describe as "scienceplanation"[46] – which they used as an authoritarian tool for preventing a pluralist debate, and for marginalizing democratic decision making. The denial of risk has constituted, in this frame, an important component of the politics of scienceplanation adopted by the government, which waited for several weeks to upgrade the gravity of the catastrophe from level 4 to level 7 (according to the *International Nuclear Event Scale*). It took TEPCO more than 2 months to announce that the Unit 1 reactor went into meltdown on the first day.[47] Overall, the risk communication discourse of the Japanese authorities and expert bodies has been framed by a very specific conception of the public/citizens as ignorant receivers of exaggerated information, as victims of "pure rumours" (rather than of radioactive risk).[48]

In the post-Fukushima context, *rumour* (*fuhyô* in Japanese) indeed has proven to be a major tool of risk governance, of compensation management, and more generally, of social management. The Japanese government considers that lay or critical evaluations of, and bad publicity about, the situation in Fukushima (renamed as "rumour") harm more than the Becquerel do. It is assumed that the importance of the latter is reinforced by the contagious and counterproductive character of the former. The government thus launched a "war against rumours," as a corollary to its politics of reclaiming and recovery of contaminated areas. This war may be seen as a powerful ignorance production machine. It serves to render invisible the complex and mostly irreversible character of the radioactive contamination that endangers the Fukushima region, its lands, resources, and food products. It also serves to claim that everything is under control (except for external threats, such as rumours) in a context in which, eight years after the accident, the stabilisation of the damaged reactors is far from being accomplished, and the uncertainties concerning the permanent circulation of caesium, iodine, strontium, as well as other dangerous radionuclides that have colonized the soil, forests, atmosphere, water, as well as people's bodies are likely to keep growing.

Far from limiting itself to pontificating over the authority of and legitimacy of "scientific" expertise – expertise which routinely denigrates citizens' science or lay research – the government's politicized "war against rumours" has expanded to include the lobbying for, and successfully obtaining, the hosting in Tokyo of the 2020 Summer Olympics as well as the vigorous "solidarity with Fukushima" campaigns that invite the public to consume food products imported from the Fukushima region in order to combat rumour. This ignorance production machine also entails harder methods such as reinforcement of the state control on media coverage, especially in relation to the critical evaluation of the consequences of the Fukushima Daichi nuclear disaster.

Another field in which the Japanese government's "war against rumours" plays a key role concerns compensations. In May 2011, the Japanese Dispute Reconciliation Committee for Nuclear Damage Compensation (JDRCNDC) forged a new category of damage to be considered in the compensation of victims, which is called "rumour-related damage." It was defined as: "concern about the risk of contaminated radioactive material in relation to products or services, due to facts that are widely known through media reports, leading consumers or trading partners to refrain from purchasing the product or service, or stop trading in the product or service, resulting in damage."[49] According to JDRCNDC, "rumour-related damage" was tightly related to the consumers' or trading partners' "psychological state of wanting to avoid the product or service."[50]

Official government publications make it a point to declare that the recognition by the state of rumour-related damage is laced with ambiguity, since such a categorization unavoidably influences the so-called damage itself (thereby reinforcing the rumour, thus the damage). That is the reason why rumour-related damage, just like mental anguish damage (due to evacuations), were categorized as temporary damages. More important however is the fact that a very significant amount of the compensation budget has been attributed to rumour-related damage. Neither large-scale land contamination or loss, nor health consequences have been considered as important as rumour-related damages.[51] What people think, feel, or judge as acceptable and safe has been framed as a major problem, and has become a priority in the political agenda. What they live (i.e., a degraded, contaminated environment), experience, and suffer from (irradiation, diseases) in their bodies, and within the nuclear anthropocene, has been categorized as a secondary concern.

Conclusion

Post-Fukushima dynamics cannot be analyzed and understood merely as a national case or an innately Japanese one. The discourses and policies that have been operational in post-2011 Japan are embedded in a national *and* transnational history. In this chapter I have presented some key elements and moments of this history. Drawing on the observation that expert consensus on the catastrophic dimensions of the damage provoked by nuclear accidents both on land and on public health was already established by the 1950s, I showed how both optimal protection and compensation for people, and optimal evacuation as a public health measure, were considered unrealistic and capacity-stretching as early as that time. Such a doctrine opened the path for exceptional economic and juridical arrangements that were vital to the launching of nuclear industries in the US and elsewhere. I argued that after Chernobyl,[52] the need to minimize land loss became central in corporate capitalist discourses and policies, while after Fukushima, the doctrine of "no zone, no sacrifice" was formalized.

The "no zone, no sacrifice" doctrine, which is a major pillar of what I propose to call the TINANT (There Is No Alternative to Nuclear Toxicity) ideology, reveals new conceptions of the relations between land and body. If, starting from the late 1990s, both health and land were gradually conceptualized as properties or assets (with financial values and costs), then the revalorization (or recovery) of contaminated land, currently in effect in Japan, further relies on the assumption that people's health is an individual resource to be exploited, in support of land (and against the sacrifice of territory). In other words, the very idea of property is extended toward the biological body as the new arena of sacrifice. The production of at least three types of ignorance plays a crucial role in this frame. With regard to geo/zone management, in a context in which zones are conceived as more and more ephemeral, the current politics of ignorance in Japan is about deleting those zones from the political agenda and from people's memories. In the field of biological/body management, the politics of ignorance production concern forgetting the state of exception (permanent exposure to 20 mSv of radiation) into which victims' bodies are inserted and governed. As to the apparatus of social/psychological management, ignorance production is about encouraging people to be deaf and blind to all alternative information (reframed as rumours) that might challenge official "scientific" expertise, and to suppress any concerns or worries other than those authorized by the state. Such a way of governing, rhetorically framed as a double war against radioactive isotopes and against rumours in order to service the reconstruction of Fukushima and the nation, represents a novel form of toxic authoritarianism and of radioactive biopolitical capitalism. The constitutive myths, ideologies, tools, and networks of this nefarious form of neoliberal governmentality should be systematically deconstructed in order to restore democracy and the potential for social and environmental justice.

NOTES

A different version of this chapter was published in Laura Centemeri, Sezin Topçu, and Peter Burgess, eds., *Rethinking Post-Disaster Recovery: Socio-Anthropological Perspectives on Repairing Environments* (London: Routledge, 2021).

1 My translation, from the French edition. The French extract is the following: "Aujourd'hui, après Tchernobyl, dans la mesure où plus personne ne peut jouer les ignorants, ses avocats en sont venus à commettre consciemment un crime. Ce crime ne s'appelle pas seulement 'genocide' – quel emploi de l'adverbe 'seulement' ! – mais 'globocide', destruction du globe." "Dix Thèses pour Tchernobyl" in Anders, *La menace nucléaire*, 317–18.
2 Beck, "The Anthropological Shock."
3 Oreskes and Conway, *Merchants of Doubt*.
4 Latour, "Why Has the Critique Run Out of Steam?"
5 The category of "nuclearists" that I mobilize here refers to a range of actors (states, industries, governmental agencies, scientific and media organizations, trade unions)

that promote nuclear energy as a "system" of government (from production-consumption to profit-making to socio-political organization and disciplining), colliding head-on with their critics, the "anti- nuclearists," that is, those who refuse precisely the generalization of such a system. See Topçu, *La France nucléaire.*

6 Topçu, "Chernobyl Empowerment."

7 Perrow, *Normal Accidents.*

8 Lagadec, États d'urgence; Guarnieri et al., *L'accident de Fukushima.*

9 Liberatore, *Management of Uncertainty*; Szarka, "From Exception"; Fujigaki, "Lessons from Fukushima."

10 Kuchinskaya, *Politics of Invisibility*; Brown, *Plutopia.*

11 Ardt, "Memories"; Kasperski, "La politique."

12 Petryna, *Life Exposed.*

13 Among rare studies on the topic, see Grandazzi, *De Tchernobyl.*

14 Gross and McGoey, "Introduction."

15 Kuchinskaya, *Politics of Invisibility.*

16 "Secrecy" refers to the confiscation of information or of its conditions of production, which has a specific history as to what concerns nuclear issues, and which continues to serve as a highly efficient tool in the management of the health and environmental impacts of catastrophes.

17 Dedieu and Jouzel, "Comment ignorer."

18 Henry, *Ignorance scientifique.*

19 Lakoff and Anderson, "Preemption."

20 Foasso, *Histoire de la sûreté.*

21 AEC, *Theoretical Possibilities.*

22 Topçu, "Organiser l'irresponsabilité."

23 The distribution of this "budget" is the following: 65 billion for the dismantling of the reactors, 64 billion for compensations; 32,5 billion for decontamination; 13 billion for temporary waste storage. *Nikkei Asian Review*, "Fukushima costs."

24 It should be further noted that no accident scenarios have yet been made public for nuclear reactors near Paris (such as the Nogent-sur-Seine power station) or for the nuclear reprocessing centre near the Hague, even though such accidents would be the most devastating (either by contaminating the densest urban zones or by generating the biggest pollution).

25 R1: urgent evacuation (within 12 hours); R2: evacuation required but more time is available to prepare it; R3: restrictions on agriculture and temporary evacuation; R4: destruction of crops and restriction of agriculture; R5: no restrictions, but regular radiologic monitoring is required. See AEC, *Theoretical Possibilities*, Appendix D.

26 IAEA, "The Fukushima Daiichi Accident," 140.

27 More precisely, rather than "returning to," "staying in" evacuated zones became a priority in zones contaminated by the Chernobyl accident, since many who needed to be evacuated were indeed never evacuated.

28 They are often divided in subcategories such as: zones with "no access allowed," "possible to pass the day but not overnight," "access authorised for public interest purposes," and so on.

29 CRIIAD, "Catastrophe de Fukushima."
30 Boilley, *Fukushima Five Years Later.*
31 Boilley, "Fukushima."
32 IAEA, "Chernobyl's Legacy."
33 Mousseau and Moller, "Conservation Consequences"; "Genetic."
34 Topçu, "Chernobyl Empowerment?"
35 Hasagewa, "Returning Home."
36 IRSN, "Examen."
37 During the emergency period, the threshold levels set by the Soviet Union temporarily were much higher (i.e., 500 mSv per year from 12 to 22 May 1986; 100 mSv per year after 22 May 1986). See Kuchinskaya, *Politics of Invisibility.*
38 IAEA, "Environmental Consequences."
39 Topçu, "Chernobyl Empowerment."
40 Brown, *Plutopia*; Kasperski and Topçu, "De Maïak."
41 World Bank, "Belarus Chernobyl."
42 UNDP, "The Human Consequences," 13.
43 For a detailed analysis, see Kuchinskaya, *Politics of Invisibility*, Chapter 4.
44 UNSCEAR, "Sources," 248; IAEA, *The Fukushima Daiichi Accident*, 14.
45 Shirabe et al. "From Risk."
46 Hirakawa, Fassert, and Hasegawa, "Rhetorical Marginalization."
47 Figueroa, "Risk Communication."
48 Ibid.
49 JDRCNDC, "Secondary Guidelines," 110.
50 Ibid., 111.
51 According to the JDRCNDC (2011), the loss of/reduction in property value (loss of value due to exposure, etc.) is estimated to be $570.7 billion, whereas loss due to "rumour-related damage" (agriculture, forestry, fisheries, manufacturing, services, etc.) is estimated to be $1.3039 trillion. See OECD, *Japan's Compensation* System, 59.
52 Krohn and Weigart, "Commentary."

BIBLIOGRAPHY

AEC (US Atomic Energy Commission). *Theoretical Possibilities and Consequences of Major Accidents in Large Nuclear Power Plants.* US Atomic Energy Commission, Wash 740 Report, 1957.

Anders, Günter. *La Menace nucléaire: Considérations radicales sur l'âge atomique.* Paris: Serpent à Plumes, 2006.

Ardt, Melanie, ed. "Memories, Commemorations, and Representations of Chernobyl." *Anthropology of East Europe Review* 30, no. 1 (2012, special issue): 1–140.

Beck, Ulrich. 1987. "The Anthropological Shock: Chernobyl and the Contours of the Risk Society." *Berkeley Journal of Sociology* 32 (1987): 153–65.

Boilley, David. *Fukushima Five Years Later: Back to Normal?* Hérouville Saint Clair: ACRO Report for Greenpeace Belgium, 2016.

– "Fukushima: l'obstination de la reconquête." *L'Acronique de Fukushima*. 2017. http://fukushima.eu.org/fukushima-lobstination-de-la-reconquete.

Brown, Kate. *Plutopia. Nuclear Families, Atomic Cities, and the Great Soviet and American Plutonium Disasters*. New York: Oxford University Press, 2013.

CRIIRAD (French Commission for Independent Research and Information on Radioactivity). "Catastrophe de Fukushima Daiichi: Un cauchemar sans fin." *Trait d'union* 73 (2017): 1–5.

Dedieu, François, and Jean-Noël Jouzel. "Comment ignorer ce que l'on sait? La domestication des savoirs inconfortables sur les intoxications des agriculteurs par les pesticides." *Revue française de sociologie* 56, no. 1 (2015): 105–33.

Figueroa, Pablo M. "Risk Communication Surrounding the Fukushima Nuclear Disaster: An Anthropological Approach." *Asia Europe Journal* 11, no. 1 (2013): 53–64.

Foasso, Cyril. "Histoire de la sûreté de l'énergie nucléaire en France (1945–2000): Technique d'ingénieur, processus d'expertise, questions de société." PhD diss., University of Lyon II, 2003.

Fujigaki, Yuki, ed. *Lessons from Fukushima: Japanese Case Studies on Science, Technology, and Society*. Cham: Springer Verlag, 2015.

Grandazzi, Guillaume. "De Tchernobyl à la Hague: La vie quotidienne entre expérience de la catastrophe et épreuve de l'incertitude." PhD diss., University of Caen, 2004.

Gross, Matthias, and Linsey McGoey. "Introduction." In *Routledge International Handbook of Ignorance Studies*, edited by Gross and McGoey, 1–14. London and New York: Routledge, 2015.

Guarnieri, Franck, Sébastien Travadel, Christophe Martin, Aurélien Portelli, and Aissame Afrouss. *L'accident de Fukushima Dai Ichi – le récit du directeur de la centrale*, vol. 1: *L'anéantissement*. Paris: Presses des Mines, 2015.

Hasegawa, Reiko. "Returning Home after Fukushima: Displacement from a Nuclear Disaster and International Guidelines for Internally Displaced Persons." *Migration, Environment, and Climate Change: Policy Brief Series* 4, no. 1 (2015): 1–8.

Henry, Emmanuel. *Ignorance scientifique et inaction publique: Les politiques de santé au travail*. Paris: Presses de Sciences Po, 2017.

Hirakawa, Hideyuki, and Masashi Shirabe. "Rhetorical Marginalization of Science and Democracy: Politics in Risk Discourse on Radioactive Risks in Japan." In *Lessons from Fukushima: Japanese Case Studies on Science, Technology, and Society*, edited by Y. Fujigaki, 57–86. Cham: Springer Verlag, 2015.

IAEA (International Atomic Energy Agency). *Chernobyl's Legacy: Health, Environmental, and Socio-Economic Impacts and Recommendations to the Governments of Belarus, the Russian Federation, and Ukraine*. Chernobyl Forum 2003–5. 2nd revised version. Vienna: 2006.

– *Environmental Consequences of the Chernobyl Accident and Their Remediation: Twenty Years of Experience*. Report of the Chernobyl Expert Forum "Environment." Vienna: 2006.

– *The Fukushima Daiichi Accident*. Report by the Director General. Vienna: 2015.

IRSN (Institut de Radioprotection et de Sûreté Nucléaire). *Examen de la méthode d'analyse coût-bénéfice pour la sûreté*. Report no. DSR 157. Paris: 2007.

JDRCNDC (Japanese Dispute Reconciliation Committee for Nuclear Damage Compensation). "Secondary Guidelines on Determination of the Scope of Nuclear Damage Resulting from the Accident at the Tokyo Electric Power Company Fukushima Daiichi and Daini Nuclear Power Plants," 31 May 2011. In *OECD: Japan's Compensation System for Nuclear Damage*, 103–16. Paris: OECD, 2012.

Kasperski, Tatiana. "La politique de la mémoire d'une catastrophe nucléaire. Les usages de l'accident de Tchernobyl en Biélorussie (1986–2008)." PhD diss., Institut d'Etudes Politiques de Paris, 2012.

Kasperski, Tatiana, and Sezin Topçu. "De Maïak à Tchernobyl, la 'guerre' radioactive: Une liquidatrice témoigne." *Mouvements* 87 (2016). mouvements.info/de-maiak-a-tchernobyl.

Krohn, Wolfgang, and Peter Weingart. "Commentary: Nuclear Power as a Social Experiment: European Political Fall-Out from the Chernobyl Meltdown." *Science, Technology, and Human Values* 12, no. 2 (1986): 52–8.

Kuchinskaya, Olga. *Politics of Invisibility: Public Knowledge about Radiation Health Effects after Chernobyl*. Cambridge, MA: MIT Press, 2014.

Lagadec, Patrick. *Etats d'urgence: Défaillances technologiques et déstabilisation sociale*. Paris: Seuil, 1988.

Lakoff, Andrew, and Ben Anderson. "Preemption, Precaution, Preparedness: Anticipatory Action and Future Geographies." *Progress in Human Geography* 34, no. 6 (2010): 777–98.

Latour, Bruno. 2004. "Why Has Critique Run Out of Steam? From Matters of Fact to Matters of Concern." *Critical Inquiry* 30, no. 2 (2004): 225–48.

Liberatore, Angela. *Management of Uncertainty: Learning from Chernobyl*. London: Routledge, 1999.

OECD. *Japan's Compensation System for Nuclear Damage*. Paris: 2012.

Oreskes, Naomi, and Eric M. Conway. *Merchants of Doubt: How a Handful of Scientists Obscured the Truth on Issues from Tobacco Smoke to Global Warming*. New York: Bloomsbury, 2010.

Mousseau, Timothy A., and Anders P. Moller. "Conservation Consequences of Chernobyl and Other Nuclear Accidents." *Biological Conservation* 144 (2012): 2787–98.

– "Genetic and Ecological Studies of Animals in Chernobyl and Fukushima." *Journal of Heredity* 10, no. 5 (2014): 704–9.

Nikkei Asian Review. "Fukushima Costs to Soar to 20 Trillion Yen." 27 November 2016.

Perrow, Charles. *Normal Accidents: Living with High-Risk Technologies*. New York: Basic Books, 1984.

Petryna, Adriana. *Life Exposed: Biological Citizenship after Chernobyl*. Princeton: Princeton University Press, 2002.

Shirabe, Masahi, Christine Fassert, and Reiko Hasegawa. "From Risk Communication to Participatory Radiation Risk Assessment." Fukushima Global Communication Program Working Paper, Series 21 (2015): 1–8.

Szarka, Joseph. "From Exception to Norm and Back Again? France, the Nuclear Revival, and the Post-Fukushima Landscape." *Environmental Politics* 22, no. 4 (2013): 646–63.

Topçu, Sezin. "Chernobyl Empowerment? Exporting Participatory Governance to Contaminated Territories." In Toxicants, Health, and Regulation since 1945, edited by

S. Boudia and N. Jas, 135–58. Studies for the Society for the Social History of Medicine, Series 6. London: Routledge, 2013.

– *La France nucléaire. L'art de gouverner une technologie contestée.* Paris: Seuil, 2013.

– "Organiser l'irresponsabilité? La gestion (inter)nationale des dégâts d'un accident nucléaire comme régime discursif." *Ecologie & Politique* 49 (2014): 95–114.

UNDP (United Nations Development Programme). *The Human Consequences of the Chernobyl Nuclear Accident: A Strategy for Recovery.* Report commissioned by UNDP and UNICEF with the support of UN-OCHA and WHO, 2002. http://www.by.undp.org/content/belarus/en/home/library/democratic_governance/publication_3.html.

UNSCEAR (Unites Nations Scientific Committee on the Effects of Atomic Radiation). *Sources, Effects, and Risks of Ionizing Radiation. UNSCEAR 2013 Report*, vol. 1: *Scientific Annex A: Levels and Effects of Radiation Exposure Due to the Nuclear Accident after the 2011 Great East-Japan Earthquake and Tsunami.* New York: 2013.

World Bank. *Belarus Chernobyl Review.* World Bank (Report no. 23883-BY), 2002. http://documents.worldbank.org/curated/en/674871468767985013/pdf/multiopage.pdf.

3

The Fukushima Process

SABU KOHSO

Event and Process

For the past ten years, since the Great East Japan triple disaster (earthquake, tsunami, and nuclear accident) of 11 March 2011, the damaged reactors at the Fukushima Daiichi Nuclear Power Plant have continued to leak radioactive nuclides into the environment, though the amount of released radiation has diminished since the accident. The Fukushima nuclear disaster has become the source of slow and long-lasting planetary radiation. It is a catastrophe without end. Not all Japanese agree about its impact, and that has led to fissures among them. This conflict has also generated strife between the pro-nuclear regime and anti-nuclear movements. This clash of perceptions is due to the attributes of radionuclides themselves, as well as the different ways by which they have affected the regime, the economy, society, and individuals.

A salient attribute of radionuclides is their virtuality: the unthinkable length of their half-lives, the complexity of their travel and accumulation patterns, and their ungraspable physical effects, all of which slip through a positivist logic of causation. Thus, their effects cannot always be contested satisfactorily in the political and judicial arenas.

To grasp the disaster's social and political impacts, we must view the catastrophe as both an event and a process. After the accident, economic and political forces – capitalism and the state – strengthened their bonds in order to collaborate in sustaining the business-as-usual regime, in terms of both industrial operations and the social order. We can observe here a dynamic process asymmetrically playing out between power operations and people's struggles. This chapter seeks to grasp this process as history repeating itself, while nevertheless offering a glimpse, just a flash, of the possibility of creating difference.

In the modern history of Japan – an earthquake-prone archipelago *and* a nation-state with intense military/economic ambitions – catastrophic events have erupted periodically, inflicting tremendous suffering and devastation. That devastation has helped drive broader infrastructural developments and economic expansion, which, in turn, have triggered still more accidents and devastation. This dynamic has been evident throughout Japan's modern social history. It is no exaggeration to say *ex post facto* that Japan's expansion as a nation-state was set in motion by catastrophes. The Great Kanto Earthquake of 1923 was followed by the development of the totalitarian state and imperialist expansion into Asia in 1937. Defeat in the Pacific War in 1945, after the US nuclear attacks on Hiroshima and Nagasaki and carpet-bombings of major industrial cities, paved the way for a postwar regime that claims to have achieved a democratic, peaceful, middle-class-oriented society, yet that same regime ultimately engendered an unending nuclear disaster. New policies of social control and remilitarization imposed by the government in the wake of Fukushima claimed to be restoring safety and bringing positive change, yet they have produced "reconstruction" projects and developments – such as Japan's hosting of the 2020 Olympics – that will generate even more environmental and social destruction.

In view of the ongoing legacies of Hiroshima and Nagasaki, Fukushima may be viewed as a tragicomic repetition (though there is a difference between a genocidal war and an accident). The "event" was perceived by many as the end of the world – a world of perceived affluence in a relatively stable, peaceful, and flourishing society. What was revealed by the nuclear disaster, however, was a war of the "peaceful nation" against itself. As experienced through the lens of everyday life, Fukushima engendered a soup of mixed emotions: grief for the tremendous losses, despair and anxiety for the future, rage against the regime that caused the disaster, and a glimmer of exhilaration over the unknown. However, an examination of the social context exposes to us a bifurcating tendency among the populace: one in favour of sustaining familiar, pre-disaster national conformity and solving problems in unity; the other wanting to confront the event and pull apart the fissures opened by the catastrophe so as to glimpse the unknown. In confrontation with government policies that have exploited the intangibility of radionuclides to deny or minimize their devastating impact, the politics of radiation exposure are playing out in two divergent directions as well: either one protests the regime by following a deeply conformist political process, or one challenges the regime by rejecting national conformity. Revealing an intention to confront the Fukushima event, a friend quipped that whereas in the heyday of the global justice movement she had single-mindedly embraced the slogan "another world is possible," now she would rather choose the slogan "staying with the trouble,"[1] because now it is the government that is seeking to mobilize the nation through a totalizing utopian vision of dystopian reconstruction and rebirth. This friend intends to carry on observing and living the Fukushima event – an event that "materializes" the breakdown of the postwar regime and its promises. Through this breach is pouring a full experience of life *and* death. This is a dreadful realization, but also a necessary impetus for action.

In what follows, I lay out the interaction between the event and the process of the unending catastrophe in the first three weeks of the state of emergency, in terms of the new formations of regime, society, people, and movement.

The State of Emergency

On 11 March 2011, two kinds of disasters arose in a chain reaction: a magnitude 9 earthquake followed by a 39 metre tsunami and a nuclear calamity. The first hit was colossal, causing 15,899 deaths, 6,157 injuries, and 2,529 missing.[2] It then triggered a series of far-reaching calamities across the Tohoku and Kanto regions, including the total blackout of the Fukushima Daiichi Nuclear Power Plant, which led to the meltdown and melt-through of nuclear cores, hydrogen explosions, and radioactive plumes and fallout. Systemic errors in emergency management – provoked mainly by authorities' conflicting priorities – led to significant shortcomings in rescue operations and disaster control. All of this created a "parallel chain crisis," in that a failure to deal with a particular crisis in one sector brought about a chain of crises in other dimensions that multiplied and intensified the disaster.[3] The catastrophe revealed fissures running through Japan's postwar regime – in the government, social milieux, and everyday life as well as in international relations. By the end of March, the worst-case scenario had been avoided: eastern Japan was not immediately made uninhabitable by a nuclear Armageddon. However, the calamitous breakdown of the reactors has not been resolved. The disaster continues, having shifted from a phase of emergency measures to one of unending crisis management. The invisible flow of radioactive nuclides is likely to cause more and more cancers and other illnesses for many years and generations to come.

The state of emergency lasted about three weeks. A number of testimonies from disaster victims attest to the Tokyo Electric Power Company's (TEPCO) miserable failure to act swiftly to protect local populations, and its flawed emergency management system at Fukushima has been exposed.[4] The responses of authorities were severely wanting. Fukushima revealed the limits of Japan's postwar constitution, though reactions to this revelation and the possibilities for action it suggested have since diverged considerably.

Rescue operations in the disaster area faced tremendous difficulties due to devastated infrastructures (transportation, communication, and electricity). Those operations were also incapacitated by serious discrepancies between the emergency instructions issued by local and regional authorities and those of the central government, as well as by problems in sharing information. Conflicting orders from the central government and Fukushima prefecture delayed evacuations and the distribution of iodine tablets to people affected by high doses of radiation. Some residents within the "crisis management district" were even left behind.

Life-and-death decisions had to be made by the leaders of local municipalities near Fukushima Daiichi, who then risked being accused of dissent and even lawbreaking by higher authorities.[5] Besides that, locals lacked the necessary information and the sophisticated technology needed to detect radioactivity – that is, they had no equipment for sampling dust, analyzing data, and predicting fallout patterns. Information from the System for Prediction of Environment Emergency Dose Information (speedi) was in the hands of the Nuclear Safety Commission, but for unknown reasons it was not shared with the prime minister's office. Crucial information repeatedly did not reach those who most needed it. The communication breakdown exposed power conflicts within the status quo, between ministries, political parties (the Democratic Party in power and the Liberal Democratic Party in opposition), government advisory agencies (Japan's Nuclear and Industrial Safety Agency and Nuclear Safety Commission, the US Nuclear Regulatory Commission, and the International Atomic Energy Agency), self-defence forces, police departments, fire departments, TEPCO, Toshiba (a nuclear apparatus supplier), and, finally, the US armed forces.

All of these disparate authorities, nevertheless, shared one concern: if they allowed a forthright assessment of the present threats posed by Fukushima, mass panic and the spectre of social unrest would arise. So the government spokesperson repeatedly announced, "There is no immediate health hazard."

Efforts to repair or seal off the crippled reactors were repeatedly set back by the constantly evolving situation on site. Decisions about how to address each critical phase were paralyzed by the conflicting priorities of nuclear advocates and safety advocates. Even as they were faced with possible apocalypse, TEPCO executives hesitated to take the decisive measure of sacrificing the plant, in favour of saving their assets (the reactors) for future use. TEPCO, like most of Japan's corporations, is organized as a strong hierarchy. The managerial class (legal and business experts) is significantly superior in rank to the engineering class (technical experts); while the control and care of highly complex devices is executed by TEPCO employees, the most dangerous work is performed by subcontractors, who lack sufficient information and protection; the employees at the reactors are strictly subservient to TEPCO headquarters in Tokyo. All of this placed tremendous pressure on those who were trying to resolve the critical conditions at the Fukushima plant even as they risked their lives by exposing themselves to high doses of radiation. It was they who made judgments as to when and how to cool off the melting fuel with seawater.

At the most critical moment, TEPCO executives asked the government for permission to evacuate all their workers from the site. Common sense dictates that the task of mending or liquidating nuclear reactors on the verge of explosion should lie beyond the ambit of wage labourers – such should only be a mission for a suicide squad, if anyone. Some commentators claim that TEPCO was desperately trying to transfer the command of its "mission impossible" to state power, aware as they were of the folly of a commercial enterprise committing itself to a project that endangered the lives of the masses.[6] But Prime Minister Kan Naoto issued an executive order that TEPCO remain at the site to halt the meltdown. If it did not do so, he said, "Japan will end." He also called for a voluntary suicide corps among the members of his administration. Some responded to it. Meanwhile, all sectors of the authority agreed that the last resort would be the military: Japan's Self-Defense Forces and, as well, the US Armed Forces, the ultimate guarantor of Japan's capitalist regime with its superior technology and manpower. However, there was reluctance within the Japanese government to invite a full foreign intervention, especially by the US. Remembering the post–Second World War occupation, they feared that Japan would be occupied again. This insistence on sovereignty was another cause of the delay and confusion in the government's response to the disaster.

The US government was losing faith in the Japanese government and TEPCO in terms of their willingness, capacity, and determination to confront the situation. But it could not abandon its most successful client-state. It ordered all American citizens living in Japan to evacuate from a fifty-mile radius of Fukushima Daiichi, and it sent forces, including the Chemical Biological Incident Response Force, to assist in disaster control and relief operations. As a consequence, some US Navy personnel have developed radiation illnesses.[7] Most other countries instructed their nationals to leave Japan immediately.

Notwithstanding nuclear power's unprecedented risks (such as nuclear accidents) and irresolvable problems (such as the disposal and storage of nuclear waste), states with access to nuclear technology have no intention of giving it up. Fukushima has shown that the use of nuclear power will never cease; our hope that it has reached a dead-end has been denied. Before the Fukushima disaster, 442 nuclear reactors in thirty countries were producing 14 percent of the world's electricity. That figure dropped to 11 percent in 2012. But by 2014, 435 reactors were operating in thirty-one countries and 68 more were under construction. Apparently, the aftermath of Fukushima amounted to little more than a momentary blip. After the disaster, Japan ordered large-scale inspections and introduced new safety regulations. In December 2011, pressured by intensifying anti-nuclear protests, the Noda administration (Democratic Party) declared denuclearization in "future Japan." But within a year, the Abe administration (Liberal Democratic Party) had reversed the proclamation and revived the policy of nuclear expansion. By August 2016, three nuclear power plants – Sendai 1 and 2 in Kagoshima prefecture and Ikata 3 in Ehime prefecture – were back online for commercial services.[8] And at no point did the Japanese government give up its project of boosting civilian nuclear exports to countries with nuclear ambitions such as Vietnam, Jordan, South Korea, Turkey, and the United Arab Emirates. In 2015 the Abe administration signed a specifically nuclear-based industrial treaty with India.[9] All of this is to say that radioactive waste will go on piling up endlessly on the planet. Its treatment is too enormous a task for current knowledge and technologies, yet nuclear states and global nuclear industries continue to claim that the problem of waste is solvable and that nuclear energy is safe, clean, and "green."[10]

The Regime

From the vantage point of statecraft, the lesson from the nuclear catastrophe was clear: Japan would need a stronger central command system, abundant resources and supplies, troops with highly advanced equipment and fearless manpower, and, finally, a strong national bond (*kizuna*).[11] The reflections of men in power on the failure of disaster control seem to have revived an old slogan from the Meiji Restoration at the dawn of Japan's modernization: *fukoku kyōhei (wealthy nation and strong soldiers)*. The term *kizuna* refers to the most intimate relations in family, friendship, workplace, local community, and so on. From the perspective of the Japanese government, applying this concept to governance implied strengthening (the presumed) totalitarian nature of Japan's modern statehood, while at the same time reviving the nationalist semantics and the anti-democratic policies that had buttressed the Meiji-era slogan *fukoku kyohei* – in other words, a vision based on the belief that Japan should once again see itself as a strong nation-family headed by the emperor.

From this perspective, the Abe administration's political reforms and policies of suppressing civil liberties inch by inch – through controversial laws and bills such as the act to protect specially designated secrets, the right of collective defence, and the conspiracy crime bill – may be seen as logical responses.[12]

In terms of the economy, it is the primary industries that have been damaged most seriously by radiation contamination. Management of the crisis triggered by the nuclear disaster will require a tremendous number of endangered workers and endless costs. TEPCO's limitless deficit will be paid by the populace through taxes and electricity bills. These will exhaust human and natural resources for years to come. Many countries have stopped importing food products from Japan. Excessive oil and liquefied natural gas imports have caused a trade deficit for the first time in thirty-one years. So-called Abenomics intervened with a double standard: to save the national economy at a time of unprecedented crisis, it reintroduced the neoliberal reforms of the Koizumi administration (2001–06). Basically, these reforms consist of public spending cuts, deregulation, and deflation via a weaker yen. By aiming at nominal economic growth, the administration seeks to hide losses in real economic growth; claiming to increase employment (especially for women), it is also accelerating the informalization of workforces. Such measures can only result in the massive transfer of the waning national wealth to a handful of ruling elites and in the deepening of the population's immiseration.

Meanwhile, earthquake "special procurements" have increased the gross production of the three prefectures most severely affected by the disaster. General contractors in Iwate, Miyagi, and Fukushima prefectures were fully engaged in the business of reconstruction, including decontamination work. Amid this construction bubble, radioactive debris and soil were accumulating endlessly, and a way had to be found to somehow manage the treatment of this hazardous waste. The central government began to require that local municipalities share the waste by incinerating it at local facilities. The Ministry of Environment considered mixing contaminated soil into compounds to be used in nationwide construction projects.[13] The Ministry of Agriculture, Forestry, and Fisheries launched a campaign called "Eat and Support Fukushima" to encourage the recovery of Fukushima's primary industries.[14] Through such policies, post-disaster governance initiated the nationalization of radiation. From the commoners' standpoint, this was a secondary and permanent disaster. The politically imposed sharing of radioactive contamination has been the most dystopic aspect of the Fukushima process.

The period of "parallel chain crisis" has passed, but what happened during that time prefigured what has been happening since. The nuclear disaster has yet to be resolved, and meanwhile, post-disaster governance has created a regime of disaster management that goes on imposing its dystopian projects on every aspect of the social system. In that same vein, governance continually seeks to maintain and reinforce its management process by covering up the fissures opened by the catastrophe.

The People

It goes without saying that it was the people of the Tohoku and Kanto regions who were most seriously harmed by the destruction and contamination. Those living in the immediate vicinity of the reactors were forced to leave their homes, and people from other parts of Tohoku and Kanto also evacuated, confronted as they were with the imminent danger of more explosions and the rise of bigger radioactive plumes. In anecdotal evidence, many of my friends living in Tokyo fled to Nagoya and other western cities, where they lived communally for a few weeks or a few months. While the majority returned to their homes once they learned that the worst-case scenario had been avoided, many of them came to intuit that Tokyo was could no longer be their permanent residence and that they needed to leave it sooner or later. Thus the first wave of evacuations was followed by a second, quieter, slower, and longer exodus to western Japan or Hokkaido.[15]

People across Japan began to express their emotions more strongly, in contrast to the rational, bureaucratic jargon of the government's discourse on reconstruction, recovery, crisis management, and national bonding. But emotions varied with the degree of physical susceptibility and sensitivity to radioactive contamination: Who among these people would be most susceptible to suffering from contamination? And who would be the ones to respond to it most radically? Physical susceptibility is epitomized by the residents of contaminated zones, farmers and fishermen, nuclear workers, sanitation workers, homeless people, and the younger population. Those who responded to the situation most sensitively and radically tended to be caregivers and care workers, mostly women, due to their engagement in reproduction on dual and conflicting registers, as both historically charged labour and historically nurtured wisdom.[16] It was the "climate" of these emotions that created a politics of life *and* struggle within the catastrophe.

In cities, this emotional climate gave rise to a multiplicity of actions, from everyday struggle (e.g., radiation monitoring and voluntary evacuation) to anti-nuclear protests (e.g., mass demonstrations and blockades against radioactive nationalization) to lawsuits against TEPCO and the Japanese government. In their interactions with the invisible flows of radionuclides, these projects were articulated as people's power or will to develop a new realm of political ontology, wherein subjectivity is decomposed into Soul or senses of relations, in order to confront the assemblage of forces operating in the spatiotemporality called the earth. This is far more complex than the conventional category of the political and does not yet have a name.

At the same time, these projects corresponded to the fissures opened up by the disaster, fissures whose effects have often been ambiguous, resulting in the break-up of families, friends, communities, and even movements. In Japan's historically congealed condition, the conformity of a homogenized nationhood consisting of smaller units of sociality – families, neighbourhoods, municipalities, and society – serves as a solid stronghold for sustaining class hierarchy and governance; while these dissolutions have often been tragic, they have also opened possibilities to create new socialities and new compositions of movements.

Society

Within a few years, the Fukushima problem had been set aside in the mainstream media. People's grief, anxiety, and anger were paid less and less attention to. The scenery of hyper-consumerist society fully returned. Program after program on TV highlighted celebrities joyfully tasting delicacies from disaster-stricken areas, in support of the recovery of local industries. A new, more aggressive nationalism was in the air. Rituals of hoisting the Japanese flag and singing the National Anthem were made mandatory in schools. Xenophobic groups abused immigrants on the street more overtly and violently than ever before. On the surface, the nation's unity was back in full stride toward a wealthier and stronger Japan.

Attention in the public discourse has shifted from the Fukushima problematic to legislative issues, namely, the Abe administration's reforms and policies. Large protests have been organized against the government with the slogan of protecting Article Nine of the "peace constitution," which bans any military intervention by Japan against other countries. Along with the symbolic emperor system in Article One, this was part of the postwar constitution drafted under the influence of the General Headquarters of the Allied Forces.[17] The constitution supposedly created a war-free enclave in the tumultuous postwar world, even as US military bases proliferated across the archipelago and the nation's economic boom was being driven by the procurement demands of the Korean and Vietnam wars.

Peace involves complications: it was made possible in synchronicity with the US military strategy in East Asia, under US protection and control. Under that regime, the status of Japan's Self-Defense Forces has been ambiguous in terms of their potential for aggression, especially among Asian countries that had suffered through Japan's wartime atrocities. Okinawa has had a large part of its land expropriated as the front line of US military operations on the Asian continent. Postwar Japan's "peaceful" regime – which has exempted the Japanese from engagement in war – has, in the context of global power contestation during and after the Cold War, never been just a matter of nationwide peace.

In the development of a peaceful enclave in Japan, nuclear power has played two vital roles: as weapon (stick) and as energy (carrot). Atomic bombs had been dropped on Hiroshima and Nagasaki in order to end a war led by the emperor-centred, militarist regime, and made the ensuing occupation quick and smooth for the Allied Forces. There have been debates about precisely why the Japanese government accepted the Potsdam Declaration that underscored Japan's surrender. Was it because of the two atomic bombs, or was it because the Russians were about to intervene in the Manchurian war? We do not know the main incentive for the government's surrender, but it is clear, at least, that the power of the new atomic weapon scared the hell out of the Imperial Council and rushed them to accept unconditional surrender.

A pacified Japan became an ideal strategic base for the US. However, in March 1954, nuclear fallout from a US thermonuclear test exploded on the Bikini Atoll in the Pacific affected a Japanese tuna boat, which shocked and enraged a populace that was still recovering from the war. In response, a mass movement arose against American imperialism, which was accelerating the development of even more destructive nuclear weapons. Initiated by a group of women in Tokyo's Suginami ward, the movement gained momentum, anticipating the 1960 uprising against the US–Japan Security Treaty (Ampo) and the decade of social unrest that followed. The rising antagonism in Japan became the first target of the Eisenhower administration's "Atoms for Peace" policy, an attempt to establish a new world order governed by the double bind of nuclear power and the entrapment of the planetary populace between two destructive scenarios – the apocalyptic Doomsday of nuclear war and the utopian dream of limitless energy in the future.

The uprisings throughout the 1960s coincided with the time of economic upswing. In this social situation, bombastic media campaigns were orchestrated by the US government and complicit Japanese interests in government, industry, and the media, which aimed to implant a utopian illusion of nuclear energy in the public mind and introduce it for commercial use. The campaigns succeeded: the Japanese anti-nuclear movement split in half between those who were only against weapons and those who were against both weapons and energy.[18] In the early 1970s, Japan began to develop nuclear energy, and by the time of the Fukushima disaster it was operating fifty-four nuclear reactors in fourteen regions. As a by-product, nuclear energy generates tons of plutonium, a radioactive substance recycled from uranium waste that can be used to make nuclear bombs. Japan's nuclear power production tacitly enabled the peaceful nation-state to export the bomb's main ingredients to its allies, and to keep the option to develop the weapon for itself in the future.

The suppression of the popular uprisings of the 1960s was the prelude to a society of high consumerism and control in the 1970s and 1980s. The gradual consolidation of the nuclear age in Japan coincided with a gigantic infrastructural reconstruction project – the 1972 "Remodeling of the Japanese Archipelago" under Prime Minister Tanaka Kakuei – which entailed massive campaigns to diffuse the flow of labour and money away from big cities to the countryside by redistributing industries and developing a national transportation and information infrastructure. This ended the hopes of the remaining revolutionary groups to build guerrilla bases in the mountains. The role of the energy and information sectors was expanded to an unprecedented degree; the Japanese archipelago was becoming a tightly controlled network of energy, information, and transportation, equipped with full security functions.

At that point, the Fukushima disaster caused an implosion of the power and control apparatuses.

The Movement

The accident was followed by a state of anarchy, a rising existential resistance. In April 2011, the first mass anti-nuclear demonstration, named Amateur Riot, was mobilized by a small neighbourhood in Koenji, a suburb of Tokyo. Led by young people living in precarious conditions, the demonstration involved heterogeneous constituencies and internalized the potential for heterogeneous actions. The crowds were not those of the old socio-political movement. At the same time, many autonomous projects were undertaken to protect daily life and reproduction from radiation. Together, these forces challenged the power and legitimacy of the establishment, the government bureaucracy, corporate industries, academia, and the mainstream media, all of which acted as proxies for the pro-nuclear regime.

Anarchic opposition faded after the Oi Nuclear Power Plant was brought back online despite a July 2012 direct action to block it and with the rise to power of the Abe administration the following December. By then, the mass protests against nuclear power had mutated into a populist movement, which claimed to be a broad movement of "normal citizens." Working with the police, this movement established a pattern of large mobilizations for strictly legal protests in Tokyo and other main cities. While keeping to the same time (Fridays until 8 p.m.) and place (the prime minister's official residence in Tokyo), the protest's theme transformed from anti-nuclear to anti-Abe, and its mode shifted from open-ended opposition to electoral campaigns for candidates from progressive political parties.

Liberal ideologues appeared in major publications and media to promote this populist movement, while disparaging the radical movements of the 1960s as vanguardist. The Japanese radical left had historically internalized the problems of authoritarian vanguardism, and the negative effects of this tendency have long been critically scrutinized by anti-authoritarians in post–New Left movements, from the citizens' movements of the 1970s to feminist and minority movements to the anti-globalization movement of the 2000s. For its part, liberal politics has turned away from the anti-capitalist, globalist, and revolutionary perspectives of previous radicals. Its discourse lacks critical awareness of its own social organization, which is, as one Marxian analysis would have it, "determined by the class interest of the people involved in it, by its relationship to capital, its historical, geographical and psychological conditions."[19] Thus, liberal discourse has returned to uncritically and unreservedly embracing the postwar constitution. Its defence of Article Nine can be viewed as an intranational pacifism based on a set of internalized beliefs: that, in the eyes of the populist movement, the desperate wish of the Japanese to sustain disengagement from war should itself automatically contribute to world peace, and that it is to achieve such a goal that all oppositional tendencies in the movement itself must be pacified. By ignoring the problematic interconnectivity of the peace constitution, the nation's entanglement with nuclear power, and Abe's post-nuclear disaster governance, the populist movement thus ended up opposing the policies of the Abe administration diametrically, within one and the same frame of national conformity. In this manner, the existential resistance that had opened a new dimension of political ontology in the immediate aftermath of the nuclear disaster has been overshadowed in public discourse.

Repetition and Difference

Around the year 2011, two series of events were set in motion that came to both embody and epitomize today's planetary reality: revolution and disaster. In December 2010, a new cycle of global uprisings began with the Arab Spring; the following March, the Fukushima nuclear disaster erupted. The wave of uprisings inspired rebellions across both hemispheres, which encouraged the early post-Fukushima struggles as well as the rise of Occupy Wall Street in New York City. Born in Japan and living in New York, I have been radically affected and deeply perplexed by these two intersecting, mutually mirroring events. They epitomize the global condition in which we are living today: the world as *logos* based on nation-states has come to constrain the desire and potential for (re)creating life, relations, and the environment, while the earth as an assemblage of forces has begun to express itself in complex entanglements and frictions. Belief in the world is collapsing, and meanwhile, the expectation of life with and as permanent struggle permeates the earth as planetary ecology.

The new cycle of global uprisings comprises waves of life-affirming struggles confronting multifaceted injustices, oppressions, and dispossessions in the name of survival, autonomy, and freedom. Interestingly, these movements have started to show a pattern that exposes both their limits and their potential for renewal. Local uprisings are either crushed immediately or quelled gradually in various ways involving violent crackdowns and institutional tampering – and in some cases, such interventions lead to civil war. At the same time, their impetuses continue to emerge in synchronicity, as if their impacts reverberate from one person to the next and travel from place to place through what George Katsiaficas calls the "eros effect."[20] This resonance seems to follow multiple flows of memes in electric signals (via media) and vital energy (via personal and local connections), rather than being under the unilateral command of some international organization. People in different places have not stopped revolting, no matter what oppression awaits.

In Japan, in the first two years after the disaster, participants in life struggles often felt that they were part of the global uprisings. This ignited exhilaration for the unknown and vitalized the will to confront the Fukushima event. All actions, from street demonstrations to everyday projects, were invigorated, fully expressing their strength to confront the unprecedented calamity and the monstrous regime. But in the past few years, this sense of being part of a planetary reverberation has been lost, and the revolutionary impetus has been subsumed by the process of reinforced social management. What has returned as a result of that process is a sense of belonging as the horizon of thinking and acting, which is confined by the apparatus of national conformity.

During the long reign of the Tokugawa Shogunate (1603–67), the isolation policy of *sakoku* (national enclosure) created popular perceptions of homogeneous ethnicity and insular territoriality. Since the advent of capitalist modernity with the Meiji Restoration in 1868, these perceptions have congealed into anti-foreignism, which facilitated the massive mobilization of Japan's imperialist expansion in the Asia-Pacific region. After the Second World War, Japan was reopened to the world, but the manufactured perception of a putatively homogeneous national body politic seems to have remained in the popular subconscious. Today, the homogenized social body is constantly being reinvented and reproduced via mass media, in concentrated fashion for both commodification and politicization. In society at large, the will to listen to, figure, or make room for otherness is easily excluded from public discourse; in this way, the sense of national belonging has been homogenized as the only horizon for thinking and acting. In the Fukushima process, the blackmailing capacity of nuclear catastrophe as an *all or nothing crisis* finds its ideal prey in the sense of belonging: for the sake of solving a problem of such magnitude, the feeling of debt to the nation must be prioritized. In this manner, the notion of Japanese insular nationhood or national belonging has been tacitly returned to during crises, time and time again, precluding both class-based politics and the aspiration for planetary becoming.

Let us return to the emotional soup that followed the disaster. Let us reactivate the will to live the event that resisted the process. In fact, as an undercurrent that has been unfolding in the past few years, more and more people have quietly changed their ways of life by developing autonomous projects, from DIY monitoring of radioactivity and minor citizens' sciences of developing and distributing knowledge on nuclear power and radiation illness, to direct actions, evacuations, and community-building. These efforts are depicted by the authorities and in the media as anomalous, as abetting mass hysteria, spreading ill-advised rumours, and causing disharmony within families, workplaces, and communities. Some evacuees from northeastern Tohoku and Tokyo to western Japan or Hokkaido are changing their subsistence and living environments. Some are engaging in farming and hunting in the countryside, in addition to part-time jobs in the city. Their ways of life deviate from the norm of the postwar society: lifetime employment, a house, two children, and a car in the suburbs. Some among these people are opposed not only to radiation exposure but also to the capitalist industrial civilization centred in Tokyo. They are discovering a new life and a new relationship with the land: a life that is ephemeral, singular, and directly connected to ecological changes, part of an archipelagic Japan that has always facilitated transcontinental Asiatic and global flows of life-energy by deterritorializing insular Japan.

With these projects, the Fukushima Event reverberated with planetary becoming for a short period. To push this reverberation further, to be part of the new global uprisings that are promoting a new political ontology, it is necessary to forget the name Japan. Radionuclides do not know borders.

NOTES

This chapter is adapted from *Radiation and Revolution (Durham: Duke University Press, 2020).*

1 This is a reference to Donna Haraway's *Staying with the Trouble.*

2 National Police Agency of Japan, "Damage Situation and Police Countermeasures," https://www.npa.go.jp/news/other/earthquake2011/pdf/higaijokyo_e.pdf.

3 "Parallel chain crisis" (*heikō rensa kiki) was* an expression used in a report by the Independent Investigation Commission on the Fukushima Daiichi Nuclear Accident (Minkan jiko cho), issued on 28 February 2012, http://park.itc.u-tokyo.ac.jp/tkdlab/fukushimanpp/minkan.html.

4 Motoyuki, "Puromeiteusu no wana" (The trap of Prometheus).

5 One example was the act of publicly sharing residents' personal information to identify those potentially unable to respond to the state of emergency because of age or physical incapacity. Risking incrimination, some local leaders shared this information in an attempt to rescue people who were left in the evacuation zone.

6 Funabashi, *Kauntodaun merutodaun (Countdown meltdown).*

7 Lynne Peeples, "US Armed Forces Sickened after Fukushima Meltdown Get Help from Online Fundraising," *Huffington Post, 8 March 2014.*

8 Fukushima Nuclear Disaster [blog], accessed 20 April 2016, http://blogs.yahoo.co.jp/fukushima_nuclear_disaster_news/33788561.html.

9 "India, Japan Reach Agreement on Nuclear Cooperation," *World Nuclear News, 14 December 2015.*

10 The most graphic example is the endless failure of the Rokkasho Reprocessing Plant. See, for example, Greenpeace, "Planning for Failure: International Nuclear Safeguards and the Rokkasho-Mura Reprocessing Plant," November 2002, http://www.greenpeace.org/international/Global/international/planet-2/report/2002/11/planning-for-failure-internat.pdf.

11 Funabashi, *Kauntodaun merutodaun.*

12 See, for example, Mina Pollman, "Japan's Controversial State Secrets Law: One Year Later," *The Diplomat,* 9 December 2015; Ayako Mie, "Security Laws Usher in New Era for Pacifist Japan," *Japan Times*, 29 March 2016; and Daisuke Kikuchi, "Controversial Conspiracy Bill Approved by Abe Cabinet," *Japan Times*, 21 March 2017.

13 "Nuclear Watchdog Questions Environment Ministry's Plans to Reuse Radioactive Soil," *Mainichi Shimbun*, 9 January 2017.

14 The campaign "Eat and Support Fukushima" (Tabete ōen shiyō) was launched by the Ministry of Agriculture, Forestry, and Fisheries and continues to this day, involving a number of fast-food chains, supermarkets, restaurants, and consumers. According

to the ministry's website. "Tabete ōen shiyō" (Eat and support Fukushima), accessed 5 February 2015, http//www.maff.go.jp/j/shokusan/eat: "We call for a wide cooperation for the campaign, in order to support the reconstruction of the disaster-stricken areas by actively consuming food products from the areas."

15 The number of forced evacuees from Fukushima Prefecture is recorded by the government's Reconstruction Agency, but this does not include the number of voluntary evacuees from the Tohoku and Kanto regions. See Reconstruction Agency, "Efforts for Reconstruction of Tohoku," accessed 10 April 2014, http://www.reconstruction.go.jp/english. For the distribution of forced evacuees across Japan, see Fukushima on the Globe, "Situation of the Evacuees," accessed 110 April 2014, http://fukushimaontheglobe.com/the-earthquake-and-the-nuclear-accident/situation-of-the-evacuees.

16 The idea is inspired by the tradition of Marxist feminism. See, for instance, Federici, *Revolution at Point Zero.*

17 In his recent book *Kenpō no muishiki* (The unconscious of the constitution), Karatani Kōjin analyzes the process: Article One was designed to preserve the emperor, which the supreme commander for the Allied Powers, Douglas MacArthur, thought necessary to quell potential communist revolution in Japan. Article Nine came along with it as a bargaining point. But the latter eventually became the unconscious of the nation, in response to its own guilt over its wars of aggression in Asia in the 1930s and 1940s.

18 Muto, "The Buildup of a Nuclear Armament Capability."

19 This phrase is cited from a critique of the anti-nuclear movement published in 1979: Midnight Notes Collective, *Strange Victories.*

20 See Katsiaficas, "Eros and Revolution"; and Thompson, "Remembering May '68."

BIBLIOGRAPHY

Federici, Silvia. *Revolution at Point Zero: Housework, Reproduction, and Feminist Struggle.* Oakland: PM Press, 2012.

Fukushima Nuclear Disaster [blog]. http://blogs.yahoo.co.jp/fukushima_nuclear_disaster_news/33788561.html.

Fukushima on the Globe. "Situation of the Evacuees." http://fukushimaontheglobe.com/the-earthquake-and-the-nuclear-accident/situation-of-the-evacuees.

Funabashi Yoichi. *Kauntodaun merutodaun* (Countdown meltdown). 2 vols. Tokyo: Bunshun bunko, 2016.

Haraway, Donna. *Staying with the Trouble: Making Kin in the Chthulucene. Durham: Duke University Press, 2016.*

Karatani Kōjin. *Kenpō no muishiki (The unconscious of the constitution). Tokyo: Iwanami shinsho, 2016.*

Katsiaficas, George. "Eros and Revolution." *Radical Philosophy Review* 16, no. 2 (2013): 491–505. https://doi.org/10.5840/radphilrev201316238.

Mainichi Shimbun. "Nuclear Watchdog Questions Environment Ministry's Plans to Reuse Radioactive Soil," 9 January 2017. https://mainichi.jp/english/articles/20170109/p2a/00m/0na/012000c.

Midnight Notes Collective. *Strange Victories: The Anti-Nuclear Movement in the U.S. and Europe*. Brooklyn: 1979.

Ministry of Agriculture, Forestry, and Fisheries. "Tabete ōen shiyō" (Eat and support Fukushima). http//www.maff.go.jp/j/shokusan/eat.

Motoyuki Maeda. "Puromeiteusu no wana" (The trap of Prometheus). *Asahi Shimbun*, 15–27 November 2011.

Muto, Ichiyo. "The Buildup of a Nuclear Armament Capability and the Postwar Statehood of Japan: Fukushima and the Genealogy of Nuclear Bombs and Power Plants." *Inter-Asia Cultural Studies* 14, no. 2 (2013): 171–212.

Thompson, A.K. "Remembering May '68: An Interview with George Katsiaficas." *Upping the Anti* 6 (October 2009): 59–74.

4

Fukushima and the Rebuild of Godzilla: Multiplying Media in an Era of Multiplying Disaster

THOMAS LAMARRE

The two great hits at the Japanese box office in the summer of 2016 were the animated film *Kimi no na wa* (*Your Name*, dir. Shinkai Makoto) and the special effects film *Shin Gojira* (*New Godzilla*, distributed internationally as *Godzilla Resurgence*, dir. Anno Hideaki). Both films evoke the tragic events of 11 March 2011, that is, the earthquake and tsunami that struck the coast of northeastern Japan, triggering the nuclear reactor meltdowns at the Fukushima Daiichi Nuclear Power Station operated by the Tokyo Electric Power Company (TEPCO). The two films could not be more different in their evocation of what has come to be known as the "triple tragedy" and glossed as "3/11." *Kimi no na wa* sticks to the natural side of the triple tragedy, in the guise of a comet that strikes the earth near a remote mountain lake, shaking the earth, turning the lake into a massive wave, and destroying a village on its shore. Nothing like nuclear meltdown enters the picture. By contrast, *Shin Gojira* repeats the entire cascade of events associated with 3/11. As Godzilla enters Tokyo, its massive size produces earthquake- and tsunami-like effects. That is only the beginning of the disaster: the ultimate source of destruction lies in the radioactive effects of Godzilla's body. The creature has ingested illegally dumped nuclear waste and through a series of mutations has absorbed the process of nuclear fission into its body. The new Godzilla acts like a nuclear reactor in meltdown at the heart of Tokyo.

The widespread popularity of both films suggests that their evocations of 3/11 struck a deep chord with audiences, and the two films were often discussed together in the Japanese press.[1] Although the focus here is primarily on *Shin Gojira*, a broad comparison with *Kimi no na wa* helps establish what is at stake.

First, both films model their disaster on 3/11, but where *Kimi no na wa* avoids nuclear destruction, *Shin Gojira* embraces it. Indeed, *Kimi no na wa* offers a potentially empowering political scenario: when government leaders refuse to act, a group of young people band together and engage in direct action, taking out the electric grid to save a small rural town from destruction. Yet its lesson with respect to the ongoing nuclear crisis appears in inverted form. The film is able to imagine what Rebecca Solnit calls "a paradise built in hell" only by stopping 3/11 short of nuclear crisis.[2] In *Shin Gojira*, such a scenario becomes unimaginable, for two reasons: everything proceeds through government mandate, and the crisis continues to escalate at such a rapid pace that even the glimmer of small-scale utopia imparted by the few brief moments of relief and camaraderie take on a grim, hellish tone.

Second, although *Shin Gojira* was the highest grossing Godzilla film to date and garnered a series of major cinematic awards in Japan, it was Kimi no na wa that turned into a runaway hit, becoming the highest grossing Japanese animated film ever, both domestically and internationally. The discrepancy between the two films in terms of reception is so striking that it is tempting to think of Kimi no na wa as a global event, whereas the impact of *Shin Gojira* remained more of a Japanese phenomenon. Some notes of caution are in order here. The distribution and reception of films are both complicated matters, and box office success does not depend on content alone. What is more, the national does not stand in opposition to the global. Both films are recognizably Japanese and global in their basic conventions. Nonetheless, the two films present very different takes on Japan. *Shin Gojira* lingers on the intricacies of governmental operations. It strives for a realistic presentation of how the government might respond to an unimaginable disaster such as Godzilla. Its presentation of bureaucracy is detailed to the point that it even strains the knowledge of Japanese audiences.3 In addition, it frequently takes the side of the government. Treating the Godzilla nuclear crisis as unforeseeable, for instance, is tantamount to treating the TEPCO reactor disaster as unforeseeable, when in fact, warnings about the Fukushima reactors had frequently been sounded, especially in the wake of the two Tōkaimura nuclear accidents of 1997 and 1999.4 What is more, because Godzilla has repeatedly been interpreted as representing a Japanese experience of nuclear trauma, the film cannot avoid engaging with issues of nationalism and militarism that Kimi no na wa deftly pushes aside. *Shin Gojira*, then, demands a different kind of interest in Japan than Kimi no na wa, implying a different investment in knowledge about Japan.

Finally, initial responses to the two films stressed the transformation of *otaku* culture. Both directors, Anno Hideaki (b. 1960) and Shinkai Makoto (b. 1973), are renowned for creating anime series and films that appeal to male fans or enthusiasts of manga, anime, games, and light novels, which are loosely characterized as otaku, often with derogatory connotations.

What surprised commentators familiar with otaku culture about the two films was the movement of otaku fare into new markets and its vast appeal to diverse audiences. Kimi no na wa proved a hit with young women, and *Shin Gojira* garnered approval from the more elite and established film community. To be sure, so-called otaku culture has long appealed to diverse audiences around the world, but largely through informal networks and through media involving more personalized platforms such as television, the internet, and the cell phone. The spectacular success of both films at the domestic box office suggested a visible, public endorsement of materials previously associated with more personalized niches and informal exchanges. Arguably, such a transformation is of a cloth with the movement of teen fiction and then fan fiction onto the big screen in the North American context.

Both films retain features associated with *otaku* culture, working within a framework that implies active participation on the part of interpretative communities. They entail a sort of intrinsic serialization, related to modes of commercialization working across multiple media, such as toys, games, comics, and television series. This intrinsic serialization invites repeated, participatory interaction, whereby interpretative communities actively engage with processes of coding and decoding, generating and transforming meanings. Although such films may allow for a one-time, stand-alone viewing, they are meant to be read within larger cross-media franchises and within certain genre parameters, and to be read repeatedly. If treated as stand-alone films, they show an affinity with what is sometimes called the puzzle film or the mind game film.[5] As such, both *Shin Gojira* and *Kimi no na wa* tend to recast the events of 3/11 into something like a cross-media game or multimedia puzzle.

This brief comparison of *Kimi no na wa* and *Shin Gojira* establishes three important lines of inquiry for discussing *Shin Gojira*: (1) the multiplication of disaster, (2) the problem of nationalizing trauma, and (3) multiple pathways of interpretation. Already a strange affinity is apparent between the multiplication of disaster and the tendency of contemporary popular culture to work across multiple media through multiple pathways. At a deeper level, flows of capital may be held responsible for producing both "mani-fold" domains (multi-plied media and multi-plied disaster), which explains their underlying affinity or semblance. Yet the two domains cannot be neatly mapped onto each other, nor can they be conflated. Something happens between them. Here I propose reading *across* them to expose the spatio-temporal paradoxes at the (capitalist) heart of the ongoing nuclear crisis, wherein energy (generation) becomes indistinguishable from weaponry (destruction).

Disaster Multiplied

The association of Godzilla with nuclear crisis and especially with the atomic bomb is a basic premise of the original 1954 film and now something of a cliché. But a Godzilla that undergoes distinct stages of transformation is a new feature within the Tōhō franchise, and the metamorphoses in *Shin Gojira* are designed to address the new nuclear situation associated with Fukushima. Promotional materials for the film explicitly linked the film to the triple tragedy. A television special on the making of the film, for instance, explains how 3/11 provided the background of the film.[6] Indeed, the new Godzilla's transformations repeat, in encrypted form, the three stages of the triple tragedy. To acknowledge these transformations and Godzilla's ability to incorporate disparate species as well as its unspecified gender, I will henceforth refer to Godzilla as they.

In the first stage, an undersea explosion occurs in Tokyo Bay. The explosion sends shock waves through the ocean, causing leaks in the commuter train tunnel that runs through the middle of Tokyo Bay, the Aqua-Line. Views of exploding plumes of water arcing skyward are captured from the Trans-Tokyo Bay Expressway, built directly over the tunnel. The government's emergency task force assumes the explosion to be due to an undersea volcano, which is say, some sort of tectonic activity. Thus, *Shin Gojira* offers a repetition of the earthquake, magnitude 9, that occurred off the northeastern shore of Honshū, the main island of the Japanese archipelago, on 11 March 2011. The initial earthquake triggered earthquakes well above magnitude 7 across eastern Japan, causing widespread damage to highways, railways, and other infrastructures. Apparently, the whole of Honshū was shifted some two and a half metres eastward. In the film, these effects are scaled down to suit both the city of Tokyo and the demands of narrative suspense. Still, the gesture is clear enough: the first stage of Godzilla recalls the first "tectonic stage" of the unfolding catastrophe of 3/11.

As the monster morphs into the second stage, they enter the Nomi River, moving deeper into the urban area. The effects of the second metamorphosis recall the second, "tsunami" stage of the triple tragedy when enormous waves struck repeatedly along the northeast coast of Honshū. Those waves, well over 10 metres in height, washed away entire villages and towns and flooded urban districts and agricultural areas. To portray Godzilla's second stage, the film shows a massive wall of debris, including boats and pieces of houses, moving up the river, with citizens running for higher ground (Figure 4.1).

These scenes plainly repeat the horrifying and widely circulated footage of villagers in northeastern Japan fleeing walls of debris carried up the rivers as the tsunami pushed inland. The scale of damage caused by the combination of the earthquakes and tsunamis was unprecedented: 500 kilometres of coastline devastated, 4,000 roads destroyed, and 71 bridges swept away, to cite some of the

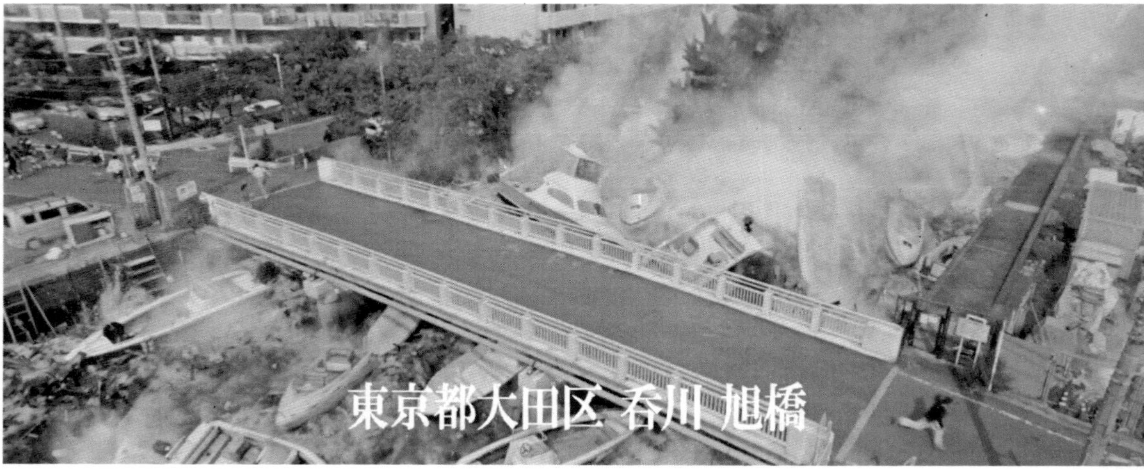

4.1 In its second stage, Godzilla swims up the Nomi River, pushing a pile of debris into the Asahibashi in manner that recalls the tsunami waters pushing walls of debris up the rivers in northeastern Japan on 11 March 2011.

salient figures. The tallies of human loss are numbing: 15,000 dead, 8,000 missing, 115,000 homeless, and more than 500,000 residents dislocated.

Shin Gojira diminishes the scale of damage, partly to suit the scale of Tokyo and partly to fit the story of an escalating threat. It would have been cinematically impossible to convey destruction on the same scale as what occurred in northeastern Japan. Nonetheless, the film makes its references to 3/11 abundantly clear, and in keeping with the triple tragedy, the calamity in *Shin Gojira* does not end with the tsunami stage.

In the third stage, Gojira takes to the land, rising on their hind legs and becoming fully visible. They topple buildings, send trains flying, and wreak destruction as they advance into the urban sprawl of greater Tokyo. This stage provides the first intimations that Godzilla is radioactive – a strange reddish glow shines forth around their dorsal spines and gills. It recalls the third, "nuclear" stage of the triple tragedy, when tsunami waves up to 15 metres in height struck TEPCO's Fukushima Daiichi Nuclear Power Plant, located some 200 kilometres north of Tokyo. Like Japan's fifty-four other nuclear reactors, those at Fukushima Daiichi were on the coast and known to be vulnerable to earthquakes. Moreover, to increase profit, TEPCO, already Japan's largest energy company, had consistently skimped on maintenance at Fukushima Daiichi, which was also one of Japan's oldest atomic power plants. Thus, when the tsunami hit, due to a complete power failure, Fukushima Daiichi lost all cooling functions, and three of the four nuclear reactors melted down, resulting in hydrogen explosions and extensive leakage of radioactive materials. Likewise, in their third metamorphosis, Godzilla begins to leak radioactive contamination and to feel like an overheating nuclear reactor on the verge of explosion. Later the film indeed makes clear that the monster's cooling system is misfunctioning and that their body is unable to withstand the thermonuclear pressures resulting from this third metamorphosis. So, the nuclear monster does an about-face, returning to the waters of Tokyo Bay to cool their meltdown.

There is a lull, with the city of Tokyo returning to normal, and mellow jazz plays over morning scenes of commuter trains running smoothly. News reports calmly announce clean-up efforts and assistance from foreign powers. The profound sense of relief following Godzilla's return to the sea recalls the collective feeling of intense relief in Tokyo when it became clear that the nuclear reactors at Fukushima Daishi would not detonate like atomic bombs and transform Tokyo into a radioactive wasteland. Yet the nuclear threat was far from over. Likewise, the first three metamorphoses of Godzilla are but a prelude to the gravest threat, the fourth metamorphosis – the inability to stop the ongoing nuclear reactor meltdowns.

In point of fact, the TEPCO disaster ranks among the worst incidents of radioactive contamination in world history. The International Nuclear Event Scale rates accidents from 1 to 7 according to their severity; the only accident rated 7 before Fukushima was Chernobyl. "Subsequent revelations suggest that Fukushima's radioactivity is actually anywhere from 2 to 4 times as high as Chernobyl's. Compared to the explosion of just one reactor at Chernobyl, which has a 1,000,000 kilowatt capacity, the explosion at Fukushima Daiichi involved 4 reactors with a combined output of 2,810,000 kilowatts."[7] A decade later, there are no conclusive or persuasive plans to control the massive ongoing radioactive leakage, particularly in the Pacific Ocean and through groundwater. What is more, it is exceedingly difficult to define and delimit the extent of radioactive contamination. The designated acute danger zone circumscribes a 20-kilometre radius around the reactors in meltdown, yet the plume of radioactive contamination from the reactor explosion stretched into zones beyond this perimeter, making some distant areas more dangerous than proximate ones. In addition, the Fukushima Daiichi reactors have released multiple forms of radioactive contamination with diverse attributes, and that contamination does not diffuse evenly or become evenly diluted.

Radioactive contamination lends itself to uneven distribution, adhering to some materials more than others, entering into air and ocean currents, flowing with groundwaters, becoming concentrated in food chains, and generating unpredictable hotspots. In the context of Chernobyl, as Kate Brown has shown, we should not fall into the trap of believing that the effects of ionizing radiation are contained with the proximity zone; "radioactive gases follow weather patterns, moving around the globe to leave shadows of contamination," so that the Cumbrian fells of England may show equally alarming levels of radioactivity due to fallout.[8] In sum, although the events of 3/11 are commonly called the triple tragedy to take into account the interrelated yet differently scaled effects of the earthquake, the tsunami, and the nuclear reactor meltdown, the ongoing disaster is not merely threefold. It is multi-plied. It is not surprising, then, that accounts of Fukushima are prone to cite Gunter Anders's phrase, "Hiroshima is everywhere."[9] The current disaster is not localizable in the manner of Hiroshima. It implies a temporal and spatial manifold in that it cannot be confined to a single place or time: it is happening in multiple sites and is far from over. It is not only that Hiroshima is potentially everywhere but also that, if the actual ongoing radioactive contamination is taken into account, Hiroshima is occurring every day.

In his account of Fukushima, Jean-Luc Nancy strives to come to terms with this manifold or multiplying effect.[10] For Nancy, it is indicative of an underlying condition, which he calls an eco-technological condition, whereby any natural catastrophe multiplies its effects through its impact on increasingly enmeshed and entangled technological systems and infrastructures. The impact of natural catastrophes is thus increasingly evaluated in terms of techno-economic loss instead of human loss. Ramona Bajema makes a similar observation: "Disasters in developed countries are not as costly in terms of lives lost as those in other parts of the world, but they are exponentially greater in terms of capital loss. Haiti lost 240,000 lives in its 2010 earthquake and suffered $14 billion worth of damage. Japan lost 20,000 lives but experienced at least $235 billion in damage."[11]

Both Bajema and Nancy strive to counter the underlying techno-rationality that encourages calculation in terms of human loss versus capital loss. Although they adopt very different disciplinary trajectories, their approaches are complementary in that both expose the irrationality operative in the supposedly rational responses to the Fukushima catastrophe. Bajema proceeds anthropologically, combining on-site observation with personal histories of survivors. Her account offers an implicit demonstration of the force of the multitude active within experiences of the nuclear disaster: instead a unified experience, a number of overlapping, intertwining experiences that result in atypical associations. Second, Bajema's sympathy for local residents who chose to return to devastated and contaminated areas to care for the land, in conjunction with her attention to the ways in which government policies frequently miss the mark, implicitly place stupidity on the side of corporate and governmental techno-rationality. She inverts the received tendency to see local residents as somehow backward and unreasonable, unable to understand scientific realities grasped by enlightened urban officials.

Both Bajema and Nancy show how the "rational" treatment of Fukushima Japan as a techno-economic problem with a techno-rational solution serves, in fact, to multiply irrational effects. Similarly, Raji C. Steinek speaks of "a vicious cycle inherent in nuclear power technology: the irresponsibility of the decision to utilize it translates in so many ways, into irresponsible forms of utilisation, magnifying its detrimental effects."[12] Entirely unreasonable projects are presented as if eminently reasonable. One example is the current plan to build a 10-metre sea wall to protect the coast and other nuclear reactors from future tsunamis, when such a barrier was in fact surpassed by the tsunami that actually happened. Another example is the recent implementation of an ice wall around reactors to prevent groundwater seepage, to which I will return.

The Rebuild of Godzilla

The new Godzilla captures something of this manifold or multi-plied disaster in very complexity of the monster's physiology and their apparently infinite capacity for metamorphosis. As such, they seem designed to introduce greater complexity into the interpretations that have become so entrenched around the original 1954 *Godzilla*, namely, that Godzilla represents the Japanese experience of the atomic bombs dropped by the United States on Hiroshima and Nagasaki in 1945. A kind of mimetic identification has gradually become solidified whereby Godzilla stands in for the experience of the atomic bomb. By the same token, it has become something of a cliché in studies of Japanese popular culture to interpret Godzilla as representing national trauma. As numerous commentators have signalled, transforming Hiroshima and Nagasaki into national trauma is an exceedingly problematic gesture, because it gives the impression that atomic trauma was evenly distributed across Japan, reinforcing the notion of the paradigm of flat, even distribution of sovereignty in the form of citizenship.[13]

Nationalizing the traumatic effects of the atomic bombs, then, makes it all too easy to ignore the uneven distribution of the actual effects, that is, the long-term suffering of victims. Ignoring unevenness makes it possible to ostracize victims, even across generations. An unthinking equation of Godzilla with the atomic bombs, Japanese trauma, or the hegemony of American military empire in the Asia-Pacific region merely reinforces the erasure of various forms of socio-historical unevenness. For instance, in the context of the 1954 *Godzilla* film, it is common practice to acknowledge the *Lucky Dragon 5* incident of 1 March 1954, when fallout from American atomic tests in the Marshall Islands severely contaminated the crew of a Japanese fishing boat, as well as their haul of tuna, with ionizing radiation. The contaminated fish had already been distributed. August Ragone writes that "when the news broke, nearly five hundred tons of fish had to be destroyed or buried. This caused a national panic and threw the fishing industry into decline. The US government later estimated that 856 such fishing ships were exposed to the test."[14] In other words, it was anxiety over the safety of food sources and the future of the Japanese fishing industry in 1954 that spurred new concern about the atomic bombs at Hiroshima and Nagasaki. Open discussion of the effects of the atomic bombs had been suppressed during the Allied occupation of Japan (1945–52). The *Lucky Dragon 5* incident then sparked a wave of nuclear fear and sense of national trauma. Evocations of this incident in the context of *Godzilla*, however, have focused largely on atomic bomb trauma, erasing the contaminated fish and ignoring ethnic and colonial unevenness: the suffering of the Marshallese people has been erased in favour of Japanese national trauma.

The new *Godzilla* risks repeating this problem of evoking a national trauma that erases unevenness. The point of departure for the new *Godzilla* is 3/11 and the ongoing nuclear crisis resulting from the nuclear reactor meltdown at Fukushima Daiichi. As such, various forms of unevenness come into play. First, there is the socio-economic unevenness that has historically subordinated the interests of Fukushima to those of the city of Tokyo, which the nuclear industry exploited in constructing reactors there. Nuclear reactors in Japan have been built only at sites where the industry has succeeded in overcoming local resistance.[15] Communities resist nuclear plants because they know that the profits and the hardships are not evenly distributed. Likewise, the hardships due to the ongoing nuclear crisis are being distributed with increasing unevenness. Norma Field puts it succinctly:

> On the one hand, the state and TEPCO have promoted differential, even discriminatory treatment in paying out compensation, while, on the other, pushed debris incineration throughout the country and more recently proposed "recycling" radioactive soil accumulated through decontamination in road construction projects throughout the country: in other words, reward differentially (on grounds that appear purely arbitrary) but burden equally, in the name of national solidarity or cost-savings (the recycling proposal).[16]

Second, while it has long been acknowledged that children and the unborn are at greater risk than adults, Field also calls attention to "a factor that is stunningly under-remarked even though it appears in the 2006 National Academy of Sciences BEIR (Biological Effects of Ionizing Radiation) VII report – the disproportionately greater risk faced by women and girls exposed to radioactivity."[17] Third, there is discursive unevenness related to who is authorized to speak, whose voice bears authority. Margherita Long notes, for instance, that "although neither the profits nor the hardship are shared, people are convinced that the government is doing its job because economic growth has become both its end and its legitimation."[18] The sense that the government's position has legitimacy and even authority is discursively shored up through a series of regimes of truth, two of which deserve mention in this context: (1) the authority of expert voices that is grounded in the image of Japan as a nation built on expertise in science and technology, which lent plausibility to the notion of "Atoms for Peace,"[19] and (2) the authority of national history over local memory, which latter, as Harry Harootunian points out, gives precedence to practices geared "toward devising protective measures on-site derived from the received experience of the household and the long tradition of care informing it."[20] "If memory is a capacity, the power of remembering is more fundamentally a figure of care."[21] Yet it is precisely care that is dismissed or discredited in favour of scientific experts and nation-centred histories, often in combination.[22]

In sum, when the new *Godzilla* evokes 3/11 and the ongoing nuclear crisis, these different yet interrelated registers of unevenness come into play: socio-economic, ethnic and colonial, gendered, age-related, technological, and historiographical or discursive forms of unevenness. These forms of unevenness ensure in advance that the risks and profits associated with nuclear power are never evenly shared. To be sure, the new *Godzilla* cannot be expected to address all of these issues as such – nor would be it desirable for it to try. The power of a film depends on it establishing its own perspective, and that perspective comes to it from without, from conventions of various kinds. Through this complex dialogue arising in the interval between the whole of the film and its constraints, a film may give rise to multiple interpretations. This was certainly true of the original *Godzilla*.

Yuki Miyamoto notes of the 1954 film that

> understandings of Gojira provide a multiplicity of interpretations: reptilian, monstrous, god-like, fallen, associated in some readings with Japan and others with the United States, and *hibakusha*. Perhaps in 1954 Japanese citizens were still in the midst of coming to terms with the lost war, the large number of soldiers' and citizens' deaths, the unprecedented nuclear attacks, and the subsequent nuclear arms race, all of which created a versatile image for Gojira to inhabit.[23]

But she also signals how "Gojira eventually lost its emotional connections to people's anxieties and fear of nuclear weapons and radiation exposure."[24] Of particular concern to Miyamoto is how the multiplicity of *Godzilla* gradually became flattened into a general trend toward trivializing the effects of ionizing radiation.

Miyamoto's account highlights the relationship between genre and gender. She situates *Gojira* within the special effects or *tokusatsu* genre, which "is almost the only venue where one finds the representation of adult men who experienced nuclear explosions and radiation exposure, as well as the depiction of their anxieties and fears about the nuclear age."[25] Yet she finds that the *tokusatsu* genre tends ultimately to downplay the fear of radiation exposure, through monsters whose exposure to radiation does not culminate in debilitating illness and prolonged suffering. On the contrary, suffering and illness are projected onto female victims exposed to ionizing radiation, whereas male figures who undergo nuclear exposure tend to power up, becoming stronger, better able to do battle. The connection with the war film is evident here. The creators of the original *Gojira* film had enjoyed great success the previous year with a remake of the Japanese war film *Hawai Mare oki kaisen* (*The War at Sea from Hawaii to Malaysia*, 1942), which they titled *Taiheiyō no washi* (*Eagles of the Pacific*, 1953).[26] Thwarted in their efforts to produce a film about postwar reconciliation in Indonesia, they opted for something more like an allegory of Japan at war – the invasion of a nuclear monster produced by the US military presence in the South Pacific.

The monster Godzilla was thus positioned to repeat the underlying ambiguity of the Japanese soldier, first evident in Japanese war films. Although the Japanese soldier was, strictly speaking, an invader, an aggressor or perpetrator of violence, war films focused greater attention on his trials and tribulations, avoiding representation of damage done to enemies.[27] Subsequently, in the postwar period, it was impossible not to acknowledge that returning soldiers were also victims in the sense that they had endured miserable, dehumanizing conditions during the war and showed symptoms of post-traumatic stress. Such facts encourage a focus on perpetrator trauma. For Miyamoto, who associates the monster with the male soldier returning home, Godzilla repeats this ambiguity: he is positioned as a traumatized victim of the atomic bomb and as a destroyer, a perpetrator, on the side of atomic bomb. The monster thus oscillates uncertainly between victim and perpetrator, martyr and destroyer.

The new *Godzilla* inherits these expressive constraints, these tropes. First, it builds directly on the war film. Especially striking are the aerial views of Tokyo on a seemingly endless series of bright, cloudless days; every detail of the city feels beautifully captured and fully exposed, vulnerable to attack (Figure 4.2).

The film revels in offering audiences a bomb's-eye view, laying out Tokyo as a potential target and a strategic theatre of operations. As Caren Kaplan notes: "Distant and abstract, requiring expert interpretation to become even marginally intelligible, the aerial or satellite image most often requires the kind of infrastructure that can only be provided by nation-states or major corporations."[28] What is more, the new Godzilla is the victim of nuclear radiation, having consumed radioactive waste, and at the same time is the embodiment of destruction, having incorporated processes of nuclear fission into their body. Finally, in keeping with the original film as well as the *tokusatsu* genre more generally, the focus in *Shin Gojira* is largely on adult male responses to the threat of nuclear radiation, which threat becomes inextricably linked to warfare and the invasion of Japan. Men confronting the new Godzilla have to power up to meet the crisis. In sum, everything about the new Godzilla feels calculated to deliver a powering up of the nationalizing of nuclear trauma now indelibly associated with the 1954 *Gojira* film, above and beyond the trivializing of radioactive exposure noted by Miyamoto.

Shin Gojira can be characterized as a "rebuild" of the 1954 *Gojira* akin to director and writer Anno Hideaki's rebuild of the animated television series for which he is best-known, *Neon Genesis Evangelion* (1995–96). To transform the original anime series into a series of four animated films (released in 2007, 2009, 2012,

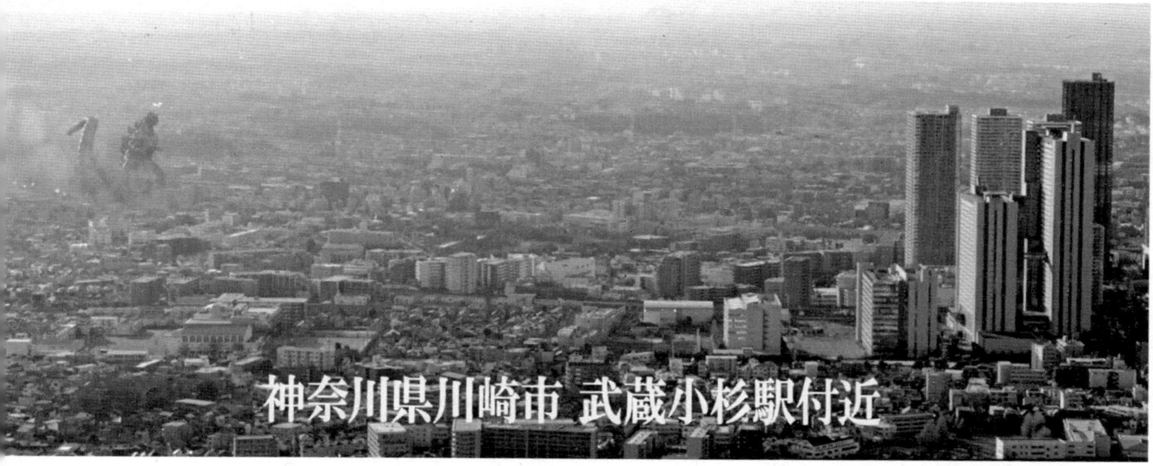

4.2 One of many aerial views, this one shows Godzilla approach the Musashi-Kosugi station in Kawasaki. Locations are consistently named throughout the film, letting audiences know precisely where the destruction is happening.

and 2020), Anno founded a new studio (Khara) and centred its activities on the rebuild of *Evangelion*. Conceptually, a rebuild differs from a remake or remix in that it uses new cutting-edge technologies alongside received techniques to power up the original, to produce an intensified version of it.[29] The concept of rebuild is not unlike that of reboot, which is often used to refer to a new version of an older film or television series. But rebuild places greater emphasis on adhering to the built environment of the prior object, on capturing its intensities instead of merely referencing them. Anno's rebuilds of *Evangelion* and *Gojira* can be compared with J.J. Abrams's reboots of the *Star Trek* and *Star Wars* franchises.

Yet where Abrams often stands accused of offering little more than a remix of the original (especially with *The Force Awakens*), Anno's rebuild of *Gojira* seems to have avoided such criticism, arguably because the rebuild strategy avoids the mimetic pitfalls of the remake, remix, and reboot. Central to rebuild is the monster, which must simultaneously capture and intensify the energies of the original.

To rebuild Godzilla, Anno worked closely with the famed special effects director Higuchi Shinji. As co-founders of Gainax Studios in the mid-1980s, Higuchi and Anno have collaborated on many projects, from their *tokusatsu* fan film of 1985, *Yamata no Orochi no Gyakushū* (The Eight-Headed Giant Serpent Strikes Back), to the rebuild of *Evangelion* in the 2000s. Promotional materials touted the technological side of the new Godzilla, especially their size (at 118.5 metres, the tallest to date) as well as the use of advanced CGI technologies for motion capture. Yet high tech alone does not make for an effective monster. The appeal and effectiveness of the original Godzilla is commonly said to lie in our

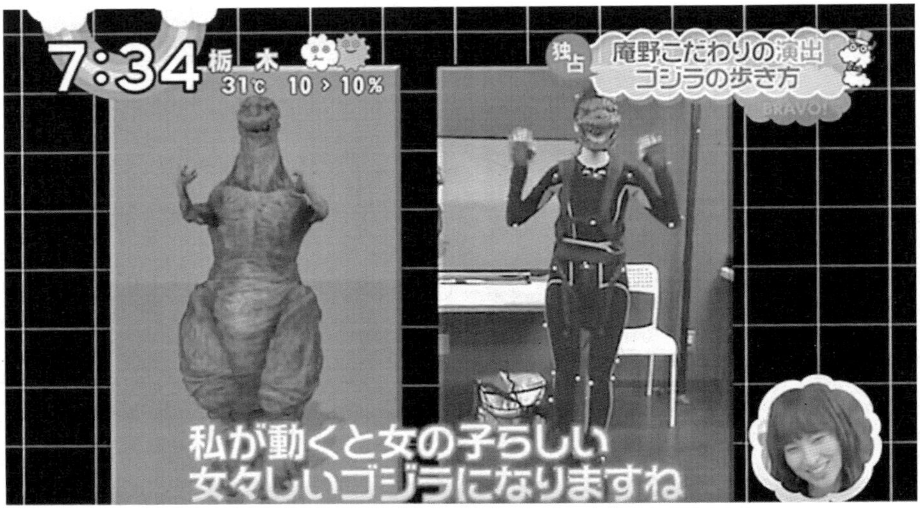

4.3 In the documentary special *The Making of Shin Gojira*, an actress shows how motion capture of her performing girlish movements can be used to make a feminized Godzilla.

intermittent awareness that the monster is, in fact, a person in a rubber suit. To rebuild such effects, *Shin Gojira* used motion capture of renowned Kyōgen actor Nomura Mansai performing Godzilla. This combination of techniques was intended to bring a sense of artifice and even a ludic quality to the monster, rather than striving for a simplistic realism or naturalism. Indeed, the television special on the making of the film gives free rein to the comic qualities of the technical set-up, allowing the young actress hosting the show to demonstrate her own feminine version of Godzilla's walk, even as the narration duly stresses how this process allows for a "more lifelike motion than CGI can depict" (Figure 4.3).

At the same time, the use of traditional drama may be discursively aligned with turning Godzilla into a Japanese thing. Higuchi, for instance, wished to distinguish this new Godzilla from the recent American incarnation in the 2014 *Godzilla*, of which he remarked: "It doesn't move viewers emotionally when it doesn't have any of the traditional feel of a person within the suit."[30] Motion capture of a Kyōgen actor was the solution, and in the film, there are moments when the Kyōgen-style movements of Nomura become palpable: in one of the key scenes, as Godzilla staggers and lumbers to a halt, the movements of their haunches are entirely those of a Kyōgen actor.[31]

The tactics of rebuild extend throughout *Shin Gojira*, radiating from the multiple image layers palpable in the monster (an open compositing of human performance, motion capture effects, CGI aftereffects, and so forth) into the series of metamorphoses through which the monster powers up their relation to nuclear energy, which metamorphoses provide the story events for the larger narrative world of the film. Rebuild thus allows for the production of a monster that is complex enough to capture something of the multiplied disaster following 3/11. Where *Gojira* captured something of the ambivalence inherent in the national reckoning with the atomic era in the wake of Nagasaki and Hiroshima (the monster as victim and perpetrator), *Shin Gojira* strives to capture the experience of Tokyo some five years into the nuclear crisis, building on the anxieties stemming from Tokyo's proximity to Fukushima during both the triple tragedy of 3/11 and the radioactive aftermath. Some of its gestures are bold, even over the top: placing a nuclear reactor in meltdown at the center of the capital of Japan. Some of its gestures are as subtle as the pauses in an actor's performance rendered in motion capture.

But can *Shin Gojira* avoid the trap of nationalizing trauma and thus of suppressing all traces of the uneven distribution of harm and profits during the ongoing nuclear crisis? A number of factors encourage audiences to read it as a national parable. Anno's brand of political realism demands extensive knowledge of the Japanese political system, especially with respect to its geopolitical position vis-à-vis the United States. Also, the figure on which the film builds has become a national icon: Godzilla. Indeed, the film proved to be the highest-grossing Godzilla film to date, primarily due to its success at the Japanese box office. It also garnered a series of Japan Academy Prizes, including Picture of the Year and Director of the Year. The critical and popular success of the film in the domestic market, along with Anno's deliberate efforts to provide a national parable, encourage a propensity to nationalize trauma. This propensity makes the task of reading the film against the grain of its majoritarian reception feel more urgent in the contemporary context, in which the Japanese government continues to propose truly outrageous procedures for nationalizing Fukushima. The Japanese government has gone to great lengths to reassure the public about the safety of food from the Fukushima area. Filmmaker Kamanaka Hitomi cites one notorious example:

> Tokyo University dining halls are currently serving 500-yen lunch specials with names like "Fukushima Plate" and "Namie Dish." They feature Fukushima rice and vegetables and are enormously popular with women students, selling out in a flash to comments from the students along the lines of "It must be safe, because it's within radiation limits set by the government!" and "If anything, it's tested food from Fukushima that we can eat with confidence!" These sorts of scenarios really hit home how resolutely the pro-nuclear energy establishment has been trivializing the accident.[32]

Again, besides trivializing radioactivity, such efforts constitute efforts to nationalize trauma. As such, even while the government declares that food from Fukushima is safe and refuses to communicate about its risks, studies have found that "consumers in Japan lack trust in their government institutions and the food industry."[33] More recently, the government has – apparently in all seriousness – proposed nationalizing the radioactive soil stockpiled in the Fukushima area, for use in road construction across the nation.[34] In such circumstances, we need to find ways of reading popular culture that do not comply with government directives to nationalize the experience of Fukushima.

I thus propose a careful examination of factors that are contributing to pushing *Shin Gojira* toward national allegory and thus toward nationalizing ionizing radiation. I also propose to read against them. The problem of nationalizing trauma is not limited to what is in the film. It is also a matter of our interpretations and interpretative communities. Arguably, the bias toward reading the original film in terms of national trauma is a legacy of a certain moment in the study of popular culture, when scholars struggled to find ways to read popular forms that conveyed their significance and complexity. For better and worse, the Japanese experience of the atomic bombs afforded a ready-made paradigm for highlighting the importance of popular culture. This legacy is palpable in *Shin Gojira*: Anno's script shows a keen awareness of how the original Godzilla has been interpreted, building prior interpretations into the dialogue and action. The rebuild is attuned to the dominant "nationalizing" interpretations of the original. Anno is well aware that we are heirs of that moment, which raises the bar for interpretation in *Shin Gojira*. The film is mindful from the outset that the monster has repeatedly been read as an allegorical expression of national trauma vis-à-vis the effects of nuclear radiation and will continue to be read in such terms. Anno knows that audiences are primed to read the rebuild of Godzilla as the rebuild of Japan – better, then, to leap head first into national allegory and make questions about the status of Japanese sovereignty central to the story.

The Rebuild of Japan

One of the more striking features of the film is its focus on national bureaucracy, which the promotional materials styled as "political realism." Anno wished to provide a realistic portrayal of how the Japanese bureaucracy would respond to an unforeseen threat like Godzilla, so the bulk of the film centres on the responses of the Japanese government to the threat. Much of the action takes place inside government buildings, in nicely appointed meeting rooms, where various committees meet and strategic initiatives and task forces are formed, dissolved, and reformulated. The film carefully tracks bureaucrats moving through the ranks in the process. Their responses are, however, removed from the actual action: information about the crisis comes to them through screens displaying television news, social media, internal memoranda, and reports redacted by committees (Figure 4.4).

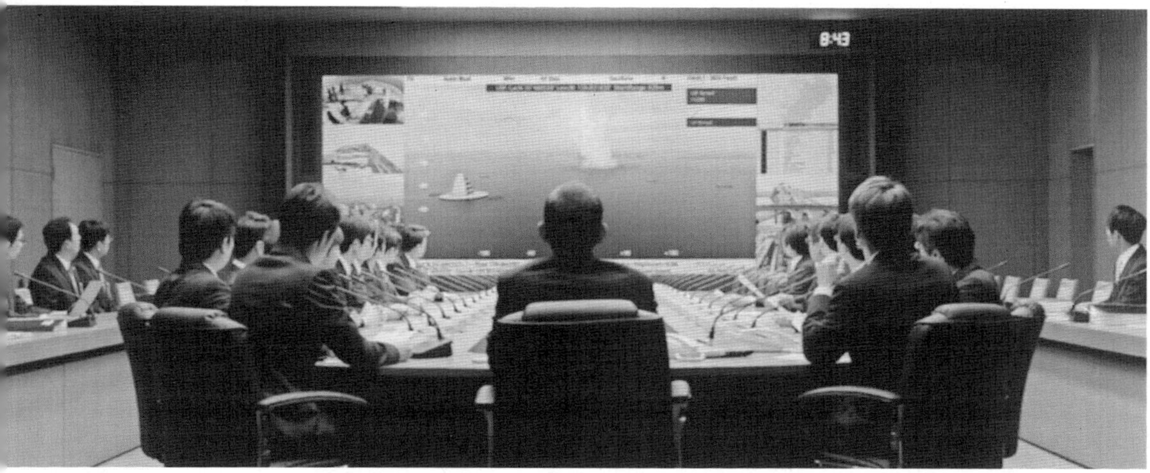

4.4 Government officials watch footage of the first events occurring in Tokyo Bay, which, from the safety of their meeting room, they take for underwater seismic activity, perhaps a volcanic vent.

A tentative schism appears. On the one hand, many of the older, established politicians in the cabinet initially refuse to believe the monster exists, despite images of it circulating on social media. Even when they must acknowledge the monster, they are so unwilling to engage with the crisis that they quickly accept reports that it will not be able to walk on land. Their desire to avoid confrontation traps the prime minister: at one point, as he is stating on television that the monster cannot walk on land, images of Godzilla walking into Tokyo are already on air.

In this presentation of the established bureaucracy and the discrediting of the prime minister, there are clear echoes of 3/11. In response to the nuclear meltdown, for instance, TEPCO hoped to discredit Prime Minister Kan Naoto, who had publicly proposed that nuclear energy be phased out. TEPCO officials plainly lied to the prime minister about what was happening, making him look incompetent and thus exacerbating the crisis rather than alleviating it. Yet the horizon of conflict in *Shin Gojira* does not extend to what is known as the nuclear village, "a chicane of relationships between power companies, government, civil servants, media, finance sector and universities, binding them in the profitable myth of safe atomic energy."[35] In the wake of 3/11, TEPCO became the supreme symbol of corporate fraudulence and unscrupulous profiteering associated with the nuclear village, with polls indicating that 77 percent of Japanese wished to be done with nuclear energy.[36] What is more, contrary to Orientalist discourses still circulating in the global media that portray Japanese as stoic and obedient vis-à-vis government policies, anti-nuclear demonstrations in Japan were nothing short of overwhelming in terms of numbers and organization. Indeed, the horizon of conflict that emerged after 3/11 might best be characterized in terms of citizens versus the nuclear village, or more simply, the people versus capitalism.[37] *Shin Gojira* takes a different tack: the horizon of conflict is the Japanese government versus the nuclear monster. In effect, the film eliminates the agency of both the nuclear village and the people. The role of people is reduced to fleeing the monster or moving quickly to evacuation sites in compliance with the government, with the aid of the Japanese Self-Defense Forces (SDF).

Because Anno's political realism settles on the government versus the monster, the film does not permit any schisms within the government to destroy the governmental machine. The timidity of the old guard prevents it from attending to the crisis in an effective manner; meanwhile, the bolder, more impetuous new guard acts *within* the system, working with and through the old guard. The division between old guard and new guard hinges on a difference in political imagination. An unwillingness and incapacity to imagine a genuine adversary characterizes the old guard; it continually disavows the monster and underestimates the crisis. For the old guard, the Second World War is long past, and the tensions with the United States do not present actual adversity or rivalry; the old guard is accustomed to their subordination to American global interests. Their leadership thus brings Japan to a dead end.

As the fourth-stage Godzilla returns to invade the city of Tokyo, the old guard mobilizes the SDF forces. The goal is to prevent the monster from crossing the Tama river, which separates Tokyo proper from the adjacent city of Kawasaki. The politicians effectively mobilize Japan's military forces, yet the battle is lost, and Godzilla enters Tokyo. Even as the prime minister and his cabinet prepare to call on the US forces for assistance, the United States has already sent out its forces to attack. When Godzilla reaches the very centre of Tokyo, that is, Tokyo Station, American bombers drop missiles on the monster. Godzilla's body responds with the display of a terrifying range of destructive powers: purple rays shoot out from their back to strike the bombers; their mouth spews jets of radioactive flames, and finally a laser-like ray erupts from the tip of their tail. Their weaponized body levels high-rises and reduces areas of central Tokyo to rubble. The US–Godzilla battle also destroys the central government buildings, and the prime minister, who chooses to remain behind, is killed.

In the end, however, the American assault merely exhausts the monster, and Godzilla goes into some kind of shutdown mode. But they will reawaken in 360 hours. The fear now is not merely that Godzilla will destroy Tokyo. The fear is that they are capable of invading the entire world. They may be capable of travelling worldwide in their present form – say, swimming to California. More importantly, because they have eight times the genetic material of humans, they are capable of reproducing asexually and of evolving into smaller forms with wings. This is why the United Nations intervenes to place the monster under US jurisdiction.

The United States, with the permission of the Japanese government, now makes plans to drop an atomic bomb on Godzilla, that is, on central Tokyo. In the end, the inability of the Japanese old guard to face up to a new adversary leaves Japan fully subordinated to global US interests. Japan's subordination has reached the point where it has agreed to its own nuclear destruction for the sake of the world.

It is precisely this subordination that the new guard within the Japanese government cannot accept. Yaguchi Rando, a young politician who initially appears in the position of deputy chief cabinet secretary, quickly emerges as the leader of a new faction.[38] From the outset, Yaguchi believes in the monster. He believes so immediately and fervently in Godzilla that he seems to have a deeper connection to the monster. The monster is the adversary, the destiny he has been waiting for, the destiny that will lead Japan out of its postwar lethargy. For the new guard led by Yaguchi, it is possible, even desirable, to imagine Japan facing an actual adversary and entering into war. Yaguchi and his allies make frequent reference to the unending postwar, referring to Japan's compromised sovereignty under the Japan–US Mutual Cooperation and Security agreement (AMPO), which has been succinctly described by Gavan McCormack as turning Japan into a "client-state" of the US.[39] Yaguchi repeatedly speaks of the Second World War. At one point, he remarks bitterly that the armchair theories of the Japanese government during the war led to the deaths of 3 million Japanese – as if the key legacy to address from the Fifteen-Year Asia-Pacific War was poor decision-making and the death of Japanese citizens, rather than the expansion of the Japanese empire, whose ambitions culminated in the military invasion of neighboring countries and massive destruction across East Asia.

When interpreted from the angle of the underlying unity of the Japanese government notwithstanding its internal differences, the rebuild of Godzilla in *Shin Gojira* appears as an apologia for the rebuild of Japan, a formal bid for renewed national unity in the context of ongoing nuclear crisis. The nuclear crisis associated with Fukushima thus appears fully nationalizable, through its transformation into national trauma. The film must eliminate so much to rebuild post-Fukushima Japan: the agency of citizens in anti-government, anti-nuclear, and anti-radiation demonstrations; the role of corporate power and capitalist profiteering, that is, the political economy of the nuclear village; and the legacy of the Japanese empire and wars of aggression in Asia. Also, sacrifices must be made: citizens must comply with evacuation orders; Japan's Self-Defense Forces must submit to governmental orders and be ready to die with selfless heroism; and the entire nation must mobilize for war, or as if for war, if it is to liberate itself from the bondage of the never-ending postwar era. In other words, national unity also requires a rebuild of war itself, in the form of a rebuild of the war film.

Yet here a paradox emerges. Although the new guard places itself on the side of war, its actions are not essentially warlike. Initially, the impasse confirms what

Aaron Gerow notes about the new wave of Japanese war films in the early 2000s: "War is transformed from the horrible space of killing to a more benevolent and less offensive practice of defending against an attack (the word used is always 'mamoru')."[40] Yet when defence, too, proves ineffectual against such a powerful, god-like, transformative monster, the film verges on paradox: the only thing that cannot be destroyed is destruction. Underlying the inability to destroy destruction looms another paradox: only destruction is truly creative. *Shin Gojira* circles frantically around this paradox.

As various attempts to destroy the monster fail, the film spirals deeper into its guiding paradox, which hinges on how destruction can be evaluated. It is no longer a matter of putting an end to destruction but of arriving at *creative* destruction, a new evaluation of destruction. But what kind of destruction may be deemed genuinely creative and valued as such? The ripples generated by this paradox begin to spread into other registers. What sort of death makes for rebirth? When is damnation salvation? In keeping with this paradox, the new guard, while on the side of war discursively, seeks creative non-warfare solutions to address the destruction that cannot be destroyed, Godzilla. During the lull between the third and fourth metamorphoses of Godzilla, Yaguchi assembles an eclectic team of researchers under the unwieldly rubric "Giant Unknown Creature Emergency Disaster Task Force" (*kyodai fumei seibutsu kinkyū saigai taisaku honbu*). This group could equally well be called the "*otaku* creative techno-science team," for its scientists, engineers, a nd programmers are characterized as *otaku*, misfits, and freethinkers. This team develops a more creative or atypical relation with the United States as well. Kayoko Ann Patterson, a special envoy from the American president, joins forces with Yaguchi and his team. Because Patterson is a Japanese American career politician whose ambition is to become president, she and Yaguchi have a good deal in common: overweening political ambition does not preclude intense loyalty to Japan. Yet the relationship between Yaguchi and Patterson is also atypical in that, despite its occasionally flirtatious tone, it is not fundamentally erotic or amorous but professional and political, with a sense of purely formal intimacy based on mutual ambition. Here US–Japan interactions are characterized by formal coordination rather than hierarchical subordination. Patterson even speaks English with a distinctly Japanese accent, suggesting a strangely Japanified America, an America for Japan.

Kayoko Patterson provides Yaguchi's atypical team with the scientific records of Maki Gorō, which allow them to decipher Godzilla. Maki's activities bring us closer yet to the central paradox of the movie. Angry over the mistreatment of his wife in Japan, who died of radiation-related illnesses, biologist Maki left Japan for the United States to work in an energy firm, where he discovered a prehistoric sea creature capable of consuming and living on nuclear waste. Maki dubbed the creature Gojira, with cryptic connotations of God Incarnate. When the US

Department of Energy prevented the publication of his research, Maki apparently took matters into his own hands. It is his boat that is found abandoned in Tokyo Bay in the opening sequence of the film, with a note saying, "do as you like" or "do what you will" (*suki ni shiro*). This phrase runs like a refrain through the movie: it seems as if Maki has sent Godzilla to Japan as a sort of test. It is as if he has foreseen how efforts to destroy Godzilla will bring Japan to the brink of nuclear annihilation. Yet he has also provided information for an alternative path, encrypted within his scientific papers. Maki is annihilator *and* saviour.

Among his research papers is a schematic diagram detailing the molecular components for the cellular functions through which Godzilla is able to transform water and oxygen into a radioactive isotope to fuel their metabolic processes. This information confirms what has already been observed: Godzilla is like a nuclear reactor that overheats when damaged or pushed to its limits. The key to stopping Godzilla, then, is freeze them, and Yaguchi's team develops plans to force-feed the monster a coagulant that will lower their body temperature, freezing them. While the film culminates in the successful implementation of this plan, it is not a solution to the crisis. In fact, the plan intensifies the paradox.

As Yaguchi coordinates with Japan's Self-Defense Forces and the US Air Force to implement his team's plan, the countdown for dropping an atomic bomb on Tokyo is reaching the point of no return. There ensues a race between those intent on freezing the monster and those intent on nuking them. Yaguchi leads a series of attacks intended to exhaust the creature. After slamming Godzilla with trains bearing bombs, assailing them with drones to sap their strength, and toppling massive buildings onto them, the monster finally collapses alongside Tokyo Station. Yaguchi's team mobilizes machinery to pump the freezing agent into their mouth. As Godzilla rises, they freeze solid, a silvery grey version of their prior version, as if in a fifth stage of metamorphosis. Again, this solution is not a solution in the sense of resolving the crisis of Godzilla: the moment the monster shows any sign of awakening, the countdown for the atomic bomb will recommence.

The freezing of Godzilla recalls TEPCO's "ice wall," first announced in 2013 with assurances that it would stop the flow of groundwater into the basements of the nuclear power plant, where it would mix with highly radioactive waste from the three reactors in meltdown. The seepage generates vast amounts of toxic water that TEPCO must remove, decontaminate, and store in tanks at Fukushima. The flow of groundwater also prevents the removal of radioactive melted fuel and the decommissioning of the plant, which means that massive amounts of radioactive materials continue to seep into the surrounding groundwater and to flow into the ocean. Every day is Hiroshima. Spending ¥34.5 billion ($324 million) in public funds, TEPCO sank brine-filled tubes to a depth 30 metres along a perimeter of about 1.5 kilometres around four of the reactors.[41] Cooling the brine to −30°C was supposed to create a frozen soil barrier or "ice wall" that would reduce the groundwater flow to a bare minimum. When the ice wall became operational in August 2018, however, the volume of water actually increased, due to unusually heavy rainfall as well as the unpredictability of the flows. Assessments of the project's success vary, but at best it may be deemed partly effective. In the meantime, TEPCO is running out of space to store the ever-growing volume of toxic water and is examining options for ocean releases.

The freezing of Godzilla in 2016 appears eerily prescient in light of the failure of the ice wall in 2018. The wall had been promoted by the Japanese government beginning in 2013 as a sure way to bring the nuclear crisis under control. Thus, at another level, it provides an index of the paralysis of the Japanese government and of its proposed rebuild of post-Fukushima Japan. The freezing of Godzilla is but a stopgap, a partial measure. While everyday life resumes in the capital of Japan, that city remains poised on the brink of destruction. Or more precisely, Tokyo is caught between two kinds of destruction – that of Godzilla, and that of the atomic bomb. Yet the two kinds of destruction have one and the same source. They are two faces of the same nuclear reality. *Shin Gojira* thus forces a confrontation with the paradox at the heart of nuclear energy: it is not just another kind of energy, for it is always already a weapon. Nuclear energy is only ever weaponry. True to form, in a manner reminiscent of the *Evangelion* series, Anno pushes conflicts into contradictions and pushes contradictions to the point where they turn into paradoxes. The film stops at a moment of nearly pure nuclear paradox, with an almost sadistic refusal to resolve it. The result is multiple perspectives, or pathways, through the film.

Multiple Pathways

The film introduces, for instance, a miraculous outcome, the possibility of a happy ending for nuclear energy itself: Godzilla has generated a new radioisotope to fuel themself, and because this isotope has a half-life of some twenty days, Tokyo will be decontaminated with three years. Such an isotope may even herald a new era of nuclear energy with manageably low radioactivity. In conjunction with the massive resignation of the prior cabinet members in the wake of the Godzilla fiasco, a happy rebuild of Japan feels possible as well: Yaguchi is poised to begin his climb to prime minister, even as Patterson, a Japanese-American, makes her bid for the US presidency. What could be happier for the political economy known as Japan?

A frozen Godzilla is not a vanquished Godzilla, however, any more than the frozen soil barrier is a solution for flows of radioactive waste from nuclear reactors in meltdown. The film thus invites a sterner, more stoic reading, reinforced by repetition of the message: we will have to learn to live with Godzilla. Which is to say, Japan will have to learn to live with the threat posed by nuclear energy/weaponry,

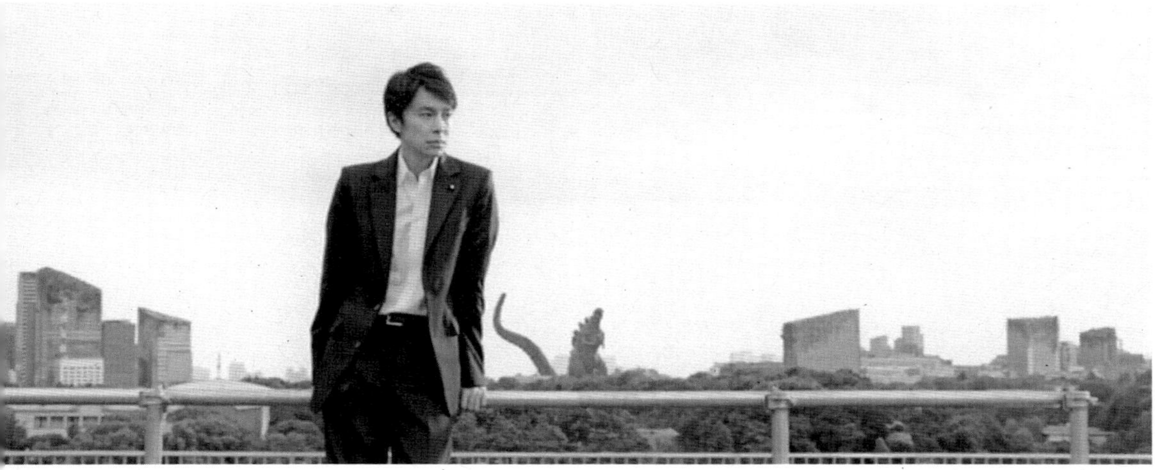

4.5 Yaguchi Randō sits atop the roof of the science museum in Kitanomaru Park, as if posing with the frozen Godzilla on the horizon.

as if perpetually mobilized for war. This register of the film encourages reading it in terms of nationalizing nuclear crisis, nationalizing nuclear trauma. The film even includes, for an instant, the iconic image of the ruins of Hiroshima, with the remains of the Prefectural Industrial Promotion Hall, now known as the Atomic Bomb Dome, gesturing toward the long history of nationalizing nuclear trauma. References are made to the history of Japan as one of repeated destruction and reconstruction. The grim tonality hovering over the idea of living with Godzilla tends to incorporate such anxieties within the project of reconstructing Japan, yet again. Thus, the stern ending encourages resignation to the fact of capitalism, in the form Naomi Klein calls disaster capitalism.[42] With each disaster, with each round of destruction, Japan will profit by reconstructing its cities, its government, its capitalist economy.

In a third, audacious movement, the film includes near the end a shot of central Tokyo seen from the rooftop of the science museum in Kitanomaru Park (Chiyoda Ward). Because Godzilla has toppled or cut down the taller buildings, the frozen monster now stands out boldly, as tall as or taller than these monuments to economic prosperity, their silvery grey sheen making them at home among the glass and metal architecture. They stand at the very heart of Tokyo, alongside Tokyo Station. The shot includes Yaguchi, as if posing in a tourist photo on a sunny day with a view of this new monument (Figure 4.5).

Godzilla is now indeed a monument, but to what? This take on Godzilla recalls the recent boom in erecting monuments to anime figures across Japan, such as the Gundam robot in Tokyo and the Gigantor (Tetsujin-28-go) statue in Kobe, which mesh with government initiatives to promote Japanese popular culture, both tourism and the "contents industry" itself.[43] Transforming Godzilla into a photo-worthy monument to the global success of Japan's contents industry, a monument more impressive than any prior architectural features of Tokyo, is tantamount to saying that Japan's future economic prosperity depends on its manga, anime, and games industries. Here Anno boldly offers us an *otaku* rebuild of Japan, an *otaku* Japan in the place of Japan, Inc., wherein the contents industry is central, not peripheral, to the nation's economic reconstruction. In keeping with this otaku pathway, Anno scatters citations of popular culture throughout the film as lures to *otaku* reading, among them references to the work of the manga artist Anno Moyoco (his spouse).

The *otaku* pathway, then, culminates in this comically audacious image of a future Tokyo in which a toy-shaped nuclear reactor dominates the cityscape. The expanded world of manga, anime, games, films, toys and so forth promises renewed prosperity, but at a price: by recapitalizing on the kernel of postwar trauma lodged within Japanese pop culture icons such as Godzilla. This may be why prominent critics associated with *otaku* culture have expressed reservations about *otaku* culture gaining such public visibility and acclaim. The well-known cultural critic Azuma Hiroki has tweeted: "To put it simply, watching *Kimi no na wa* and *Shin Gojira*, I felt the *otaku* era had ended."[44] For Azuma, the problem lay in a new degree of boldness, and he adds, "the characteristically half-hearted, irresolute side of the *otaku* has vanished entirely."[45] It is as if the contents industry, with the aid of a younger generation of politicians, could successfully replace the previous military-industrial complex known as Japan, Inc. that wholeheartedly promoted the nuclear industry. But then a certain uneasiness looms behind this ludic gesture. Is the contents industry replacing the military-nuclear-industrial complex or fusing with it? Is Godzilla fusing with disaster capitalism, albeit in a playful manner?

Here, in an offhand way, the *otaku* pathway reveals something about nuclear energy. Although the nuclear industry is typically justified in terms of its capacity to provide clean, abundant, and cost-effective energy for the "new" information economy, its scale and organization align it squarely with the "older" industrial formation. In the case of Japan, for instance, nuclear reactors are among the few remaining major industrial constellations. If the nuclear industry remains attractive to government and commercial interests despite abundant evidence that it is not cost-effective financially or socially, it is because it promises to sustain the "older" military-industrial complex and those who profit from it, amid the information revolution. As such, the nuclear industry appears poised between the two formations of capitalism, in terms of both labour and expropriation: nuclear workers range from lifetime loyal company employees to the most precarious, abused, and exploited workers imaginable; resources must be expropriated from around the globe, but more importantly, radioactive waste demands the expropriation of lands over thousands of years. As the former chairman of the US Nuclear Regulatory Commission, Gregory Jackzo, wryly comments: "I think a reasonable standard for any source of electricity should be that it doesn't contaminate your community for decades."[46]

For Jackzo, the Fukushima crisis was a game changer. He had been asked to downplay the dangers made evident by the reactor meltdowns in Fukushima; careful study, however, led him to a different conclusion: "My journey, from admiring nuclear power to fearing it, was complete: This tech is no longer a viable strategy for dealing with climate change, nor is it a competitive source of power. It is hazardous, expensive and unreliable, and abandoning it wouldn't bring on climate doom."[47] Why then does the nuclear village in Japan continue to promote nuclear energy? There are the current profits, of course, but more important is the future of profit itself: the stabilization (the freezing) of the present political economy of energy ensures that those who currently profit will continue to profit.[48] Because nuclear reactors and nuclear waste demand constant management over periods of time well beyond anything previously envisioned, their very existence constitutes a bid to keep the contemporary socio-economic and political configurations firmly in place, seemingly for all eternity. Nuclear reactors, like nuclear weapons, leave us poised in an unsustainable transition between industrial and computational configurations of capitalism, as if eternally.

Shin Gojira deliberately strands its audience in such a situation, in which the city of Tokyo is left hovering between a nuclear reactor about to explode and the imminent detonation of a nuclear bomb. For a moment, in the image of a calm and intrepid Yaguchi posing with a giant nuclear toy, it feels as if we might dwell for all eternity on the brink of disaster. The brink of disaster begins to feel like an eternity.

Ultimately, however, we know it is impossible to recapitalize forever on radionuclides. Profits may be made from ventures as varied as toxic waste disposal, dark tourism, medical treatment of victims, news reportage, and packaging of data derived from this inadvertent large-scale random clinical trial of the effects of radiation, to name a few. Yet, the protean, divagating radionuclide continues to evade the extractive grasp of capitalism, and at some point, what appeared to be disposable waste turns out to be pure anti-production. If capitalism is not abandoned, humanity will have to be. Is it easier to imagine the end of humanity than the end of capitalism?

The final images of the film confirm as much: what are those grotesque humanoid skeletons sprouting from the tip of Godzilla's tail (Figure 4.6)? This parting shot has inspired a good deal of speculation in fan communities. Is Godzilla giving birth parthenogenetically to a new species of human? The possibility of new hybrid species intensifies the film's central paradox, for it is not

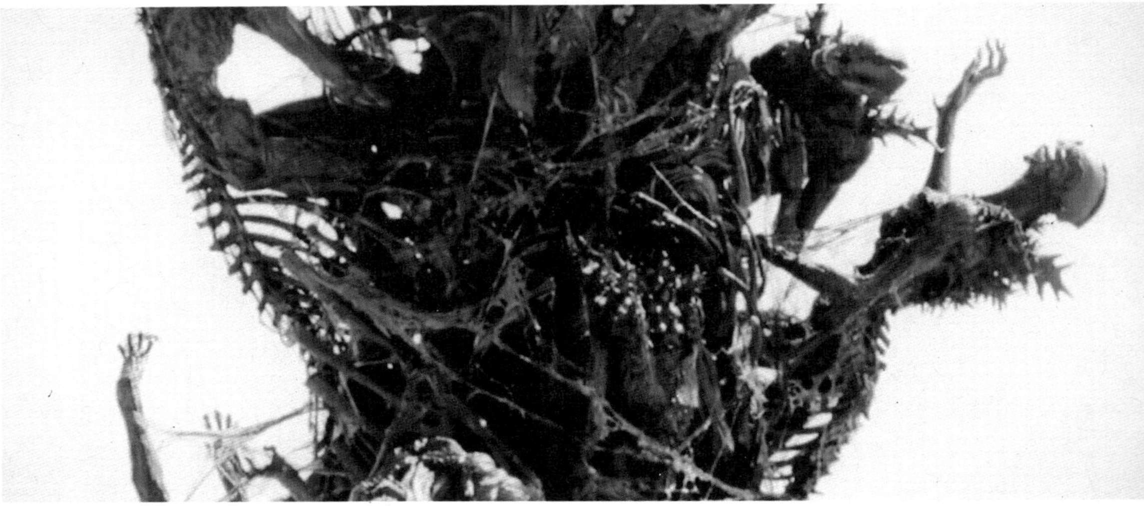

4.6 At the end of the film, a quick pan shows grotesque humanoid forms apparently sprouting from Godzilla's tail.

clear whether the human-Godzilla fusion marks the end of humans, or whether humans are somehow preserved within the monster. Is Godzilla the annihilator or the saviour of humanity? Then again, these human forms may be people who have come somehow to inhabit Godzilla. Maki Gorō may have abandoned his boat in Tokyo Bay in order to fuse with Godzilla. Is Maki a sort of puppet master, a ghost in the shell?

These eerie scenarios of asexual reproduction through the sprouting of grotesque post-human forms imply a fusion with radioactivity at the deepest level, a fusion with nuclear fission itself, which becomes expressed in modes of self-(re)production like grafting, splitting, and budding. Such an embodiment of nuclear fission presents a new variation on what Miyamoto characterized as a masculinist imagination of nuclear exposure in terms of "powering up" instead of merely suffering illnesses. Reproduction becomes indistinguishable from non-gendered self-expression. It is not a coincidence that few provisions are made for femininity in the film and that the three main female characters work with men and expose themselves to the same dangers as men. Neither human sexual reproduction nor human biology is presented as special, of value in itself, even though the film hinges on the protection of human life. Along this uncanny pathway, humans are already non-biological life forms; their lives do not centre on traditional reproductive concerns. This pathway, then, not only evokes long-standing male *otaku* fantasies about non-reproductive sexuality, planetary rebirth, and new-type humans but also opens into a sort of speculative practice on the part of fan communities to decipher the brain teasers and to extend the enigmatic moments into new fictions. Significantly, just as it is through auto-reproduction that Godzilla threatens to leave Japan and become a global concern, the enigmas surrounding Maki and the human forms sprouting from Godzilla have inspired speculation among fans around the world.

In sum, there are at least four pathways through *Shin Gojira*, which imply different relations to radioactivity in the context of the ongoing nuclear crisis. The happy pathway allows viewers to imagine overcoming radiation, while the stoic pathway enjoins us to accept the difficult fact that we will have to find a way to live with the monster, with nationalizing radiation proposed as a path of least resistance. The ludic pathway presents some overlap with the happy and stoic pathways yet introduces a third possibility: recapitalizing on radioactivity within the global content industry. While such a proposition is so preposterous as to defy belief, its mocking tone also exposes the outlandish nature of many current initiatives striving to profit from the nuclear crisis. Fourth and finally, the uncanny pathway hints at dark and strange forms of living with the nuclear crisis based on situating the human body within nuclear fission while incorporating nuclear fission into the body.

The result of these multiple pathways or perspectives is something akin to what Thomas Elsaesser dubs the mind game film. In contrast with the classical film form, which he characterizes as "excessively obvious," the mind game film is "excessively enigmatic."[49] Elsaesser notes that the tendency toward excessively enigmatic cinema follows from the transformation in the production and reception of films: increasingly, films not only are displayed in multimedia formats (in theatres, on televisions, on computer screens and tablets, and so forth) but also are read as one entry within a larger multimedia franchise comprising products as diverse as console games, iOS games, comics, television series, toys, and many others. Here another question arises: how do these pathways hold together?

Shin Gojira provides one response: each pathway entails a conflict or contradiction that is intensified to the point of paradox. The stoic pathway, for instance, reaches the point where destruction is reconstruction; the nation must be destroyed in order to be built, and the only way to avoid war is total mobilization for it. Thus one pathway resonates with and potentially flips into other pathways at the point of paradox. Once destruction turns into construction, annihilation becomes salvation, and death is rebirth. If death is rebirth, then reproduction is unnecessary, because destruction is generative and creative, and so forth. Thus, *Shin Gojira* circles around the paradox that it cannot name or represent but that serves a transcendental principle that is experienced throughout the film: nuclear energy is weaponry. As such, whether it intends to do so or not, the film stages the "deep link between nuclear energy (Fukushima), nuclear weapons (Hiroshima) and the military bases (Okinawa)."[50]

Conclusion: The Disaster Complex

In modelling its political realism on 3/11 and the resulting nuclear crisis, *Shin Gojira* provides some insight into two contemporary domains in which the entanglement of infrastructures and networks results in a continuous multiplication of effects. On the one hand, the ongoing release of massive amounts of radionuclides from the nuclear reactors in meltdown at the TEPCO plants in Fukushima was a direct result of the enmeshing of various technological systems. Also, the inability to rectify the situation technologically is due both to the multiplication of aleatory natural factors such as groundwater seepage, weather patterns, ocean currents, gradients, and so forth; and to the enmeshing of political and economic interests known as the nuclear village or the military-industrial complex. The result is a sort of "disaster complex," a mangle of natural and technological systems that ground the nuclear political economy. At the same time, popular culture in Japan has developed various creative strategies and business models for the serialization of its products or objects across media. These objects are like what Timothy Morton calls hyperobjects, or "things that are massively distributed in time and space relative to humans."[51] These media objects oscillate between a discrete and non-discrete existence: *Shin Gojira*, for instance, is at once a stand-alone film and part of genre systems (the Godzilla studio franchise and the larger *tokusatsu* mode) and a multiple franchise that involves mixed-media strategies for reception in diverse media forms across a plethora of platforms. The film thus includes multiple pathways harkening to interpretative communities. Let us call this serialization effect, somewhat reductively, an "*otaku* complex."

These two domains of multiplication act in different registers, bringing into play distinctive modes. The one cannot be reduced to the other. But, increasingly in the context of ongoing nuclear crisis, something is happening between them. As *otaku*-related commentators on both *Shin Gojira* and *Kimi no na wa* noted, even though the *otaku* complex has long shown an interest in disaster, catastrophes, and apocalyptic scenarios, something shifted after 3/11. When Azuma speaks of the two films in terms of the end of *otaku* culture, he calls attention to a shift in a structure of feeling. Likewise, Uno Tsunehiro remarked how the relation between the domain of the ordinary (everyday life) and that of the extraordinary (disastrous or apocalyptic scenarios) had shifted.[52] The extraordinary is too close to everyday life. Put another other way, the subjectivity associated with the *otaku* complex is increasingly in communication with the disaster complex. But communication is something of a misnomer, for the two complexes are not engaging in a dialogue constrained by procedures of consensus, which might allow for a critical relation between the two in the form of critical reason. As this account of *Shin Gojira* demonstrates, even if the film includes moments of criticism of government policies that verge on a critical rationality, the overall effect is not what is normally deemed critical.[53] Instead, the film explores the feel of the disaster complex, drawing on the *otaku* complex to gauge how it feels. The disaster complex and *otaku* complex begin to explore this feeling for each other. A new subjectivity starts to emerge across them, one that is neither of the otaku complex nor complicit with the military-industrial disaster complex.

What then does such a film give us, if we cannot expect a critical stance of it? On the one hand, it makes us aware of the emergence of a new structure of feeling, which, as it emerges, will surely begin to expose the fault lines of a deeper history. On the other hand, with its emphasis on multiple pathways, this recomposition of subjectivity shows signs of a profound engagement with imagining other possible futures. Such a gesture may easily collapse into a static formation of somewhat perverse consumer choices. Yet its feel also conveys an involuntary movement toward another possible world, a non-conscious worlding. To realize such a world will require an effort on our part, for building on this recomposition of subjectivity can only begin through a reconsideration of the entrenched socio-historical and geopolitical categories that so avidly make Fukushima a Japanese crisis. Even though it is abundantly evident that "Fukushima" radionuclides are as much a global event as Japanese popular culture is, this recomposition of subjectivity is currently forced to implode back into its capital, lest its feeling contaminate the rest of the world with longings for another future.

NOTES

1 The two films are consistently linked in scholarship as well. Nearly all of the Japanese sources cited here connect the two films, and in his analysis, Asano Shun'yū captures the enthusiastic tone of reception in Japan, speaking of a veritable Renaissance in Japanese cinema.

2 Solnit, *A Paradise Built in Hell*, 7.

3 Hayashi Nobuya's quantative analysis of reviews based on text mining shows the degree to which the complexity of the storyline affected reviewers' overall assessment of the film.

4 See Katsuya Hirano's interview with the former mayor of Tōkaimura: Murakami Tatsuya and Katsuya Hirano, "Fukushima and the Crisis of Democracy."

5 See Watanabe Daisuke, "*Kimi no na wa* no dai-hitto wa naze 'jiken' na no ka?"; Azuma Hiroki, *Gēmu teki riarizumu no tanjō*; and Thomas Elsaesser, "The Mind-Game Film."

6 See the special feature *The Making of Shin Gojira* included in Anno Hideaki, *Shin Gojira*, DVD.

7 Yagasaki Katsuma, "Internal Exposure Concealed," citing Watanabe Etsushi et al., *Hōshasen hibaku no sōten*, 170.

8 Kate Brown, "Chernobyl's Disastrous Cover-Up."

9 See, for instance, Cécile Asanuma-Brice, "From Atomic Fission," 95.

10 Nancy, *After Fukushima*.

11 Bajema, "Brave New Sanriku," 74–5.
12 Steineck, "Nuclear Power," 35.
13 See, for instance, Yoshikuni Igarashi, *Bodies of Memory*, esp. Chapter 1, "The Bomb, Hirohito, and History"; as well as Lisa Yoneyama, *Hiroshima Traces*, esp. Chapter 5, "Ethnic and Colonial Memories."
14 Ragone, *Eiji Tsuburaya*, 34.
15 Dusinberre and Aldrich, "Hatoko Comes Home," 1–23. Kamanaka Hitomi's documentaries provide a long chronicle of such unevenness from a global perspective. See her interview with Katsuya Hirano, "Fukushima, Media, Democracy."
16 Field, "From Fukushima," note 10.
17 Ibid., n24, citing Olson, "Atomic Radiation Is More Harmful to Women."
18 Long, "Japan's 3.11 Nuclear Disaster."
19 Penney, "Nuclear Nationalism and Fukushima."
20 Harootunian, "Reflections from Fukushima," 34.
21 Ricoeur, *Memory, History, Forgetting*, 508, cited in Harootunian, "Reflections from Fukushima," 34.
22 See Long's discussion of Yamashita Shun'ichi in "Japan's 3.11 Nuclear Disaster."
23 Miyamoto, "Gendered Bodies in *Tokusatsu*," 1092.
24 Ibid.
25 Ibid., 1088.
26 Ragone, *Eiji Tsuburaya*, 34.
27 See Dower, "Japanese Cinema Goes to War"; and Hirano, *Mr. Smith Goes to Tokyo*, 24–7.
28 Kaplan, *Aerial Aftermaths*, 181. For an account of aerial photography and 9/11 germane to Anno's vision of Tokyo, see Kaplan's "'A rare and chilling view': Aerial Photography as Biopower in the Visual Culture of '9/11.'"
29 A fuller account of Anno's conceptualization of rebuild appears in Thomas Lamarre, "The Rebuild of Anime."
30 Anno Hideaki and Tōhō kabushiki kaisha, eds., *Ji āto obu Shin Gojira*, 484; also cited in Hirabayashi, "*Shin Gojira* ni okeru kitsune to hebi," 183.
31 In "*Shin Gojira* ni okeru kitsune to hebi," Hirabayashi provides a careful account of how Nomura Manzai plays the role of Godzilla, with an emphasis on how the use of Kyōgen brings connotations of a six-hundred-year tradition to the monster. But it is worth noting that, as is true of Nō and Kyōgen, temporality is already out of joint, oddly layered; a linear view of the past and thus a nation-centred account becomes impossible to sustain.
32 Hitomi and Hirano, "Fukushima, Media, Democracy."
33 Reiher, "Food Safety and Consumer Trust," 54.
34 Brennan, "Fukushima." See also Field, "From Fukushima," n10.
35 Andrews, *Dissenting Japan*, 288.
36 Ibid., 288–9.
37 See Sabu Koso's contribution to this volume.
38 The potential conflict between what I am calling the old guard and the new guard gradually transforms into an amical tension between Yaguchi Randō and Akasaka

Hideki, whose dialogues about postwar Japanese sovereignty clearly announce the film's geopolitical concerns. Asano Shun'yū cites their dialogue in full in "Eiga *Shin Gojira* ni miru bunmyaku bunseki," 45.

39 Gavan McCormack coins this term in *Client State*. For an overview, see McCormack, "Japan's Client State (Zokkoku) Problem."

40 Gerow, "Fantasies of War and Nation."

41 A multitude of reports are found online. The paragraph here is largely based on the Reuters report: Sheldrick and Foster, "Tepco's 'Ice Wall.'"

42 Klein, *The Shock Doctrine*.

43 As Aaron Gerow shows in "Wrestling with Godzilla" through his discussion of intertextuality and markets in the context of the original *Gojira*, there is a history to this *otaku* dimension, which Anno effectively brings to fruition in his rebuild.

44 Azuma Hiroki, "*Shin Gojira* to *Kimi no nawa*."

45 Ibid.

46 Jackzo, "I Oversaw the U.S. Nuclear Power Industry."

47 Ibid.

48 Steineck, in "Nuclear Power," beautifully traces the logical connections between pro-nuclear parties and "social structures that sustain the inequitable distribution of benefits and risks" (35).

49 Elsaesser, "The Mind-Game Film,"

50 Aramaki Osamu, "The Japan–US–Okinawa Relationship," 16.

51 Morton, *Hyperobjects*, 1.

52 Uno Tsuneohiro, "Imagination After the Earthquake."

53 Kosuke Shimizu provides an insightful account of the general question of critical thinking with specific reference to its waning in the context of dealing with Fukushima, in "Reflection, the Public, and the Modern Machine."

BIBLIOGRAPHY

Andrews, William. *Dissenting Japan: A History of Japanese Radicalism and Counterculture, from 1945 to Fukushima*. London: Hurst, 2016.

Annō Hidekai and Tōhō kabushiki kaisha, eds. *Ji āto obu Shin Gojira* (Art of Shin Godzilla). Tokyo: Kabushiki Kaisha Karā, 2016.

Annō Hideaki et al. *Shin Gojira*. Tokyo: Tōhō Kabushiki Kaisha, 2017. DVD.

– "The Making of *Shin Gojira*." Television special. https://www.youtube.com/watch?v=YDlMH4foXdM. Subsequently included in *Shin Gojira*. Tōhō Kabushiki Kaisha, 2017. DVD.

Aramaki Osamu. "The Japan–US–Okinawa Relationship with Respect to the Military Bases: The Discourses since 3/11." In *Japan's 3/11 Disaster As Seen from Hiroshima: A Multidisciplinary Approach*. Hiroshima City University, Faculty of International Studies Book Series, vol. 5. Tokyo: Soeisha/Sanseido Shoten, 2013.

Asano Shun'yū. "Eiga *Shin Gojira* ni miru bunmyaku bunseki: monogatari no naibu kōzō to monogatari dōshi no kakawariai." *Ronbunshū*. Edited by Kanazawa daigaku ningen shakai iki keizai gakurui shakai gengogaku enshū. No. 13 (22 March 2018): 33–49.

Asanuma-Brice, Cécile. "From Atomic Fission to Splitting Areas of Expertise: When Politics Prevails over Scientific Proof." In *Planetary Atmospheres and Urban Society after Fukushima*, edited by Christophe Thouny and Mitsuhiro Yoshimoto, 95–112. Singapore: Palgrave Macmillan, 2017.

Azuma Hiroki. *Gēmu teki riarizumu no tanjō: dōbutsuka suru posutomodan 2* (The Birth of Gamic Realism: Animalized Postmodern 2). Kōdansha gendai shinsho 1883. Tōkyō: Kōdansha, 2007.

– "*Shin Gojira* to *Kimi no nawa*. o mite omotta no wa, otaku no jidai wa owatta n da na to iu koto" (What I thought watching *Shin Gojira* to *Kimi no nawa*. was that the era of otaku is over and done.) https://twitter.com/hazuma/status/773901267173117952.

Bajema, Ramona. "Brave New Sanriku: Recovering from 3.11." In *Planetary Atmospheres and Urban Society after Fukushima*, edited by Christophe Thouny and Mitsuhiro Yoshimoto, 73–94. Singapore: Palgrave Macmillan, 2017.

Brennan, David. "Fukushima: Radioactive Soil Might Be Used to Build New Roads – and Residents Are Not Happy." *Newsweek*, 30 April 2018. https://www.newsweek.com/fukushima-radioactive-soil-might-be-used-build-new-roads-and-residents-are-not-906184.

Brown, Kate. "Chernobyl's Disastrous Cover-Up Is a Warning for the Next Nuclear Age." *The Guardian*, 4 April 2019.

Dower, John W. "Japanese Cinema Goes to War." In *Japan in War and Peace: Selected Essays*. New York: New Press, 1993.

Dusinberre, Martin, and Daniel P. Aldrich. "Hatoko Comes Home: Civil Society and Nuclear Power in Japan." *Journal of Asian Studies* (2011): 1–23.

Elsaesser, Thomas. "The Mind-Game Film." In *Puzzle Films: Complex Storytelling in Contemporary Cinema*, edited by Warren Buckland, 13–41. Oxford: Blackwell, 2010.

Field, Norma. "From Fukushima: To Despair Properly, to Find the Next Step." *Asia-Pacific Journal: Japan Focus* 14, no. 3 (2006).

Gerow, Aaron. "Fantasies of War and Nation in Recent Japanese Cinema." *The Asia-Pacific Journal: Japan Focus* 14, no. 3 (2006).

– "Wrestling with Godzilla: Intertextuality, Childish Spectatorship, and the National Body." In *In Godzilla's Footsteps: Japanese Pop Culture Icons on the Global Stage*, edited by William M Tsutsui and Michiko Itō. New York: Palgrave Macmillan, 2006.

Harootunian, Harry. "Reflections from Fukushima: History, Memory, and the Crisis of Contemporaneity." *boundary 2* 42, no. 3 (2015): 23–35.

Hirabayashi Kazunari. "*Shin Gojira* ni okeru kitsune to hebi: Nomura Manzai no 'nessen' no engi o chūshin ni." *Kindai daigaku Nihon bunka kenkyūsho kiyō* 2 (March 2019): 179–98.

Hirano, Kyōko. *Mr. Smith Goes to Tokyo: The Japanese Cinema under the American Occupation, 1945–1952*. Smithsonian Studies in the History of Film and Television. Washington, DC: Smithsonian Institute, 1992.

Igarashi, Yoshikuni. *Bodies of Memory: Narratives of War in Postwar Japanese Culture, 1945–1970*. Princeton: Princeton University Press, 2000.

Jackzo, Gregory. "I Oversaw the U.S. Nuclear Power Industry. Now I Think It Should Be Banned." *Washingon Post*, 20 May 2019. https://www.washingtonpost.com/

outlook/i-oversaw-the-us-nuclear-power-industry-now-i-think-it-should-be-banned/2019/05/16/a3b8be52-71db-11e9-9eb4-0828f5389013_story.html.

Kamanaka Hitomi and Katsuya Hirano. "Fukushima, Media, Democracy: The Promise of Documentary Film. An Interview with Kamanaka Hitomi with Introduction by Katsuya Hirano." *Asia-Pacific Journal: Japan Focus* 16, no. 3 (August 2018).

Kaplan, Caren. *Aerial Aftermaths: Wartime from Above*. Durham: Duke University Press, 2018.

Lamarre, Thomas. "The Rebuild of Anime." *Mechademia 5: Fanthropologies* (2010): 349–53.

Long, Margherita. "Japan's 3.11 Nuclear Disaster and the State of Exception: Notes on Kamanaka's Interview and Two Recent Films." *Asia-Pacific Journal: Japan Focus* 16, no. 4 (August 2018).

McCormack, Gavan. "Japan's Client State (Zokkoku) Problem." *Asia-Pacific Journal: Japan Focus* 11, no. 2 (June 23, 2013).

Miyamoto, Yuki. "Gendered Bodies in *Tokusatsu*: Monsters and Aliens as the Atomic Bomb Victims." *Journal of Popular Culture* 49, no. 5 (2016): 1086–106. doi:10.1111/jpcu.12467.

Morton, Timothy. *Hyperobjects – Philosophy and Ecology after the End of the World*. Minneapolis: University of Minnesota Press, 2013.

Murakami Tatsuya and Katsuya Hirano. "Fukushima and the Crisis of Democracy: Interview with Murakami Tatsuya." *Asia-Pacific Journal: Japan Focus* 13, no. 1 (May 2015).

Nancy, Jean-Luc. *After Fukushima: The Equivalence of Catastrophes*. New York: Fordham University Press, 2015.

Olson, Mary. "Atomic Radiation Is More Harmful to Women." *Nuclear Information and Resource Service* (2011). https://www.nirs.org/wp-content/uploads/radiation/radhealth/radiationwomen.pdf.

Penney, Matthew. "Nuclear Nationalism and Fukushima." *Asia-Pacific Journal: Japan Focus* 10, no. 2 (March 2012).

Ragone, August. *Eiji Tsuburaya: Master of Monsters: Defending the Earth with Ultraman, Godzilla, and Friends in the Golden Age of Japanese Science Fiction Film*. San Francisco: Chronicle Books, 2014.

Reiher, Cornelia. "Food Safety and Consumer Trust in Post-Fukushima Japan." *Japan Forum* 29, no. 1 (2017): 53–76.

Ricoeur, Paul. *Memory, History, Forgetting*. Translated by Kathleen Blamey and David Pellauer. Chicago: University of Chicago Press, 2004.

Sheldrick, Aaron, and Malcolm Foster. "Tepco's 'Ice Wall' Fails to Freeze Fukushima's Toxic Water Build-Up." *Reuters*, 8 March 2018. https://www.reuters.com/article/us-japan-disaster-nuclear-icewall/tepcos-ice-wall-fails-to-freeze-fukushimas-toxic-water-buildup-idUSKCN1GK0SY.

Shimizu, Kosuke. "Reflection, the Public, and the Modern Machine: An Investigation of the Fukushima Disaster in Relation to the Concepts of Truth and Morality." *Japanese Journal of Political Science* 18, no. 4 (2017): 536–51.

Solnit, Rebecca. *A Paradise Built in Hell: The Extraordinary Communities That Arise in Disasters*. New York: Viking, 2009.

Steineck, Raji C. "Nuclear Power: A Techno-ethical Perspective." *The Impact of Disaster: Social and Cultural Approaches to Fukushima and Chernobyl* 9 (2015): 25–49.

Uno Tsunehiro. "Imagination after the Earthquake: Japan's Otaku Culture in the 2010s." Translated by Jeffrey C. Guarneri. *Verge: Studies in Global Asias* 1, no. 1 (2015): 114–36.

Watanabe Daisuke. "*Kimi no na wa* no dai-hitto wa naze 'jiken' na no ka? Sekai-kei to bishōjo gēmu no bunmyaku kara yomitoku" (Why was the box office hit of *Your Name* an "event"? A close reading based on the contexts of world type genre and beautiful girl games). *Real Sound,* 8 September 2016. http://realsound.jp/movie/2016/09/post-2675.html.

Watanabe Etsushi, Endo Junko, and Yamada Kosaku. *Hōshasen hibaku no sōten: Fukushima Genpatsu Jiko no kenkō higai wa nai no ka.* Tokyo: Ryokufū shuppan, 2016.

Yagasaki Katsuma. "Internal Exposure Concealed: The True State of the Fukushima Nuclear Power Plant Accident." *Asia-Pacific Journal: Japan Focus* 14, no. 3 (2016).

Yoneyama, Lisa. *Hiroshima Traces: Time, Space, and the Dialectics of Memory.* Berkeley: University of California Press, 1999.

Afterword to Part One

Repeating, Multiplying: The Ongoing Now of Nuclear Aftereffects

DANIEL C. O'NEILL

The chapters in the section are engaged with an impasse, encountered across divergent affective and analytical registers. Each chapter takes up the notion of aftereffects by identifying a series of enduring problems linked to the nuclear, problems rendered as ideological contradictions, historical formations, or anthropogenic forces that continue to produce harm or the threat of harm. They are imagined to be enduring habits of thought, repeating historical formations, disciplining of bodies and minds, and multiplying through the manifold relations of history, media, and capital. Each chapter is engaged with complexities that resist an overview. As such, I will briefly discuss each work and attempt to draw out some of the broader themes and conceptual insights that invite further elaboration and reflection. I will explore what these chapters are doing collectively to transform our understanding of the nuclear and its aftereffects, which are toxic, enduring, and mutating. The speculative and open-ended nature of the comments I offer reveal a certain limit or an impasse in my own work on the 3/11 disasters, as I continue to think through the question of critique in relation to the ongoingness of the nuclear, even as the coordinates of agency and resistance are already becoming something else as the non-human and more-than-human turns in recent theory are being absorbed into environmental studies.[1] It is my hope that these comments and questions will allow us to relate to these chapters as a series of reciprocal determinations and interferences through which we can begin to glimpse our relatedness to forms of nuclear toxicities that carry consequences for imagining other possible futures.

Sharae Deckard, "The Future Is Behind Them!": Post-Apocalypse and the Enduring Nuclear in Post-Soviet Russian Fiction"

Through a reading of the post-apocalyptic novel *Metro 2033* (2002), Deckard invites us to consider what endures as a nuclear aftereffect. In her analysis, the "enduring nuclear" refers to the enduring presence of the nuclear state in our own time, a persistence made legible by the novel's symptomatic failures to imagine a world without nuclear. More pointedly, these symptoms expose our perverse attachments to the habits of thought and modes of living organized around extractive energy regimes. Mobilizing the trope of the uncanny, Deckard extends the meaning of the enduring nuclear to draw our attention to the problem of time and, specifically, to the long temporality of radioactive decay to which we are all exposed, though unevenly so. Deckard makes the point that the toxicities staged as a future reality in the novel are in fact taking place in the present as an ongoing disaster for many (from Ukrainians to those living in the Urals to downwinders elsewhere).[2] The dissonance between a postcolonial disaster unfolding in the present and a post-apocalyptic disaster that is held affectively as a future threat gives rise to the many figurations of the uncanny that she examines with ethical urgency. In Deckard's reading, *Metro 2033* becomes an evocation of uncanny returns, a testimonial in which nuclear futurity unravels to reveal a traumatic logic where the past continues to make claims on the present, thus rendering transition to the future nearly impossible.

The trope of failure takes the form of a limit that is repeated throughout the novel: the failure to live differently, the failure to move into a post-nuclear future, the failure to imagine otherwise. Though the novel appears to produce a cyclical model of history in which the future is grasped only as a repetition of the past, its closing moment offers an opening and transforms the limit into a revelation. Deckard notes that while the protagonist is finally unable to think beyond the nuclear potential, an "epiphany" does transpire in his final moments in which the dream of a multispecies commons is momentarily evoked only to be destroyed. If, as Deckard compellingly demonstrates, the novel is preoccupied, if not obsessed, with the blockage of the capacity to think beyond the nuclear, how might we account for the epiphany? The epiphany feels different from the uncanny, where what is suppressed returns as its own constitutive occlusion. If we were to understand epiphany as a sudden revelation residing outside the economy of repression, how might we account for the appearance of the revelation? How might the epiphany (or other novel figurations of time) capture something of the materiality of the nuclear already immanent in our own present? I raise these questions not to contest the insights offered by Deckard's luminous reading but to extend them in the direction expanding the novel's speculative capacity.

Sezin Topçu, "From Toxic Lands to Toxic Rumours: Nuclear Accidents, Contaminated Territories, and the Production of (Radio)active Ignorance

Topçu tracks with great analytical clarity the means by which the nuclear state proffers an obfuscating notion of nuclear harm that is temporary and even immaterial, thereby diverting public attention from the long-term harm caused to properties and bodies. If Kohso is engaged with the larger historical forces that consolidate the nuclear state, then Topçu examines the exercise of biopower that is performed at the level of everyday life where the perception of nuclear harm is managed through precise and detailed norms.

For Topçu, nuclear accidents produce opportunities. She provides an account of these opportunities and shows how nuclear regimes, in different historical moments, have responded to disasters by exerting a power that endeavours to administer life by subjecting it to precise controls. Here, the mechanisms of power are multiplied, through territorial decisions, bodily practices, and the production of ignorance, in order to prioritize remediation, remedies, and compensation. Contaminated lands are zoned to preserve territorial capital. Nuclear refugees are encouraged to return to their normal lives in the contaminated zones for as long as they learn to "optimize their food intake, control their movements, and discipline their bodies."[3] This management of bodies not only shifts the emotional and financial burden of care onto the individual but also banalizes nuclear harm, ensuring its naturalization. Among the strategies mobilized by the nuclear state, the most pernicious is the conversion of harm into hearsay in which the health effects and risks linked to radiation exposure are rendered as misinformation or mere rumours that distract the mind from rational thought.

Topçu's analysis demonstrates with great compelling force the ways in which the bodily and mental well-being of the population are managed to align with determining interests that ensure the nuclear regime's economic well-being. In curating the ensemble formed by institutions, procedures, calculations, and tactics that allows for the exercise of this biopower, land becomes displaced as the object of control. In her analysis, the consideration of land risks becomes nothing else but the effects of a perpetual zoning or a series of territorializations, where environmental relations, even those induced by harm, become merely the mobile effects of a regime of governmentalities. How might the relation between land and bodies as well as other forms of environmental interdependence be further articulated so that we may understand the full force of her critique? How might radiation generate forms of environmental variability not moulded to the political administration of life in the nuclear exclusion zones?

Sabu Kohso, "The Fukushima Process"

Fukushima is an event, a breaking point in the status quo at which radical action becomes a necessity. It also serves as an opportunity for the nuclear state to capitalize on the disasters. Here, Fukushima as an event is subsumed under a "process" in which radiation is normalized and pre-existing political preferences are strengthened in the name of national belonging. Fukushima is thus captured both as an event and a process; their interaction is described by Kohso as a historical pattern that is fated to repeat itself for as long as the iron grip of the nation-state on our political imagination is left unchallenged. To highlight the ideological character of the process and submit it to critique, Kohso brings in a transnational perspective to keep track of the discursive formations that are mobilized across a horizon of interests only to shore up the threatened geopolitical arrangements established in the Cold War. Hence, the chapter is committed to a critique that enlarges the analytical scope, drawing our attention to the interconnectivity of the transnational entanglements of nuclear power and Abe's post-nuclear disaster governance. For Kohso, the "nationalization of radiation" narrows the historical vantage point and forecloses the conflict and opposition necessary for the emergence of transformative politics. As such, any social movement failing to account for the historical entanglements of peace with nuclear power is a movement doomed to produce national conformity. In a sense, this history of conformity, a history that repeats itself, is prefigured.

Though the critical thrust of Kohso's chapter leaves us with a compulsive notion of historical repetition, it does offer a faint note to suggest the contrary. Kohso writes, "We can observe here a dynamic process asymmetrically playing out between power operations and people's struggles. This chapter seeks to grasp this process as history repeating itself, while nevertheless offering a glimpse, just a flash, of the possibility of creating difference."[4] Here, Kohso offers us a tantalizing notion of difference, as though repetition itself becomes the condition of possibility for its generation. This note, however erratic, invites broader reflections on the figurations of history and nuclear crisis. The tense combination of the post-nuclear and its uncanny re-enactment (in Deckard) and of Fukushima as event and process (in Kohso) produces an uneasy hybrid in which history, here divested of the certainty of a redemptive conclusion, is marked with a certain overdetermination that compels each thinker to reach for alternative forms of historical time based on notions such an epiphany or a flash. If we were to understand their critical engagements with the socio-political aftereffects of nuclear disasters only within a model of history as repetition, then we would run the risk of producing only circular futures. How, then, is the flash, as a figuration of historical time, related to a history figured as repetition? If the flash is not reducible as an effect of normalization, then what kind of historical experience does it afford?

In Kohso's case, it is perhaps the planetary perspective he mobilizes – one that allows him to track the larger forces that converge to reproduce the nuclear status quo as well as the uneven distribution of nuclear aftereffects – that becomes the vehicle for generating a critical standpoint with which we can begin to imagine political change. Still, what do we make of epiphanies, glimpses, and flashes, the unanticipated figurations of time inserted into a narrative dominated by repetition? If the repetition compulsion in Deckard's analysis corresponds to a failure that is doomed to repeat itself well into the future and if the historical repetition in Kohso's diagnosis reveals a process in which existing political arrangements are disrupted only to be reinforced, how might a reorientation to the "everyday" change the narrative? How might a shift in focus to what Kohso calls a "politics of life and struggle within the catastrophe" allow for different forms of repetition to come into view, forms claimed by the processes of ongoingness in which repetition becomes indistinguishable from the rhythms of daily survival? How might we make sense of a crisis experience that is embedded in ordinariness, where the possibilities of rupture are repeated within the minute, tiny gestures of living with toxic effects? How might we account not only for the "virtuality of radionuclides' invisible flows" but also for the felt presence of their toxic effects on affected bodies as well as on those who are caring for them, effects that do not hold well by any temporality in particular? How might these immanent forms of repetition generate new critical horizons that might help us recognize the agential resources of non-political beings and emerging commons of well-being?

Thomas Lamarre, "Fukushima and the Rebuild of Godzilla: Multiplying Media in an Era of Multiplying Disaster"

The concern in Lamarre's chapter is not with the exposure of ideological contradictions, historical formations, or strategies of governance. Instead, it captures a paradox where nuclear energy becomes indistinguishable from nuclear weaponry. *Shin Gojira* becomes an object of analysis; and as an object it recalls the first Godzilla film as much as the social-political matrix of the Cold War era. It is also an object that responds to the new circumstances generated by the 3/11 disasters, a response that is at the same time a movement in which the relations between popular culture and nuclear power are reconfigured. *Shin Gojira* may recall Godzilla in ways to invite a reading of the film as an allegory of national trauma. Lamarre makes the point, however, that the allusion does not simply replay a traumatic scenario in which postwar Japan rises from the ashes of nuclear destruction, as though history repeats itself as an eternal cycle of destruction and creation. For Lamarre, the critical task is to provide a theoretical account of how *Shin Gojira* transforms the trope of repetition in order to capture something of the ongoingness of nuclear harm without erasing the social-political unevenness that has historically securitized some lives by rendering other lives precarious.

Like Deckard, Lamarre offers a model of theorizing at the threshold, at the end of the world. Here, rather than a failure to transcend the nuclear imaginary that Deckard identifies as the traumatic core of speculative fiction, Lamarre recasts the limits of the nuclear imaginary as a paradox, drawing our attention to the fact that the dissemination of nuclear energy is inescapably entangled with the proliferation of nuclear weapon capabilities. In short, nuclear power is a form of energy that is "always already a weapon."[5] It is a paradox whose effects render acts of destruction and creation indistinguishable. While Anno's film does not shy away from capitalizing on these effects, Lamarre's reading intensifies the paradox, enlarging the scene of analysis so that we may forge multiple pathways out of the traumatic discord that capitalist production has traded on for its cycles of growth and retraction.

Post-Nuclear Lives and Critique

The essays collected here are galvanized by a refusal to think of the "aftereffects of Chernobyl and Fukushima" as localized problems whose resolution is subjected to national security or legal frameworks. They approach the problem of the nuclear as a turning point for broader historical reflections and imagining multiple futures. They do so in ways that recontextualize the problem and enlarge the analysis to account for the matrix of power relations, epistemologies, and technologies of disciplining and control as well as the attachments and identifications that continue to consolidate nuclear regimes and their toxicities. The chapters all work through and beyond standard narratives to attend to the incommensurate forces that converge on the nuclear. Variously described as ideological contradictions, historical formations, impossibilities and paradoxes, these forces are encountered as an impasse, suspended in a condition where time seems out of joint, where things relate and do not quite cohere, where phenomena overlap and interfere with one another. As such, the aftereffects they examine require and demand a reinvention of the genres of theorizing adequate to the impasse – a different mode of critique that attends to their uncertainties, mixed feelings, and intensities.[6]

Kohso's essay teaches us that any attempt to understand nuclear regimes and their historical resilience (their ability to survive disasters) must begin from a broader perspective that takes into view the global contestations of power during and after the Cold War. The historical perspective broadens the scope of the critique. Topçu's chapter exposes an array of managerial strategies that has allowed an unrecognized form of nuclear power to obfuscate the incalculable harm to which life and populations have been subjected. The critique focalizes on the micro-political analytic of nuclear power whose multiplying effects are both overdetermined and dynamic. Critique comes in many forms, but if we were to take critique to be a movement of thought compelled to produce a clarifying perspective (a broadening or a zooming in), what happens to critique when it is confronted with an impasse, with its tensions and contradictions? How might the confrontation produce another impasse from which the necessity of a new mode of critique is made legible? The critiques these chapters perform, while targeting the nuclear regimes as a practice, discourse, and institution, are pertinent and necessary. And while we rightfully denounce the harmful effects of the radioactive materials produced by these regimes, the effects we capture and document as the means of materializing our verdicts, questions remain: How do we live with these harmful effects, effects that will outlast regimes? How might we live with these effects while disrupting the processes of their normalization? How might we reinvent critique to allow us to stay with the trouble and build worlds?

Deckard's reading offers a critique of our inability to extricate ourselves from the extractive practices of capitalism, an inability that also reveals the positive capacities of nuclear power to assert its totalizing presence in the historical present. While Deckard encapsulates these relations as the uncanny, I take the tension she generates between the post-apocalyptic and the postcolonial to be a kind of paradox that invites reassessment, a formation that paradoxically leads us from the future back into the present. Here, the novel's inability to imagine a future without nuclear becomes instructive. For one, a nuclear-free future is not the same thing as a radiation-free future; a radiation-free future is, in fact, an impossibility. Faced with this impossibility, we are left with no options but to confront the question of how we might live with radiation now (and not only in the future). This means articulating a form of critique or a decolonizing pedagogy that would allow us to imagine otherwise while maintaining traction in the various modes of living that are transformed and continue to be transformed by the enduring presence of radiation.

Whether or not we can disentangle ourselves from nuclear power, a critique of the nuclear that entails a return to a pre-toxic state is not tenable, nor is it desirable. As Livia Monnet points out in the introduction, "a complete nuclear decolonization or denuclearization is impossible"[7] (i.e., radioactive wastes from twentieth- and twenty-first-century nuclear weapons and nuclear energy industries will remain in the earth systems for millennia if not hundreds of thousands of years). A decolonizing pedagogy of the nuclear will not indulge in the fantasy of a pre-toxic futurity and erase the history of those left in the wake of the nuclear arms race. It will reveal past injustices, but it will also provide an account of how people sustain themselves in the present when they feel their health is no longer a given, but is something they must fight for, how life may be held together in a precariousness that endures.[8] Our conventional modes of critique exhibit a great capacity to attend to the brutalities that constrain life. Even as we rely on dreams, flashes, and epiphanies as exit signs that would free us from the prison house of the nuclear, radiation's ongoing material presence in our everyday life, as well as the uneven distribution of its effects through time and space, remains a brute reality we have yet to fully address. How might we account for forms of human and non-human existences that flourish or wither among the enclosures we now occupy? What new language of critique can we construct that may be more affirming of these life forms? How might critique propose a sustaining sense of the commons that calls for our attention now?

Lamarre's chapter expands his diagnosis of the nuclear regimes beyond their industrialism and nationalism. In engaging with an *otaku* cultural imaginary, he seems to be doing something different with critique. His is an attempt to track the dynamics that are generated when a series of "complexes" is put into a communicative conversation, dynamics he encapsulates as a nuclear paradox, that, upon its intensification, mutates into multiple pathways of destruction, creation, and suspended animation, leading to futures we cannot determine in advance. Here he relies on a notion of multiplying to unleash a certain kind of potentiality that is otherwise held captive in the paradox. In his analysis, Lamarre explores the consonances as well as the dissonances generated by the linguistic assemblage of "recall," "remix," "rebuild," "building up," and "powering up," reconfiguring the relations among these words, intensifying them so that the word "repetition" is made to do something else. The monstrosity that haunts Deckard's essay returns in Lamarre's as a monstrous fusion and an uncanny pathway that "hints at dark and strange forms of living with the nuclear crisis."[9] I wonder, if we were to rethink these uncanny relations in terms of kinship or what Haraway describes as an "obligatory enduring relatedness," how might such a recontextualization renew attention to a different kind of living with the nuclear, a living that feels like an event but not quite so, an ongoingness that is not necessarily strange but whose banality is interrupted.[10]

In engaging with the nuclear as an impasse that feels as overwhelming as it is inevitable, these chapters transform critique into something else, a critique that is also a decolonizing pedagogy of the nuclear, a mode of writing and thinking that redirects attention to multiple forms of impasse and passing, dramatic and undramatic. From these chapters, we have reasons enough to renew the conviction that we must reject the trope of the nuclear as a custodian of global security and to contest the notion that nuclear energy is clean and sustainable. Though these chapters do not offer us a smooth or speedy arrival at a denuclearized future, they do enjoin us to stay with the trouble, and to live the precarious life, otherwise.

NOTES

1 For a reconceptualization of agency in new materialisms, see Coole and Frost, *New Materialisms*, 1–46; and Barad, *Meeting the Universe Halfway*, 132–88. On multispecies collaboration, see Tsing, *The Mushroom at the End of the World*, 17–27.
2 Deckard, "The Future Is Behind Them!," in this volume.
3 Topçu, "From Toxic Lands to Toxic Rumours," in this volume.
4 Kohso, "The Fukushima Process," in this volume.
5 Lamarre, "Fukushima and the Rebuild of Godzilla," in this volume.
6 For recent conversations that enliven the question of critique, see Cauwer, *Critical Theory at a Crossroads*, xvii–xxx.
7 Monnet, "Introduction," in this volume.
8 On the question of health in the atomic age, see Masco, "Atomic Health," 133–56.
9 Lamarre, "Fukushima and the Rebuild of Godzilla," in this volume.
10 Haraway, *Staying with the Trouble*, 1–8; Paulson, interview.

BIBLIOGRAPHY

Barad, Karen. *Meeting the Universe Halfway: Quantum Physics and the Entanglement of Matter and Meaning.* Durham: Duke University Press, 2007.

Cauwer, Stijn De. *Critical Theory at a Crossroads: Conversations on Resistance in Times of Crisis.* New York: Columbia University Press, 2018.

Coole, Diana, and Samantha Frost. *New Materialisms: Ontology, Agency, and Politics.* Durham: Duke University Press, 2010.

Haraway, Donna. *Staying with the Trouble: Making Kin in the Chthulucene.* Durham: Duke University Press, 2016.

Masco, Joseph. "Atomic Health, or How the Bomb Altered American Notions of Death." In *Against Health*, edited by Jonathan Metzl and Anna Kirkland, 133–56. New York: NYU Press, 2010.

Morton, Timothy. *Hyperobjects: Philosophy and Ecology after the End of the World.* Minneapolis: University of Minnesota Press, 2013.

Paulson, Steve. "Making Kin: An Interview with Donna Haraway." Interview in *Los Angeles Review of Books*, 6 December 2019. https://lareviewofbooks.org/article/making-kin-an-interview-with-donna-haraway.

Tsing, Anna Lowenhaupt. *The Mushroom at the End of the World: On the Possibility of Life in Capitalist Ruins.* Princeton: Princeton University Press, 2015.

TWO

The Cold War and Post–Cold War Nuclear State and Its Geopolitics: Imaginaries and Contestations

5

Shaking, Trembling, Rattling, Shouting: Seismic Politics in the Nuclear Age

JOSEPH MASCO

Vibration is the most basic form of energy; it can be both destructive and sympathetic, the oscillating source of destruction, or sound, or social effervescence. In this chapter, I consider how the atomic bomb has functioned as a vibrational force since 1945, one that not only has unlocked and transformed matter into energy, remaking the material world as well as the international order, but also has become a vehicle for lively social engineering in both negative and positive keys. For how a military and technoscientific revolution plays out in the *longue duree* of history, in the process of remaking earth, society, and consciousness, is a matter of institutions and logics as well as affects and atmospheres.

In the early days of the Manhattan Project, Los Alamos theoretical physicists postulated, dreamed of really, a massively slowed down nuclear event, attempting to visualize fission one atom at a time. Their idea was to chart atomic fission from the first splitting of a nucleus through each subsequent generation, tracking at the subatomic level the development of a chain reaction leading to a massive release of energy in the form of an explosion. They invented an informal unit of measure to describe the time they theorized it took one atom of uranium to fission – one-hundred-millionth of a second (.01 microsecond): they called it a "shake," slang for "shake of a lamb's tail."[1]

A nuclear explosion, perhaps the single most destructive event people have engineered, occurs in less than one hundred shakes. This means that all the nuclear detonations in human history – the 2,100 or so nuclear events that constitute an unprecedented human intervention in the biosphere, remaking geological and human time in perpetuity – do not collectively add up to a single second of explosive linear time.[2] Each of these detonations was a planetary seismic event, producing the sharp spike of a shock wave watched by geoscientists and security agents of all kinds. Indeed, the Cold War, from one key perspective, was

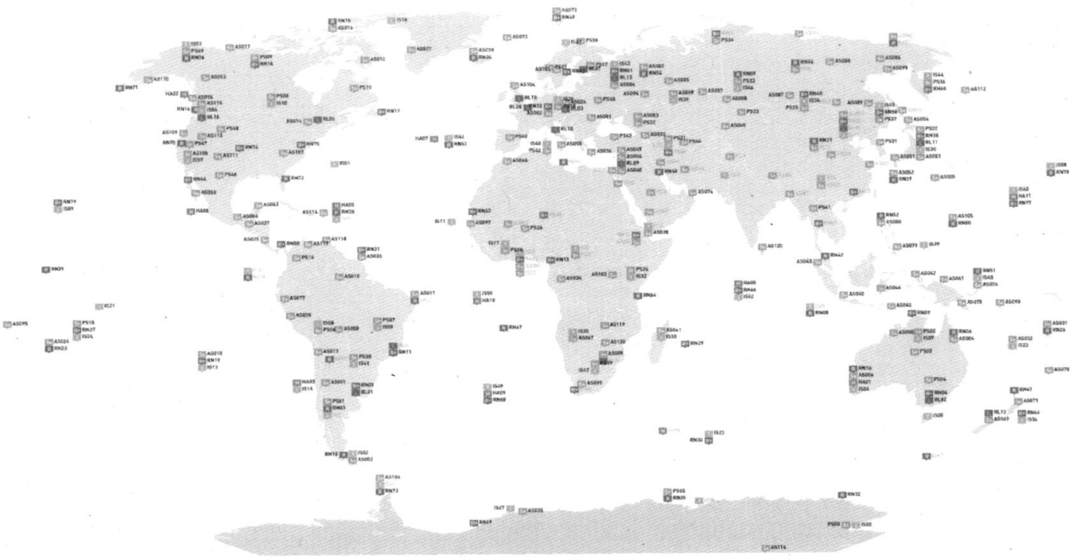

5.1 International monitoring system for nuclear detonations.

fought on a shake-by-shake basis not only in laboratories and testing sites but also seismically at planetary scale. The test programs of the United States, the Soviet Union, and, ultimately, England, France, China, India, Pakistan, Israel, and North Korea were theatrical displays of technoscientific power and national commitment written into the geology of Earth. Each detonation was an experiment but also a political statement in a global theatre of nuclear messaging. In the 1950s, a global network of seismic sensors was created to listen for and record the distinctive wave of a nuclear detonation, becoming a surveillance tool, a scientific research platform for geologists, and a new planetary mode of threat communication between nuclear powers (see Figure 5.1).[3] Nuclear states, in other words, used their research and development programs not only to create nuclear arsenals but also to communicate with one another via nuclear testing itself. They used the vibrational force of each nuclear detonation, transmitted via the global seismic sensor array to all interested parties, as an illustration of their technological capability, political resolve, and ongoing research commitments.

Researching the shake, and studying the resulting seismic quakes of nuclear detonations, became a total project in the US nuclear complex, one that evolved across radically shifting experimental regimes: above-ground testing (1945–62), underground testing (1962–92), and computer simulations (1993 to today).[4] Understanding the bomb in a shake-by-shake manner energized a phenomenal expenditure of money, energy, and technoscientific and military-industrial power in the United States – just under a $6 trillion investment in the twentieth century

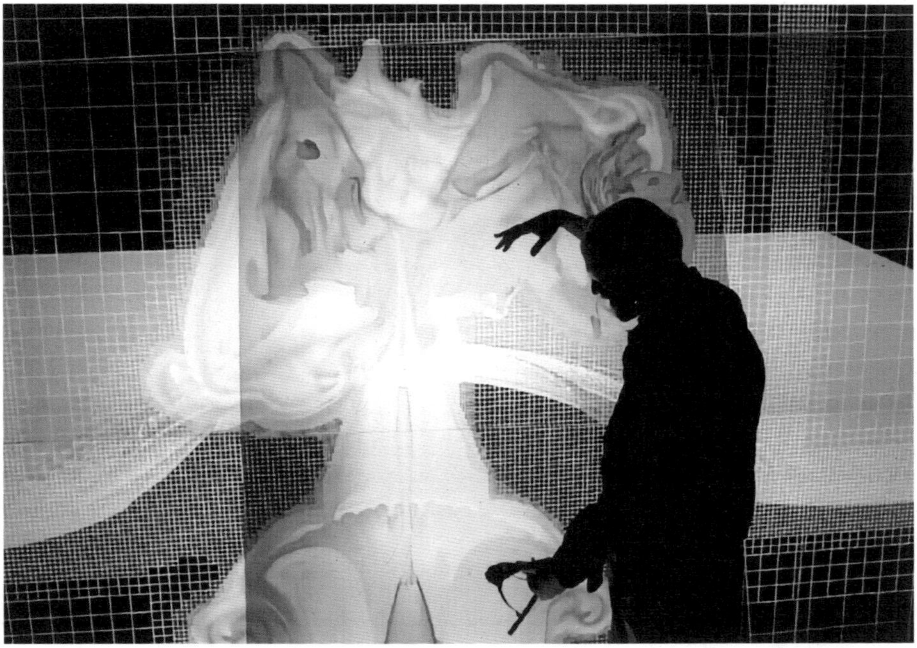

5.2 Manipulating high-resolution data in a virtual cave simulation at Los Alamos National Laboratory.

alone.[5] Indeed, the technoscientific challenge of the shake has only recently moved from a concept to a capacity, as the cumulative knowledge of the US nuclear program can now be projected into virtual reality spaces, allowing weapons scientists to interact with the exploding bomb one simulated shake at a time (see Figure 5.2). Moreover, there is increasing enthusiasm in the US for completing the Manhattan Project not by getting rid of the bomb but by rejuvenating and rebuilding the US nuclear complex, producing twenty-first-century nuclear weapons that could sit on the shelf for decades without degradation and still be used militarily at a moment's notice.

The Department of Energy has committed to rebuilding the US nuclear complex over the next twenty years to the projected tune of at least $1.5 trillion. This plan involves the first entirely new weapons designs since the 1980s and is an effort to create a nuclear arsenal and production program that can last through the end of the twenty-first century. These planned weapon systems will be less complex mechanically and more robust than the Cold War designs in the current arsenal (which have been painstakingly maintained part by part since 1992). They will also employ a new generation of weapon scientists through mid-century and will be loaded into a new triad of bombers, missiles, and submarines. Understanding the shake in a serious way means these designs will not have to be detonated, as did all Cold War systems, before being deployed into US military arsenals.

The promise of the virtual laboratory now points to a permanent nuclear production capacity in the US, one that can maintain a *de facto* test ban while also introducing new weapons. As the DOE's yearly programmatic report on "nuclear modernization" to the US Congress puts it: "By 2038, a new generation of weapons designers, code developers, experimentalists, and design and production engineers must demonstrate an understanding of nuclear weapons functionality using more predictive and more precisely calibrated computer-aided design and assessment tools than are possible today. High-fidelity experimental capabilities will produce quantitative data that preclude resumption of underground nuclear testing."[6]

Still shaking, after all these years. This twenty-first-century commitment to building new nuclear weapons and maintaining a state-of-the-art nuclear warfighting capacity is at odds with the Treaty on the Prohibition of Nuclear Weapons, which entered into force on 22 January 2021. Signed exclusively by non-nuclear states, the treaty adds nuclear weapons to the list of illegal weapons in international law. The contrast between this demand from non-nuclear states for immediate denuclearization and the modernization plans of existing nuclear states also reveals the continuing conceptual hold the bomb has on American society. For across seemingly opposed political regimes, four US presidents (Bush, Obama, Trump, and Biden) have each committed to re-energizing nuclear nationalism as the basis of American power, revealing, once again, the serious affective, technoscientific, and political recommitment to the shake.

In addition to the technoscientific and political investment in the shake, an equally serious military engineering commitment in the nuclear age has been to understand vibration and shock. Beginning in the mid-twentieth century, the test regimes from New Mexico to Nevada to Alaska and the Marshall Islands were orchestrated to a large extent around exploring the physical power of vibration and shock as a destructive force. Each experimental nuclear detonation was designed to simultaneously explore how blast, heat, and radiation would affect machines, structures, and biological specimens of all kinds – studied for both nuclear warfighting and civil defence purposes. Here vibration and shock were carefully engineered, examined, theorized, and managed by developing high-speed test cables capable of moving experimental information from the explosion itself to distant computers just slightly ahead of the destructive force itself. Moreover, diagnostic equipment was installed on giant springs throughout test sites so as to allow sensor arrays and equipment rooms to literally absorb the shock of the blast by bouncing in the air and riding the shifting land mass underneath, thereby gaining the precious milliseconds needed to transmit more data off-site. Indeed, the entire nuclear Cold War system could be read as a multigenerational project to predict and counter shock by engineering hardened but flexible structures. From nuclear test facilities to nuclear bunkers to ICBM silos scattered across the United States – all were built to shake and bounce, absorbing a high degree of physical shock during a nuclear attack. The most fortified structures ever engineered were thus also designed to be highly flexible: steel and concrete were assembled not only to protect but also literally to rock and roll. The North American Aerospace Defense Command (NORAD), for example, committed to 25-ton nuclear-blast-proof doors to protect its underground command and control facility in Colorado (see Figure 5.3); indeed, the entire structure was built on springs and could shift and sway to the hyperviolent tune of a nuclear attack (see Figure 5.4). In other words, the experimental politics of the shake produced a national infrastructure designed literally for shaking – with machines, and springs, and vibrational force the organizing principles of the nuclear state.

It is almost impossible today to describe the extraordinary conceptual, financial, and industrial energy empowering this vibrational system. Literally millions of experts, designers, engineers, military personnel, and politicians were coordinated through nuclear logics, need-to-know command-and-control systems, and the politics of shock. Put differently, the nervous system of the nuclear state has always also been quite nervous – a quivering universe of experts building systems for, and imaginatively rehearsing over and over again, apocalyptic scenarios. The serious study of shakes and shaking also involved quivering and worrying, threatening, and bullying on a global scale. As a total social formation, nuclear nationalism remade politics, science, environment, and public emotions, both informing and distorting how business, defence, education, geopolitics, and danger itself are understood.[7]

The affective universe produced by the nuclear revolution has always been one of its most powerful achievements. In 1945, the future became radically split for many between apocalyptic visions of nuclear war and utopian hopes for advanced technology and a world without war. Key moments of global nuclear terror – 1955, 1962, 1983, and 2001, to locate a few famous geopolitical emergency moments – turned the international order into a quivering system, beset by worst case scenarios and nightmare visions of the end. However, we tend to forget today how energizing the bomb was for activists of all political agendas. From the late 1950s on, the fight to denuclearize the world empowered and linked social movements devoted to peace, human rights, and the environment. Nuclear fear and the critique of Cold War militarism enabled vast networks of actors to see one another as allies, mobilizing alternative notions of society and the future in ways that have also foundationally remade American society across race, gender, justice, and the environment.[8]

Activists marched by the millions from the late 1950s through to the end of the Cold War, sympathetically linked by national security affects and the desire for a different and better world.[9] Some of these protests started small, as when a few women took their children to the park in New York in the late 1950s instead of practising the end of the world in civil defence drills. The Mothers Against Bombs

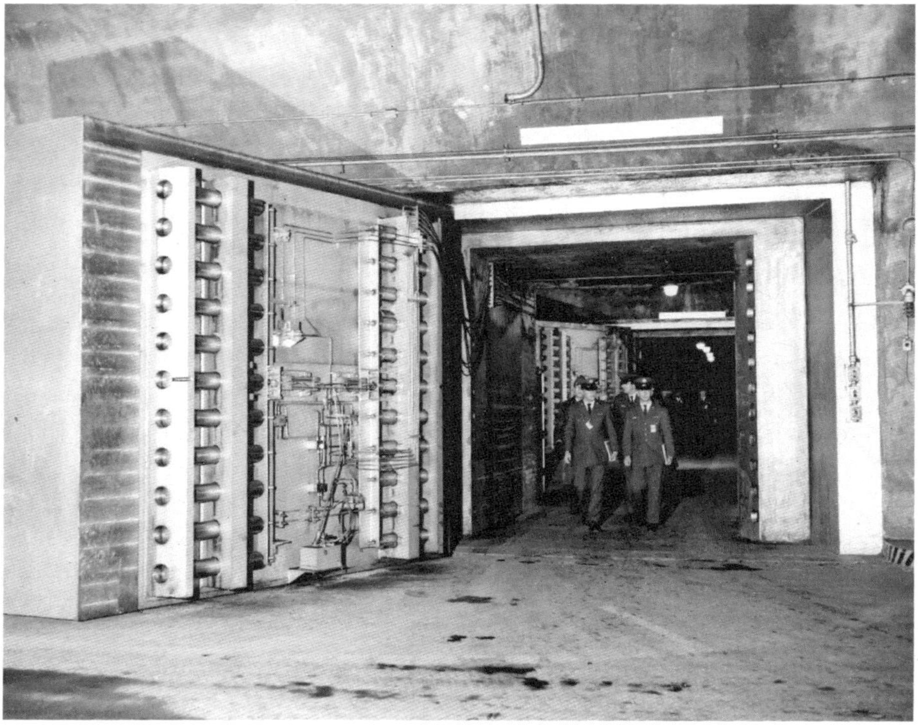

5.3 25-ton blast-proof security doors at NORAD, Cheyenne Mountain, Colorado.

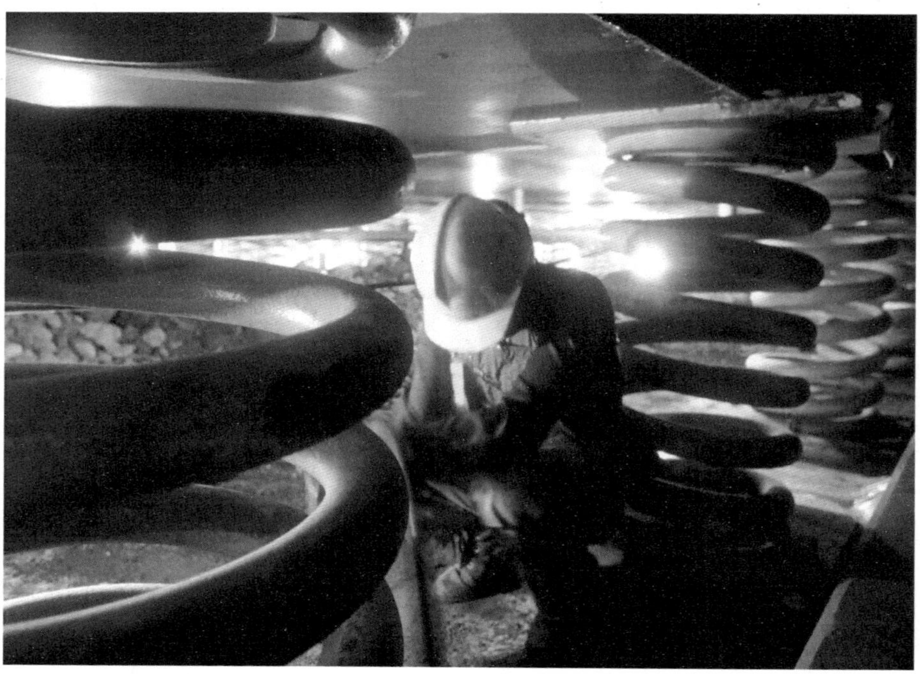

5.4 Springs supporting the NORAD facility in Colorado.

movement brought gender into anti-nuclear activism immediately, as did protests from inner-city groups that had noticed that atomic civil defence was focused almost exclusively on saving white suburban families rather than on a protecting a multiracial America, be it urban or rural. Indigenous, non-White and colonial subjects organized globally as their homelands were threatened by nuclear tests, uranium mining, and military expansion. Sexuality was always central to the Cold War system: officials prosecuted homosexuality as a national security threat, thus weaponizing sex in a world in which competing intelligence agencies not only sought information but also engaged heavily in espionage and covert actions. The politics of radioactive fallout generated similarly foundational concerns about the stability and toxicity of the global environment, energizing activists to protect not only Americans from total nuclear war but all living beings from the cumulative biological effects of the nuclear test programs themselves.

The nightmarish world of Cold War nuclear danger – which promised to end everything in a radioactive flash – invited everyone to think about the qualities of life, and the nature of politics and power, and to articulate their commitments to one another and the future. The bomb became both a symbol of American military power and a direct challenge to a political system increasingly organized around the official secrecy, covert actions, and end-times visions of nuclear nationalism. People from all walks of life sought to articulate the terms of a better way of living in the face of the Cold War system, a state-based emotional management project seeking to ground American-ness in a specific kind of naturalized militarism, technological determinism, and apocalypticism. Camping out in front of nuclear production sites and military bases, some activists devoted their lives to being visibly in opposition to the nuclear state on precisely these terms. Others sought to build alliances capable of supporting alternative visions of daily life that were not based on managing apocalyptic potentials as a standard mode of politics – opting out, tuning in, mobilizing, shouting, critiquing.

In addition to the protests and marches, and the arrests and acts of sabotage, there were explicit opportunities to shake it up. Consider the Give Peace a Dance project, an annual twenty-four-hour dance marathon organized in Seattle in the mid-1980s to raise money for local anti-nuclear groups. It was formally a "legs against arms" campaign, fighting Cold War doom and gloom with activism, social energy, and humour. Participants organized into costumed teams, collected sponsors, and danced through the night, trading turns on the dance floor to keep their team (and, one could say, anti-nuclear activism itself) alive. The project soon went global, with dance-a-thons for a nuclear-free world on multiple continents. Instead of offering end-of-the-world images, Give Peace a Dance promoted the idea of a denuclearized planet, one that could be free of the minute-to-minute danger of nuclear war.

Art Chantry's posters for the event immediately became iconic Cold War political statements and constitute the happiest anti-nuclear art ever produced.

For the 1986 event (see Figure 5.5), he created a montage of dance step diagrams (the fox trot, monkey, twist, polka) overlaid against an atomic bomb (marked "the big reason"). Playing off of John Lennon's antiwar song "Give Peace a Chance" (introduced during his famous "Bed-Ins for Peace" protests with Yoko Ono in 1969), the dance marathon suggested that more collective agency is needed to remake the world, inviting participants to dance their way to a better geopolitics. For the following year's dance marathon (see Figure 5.6), Chantry notoriously offered a waltzing, cross-dressing Reagan and Gorbachev, presenting them as a couple in a spectacular (anti)nuclear dance. These posters play on the October 1986 summit in Reykjavik, Iceland, where Reagan and Gorbachev, as leaders of the United States and Soviet Union, discussed a radical end to the Cold War arms race – namely, the complete elimination of nuclear weapons. Reagan's stubborn commitment to an imaginary space-based laser shield against intercontinental ballistic missiles (known as the Strategic Defense Initiative, or Star Wars) killed the possibility not only of an end to the arms race but also of an affirmative global project of immediate denuclearization. The US has spent hundreds of billions of dollars on space-based missile defence over the past three decades but has not yet produced a viable technology, leaving the world with 14,000 known nuclear weapons today. Chantry's posters are subversive not only for presenting Reagan and Gorbachev as cross-dressers but also for suggesting that an arms race requires a partner, just as peace requires a dance team.

For his 1988 poster, Chantry literally changed the frame on everyday reality (see Figure 5.7). A jitterbugging couple, ecstatic in their physical and musical connection, literally kick the bomb out of the frame, creating a disjointed view that points to a new world almost in sight. Rejecting the rectangular format of the standard poster, Chantry here demands that we change our perception, break with the convention, and ignore the rules of the Cold War system. Inviting viewers to "kick the habit!" and "vote with your feet!," the poster promises not just a less violent world but also a happier one without nuclear weapons. The kinetic energy of dancing and denuclearizing at the same time is beautifully illustrated as a world-view-altering process, one in which participants are not isolated in the despair of Cold War apocalypticism but rather animated by social action and an elimination of collective dangers. Here the good vibrations involve the coming together of people across class and race, gender and sexuality, to confront military states and argue for a better social contract and an alternative notion of security. The Give Peace a Dance project is the flipside to the international seismic signalling of nuclear test detonations, a world of technologies, bodies, and affects remade by both fear and promise, sympathetically organizing on behalf of positive futurities. The excitability of the national subject in the nuclear age – under forms of terror, outrage, and hope – has been as powerful as the technology itself in reshaping American society.

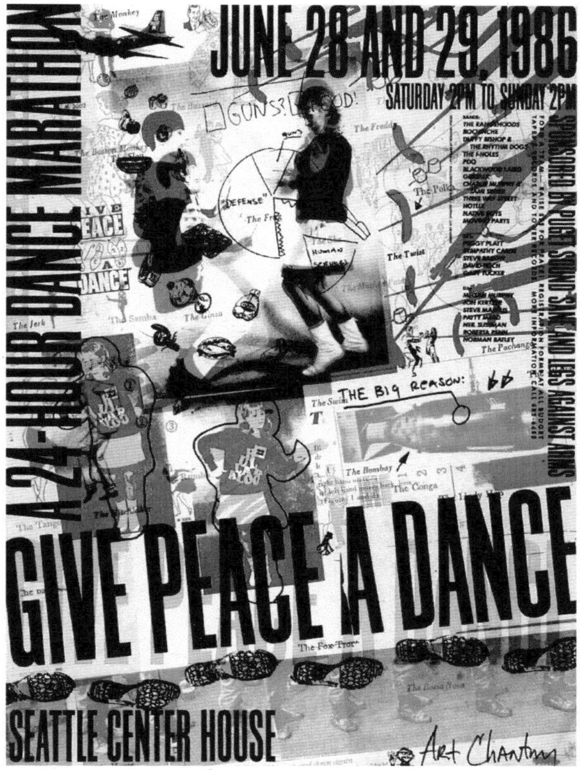

5.5 Art Chantry's 1986 Give Peace a Dance nuclear disarmament poster.

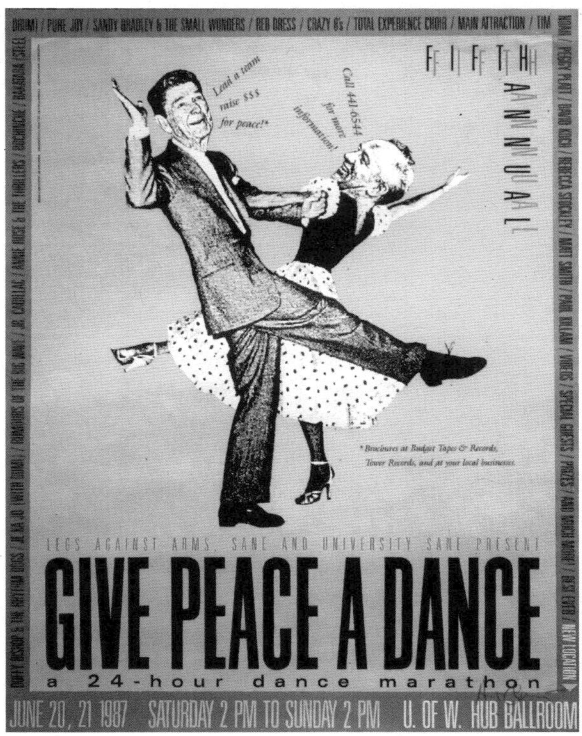

5.6 Art Chantry's 1987 Give Peace a Dance posters with Reagan and Gorbachev dancing nuclear weapons away.

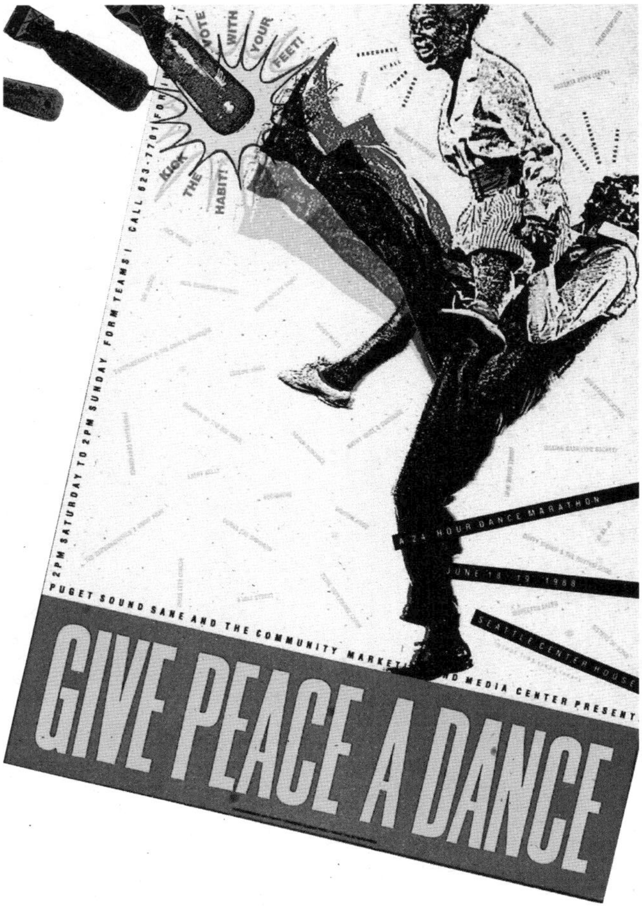

5.7 Art Chantry's 1988 Give Peace a Dance poster, busting out of the nuclear warfare frame.

Well into the third decade of the US war on terror, we might well ask about the state of our national security excitations. A vast new national security apparatus has been built over the past generation in the name of ending terror. The investment in counterterror has not only created multiple war fronts and a high body count but also entirely new departments and disciplines (homeland security, cybersecurity, biosecurity, to name only a few). Indeed, the counterterror state has expanded its reach on a scale only matched by that of the early Cold War nuclear state. But what now of the activisms? In February 2003, a synchronized global march to protest the imminent US invasion of Iraq showed that the political energy is still there, as is the capacity of people to unify across social and national divisions, and to do so on record terms. However, while linking a vast set of organizations, this antiwar march produced a profound moment but not a lasting political movement. But then, the massive protests of the Cold War system only really emerged about fifteen years into the US–Soviet nuclear confrontation. So perhaps it takes a new generation of people raised under the radical conditions of a new kind of war to see the emerging future and decide they want a different one. If the twentieth century offers any guide, we know that these kinds of social vibrations can start small and amplify to become a serious political force (of the kind that drove Reagan, the ultimate Cold Warrior, to meet with Gorbachev in Iceland in the first place to consider global denuclearization).[10] Perhaps then, just around the corner but coming into view, the negative affects of counterterror and resurgent nuclear nationalisms will be matched by a renewed commitment to social organizing and peaceful future-building. In other words, to shaking it in a positive key for a less violent world – now, wouldn't that be a good vibration?

NOTES

This chapter is adapted from Joseph Masco, *The Future of Fallout, and Other Episodes in Radioactive World-Making* (Durham: Duke University Press, 2021).

1 McPhee, *The Curve of Binding Energy*, 115; Glasstone and Dolan, *The Effects of Nuclear Weapons*, 17.
2 Waters et al., "The Anthropocene."
3 Heller, "Seismic Monitoring."
4 Masco, *The Nuclear Borderlands.*
5 Schwartz, *Atomic Audit*
6 DOE, *Fiscal Year 2014*, 1–6.
7 Masco, *The Theater of Operations.*
8 Zaretsky, *Radiation Nation.*
9 Wittner, *Confronting the Bomb.*
10 Fitzgerald, *Way Out There.*

BIBLIOGRAPHY

DOE (US Department of Energy). *Fiscal Year 2014: Stockpile Stewardship and Management Plan*. Washington, DC: GPO, 2013.

Fitzgerald, Frances. *Way Out There in the Blue: Reagan, Star Wars, and the End of the Cold War.* New York: Touchstone, 2000.

Glasstone, Samuel, and Philip J. Dolan, eds. *The Effects of Nuclear Weapons*. Washington, DC: Department of Defense, 1977.

Heller, Arnie. "Seismic Monitoring Techniques Put to a Test." *Science and Technology Review*, April 1999, 18–20.

Masco, Joseph. *The Nuclear Borderlands: The Manhattan Project in Post–Cold War New Mexico*. Princeton: Princeton University Press, 2006.

– *The Theater of Operations: National Security Affect from the Cold War to the War on Terror.* Durham: Duke University Press, 2014.

McPhee, John. *The Curve of Binding Energy*. New York: Farrar, Straus and Giroux, 1973.

Schwartz, Stephen. *Atomic Audit: The Costs and Consequences of U.S. Nuclear Weapons since 1940*. Washington, DC: Brookings Institution Press, 1998.

Waters, Colin N., et al. "The Anthropocene Is Functionally and Stratigraphically Distinct from the Holocene." *Science* 351, no. 6269 (2016): 137–47.

Wittner, Lawrence S. *Confronting the Bomb: A Short History of the World Nuclear Disarmament Movement*. Stanford: Stanford University Press, 2009.

Zaretsky, Natasha. *Radiation Nation: Three Mile Island and the Political Transformation of the 1970s*. New York: Columbia University Press, 2018.

6

What Is the Matter with Nuclear Weapons Communication?

BRYAN C. TAYLOR

> Bodies will have to pay for their excesses of language.
>
> ALAIN BADIOU

> Mama always told me not to look into the eyes of the sun
> But Mama, that's where the fun is.
>
> BRUCE SPRINGSTEEN

Future historians will not cite the 2016 US presidential election as a high-water mark of eloquent political oratory or robust democratic deliberation. Instead, that campaign unfolded in a series of increasingly disturbing eruptions of populist authoritarianism. Its victor voiced a potent vernacular discourse, marked by exceptionalism, misogyny, racism, fearmongering, and an arrogant imperviousness to accountability. Remarkably, this discourse transcended mere partisanship to offend norms associated with the American political traditions of both liberal democracy and civic republicanism.

Amid this furore, the concerns of nuclear disarmament, non-proliferation, and environmentalism languished. In their primary and post-nomination debates, the presidential candidates consistently failed to address important issues surrounding current US policies, such as weapons modernization and radioactive waste disposal. Instead, they invoked nuclear weapons in an offhand, clichéd manner – as a measure of their gritty reliability in managing security threats, and an illegitimate enterprise pursued by *other* "rogue" nations. Additionally, nuclear experts likely winced at Donald Trump's steep learning curve concerning the structured qualities of nuclear deterrence. In a revealing series of naive statements,

Trump expressed amazement concerning any need for US nuclear restraint: "*If we have them, why can't we use them?*"[1] He further threatened shibboleths of the global nuclear order by questioning continued US participation in the NATO alliance and by issuing vague declarations concerning a renewed arms race with Russia.[2]

I offer these events as a means of approaching the central concern of this chapter: *the relationship between materiality and symbolism in nuclear weapons communication*. That is, these events remind us that that this relationship is thoroughly entangled, highly fraught, and underconsidered.

To illustrate, in the first case we see Trump stumbling over the central paradox of deterrence: his nation's most powerful weapons are militarily useless. However, the significance of this exceeds the surface concern of whether a speaker has accurately understood and represented a policy. Instead, Trump's statements were controversial because they disrupted cultural conventions for the public expression by political authorities of nuclear knowledge. That is, it was not merely that Trump expressed this frustration: it was that he did so openly. In doing so, he violated traditional protocol dictating that US presidents speaking to a general audience should neither idly endorse nuclear warfighting nor decline the nation's entitlement to first use of nuclear weapons. Commentators noted that this breach was also disturbing because, since the declared end of the Cold War, American politicians and citizens have largely maintained a regime of apathetic silence and inarticulateness surrounding the persistence of mutually assured destruction.[3] Or, to use a (materialist) metaphor: The capacity for nuclear deliberation among the body politic has atrophied alarmingly.

In the second case, Trump's impulsive promise to "expand [America's] nuclear capability" (thus countering the work of previous administrations to reduce the numbers and salience of nuclear weapons in US defence policy) was controversial in part because of the means by which it was delivered: via Twitter. Here, public perception of nuclear decorum was tied to the properties of social media: Trump's (apparently unvetted) posting of a nuclear policy vision in an electronic, 140-character-limit format was, to say the least, novel.

What both of these examples illustrate is how material concerns such as *format*, *time*, *place*, *design*, and *effectivity* saturate the performance of nuclear weapons communication. In this chapter, I seek to advance understanding of these issues, in the service of two goals. The first is to clarify the distinctive concerns of communication studies within the nuclear-environmental humanities. The second is to empower related critique by advancing current modes of conceptualizing, evaluating, and intervening in entangled nuclear weapons ontology.

I do so in the following sequence of portraits. First, I offer two stories of *communication* – one emphasizing its discursive qualities, the other its materiality. Subsequently, I do the same for nuclear weapons. I then extend this pattern by presenting two accounts of the relationship *between* the hybrid ontologies of communication and nuclear weapons. The first account emphasizes their relative *independence*, the second their *interdependence*. My goal in presenting this account is to provide nuclear critics with resources for engaging the onto-political complexity of public nuclear discourse. That complexity requires a transcendence of traditional orientations toward communicative registers of "mere" semantics, grammar, or pragmatics. As an alternative, we require enhanced critical means to engage how materiality and symbolism suffuse both communication and nuclear weapons and thus structure their articulation.

Two Stories of Communication

Conventional images of communication in modern Western society have been shaped by diverse theoretical, philosophical, and theological legacies, ranging from Christian spiritualism and liberal individualism to the technocratic engineering of information systems.[4] While it is possible to distinguish these contributing components, actual cultural communication proceeds under the influence of their amalgamation.

THE DISCURSIVITY OF COMMUNICATION

One hegemonic cluster of onto-epistemologies in this history blends the premises of idealism, representationalism, subjectivism, and transcendentalism. Depicted in this tradition, communication is the primarily linguistic process of connecting human minds (which are presumed distinct from bodies) through the exchange of verbal and written symbols. It emphasizes the reliance of communicators on these symbols to express their private understandings and intentions. It also situates them as users of pre-existing cultural codes, whose rules for the contextualized linkage of signifiers and signifieds structure communicators' access to – and competent use of – those symbols.

The primary concern (both practical and theoretical) in this tradition involves the collaborative accomplishment by individual speakers of intersubjectivity, defined as their co-orientation of personal meanings. While this tradition idealizes the authenticity of immediate (i.e., oral, face-to-face) communication, it also acknowledges the saturation of modern society in mediated communication by appropriating semiotics and cybernetics. It uses these traditions to problematize the contingency of messages transmitted in mechanical and electronic channels. It cautions that the fidelity of meaning encoded by sources in those messages may be compromised by poor design, systemic noise, or incompetent decoding by receivers. Uncertainty and ambiguity arising from these contingencies are viewed as deficiencies and obstructions to successful communication.

Additionally in this view, the voluntary and undistorted expression of personal meanings between speakers (e.g., in dialogue) is considered to serve the beneficial (if not therapeutic) goals of affirmed identities and enhanced relational bonds. As communication expands in scale, this process changes in degree but not kind. Audience members are affected by sources' design and delivery of messages, which offer them new information and psychologically influence their attitudes, values, and beliefs. In this view, the integrity and viability of a liberal democratic society are inextricable from the quality of its communication processes (e.g., the ability of citizens to provide "feedback" to official messages).

Undoubtedly, this logocentric and hermeneutic account of communication offers benefits. Indeed, it has elevated the status of "communication" to a persistent, defining (although perhaps irresolvable) concern of popular culture. Nonetheless, it also creates significant liabilities. This is partly because its concerns (i.e., with meaning) are commonly associated with intangible, ideational phenomena as well as qualities of transience, arbitrariness, and inconsistency. In Western metaphysics, which has historically been dominated by the logics of empiricism, dualism, and rationality, these concerns have been opposed and subordinated to those of *materiality*. Used here, this term designates those worldly forms and processes that appear to enjoy a *prior, objective, tangible*, and *embedded* existence in our everyday lives. We typically characterize material phenomena as being relatively *solid* (persistent in form), *independent* (not reliant for their being on human perception), *obdurate* (resistant to human will), and *determinate* (i.e., externalizing an irresistible force, and imposing it on human activity). In these logics, materiality is prioritized as an enduring realm of the immanent Real that may be indexed – but not fully manifested – by mere derivative discourse.

THE MATERIALITY OF COMMUNICATION

So let us now consider a different story of communication – one that incorporates the eclectic premises of socio-materialism, including agential realism, biopower, flat ontology, media ecology, political economy, feminist psychoanalysis, and transhumanism.[5] Space here does not permit me to delineate the alternately competitive and complementary relations between these traditions. Instead, I seek to synthesize their central themes to develop a plausible construct that facilitates the purpose of this chapter. In this materialist story of communication, subsequently, we are called upon to reject entrenched habits of privileging disembodiment and humanism. We are reminded that communication is *always* material, in that it must assume some physical form, requires some kind of biological, mechanical, or electrical infrastructure for its production and circulation, and is thus organized by natural, institutional, and technological forces.

In this view, our apparently human-authored performances of communication *actually* depend upon, and coexist with, the plural and simultaneous agencies of other entities, including *objects, bodies, species, places, technologies, environments*, and even *spirits*. Communication is also shaped by extradiscursive phenomena of the human psyche (e.g., unconscious desire), which both predate and haunt the subject's entry into cultural regimes of language and discourse. These agencies manifest themselves as these entities *participate* in our communication through organic, unpredictable, and influential forms of *sentience*, *expression*, *mediation*, and *collaboration*. Thus, we should no longer suppress and minimize the significance of this materiality in explaining communication, as if it was mere "context" for more important "textuality." We should instead depict ongoing "assemblages" of materiality and discourse, which emerge among and between heterogeneous networks and economies, and which may serve to temporarily stabilize their flux, thus facilitating coherent and productive exchange. That communication subsequently appears as a dynamic and deeply contingent phenomenon bristling with intensity, relationality, temporality, and imbrication. In this view, communication may be a kind of construction, but not one that is singular, autonomous, determinate, or complete. Instead, communication may be considered "the ongoing, situated and embodied process whereby human and non-human agencies interpenetrate ideation and materiality toward meanings that are tangible and axial" for specific identities, relationships, and communities.[6]

Obviously, the distinction developed in these two stories between communication's discursivity and materiality emphasizes their incongruity. As new materialist theory reminds us, however, actual phenomenal entanglements have no need for such distinctions, and we should thus reflect on how and why we make them. Here, I use these stories to serve the heuristic purpose of conceptualizing phenomena that are entangled in nuclear weapons communication.

Two Stories of Nuclear Weapons

We proceed now to apply this heuristic to the phenomena of nuclear weapons. Here, we reverse the previous order of the stories and consider first the conventional account of nuclear *materiality*.

THE MATERIALITY OF NUCLEAR WEAPONS

At its most basic level, the development of nuclear weapons relies upon physical manipulation of the energetic potential in matter. In different ways, the processes of nuclear fission and fusion are stimulated by human penetration of the elemental structure of uranium and plutonium. These interventions may generate a state of "critical mass" that liberates certain types of subatomic particles at a rate sufficient to produce a self-sustaining reaction. Under the right conditions, this reaction releases enormous amounts of energy, which manifests itself in three hellish registers: ionizing radiation in the form of gamma- and X-rays; incinerating heat capable of generating wide-area wildfires; and a concussive blast. The metrics of this power are astonishing: a nuclear detonation occurs with blinding speed and explosive force that are typically measured on a scale of *millions of tons* of conventional TNT. Currently, the global nuclear arsenal is estimated to include around *15,000* warheads; their combined explosive yield is close to *4,000* megatons of TNT. For this reason, it is projected that total nuclear war would permanently destroy the planet as a site of human habitation – never mind civilization (e.g., by creating long-term high-altitude dust clouds that block sunlight, thus ensuring the failure of food production systems). Here, evidence of the somatic and environmental effects gathered from the only cases of actual nuclear-military usage – the US attacks on Hiroshima and Nagasaki conducted in August 1945 – has played a crucial role.

Let us now turn outward to another type of nuclear materiality. As theorized by structural realists, the international political system is fundamentally anarchic, in that it possesses no overarching authority to regulate the security of nation-states.[7] Those states must therefore remain suspended in the imperative of self-help, by which they must develop capabilities to secure both their home territory and their international projection of valued interests (military bases, alliances, etc.). As nations arm themselves for this purpose, they are subject to an inherent "security dilemma."[8] This is because most military technology can be used for either offensive or defensive purposes. So it is not possible for nations to use those weapons to unambiguously express – or reliably interpret – their intentions toward one another. As a result, nations engaged in assessing threats must favour concrete evidence of their adversaries' *motives* and *capabilities* (i.e., rather than their unreliable expressions of intention). Insecure nations will subsequently prioritize the imperatives of technological innovation and escalation. So as not to be victimized by an adversary's adoption of technological innovation, they must do so first, even though it may provoke that adversary to respond in kind, and even when they did not originally intend to do so.

Continuing this realist narrative of nuclear materiality, states will develop military technology that corresponds with their perception of threats and their motives for projecting power. Nations that exist under high degrees of threat, and/or who desire to project global stature, will thus seek to develop the most powerful military weapons possible. Historically, at least nine states have been able to cultivate the various capabilities (e.g., scientific, technological, engineering, and industrial) required for producing nuclear weapons – including their design, testing, manufacturing, and disassembly. This undertaking is massively complex, intrusive, dangerous, and costly. It requires nuclear governments to traumatically disrupt both natural environments and Indigenous cultures, develop new institutions (e.g., of logistical support and regulatory oversight), expend considerable national treasure at the expense of other needs, and systematically manage domestic and international public opinion in order to maintain their presumption of authority and legitimacy.

Correspondingly, nuclear states must also develop specialized military and defence-related infrastructures for deploying and maintaining these weapons, for strategizing their usage, and for the reliable exercise of command that balances conventional needs for civilian control with military needs for operational autonomy. A significant function of that infrastructure involves remote monitoring and timely response to the physical signs of adversaries' nuclear operations, including infrastructure engineering, airborne emissions, unnatural geological activity, military unit deployments, and – most feared – actual launch of an attack. While nuclear states may or may not choose to declare their nuclear capability, this surveillance provides evidence supporting official claims concerning whether and when they have "crossed the threshold."[9]

The possession of viable nuclear weapons by a nation enrols it in the condition of strategic deterrence, whereby adversaries are discouraged from making or acting on threats for fear of nuclear retaliation. Sufficient development by a nation of large varieties and numbers of nuclear weapons extends this condition to create an "in-kind" stalemate with other nuclear nations. Specifically, they are discouraged from launching a nuclear attack for fear that their first strike will not overwhelm an adversary's defences and destroy its nuclear forces before that adversary can use its surviving nuclear capabilities to inflict unacceptable levels of damage on that attacker. The projected speed and force of these attacks has required nuclear states to create semi-automated systems for threat assessment and related decision-making. Nuclear states must thus remain vigilant against accidental nuclear launch or other kinds of mistakes that could trigger a disastrous cascade of irreversible actions among densely interconnected human and non-human systems. An equally urgent concern involves unauthorized diversion or theft of stored nuclear materials that increases the number and range of nuclear-capable actors – particularly, sub-state and terrorist groups.

Historically, the international development of nuclear weapons has created numerous material legacies for the global spheres of worker safety, public health, and environmental integrity.[10] In North America, for example, that development relied on the dangerous labour of Indigenous groups, who mined radioactive materials and transported them to top-secret industrial sites for refining and weaponization.[11] In the Soviet Union, those facilities were built with the forced labour of prisoners. Combined with the radioactive fallout created by thousands of aboveground nuclear tests, those activities have produced generally undeniable effects, albeit highly contested ones in the specific. These effects arise from enormous, uncontrollable (and thus unknown) emissions of toxic and radioactive materials into the surrounding air, soil, and water. These emissions have subsequently been deposited in those natural ecosystems, continuously contaminating a diverse, ever-growing range of bodies and sites. Nuclear weapons development has also produced more formally processed streams of waste, which have been variously packaged and transported to temporary storage sites, where their installation initiates yet another set of effects.

Since the end of the Cold War, affected groups of nuclear workers and "downwinders" have been increasingly outspoken in asserting their claims for recognition and compensation, leading to scientific and political controversies surrounding the collection, analysis, and litigation of related evidence (e.g., dosimetry models of radiation exposure thresholds). Most nuclear governments have subsequently developed health and safety regimes surrounding nuclear weapons production – partly because of the staggering costs they may have to pay in mitigating its environmental and health-related impacts. These natural effects are both distinct from and related to the *political* effects of nuclear weapons development. Here, the Cold War history of nuclear deterrence instructs international audiences in a particular lesson: that nuclear weapons are a terrifying, effective means of inhibiting both direct nuclear attack as well as most forms of conventional conflict.

To summarize, this materialist story of nuclear weapons emphasizes conditions of necessity and relations of causality. It depicts nuclear weapons both as the *effect* of objective, inexorable forces and as their *agent* in producing effects. These effects are depicted as forceful and visceral – as sources of both private fear and public urgency. Their morbid significance is thickened by the evolving attribution of empirical qualities, such as number, size, weight, location, motion, speed, rate, accuracy, and explosive yield. The political gravity of nuclear weapons is subsequently tied to their determination of foreign policy and military strategy, as well as their apparent irreversibility as a condition of global (in)security.

THE DISCURSIVITY OF NUCLEAR WEAPONS

Alternatively, the *discursive* story of nuclear weapons emphasizes their qualities of ambiguity, contingency, and undecidability. It rejects the pursuit of nuclear certainty through objective analysis and embraces the fullness of nuclear weapons' existence across multiple modes of reality. Ironically, this story seems counter-intuitive precisely because the dreadful, numinous aura surrounding nuclear weapons stimulates in audiences a defensive preference for hyper-rationality.

The discursive story of nuclear weapons positions humans differently: as beings endowed with the capacity of speech and expression, who must call upon those resources to understand, interact with, and "manage" a radically defiant, existential threat of our own making. A principal intellectual grounding of this story is Jacques Derrida's famous 1984 manifesto for "nuclear criticism."[12] In that essay, Derrida argues that the apocalyptic quality of full-scale nuclear war sits in paradoxical relationship to cultural communication. Specifically, because that rupture-event of "remainderless destruction" has not (ever/yet) occurred, its actuality is not known in any comprehensive or reliable fashion. Indeed, it is arguably unrecordable, in that any attempt at documenting its occurrence would lead either to the destruction of the medium employed or to the production of survivable texts that remain forever unread because they lack surviving audiences. As a result – and while Derrida's position here is often exaggerated – nuclear war may be considered "fabulously textual." It is an artifact of ongoing discursive activity, organized through genres such as technical simulation, aesthetic lamentation, social movement protest, legislative deliberation, diplomatic negotiation, and news-journalistic coverage.

These discursive premises direct our attention to the various vocabularies, grammars, practices, and formations exchanged between speakers and audiences that establish the significance and implications of nuclear weapons for human existence. Here, we are plunged into a history of controversy in which various social and political groups have developed and deployed discourse to normalize their preferred images of a nuclear world (which has in turn advanced their general visions of order, governance, and security). While there can be little doubt that the nuclear states have succeeded in normalizing their development of nuclear weapons as a political necessity, that accomplishment has not always been inevitable or stable. Rather, it has been perpetually shadowed by popular protest, advanced in narratives whose plots emphasize the immorality of omnicide; the regressive politics of siting environmentally dangerous nuclear waste facilities among powerless communities; the tragic irony of states contaminating their citizens through the testing of nuclear weapons; and the needless cost and impracticality of maintaining a nuclear arsenal. Here, speakers *animate* nuclear culture by employing discursive strategies and techniques to secure the identification of audiences with their nuclear visions. In this ontology, successful outcomes of nuclear communication do not depend on a valid depiction of pre-existing, objective truth. Instead, speakers *actively produce* outcomes *with* situated audiences by drawing upon cultural conventions and rhetorical commonplaces to create nuclear stories that seem both plausible and compelling.

This insight leads us to investigate a variety of idealist nuclear phenomena, focusing on how they are generated *in* and *through* communication. Here, the hegemony of nuclear materialism is disrupted by counter-intuitive arguments – for example, that nuclear deterrence has sublimated the physical potential of nuclear weapons in favour of their symbolic function as a messaging apparatus whose only useful function is to be manipulated for communicative purposes.[13]

Nonetheless, the ambiguous, indeterminate qualities of this symbolism inspire a range of nuclear "thought-styles."[14] Among military and political officials, these styles are distinguished by varying attributions of utility to the weapons themselves, of rationality to opponents, and of communicability to nuclear policies and strategies. In these systems of cognition, particular displays of force may accomplish the didactic "coupling" of potential and actual conflicts for intended audiences and stabilize the outbreak of crisis. Related studies utilize constructivist and post-structuralist theories of language to emphasize the social and cultural underpinnings of nuclear phenomena such as *myths*[15] (i.e., non-rational but compelling belief systems held by nuclear policy-makers), *norms*[16] (e.g., of "non-use), *constructs*[17] (e.g., "proliferation"), *policies*,[18] and *valuing*.[19]

As a result, key claims concerning the identity and status of nuclear actors (e.g., as "weapons-possessing states") and events (e.g., "crisis"[20]) may be understood as motivated attributions. In this view, speakers' performance of these attributions is conditioned by several factors. These include their participation in the historical discourses of international politics, their positioning in cultural and institutional hierarchies, their use of arbitrary professional and technological conventions to resolve ambiguous evidence, and their investment of identity and interests in related narratives.[21] Indeed, this critique problematizes the apparent facticity of nuclear weapons "themselves" in arguing that their conception and development are predicated on ideological discourses that pre-establish their value and utility.[22] In this story, no *symbolic* dimension of nuclear weapons exists entirely *before*, *outside of*, or *apart from* communication's powers.

Two Stories of the Nuclear Weapons – Communication Relationship

It is not possible here to survey fully the modes of relationship between the materiality and discursivity of communication and of nuclear weapons. Potentially, those modes span an extraordinary range, from positivistic images of *causality* and *correlation*, to neo-Marxist images of *dialectics* and *(over)determination*, to constructivist images of *(co-)constitution*, *embedding*, and *mutual shaping*, to more postmodern and new materialist images of *articulation*, *assemblage*, and *hybridity*. Each of these images, of course, possesses both strengths and limitations in its ability to explain related entanglements. In the remainder of this chapter, however, I begin to develop that survey by employing yet another heuristic.

Specifically, I propose here a distinction between two ontologies of the communication/nuclear weapons relationship. The first ontology emphasizes the relative *independence* of their respective phenomena. The second emphasizes their relative *interdependence*. As we shall see, these ontologies organize distinct accounts of related research and provide different answers to critical questions. So oriented, I turn now to selectively review thirty years of scholarship on "nuclear weapons" published in the fields of rhetorical and communication studies.

THE RELATIVE INDEPENDENCE OF COMMUNICATION AND NUCLEAR WEAPONS

Considered according to an ontology of *independence*, communication and nuclear weapons comprise *separate* sets of phenomena. The associated epistemology here is one of *representation*: nuclear weapons are a worldly object whose distinctive qualities are depicted and interpreted by cultural actors through their use of communication codes. Key critical concerns involve the conditions surrounding those actors' practices of selecting, arranging, circulating, and consuming nuclear messages and texts. Here, we see three related patterns in the literature.

First, nuclear weapons are ominous and paradoxical *objects of representation* in the discourses of public culture. This is because the development of nuclear weapons has confounded binary distinctions underpinning key structures of collective meaning in the modern West. These distinctions, which guide conventional conceptions of governance and citizenship, include those made between *war* and *peace*, *attack* and *defence*, *rationality* and *irrationality*, *suicide* and *homicide*, *security* and *danger*, *military* and *civilian*, and *battlefield* and *home*. In stimulating this public disorientation, nuclear weapons *compel* agonistic communication by social and political groups, who seek alternately to rationalize and challenge their continued existence. This communication may be characterized and judged according to its use of particular forms of reasoning, as well as other cultural-symbolic resources (genres, archetypes, etc.). Nuclear speakers utilize these resources to induce identification by audiences with their underlying interests and to stimulate changes in audience members' thought, affect, and behaviour.[23] Nuclear weapons thus form *symbolic sites of struggle* in which their features, functions, and capabilities are highlighted and moralized through cultural discourses, which promote idealized, competing visions of governance and security.[24]

In a second scholarly tradition, nuclear weapons exemplify *the modern-era colonization of public deliberation* through instrumental discourses that promote the nominally rational – but thoroughly amoral – ends of *efficiency, effectiveness, certainty*, and *control*. Here, critics engage the documented communication of military and political elites to clarify and challenge their practices of nuclear administration. Five examples of these practices may serve to illustrate. The first involves the *teleological compulsion* (e.g., cultivated among nuclear scientists and defence planners) to "perfect" nuclear weapons as a technology of foreign policy and military strategy.[25] Our second example involves the use by nuclear officials of *technical reasoning*, the validity of which is premised exclusively on the superiority of professional knowledge, specialized techniques, and the didactic assertion of abstract "information."[26] A third example involves the official dialects of *domestication* (i.e., which utilize familiar metaphors and euphemisms to render nuclear objects innocuous) and *bureaucratization* (which employs alienating devices such as acronyms, maxims, and jargon to inhibit public reflection).[27] Fourth, nuclear officials commonly utilize an "engineering" paradigm in managing public deliberation as a hyperformal process of "transmission," in which citizen "input" on proposed policies is passively tolerated, routinely devalued, and targeted for corrective "education."[28] Fifth and finally, official nuclear discourse reflects a foundational attitude of Heideggerian *enframing*, which appropriates the integrity of the natural world as if a resource for human exploitation.[29] These critical findings establish that the historical infiltration and steering of public deliberation by technocratic discourse has succeeded in framing the ongoing development of nuclear weapons as both necessary and virtuous – and thus largely unassailable. As a result, nuclear weapons have achieved a quasi-autonomous status in contemporary society.[30]

In a third tradition of communication research, nuclear weapons exemplify cultural struggle over the *scientific* development of military technology. Here, critics focus on the use of ideological discourse that that frames the role of science in supporting the coercive projects of neoliberal globalization and US neo-imperialism.[31] The significance of nuclear weapons emerges from discourse that frames the scientific enterprise as alternatively noble, compromised, and redeemable. Nuclear officials, for example, commonly draw upon religious and cultural *mythos* to dramatize nuclear weapons as emblems of supernaturally bestowed favour and of national power and progress.[32] A related concern involves cultural commemoration of the historical figures involved in nuclear weapon production as instructive (i.e., both exemplary and cautionary) lessons concerning individual character, organizational culture, and the unforgiving politics of nuclear institutions.[33] This rhetoric, critics conclude, frequently distorts conditions of ambiguity, complexity, and uncertainty surrounding nuclear science in order to assert hegemonic narratives of loyal, compliant, and patriotic nuclear citizenship.[34] As David J. Tietge observes: "Nuclear devices of war are technological ends in themselves that ultimately cannot be used. As a result, *they take on a linguistic property* rather than an operational one, and *they manifest the symbolic heritage they draw forth* because of their destructive function."[35] Tietge's statement here represents an exemplary formulation of relative independence between nuclear weapons and communication. Specifically, while this formulation appears to suggest an entanglement of discourse and materiality, Tietge's analysis subsequently defaults to critique this "predominantly rhetorical weapon."[36] Nonetheless, we can sense in this formulation the instinctive striving of communicative explanation to transcend the constraints of a discursive paradigm.

To summarize, our first ontology reflects a traditional orientation in communication scholarship toward nuclear weapons as primarily objects, topics, and contexts of communication. Depicted thus, nuclear weapons are controversial "problems" of security whose potential destructiveness stimulates the strategic use of communication for purposes such as information and deliberation. That usage proceeds in distinctive practices such as *arguing, defining, framing, naming,* and *narrating.* Configurations of political and economic interests surrounding this process may *shape* (and be shaped by) the character of related communication (e.g., as spectacle or propaganda). In this view, however, nuclear weapons exist largely *apart* from communication, which remains a resource mobilized by nuclear society to serve its aesthetic and pragmatic needs.[37] Speakers subsequently *infuse* nuclear weapons phenomena with symbolic qualities such as *necessity, inevitability, legitimacy, priority, authority,* and *validity* (as well as their opposites).

While the power relationships surrounding this process are clearly not equitable, critique may nonetheless serve to deconstruct hegemonic discourses such as realism, exposing their reliance on rhetoric to create compelling – and logically self-sealing – stories of nuclear governance.[38] It may empower both non- and oppositional-scientific groups[39] as well as Indigenous stakeholders by modelling alternatives that promote inclusion, diversity, and mutual accountability. Their communication may subsequently restore the validity of those interests for undistorted policy deliberation (e.g., surrounding the siting of waste storage[40]).

THE RELATIVE INTERDEPENDENCE OF COMMUNICATION AND NUCLEAR WEAPONS

Let us now consider our second, contrasting ontology, which rejects the conventional separation of nuclear weapons and communication phenomena. It emphasizes, instead, their inextricable *connectedness* – if not fusion – based on demonstrable convergence in their infrastructures, functions, and politics. In published scholarship, these arguments have historically been the minority view, expressed in occasional and implicit fashion (e.g., the concession that nuclear test-verification technologies *themselves* function as symbolic tokens of political resolve, deployed by states pursuing arms control[41]). As we shall see, however, the critical use of (new) materialist traditions such as agential realism[42] and actor-network and practice theories[43] to explore this relationship is growing, both within and beyond communication studies. Here, we could draw for illustration on many examples, including animist intersubjectivity voiced in the deliberation of nuclear waste siting,[44] accretions of everyday life performance that constrain the emergence of nuclear publics,[45] and the intertwined decomposition of semiotic and radioactive materials that haunts official efforts to warn future humans concerning nuclear risk.[46] I will focus, however on two sites related to the evolving apparatus of US nuclear deterrence.

Inevitably, this "interdependent" ontology is drawn to that apparatus because (as discussed earlier) it has historically formed a vast, complex, and dangerous signalling enterprise. Of particular interest, however, is how this apparatus has recently expanded its scope to enrol new and different phenomena as relevant actors. Specifically, I am referring to a brief post–Cold War period in which nuclear officials contemplated redesign of the US nuclear "strategic triad," which had previously included the traditional "delivery platforms" of air force bombers, intercontinental missiles, and naval ballistic missile submarines. In these proposals, however, nuclear officials reformulated that triad to include (in addition to strategic missile defences and global-strike conventional weapons) the network of mines, factories, laboratories, and testing facilities that constitute the US nuclear weapons production complex.[47] Significantly, these proposals elevated that complex to the status of a "robust and responsive infrastructure,"[48] establishing its perceived viability as instrumental for the effectiveness of "deterrence by capability." In this mode of deterrence, a nation's *demonstrated potential* to reliably produce nuclear weapons forms a latent, but significant, influence on adversaries. As a result, a series of scandalous revelations during the late–Cold War and early post–Cold War periods concerning the decrepit and contaminated state of these facilities served to energize calls for their ongoing "modernization."

Importantly, these proposals were not endorsed to produce official redesign of the nuclear triad. Nonetheless, they show how official efforts to understand and manage the complex dialectic between nuclear discursivity and materiality continue to shape ongoing deliberations of nuclear deterrence strategy. In response, nuclear criticism can revise simplistic images of nuclear deterrence (e.g., as-if linear and direct "messaging") to more fully represent communicative concerns of situated interaction, performance, and relationship. One example here involves reconceptualizing nuclear weapons deployment as "non-linguistic communication" that theatrically "stages" preferred impressions of nuclear states and their adversaries for multiple audiences, including remote third parties.[49]

Additionally, we should consider how, as nuclear deterrence claims new phenomena for communicative relevance and functionality, its logical imperatives penetrate ever more deeply into the mundane forms and practices of institutional and cultural communication, restricting their possibilities. That is, this condition ensures that developments and activities (infrastructural decay; public protest, etc.) appearing to compromise the integrity of these components will be rearticulated by opponents from conventional frames such as fiscal prudence or free expression to assume the stigma of *facilitating existential threat*. Expressed more directly, nuclear officials cannot allow any of the signifying components of the nuclear deterrence apparatus to deviate from their ideal functioning, lest adversaries interpret this variance as compromising the projected appearance of capability and willingness required for all successful deterrence. Adopting this perspective helps explain why, for example, the severe legal punishments imposed in the US on nuclear activists who symbolically deface weapons sites and facilities (e.g., using blood, urine, and hammers) are so disproportional to their actual, negligible impairment of operations.[50] Indeed, some commentators argue that this historical dynamic extends to the very core of democratic deliberation in nuclear societies. Here, public anti-nuclear dissent has been demonized by the state as contributing to adversaries' perceptions of its reduced resolve to engage in nuclear warfare, and hence, to national insecurity.[51]

Let us return now to the case of the US nuclear weapons production complex, but with a different focus. Instead, we may consider recent interdisciplinary scholarship that depicts the conflation of nuclear weapons and communication in the nation's post–Cold War nuclear laboratories.[52] To review, the declared "end" of the Cold War – with its attendant development of treaties and policies constraining future nuclear weapons production – plunged these institutions into a period of extended, traumatic change, requiring nuclear weapons scientists to adapt their capabilities to new missions (e.g., counterproliferation). Collectively, these studies depict how this process dispelled the "black box" image of nuclear weapons as a discrete, *a priori*, autonomous technology. Instead, they depict "nuclear weapons" as a contingent, temporary – and primarily *communicative* – stabilization by techno-political actors of contingent, fluctuating, and inter-related material and discursive phenomena.

Specifically, these studies emphasize the role of communication media and practices in materializing the "tacit," embodied, and craft-related knowledge of nuclear weapons designers (whose numbers are rapidly dwindling). During the Cold War, this professional knowledge was maintained through ongoing oral and written communication between colleagues and by the spectacular ritual of explosive nuclear weapons testing. It is something of a postmodern irony here that, even as they lethally contaminated surrounding landscapes and bodies, these tests partly remained a form of simulation that both evoked and deferred the culminating "reality" of nuclear war. Nonetheless, the documentation and preservation of this expert knowledge has become an urgent matter for the US government, which is currently engaged in a program of "stockpile stewardship," wherein the safety, effectiveness, and reliability of the nuclear arsenal must be certified in the absence of verification traditionally obtained via explosive nuclear testing.[53]

In this process, the role of communication in materially formatting the recording, storage, transmission, and understanding of nuclear "information" (e.g., as videotaped oral history, algorithmic extrapolation of historical data) remains crucial to the state's management of two security issues. The first issue involves the undesirable possibility that, in the absence of explosive testing, nuclear weapon design knowledge might decay, gradually and irrevocably, to produce their *de facto* "uninvention." Here, "communication" becomes that set of evolving, hybrid processes by which professional knowledge is communally cultivated and supervised among nuclear weapons designers. As a result, ongoing "stockpile stewardship" is not merely an engineering project that maintains nuclear weapons, but one that is deeply implicated in theoretical debates concerning the co-constitutive relationships among discourse, knowledge, and technology.[54] Put more simply, communication developed in the stockpile stewardship regime does not so much *refer* to nuclear weapons as *constitute* them by *speculating about* and *projecting* their viability under scientific and political conditions that *prohibit* the generation of final, reliable nuclear knowledge. A second security issue here reverses the critical concern with insufficient communication to problematize the condition of *undesirable sufficiency*. This issue poses the question of *how much* of *which type* of formatted information – acquired *when*, *how*, and by *whom* – might facilitate the construction and use of even a rudimentary nuclear weapon design by unauthorized groups. Viewed through this prism of counterproliferation and anti–terrorism programs, "communication" becomes the totality of hybrid practices by which nuclear actors may access, acquire, appropriate, commodify, configure, innovate, and translate warhead design information, across space and time, to serve their purposes.

Conclusion

This chapter has covered considerable ground in arguing for enhanced critique of the entanglements of nuclear weapons and communication. It began by discussing events in the 2016 US presidential election as illustrations of this entanglement, and proceeded to develop heuristic narratives of the material and symbolic dimensions of communication and then nuclear weapons. I subsequently proposed two contrasting accounts of the relationship between nuclear weapons and communication, emphasizing their relative independence and interdependence.

Throughout, I have sought to provide a portrait of these relationships complementing recent discussion of communication and nuclear *power*.[55] As a "theoretical problematic," these scholars argue, that site distinctively illustrates how "nuclear phenomena challenge core concepts of social, rhetorical and discursive construction, defying both representational and constitutive models of communication."[56] Nonetheless, the entangled phenomena of nuclear weapons and power manifest "relationships among the recalcitrant material world, human institutions, and language."[57] As a result, these authors conclude, both sets of nuclear phenomena "warrant closer attention by communication scholars."[58] That attention may usefully be directed toward "dual-use" (i.e., civilian/military) entanglements, including regimes of "nuclear colonialism" in radioactive waste siting,[59] the nuclear fuel cycle,[60] anti-nuclear social movements,[61] the assessment and regulation of nuclear risk,[62] and issues of nuclear technology transfer and proliferation. In all these cases, the social, political, technological, and discursive aspects of nuclear phenomena, "are neither separable nor self-contained; they are complex, interconnected, recursive and always negotiated."[63]

Hopefully, this chapter has provided readers with resources to engage these concerns in critiquing the – admittedly, artificially delimited – sphere of *nuclear weapons*. That concession, of course, should not be misread as an attribution of either falsehood or triviality to that delimited sphere. Instead, it signals the productive work of human and non-human agencies in shaping nuclear communication to create significant and consequential differentiation. Future work here might proceed along at least three paths. The first involves rigorous comparison and contrast of theoretical positions in (new) materialism to depict their logics in conceptualizing material-symbolic articulations involving nuclear weapons and communication. A second path involves examining evolving paradigms of post–Cold War nuclear strategy (e.g., "tailored deterrence"[64]) to analyze how they conceptualize and administer the entanglements of nuclear weapons and communication. A final path mobilizes the primary theme of this volume by decolonizing traditional onto-epistemologies through which Western nuclear states dubiously assert the nuclear status of developing nations as a premise for sanctioning, containing, assimilating, and invading them. As a result, we may identify critical handholds and pressure points that allow us to disrupt the communicative politics that currently suspend us in the seemingly endless – and endlessly terrifying – condition of nuclear insecurity.

NOTES

1 Fisher, "Donald Trump."
2 Shear and Glanz, "Trump Says."
3 Fisher, "Donald Trump"; R. Meyer, "The Difficulty," 173–92.
4 Anderson, *Communication Theory*, *passim*; Cobley, "Communication"; Peters, *Speaking into the Air*, *passim*.
5 Aakhus et al., "Communication and Materiality"; Angus, "The Materiality"; Fox and Allred, "New Materialist"; Gillespie, Boczkowski, and Foot, *Media Technologies*, *passim*; Leonardi, Nardi, and Kallinikos, *Materiality and Organizing*, *passim*; Kuhn, "Introduction"; Ott and Domenico, "Conceptualizing Meaning"; Packer and Wiley, "Introduction"; Pfeiffer, "The Materiality."
6 Ashcraft, Kuhn, and Cooren, "Constitutional Amendments."
7 Martin, "The Continuing Value."
8 Herz, "Idealist Internationalism."
9 Hymans, "When Does a State."
10 Taylor and Kinsella, "Introduction."
11 Van Wyck, *Highway of the Atom*, *passim*.
12 Derrida, "No Apocalypse."
13 Dauber, "Better Red or Dead."
14 O'Neill, *Honor, Symbols, and War*, *passim*.
15 Lavoy, "Nuclear Myths."
16 Frey, "Nuclear Weapons"; Tannenwald, "The Nuclear Taboo," *passim*.
17 Mutimer, "Testing Times"; Woods, "Inventing Proliferation."
18 Moshirzadeh, "Discursive Foundations."
19 Ritchie, "Valuing and Devaluing."
20 Weldes, "The Cultural Production."

21 Bashi, "Examining the Discourse"; Hecht, "Globalization Meets Frankenstein?"; von Meier, Miller, and Keller, "The Disposition."
22 Angus and Cook, "Nuclear Technology."
23 Hogan, *The Nuclear Freeze Campaign*, *passim*; Kinsella, "One Hundred Years"; Mechling and Mechling, "The Campaign"; Taylor, "Nuclear Weapons"; Taylor and Kinsella, "Introduction."
24 Bjork, *The Strategic Defense Initiative*, *passim*; D. Meyer, "Framing National Security."
25 Brummett, "Perfection and the Bomb."
26 Bazerman, "Nuclear Information"; Farrell and Goodnight, "Accidental Rhetoric."
27 Schiappa, "The Rhetoric."
28 Katz and Miller, "The Low-Level Radioactive Waste."
29 Kinsella, "Heidegger and Being."
30 Krug, "Silence."
31 Mosco, "Star Wars"; Mosco and Foster, "Cyberspace."
32 Goodnight, "Ronald Reagan's Reformulation"; Rushing, "Ronald Reagan's 'Star Wars' Address."
33 Taylor, "The Politics" and "Organizing."
34 Mechling and Mechling, "The Atom"; Taylor, "Shooting Downwind"; Taylor and Hendry, "Insisting."
35 Tietge, *Flash Effect*, 148, emphasis added.
36 Ibid., 149.
37 Lukens-Bull and Woodward, "Israeli Nukes."
38 Beer and Hariman, "Realism and Rhetoric."
39 Slayton, "Discursive Choices.".
40 Endres, "From Wasteland" and "Sacred Land."
41 Schiappa, "The Rhetoric."
42 Anaïs and Walby, "Secrecy."
43 Bourne, "'Guns don't kill people'"; Pouliot, "The Materials."
44 Endres, "Sacred land," 328–45.
45 Pezzullo and Depoe, "Everyday Life."
46 Van Wyck, *Signs of Danger*, *passim*.
47 Guthe, "The Nuclear Posture."
48 Taylor, "A Hedge."
49 A. Gillespie, "Nuclear Brinkmanship."
50 Schlosser, "Break-in."
51 Taylor, "The Means."
52 Gusterson, *Nuclear Rites*, *passim*; MacKenzie and Spinardi, "Tacit Knowledge"; Masco, *The Nuclear Borderlands*, *passim*; McNamara, "TRUTH Is Generated HERE."
53 Taylor and Hendry, "Insisting."
54 Masco, "Nuclear Technoaesthetics," 367.
55 Kinsella et al., "Communicating Nuclear Power."
56 Ibid., 278.
57 Ibid.
58 Ibid.

59 Ibid., 286.

60 Ibid., 290.

61 Ibid., 294.

62 Ibid., 295.

63 Ibid., 291.

64 Lantis, "Strategic Culture," 467–85.

BIBLIOGRAPHY

Aakhus, Mark, Dawna Ballard, Andrew J. Flanagin, Timothy Kuhn, Paul Leonardi, Jennifer Mease, and Katherine Miller. "Communication and Materiality: A Conversation from the CM Café." *Communication Monographs* 78, no. 4 (2011): 557–68.

Anaïs, Seantel, and Kevin Walby. "Secrecy, Publicity, and the Bomb: Nuclear Publics and Objects of the Nevada Test Site, 1951–1992." *Cultural Studies* 30, no. 6 (2016): 1–20.

Anderson, James A. *Communication Theory: Epistemological Foundations*. New York: Guilford Press, 1996.

Angus, Ian. "The Materiality of Expression: Harold Innis' Communication Theory and the Discursive Turn in the Human Sciences." *Canadian Journal of Communication* 23, no. 1 (1998). https://cjc-online.ca/index.php/journal/article/viewArticle/1020/926.

Angus, Ian, and Peter G. Cook. "Nuclear Technology as Ideology." *Canadian Journal of Political and Social Theory* 11, nos. 1–2 (1987): 186–97.

Ashcraft, Karen Lee, Timothy R. Kuhn, and François Cooren. "Constitutional Amendments: 'Materializing' Organizational Communication." *Academy of Management Annals* 3, no. 1 (2009): 1–64.

Bashi, Roya Ghafele. "Examining the Discourse on Nuclear Weapons." *Disarmament Forum* 2, (2000): 77–86.

Bazerman, Charles. "Nuclear Information: One Rhetorical Moment in the Construction of the Information Age." *Written Communication* 18, no. 3 (2001): 259–95.

Beer, Francis, and Robert Hariman. "Realism and Rhetoric in International Relations." In *Post Realism: The Rhetorical Turn in International Relations* (1996): 1–30.

Bjork, Rebecca S. *The Strategic Defense Initiative: Symbolic Containment of the Nuclear Threat*. Albany: SUNY Press, 1992.

Bourne, Mike. "'Guns Don't Kill People, Cyborgs Do': A Latourian Provocation for Transformatory Arms Control and Disarmament." *Global Change, Peace, and Security* 24, no. 1 (2012): 141–63.

Brummett, Barry. "Perfection and the Bomb: Nuclear Weapons, Teleology, and Motives." *Journal of Communication* 39, no. 1 (1989): 85–95.

Cobley, Paul. "Communication: Definitions and Concepts." *The International Encyclopedia of Communication*. Edited by Wolfgang Donsbach. 2008. Blackwell Reference Online. http://www.blackwellreference.com.colorado.idm.oclc.org/subscriber/tocnode.html?id=g9781405131995_chunk_g97814051319958_ss72-1.

Dauber, Cori. "Better Red or Dead: Risk in a Nuclear Age." *Proceedings of the SCA Alta Conference on Argumentation* (1987), 137–42.

Derrida, Jacques. "No Apocalypse, Not Now (Full Speed Ahead, Seven Missiles, Seven Missives)." *diacritics* 14, no. 2 (1984): 20–31.

Endres, Danielle. "Animist Intersubjectivity as Argumentation: Western Shoshone and Southern Paiute Arguments against a Nuclear Waste Site at Yucca Mountain." *Argumentation* 27, no. 2 (2013): 183–200.

– "Sacred Land or National Sacrifice Zone: The Role of Values in the Yucca Mountain Participation Process." *Environmental Communication: A Journal of Nature and Culture* 6, no. 3 (2012): 328–345.

– "From Wasteland to Waste Site: The Role of Discourse in Nuclear Power's Environmental Injustices." *Local Environment* 14, no. 10 (2009): 917–37.

Farrell, Thomas B., and G. Thomas Goodnight. "Accidental Rhetoric: The Root Metaphors of Three Mile Island." *Communication Monographs* 48, no. 4 (1981): 271–300.

Fisher, Max. "Donald Trump, Perhaps Unwittingly, Exposes Paradox of Nuclear Arms." *New York Times*, 3 August 2016. https://nytimes.com/2016/08/04/world/donald-trump-nuclear-weapons.html.

Fox, Nick J., and Pam Alldred. "New Materialist Social Inquiry: Designs, Methods, and the Research-Assemblage." *International Journal of Social Research Methodology* 18, no. 4 (2015): 399–414.

Frey, Karsten. J. "Nuclear Weapons as Symbols: The Role of Norms in Nuclear Policy Making." IBEI Working Papers no. 3. Barcelona Institute of International Studies (2006): 1–19.

Gillespie, Alex. "Nuclear Brinkmanship: A Study in Non-Linguistic Communication." *Integrative Psychological and Behavioral Science* 47, no. 4 (2013): 492–508.

Gillespie, Tarleton, Pablo J. Boczkowski, and Kirsten A. Foot. *Media Technologies: Essays on Communication, Materiality, and Society.* Cambridge, MA: MIT Press, 2014.

Goodnight, G. Thomas. "Ronald Reagan's Re-Formulation of the Rhetoric of War: Analysis of the 'Zero Option,' 'Evil Empire,' and 'Star Wars' Addresses." *Quarterly Journal of Speech* 72, no. 4 (1986): 390–414.

Gusterson, Hugh. *Nuclear Rites: A Weapons Laboratory at the End of the Cold War.* Berkeley: University of California Press, 1998.

Guthe, Kurt. "The Nuclear Posture Review: How Is the 'New Triad' New?" Washington, DC: Center for Strategic and Budgetary Assessments (2002). http://www.militaerbibliothek.de/NRANEU/docs/R.20020729.Nuclear_Posture_Review.pdf.

Hecht, Gabrielle. "Globalization Meets Frankenstein? Reflections on Terrorism, Nuclearity, and Global Technopolitical Discourse." *History and Technology* 19, no. 1 (2003): 1–8.

Herz, John H. "Idealist Internationalism and the Security Dilemma." *World Politics* 2, no. 2 (1950): 157–80.

Hogan, J. Michael. *The Nuclear Freeze Campaign: Rhetoric and Foreign Policy in the Telepolitical Age.* East Lansing: Michigan State University Press, 1994.

Hymans, J.E C. "When Does a State Become a 'Nuclear Weapon State'?" *Nonproliferation Review* 17 (2010): 161–80.

Katz, Steven B., and Carolyn R. Miller. "The Low-Level Radioactive Waste Siting Controversy in North Carolina: Toward a Rhetorical Model of Risk Communication." *Green Culture: Environmental Rhetoric in Contemporary America* (1996): 111–40.

Kinsella, William J. "Heidegger and Being at the Hanford Reservation: Standing Reserve, Enframing, and Environmental Communication Theory." *Environmental Communication* 1, no. 2 (2007): 194–217.

– "One Hundred Years of Nuclear Discourse: Four Master Themes and Their Implications for Environmental Communication." *Environmental Communication Yearbook* 2 (2005): 49–72.

Kinsella, William J., Dorothy C. Andreas, and Danielle Endres. "Communicating Nuclear Power: A Programmatic Review." *Communication Yearbook* 39 (2015): 277–309.

Krug, Gary. "Silence, Oblivion, and Nuclear Criticism." *Journal of Communication* 45, no. 2 (1995): 204–9.

Kuhn, Timothy. "Introduction: Engaging Materiality, Communication, and Social Problems." In *Matters of Communication: Political, Cultural, and Technological Challenges to Communication Theorizing*, 1–10. New York: Hampton Press, 2011.

Lantis, Jeffrey S. "Strategic Culture and Tailored Deterrence: Bridging the Gap between Theory and Practice." *Contemporary Security Policy* 30, no. 3 (2009): 467–85.

Lavoy, Peter R. "Nuclear Myths and the Causes of Nuclear Proliferation." *Security Studies* 2, nos. 3–4 (1993): 192–212.

Leonardi, Paul M., Bonnie A. Nardi, and Jannis Kallinikos. *Materiality and Organizing: Social Interaction in a Technological World.* Oxford: Oxford University Press on Demand, 2012.

Lukens-Bull, Ronald, and Mark Woodward. "Israeli Nukes versus Palestinian Slingshots." *Consortium for Strategic Communication, Report 901*, no. 6 (2009): 1–13.

MacKenzie, Donald, and Graham Spinardi. "Tacit Knowledge, Weapons Design, and the Uninvention of Nuclear Weapons." *American Journal of Sociology* 101, no. 1 (1995): 44–99.

Martin, Susan B. "The Continuing Value of Nuclear Weapons: A Structural Realist Analysis." *Contemporary Security Policy* 34, no. 1 (2013): 174–94.

Masco, Joseph. *The Nuclear Borderlands: The Manhattan Project in Post-Cold War New Mexico.* Princeton: Princeton University Press, 2006.

– "Nuclear Technoaesthetics: Sensory Politics from Trinity to the Virtual Bomb in Los Alamos." *American Ethnologist* 31, no. 3 (2004): 349–73.

McNamara, Laura A. "TRUTH Is Generated HERE: Knowledge Loss and the Production of Nuclear Confidence in the Post–Cold War Era." *Nuclear Legacies: Communication, Controversy, and the US Nuclear Weapons Complex* (2007): 167–98.

Mechling, Elizabeth Walker, and Jay Mechling. "The Atom According to Disney." *Quarterly Journal of Speech* 81 (1995): 436–53.

– "The Campaign for Civil Defense and the Struggle to Naturalize the Bomb." *Western Journal of Communication* 55, no. 2 (1991): 105–33.

Meyer, David S. "Framing National Security: Elite Public Discourse on Nuclear Weapons during the Cold War." *Political Communication* 12, no. 2 (1995): 173–92.

Meyer, Robinson. "The Difficulty of Talking about Doomsday." *The Atlantic*, January 2016. https://www.theatlantic.com/technology/archive/2016/01/the-difficulty-of-talking-about-doomsday/431676.

Mosco, Vincent. "Star Wars Is Already Working." *Science as Culture* 1, no. 1 (1987): 12–34.

Mosco, Vincent, and Derek Foster. "Cyberspace and the End of Politics." *Journal of Communication Inquiry* 25, no. 3 (2001): 218–36.

Moshirzadeh, Homeira. "Discursive Foundations of Iran's Nuclear Policy." *Security Dialogue* 38, no. 4 (2007): 521–43.

Mutimer, David. "Testing Times: Of Nuclear Tests, Test Bans, and the Framing of Proliferation." *Contemporary Security Policy* 21, no. 1 (2000): 1–22.

O'Neill, Barry. *Honor, Symbols, and War.* Michigan: University of Michigan Press, 2001.

Ott, Brian, and Mary Domenico. "Conceptualizing Meaning in Communication Studies." In *A Century of Communication Studies: The Unfinished Conversation*, edited by Pat J. Gehrke and William M. Keith, 234–60. London: Routledge, 2014.

Packer, Jeremy, and Stephen B. Crofts Wiley. "Introduction: The Materiality of Communication." *Communication Matters: Materialist Approaches to Media, Mobility, and Networks* (2012): 3–16.

Peters, John Durham. *Speaking into the Air: A History of the Idea of Communication.* Chicago: University of Chicago Press, 1999.

Pezzullo, Phaedra, and Stephen Depoe. "Everyday Life and Death in a Nuclear World: Stories from Fernald." *Public Modalities: Rhetoric, Culture, Media, and the Shape of Public Life*, 85–108. Tuscaloosa: University of Alabama Press, 2010.

Pfeiffer, K. Ludwig. "The Materiality of Communication." In *Materialities of Communication,* edited by T. Lenoir and H.U. Gumbrecht, 1–12. Stanford: Stanford University Press, 1994.

Pouliot, Vincent. "The Materials of Practice: Nuclear Warheads, Rhetorical Commonplaces, and Committee Meetings in Russian–Atlantic Relations." *Cooperation and Conflict* 45, no. 3 (2010): 294–311.

Ritchie, Nick. "Valuing and Devaluing Nuclear Weapons." *Contemporary Security Policy* 34, no. 1 (2013): 146–73.

Rushing, Janice Hocker. "Ronald Reagan's 'Star Wars' Address: Mythic Containment of Technical Reasoning." *Quarterly Journal of Speech* 72, no. 4 (1986): 415–33.

Schiappa, Edward. "The Rhetoric of Nukespeak." *Communications Monographs* 56, no. 3 (1989): 253–72.

Schlosser, Eric. "Break-In at Y-12." *The New Yorker*, 9 March 2015. http://www.newyorker.com/magazine/2015/03/09/break-in-at-y-12.

Shear, Michael D., and James Glanz. "Trump Says the U.S. Should Expand Its Nuclear Capacity." *New York Times*, 22 December 2016. https://nytimes.com/2016/12/22/us/politics/trump-says-us-should-expand-its-nuclear-capability.html.

Slayton, Rebecca. "Discursive Choices: Boycotting Star Wars between Science and Politics." *Social Studies of Science* 37, no. 1 (2007): 27–66.

Tannenwald, Nina. "The Nuclear Taboo: The United States and the Normative Basis of Nuclear Non-Use." *International Organization* 53, no. 3 (1999): 433–68.

Taylor, Bryan C. "'A Hedge against the Future': The Post–Cold War Rhetoric of Nuclear Weapons Modernization." *Quarterly Journal of Speech* 96, no. 1 (2010): 1–24.

– "'The Means to Match Their Hatred': Nuclear Weapons, Rhetorical Democracy, and Presidential Discourse." *Presidential Studies Quarterly* 37, no. 4 (2007): 667–92.

– "Nuclear Weapons and Communication Studies: A Review Essay." *Western Journal of Communication* 62, no. 3 (1998): 300–15.

– "Organizing the 'Unknown Subject': Los Alamos, Espionage, and the Politics of Biography." *Quarterly Journal of Speech* 88, no. 1 (2002): 33–49.

– "The Politics of the Nuclear Text: Reading Robert Oppenheimer's Letters and Recollections." *Quarterly Journal of Speech* 78, no. 4 (1992): 429–49.

– "Shooting Downwind: Depicting the Radiated Body in Epidemiology and Documentary Photography." *Transgressing Scientific Discourses: Communication and the Voice of Other* (1997): 289–328.

Taylor, Bryan C., and Judith Hendry. "Insisting on Persisting: The Nuclear Rhetoric of 'Stockpile Stewardship.'" *Rhetoric and Public Affairs* 11, no. 2 (2008): 303–34.

Taylor, Bryan C., and William J. Kinsella. "Introduction: Linking Nuclear Legacies and Communication Studies." In *Nuclear Legacies: Communication, Controversy, and the US Nuclear Weapons Complex,* edited by Bryan C. Taylor, William J. Kinsella, Stephen P. Depoe, and Maribeth S. Metzler, 1–37. Lanham: Lexington Books, 2007.

Tietge, David J. *Flash Effect: Science and the Rhetorical Origins of Cold War America.* Athens: Ohio University Press, 2002.

Van Wyck, Peter. *Highway of the Atom.* Montreal and Kingston: McGill-Queen's University Press, 2010.

– *Signs of Danger: Waste, Trauma, and Nuclear Threat.* Minnesota: University of Minnesota Press, 2004.

Von Meier, Alexandra, Jennifer Lynn Miller, and Ann C. Keller. "The Disposition of Excess Weapons Plutonium: A Comparison of Three Narrative Contexts." *Nonproliferation Review* 5 (1998): 20–31.

Weldes, Jutta. "The Cultural Production of Crises: US Identity and Missiles in Cuba." *Cultures of Insecurity: States, Communities, and the Production of Danger* 14 (1999): 35–62.

Woods, Matthew. "Inventing Proliferation: The Creation and Preservation of the Inevitable Spread of Nuclear Weapons." *Review of International Affairs* 3, no. 3 (2004): 416–42.

Annotations in the margins provide an overlay of personal and academic reflections on the occasion of the agency's ten-year anniversary.

ABOUT THIS DOCUMENT

The National Toxic Land and Labor Conservation Service ("National TLC Service") is pleased to deliver this report summarizing the history of the agency, discussing our operations within the current state of atomic commemoration, and reporting real progress in fulfilling the long-term goal of collaboratively rerouting Cold War public memory. The last decade has witnessed periodic spikes in public concern over nuclear security and a steady increase in awareness of the toxic legacies faced by Indigenous communities alongside the largest new investments in warheads weapons delivery systems since the end of the Cold War.

OPERATIONAL DISCLOSURE

In a rare display of bipartisan cooperation and as testament to the value placed in our mission, U.S. Congress maintained the National TLC Service's annual appropriation of approximately $0 across three presidential administrations, from the agency's founding in 2011 through fiscal year 2021. This budget, supplemented by personal funds, the labor of underpaid research assistants and understanding spouses/partners, and myriad small grants and lecture funds from several U.S. universities,[1] permitted us to meet, and in some cases, exceed our organizational goals over the first years of operations. The improvisational skills honed through years of operating in the interstices between service provision, artistic expression, and scholarly research allowed agency personnel to work sporadically and opportunistically even through the depths of the Trump administration. The historic appointment of Deb Haaland (Laguna Pueblo) has placed a member of a uranium-affected Indigenous community at the helm of the Department of Interior. While it remains to be seen how much the inertia of statecraft might resist Haaland's agenda, the time is right for the non-Native provisional co-directors of the National TLC Service to step aside and offer the research and conceptual framework we have developed as offerings toward intellectual and material decolonization. This will therefore be our final report as interim co-directors, as we look toward handing the reins to an unknown successor.

Above: Logo of the National Toxic Land/ Labor Conservation Service.

7. THE NATIONAL TOXIC LAND/ LABOR CONSERVATION SERVICE

10-Year Final Report on Public Agency Organizing and Operational Responses to Cold War Legacies and the Nuclear Stockpile

SARAH KANOUSE & SHILOH KRUPAR,
INTERIM CO-DIRECTORS

BACKGROUND

The National TLC Service is a satiric but also sincere federal agency dedicated to the vigilant detection and exposition of military ecologies. Since the service was established in 2011, we have been hard at work collaboratively developing justice-oriented forms of public memory surrounding the staggering domestic impacts of the Cold War and U.S. atomic arsenal.

Few Americans realize the degree to which Cold War defense priorities live on through nuclear deployments, ongoing stockpile stewardship, and the containment and monitoring of toxic military residues.

- → The United States nuclear arsenal currently stands at around 3,700 – more than it possessed in 1955, with much greater lethality;[2]
- → The global active nuclear arsenal still stands at 9,500, with another 3,650 awaiting dismantling;[3]
- → The Obama administration began, and Trump intensified, a new era of investments in nuclear weapons and delivery systems, estimated in 2017 to total $1.2 trillion over 30 years;[4]
- → Since taking office President Biden has continued to support these programs, requesting $43 billion for nuclear weapons in his latest annual budget;[5]
- → Nuclear weapons bases and facilities cover more than 15,000 square miles of land in the U.S.;[6]
- → The global stockpile of fissile materials includes over 500 tons of plutonium-239, which is hazardous to humans for ~240,000 years.

Though it flies in the face of modernism's celebration of artistic autonomy, the practice of placing artists in organizational contexts is at least 75 years old and has expanded, in ambition if not in size, since the 1970s. Under the New Deal, the U.S. government's work relief programs hired artists to document economic conditions, decorate new public buildings, and produce performances to celebrate working people – an emergency measure that seeded the community arts movement a generation later. Other often-cited precedents include the Artist Placement Group, episodic efforts to place avant-garde artists in public school contexts, and the forty-year tenure of artist Mierle Laderman Ukeles as the official, unsalaried artist-in-residence at the New York Department of Sanitation. While the history of scholars working in such "embedded" contexts is less established, William Bunge's/Gwendolyn Warren's Detroit Geographical Expedition and Institute is a widely admired model. Today's uptick in municipal government artist-in-residence programs closely tracks the institutionalization of socially engaged art as a distinct field rendered newly relevant by the predations of neoliberal academia and urban governance. Analogs for scholars can be seen in new graduate programs in "public humanities" and Mellon-funded initiatives that encourage humanists to explore the policy implications of their research.[7]

Alongside efforts to embed artists within pre-existing agencies, artists themselves have been organizing collectives along pseudo-institutional lines. Called "mockstitutions" by artist and critic Gregory Sholette and "parafictions" by art historian Carrie Lambert-Beatty, these self-authorizing agencies, corporations, bureaus, laboratories, and coalitions exist as both parodies of organizational culture under neoliberalism and as grassroots correctives that "intersect with the world as it is being lived."[8] These two modalities – the embedded artist and the mockstitution – express what critic Grant Kester describes as modern art's internal conflict between "the desire for solidarity and reconciliation and the imperative of resistance and critique."[9] As these emerging practices mature – and as practitioners and sponsors move between modalities – the possibilities and limitations of each approach may come to exist as part of a broader ecology of strategies by which the traditional boundaries around art as an aesthetic and critical activity are redrawn.

The yet-to-be-fulfilled potential for artists and humanities scholars to put their unique skills and talents toward the service of government accountability inspired the wishful establishment of the National TLC Service, with Sarah Kanouse and Shiloh Krupar self-appointed as provisional co-directors. Our mission includes taking stock of the ways nuclear militarism has shaped our domestic landscapes, historical narratives, national priorities, and global politics. The agency is also responsible for organizing participatory cultural programming that addresses the historical geographies of U.S. warmaking and nuclear colonialism.

After an audit of actually existing bureaucracies, the U.S. Department of the Interior (DOI) was determined to be the most appropriate administrative home of the National TLC Service. With the most heterogeneous mission of any Cabinet office, the DOI is forced to cope with contradiction on a day-to-day, operational basis and embodies the deepest paradoxes of the American experience. Moreover, the DOI has long served as an institutional dumping ground, and it continues to serve in this capacity for the lands most directly affected by the U.S. Department of Defense (DOD) and U.S. Department of Energy (DOE). After undergoing expedited clean-up, former military facilities are often transferred to the DOI's Fish and Wildlife Service to be administered as part of the national wildlife refuge system. The "military to wildlife" or "bombs to birds" conversion process has become a preferred way to dispose of obsolete facilities deemed too toxic to return to any other use. The cabinet office long derided as the "Department of Everything Else," therefore, emerged as a most fitting home for the people and ecologies remaindered by the U.S. nuclear program.

The resolution establishing the National TLC Service outlines these historical and continuing practices, explicitly positioning the agency as a response to previous governmental failures and evasions of responsibility.[10] The resolution's text contests the widespread but inaccurate perception that the Cold War is safely past and that its toxic legacies have been resolved. Indeed, we live in a world in which toxicity and exposure are not the exception but the rule: In what has become our motto, "the Cold War isn't over; it's not even past." The National TLC Service was designed to confront the U.S.'s ongoing toxic open door by challenging dominant ways of knowing and ignoring the remains of the Cold War. Our very existence signals the charge to perform government differently, to diverge from the bureaucratic production of secrecy and unaccountability that has marked U.S. atomic policy for over seventy years. In support of this goal, we approached the first years of operations in three distinct phases: Establishing Operations, Surveying the Terrain, and Developing Public Programs.

Our motto riffs on a now familiar idiom, originally penned by William Faulkner, about the past not really being past.[11] At the most literal level, it signals the complex ways that the Cold War lives on through environmental contamination and health effects, military-industrial economies, the continued existence of nuclear weapons, and the unsolvable problem of nuclear waste, to name only a few examples. Conceptually, it draws upon the creative historical-materialist methodology of Walter Benjamin, who juxtaposed fragments from the past with the present in order to debunk the idea that history proceeds in a natural step-by-step or progressive linear way. We also find that the motto speaks to the ongoing renewal of the nuclear arms race, with over $72 billion in global funding in 2020 alone.[12] Defense policy analysts deem nuclear weapons stockpiling to be back on the agenda — this time for the purposes of blocking rising nuclear state powers and non-state nuclear actors. Cavalier pro-proliferation wonks in the Department of Defense lack on-the-ground knowledge about the conditions of facilities and communities across the country that have been part of the U.S. national nuclear assembly line for more than 75 years. Either that, or they cynically hope that the risks of the new nuclear arms race can be more fully offshored in the global extractive economy beyond the settler colonies of North America, with Namibia and Kazakhstan now operating some of the largest mines in the world.[13]

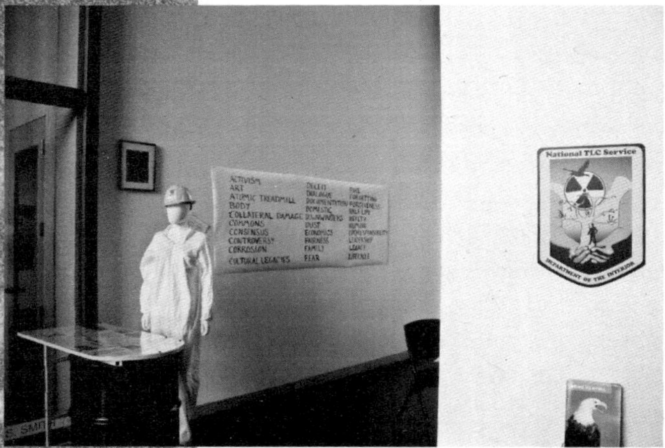

The National TLC Service opened a temporary field office to the public in the IDEA Space at Colorado from March to May 2016.

PHASE 1: Establishing Operations

In order to *behave* governmentally, the National TLC Service must first *appear* governmental. Creating a visual identity and communications strategy was an essential first step in signaling our critique and constituting our public. Our appearance must reference the visual culture of governmentality while suffusing it with contradictory, challenging, and uncanny imagery. We quickly released a logo, launched the official National TLC Service website, developed informational brochures about the agency, and produced a short video to communicate our mission.[14]

Our visual identity was carefully developed to signal our mission of listening to and caring for the uncanny byproducts and "collateral damage" – human and non-human – of the atomic arsenal. Our logo represents our commitment to conservation by echoing the chevron-shaped badges of the U.S. Fish and Wildlife Service, National Park Service, and National Forest Service. On the other hand, what is being conserved here is distinct/unusual. Two cupped hands carefully hold a gnarled, toxic green tree and the silhouette of a stiffly posed, hard-hatted worker. The hands belong to a figure whose shoulders merge into the shape of mountains. Above these hills, a lone goose flies across the hazardous waste-symbol sun. The agency's website echoes the brown-and-green palette long favored for environmentally themed organizations but replaces soothing earth tones with contrasting bright yellow and neon blue. These choices of color and iconography acknowledge that "nature" and "waste" are not mutually exclusive or eternal categories, with military and industrial actors both utilizing and producing new forms of nature – from transuranic elements to disposal cell dog parks – in ways that cannot be entirely predicted or controlled.

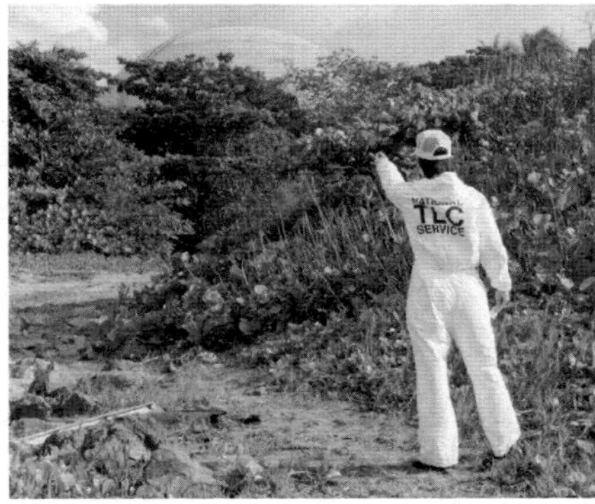

National TLC Service Agent Shiloh Krupar conducts a site visit to the grounds surrounding the shuttered Boiling Nuclear Superheater (BONUS) Reactor Facility, now a popular surfing spot, in Rincón, Puerto Rico.

In a similar vein, our field uniform replaces the reassuring, retro brown workwear of the park ranger with the hazmat coverall – an outfit whose brilliantly white appearance signals the frightening invisibility of the contamination it protects against. Where the ranger reassures, the TLC agent disturbs. The hazmat suit announces, "This is no ordinary dirt," and our appearance in uniform far from known hazardous sites suggests that zones of exposure extend well beyond the confines of the nuclear weapons complex. The replacement of the soaring American bald eagle with a taxidermied specimen – beak agape, tongue visible, and frozen mid-squawk – lends an uncanny quality of suspended animation to the cover of our brochure. Our official agency video references classic moments of Cold War culture, restaging or compositing them with present-day footage to demonstrate that their residues live on in many forms. These visual strategies call into question both the public face of government and the authority and plausibility of the TLC Service itself. If the rhetoric of governmentality typically performs a reductive function that assumes legitimacy, suppresses controversy, and flattens complexity, the National TLC Service marshals the language and visual culture of bureaucracy to critique and place it in question. Our goal is not to deceive but to productively confound, inviting our public to imagine alternatives to the insufficiency, limited efficacy, and impoverished public engagement of the actually existing governmental response to the damaging effects of military and nuclear toxics.

As someone whose political identity has long slid leftward on the anarchist spectrum, I initially conceived the National TLC Service as a purely satiric, critical project. The government's handling of the legacies of its nuclear program has been so slipshod, thoughtless, and downright callous that "exposing" it to the mutations of negative publicity is an obviously important first step. Yet lambasting the government for its paternalistic, technocratic blindspots is a popular spectator sport on both the left and the right. In the absence of a prefigurative position, parody may unintentionally strengthen the already hegemonic power of what Ruth Wilson Gilmore calls the "antistate state."[15] At first, I thought of the National TLC Service as an honest forgery: In the web of obfuscation that surrounds the "greening" of the U.S. military, only a fake agency can tell the truth. What we initially considered a one-off thought experiment became an idea people wanted to believe in, and we soon found ourselves invited to expand the project in real life.

The move from idea to performance allowed us to modulate the satiric affect and interject a note of sincerity to propose, through enactment, what a government rooted in mutual accountability, vulnerability, and transparency might look like. Our ability to pursue this mission is necessarily ad-hoc given our funding, personnel, and day jobs. Indeed, the problems stemming from nuclear militarism overwhelm the capacity of the actually existing Department of the Interior. Clearly, nuclear crimes cannot be addressed by either the private markets of neoliberal capitalism or the voluntary associations of pirate utopias. The TLC Service seeks to use our amateur status to test other avenues for action on a local level that could be cultivated by the state on the geographical and temporal scale on which it only can act.

Of course, on the time scale of the atomic, all governments are ad-hoc. (Sarah)

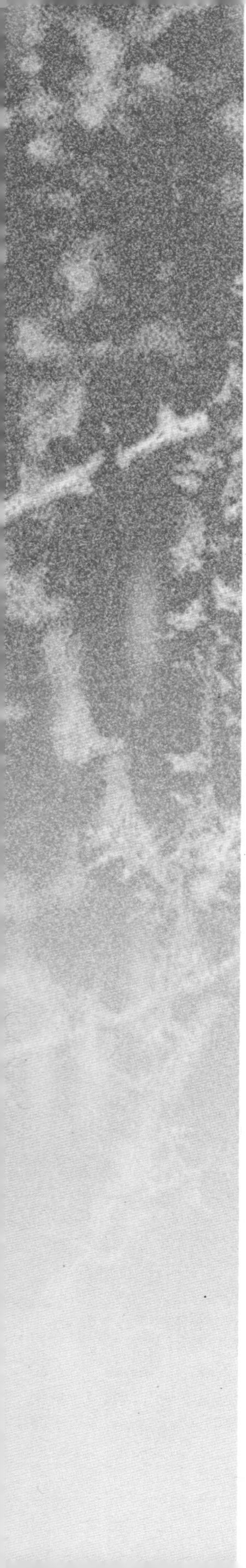

PHASE 2: Surveying the Terrain

To better understand the current state of nuclear commemoration, the National TLC Service completed an audit of the rhetorics of atomic memory. We conducted site visits to key locations for documentary purposes as well as carefully examined promotional materials, media coverage, and audience commentary posted to various Internet sites. Our field trips included a wide range of sites from famous Manhattan Project locations to obscure waste disposal cells, to understand the extent and diversity of atomic sites. We also surveyed contemporary popular media, including video games and television shows. While these sites and products vary greatly in terms of audience size, demographics, and profile, they demonstrate that the Cold War era has experienced a resurgence of popular interest thirty years after its ostensible end. Our investigations found that the former nuclear complex increasingly functions as an exhibitionary complex, as the domestic spaces and ruins of war are repurposed for atomic commemoration. Simultaneously, existing atomic museums and televisual nuclear nostalgia bolster public awareness of the Cold War both as a consumable bygone era and as infotainment.

The most ambitious and high-profile of these commemorative efforts is the Manhattan Project National Historical Park, established as a provision within the National Defense Authorization Act of 2015. The park resulted from over a decade of conversations between the National Park Service (NPS) and the DOE, with sustained advocacy from a number of non-governmental organizations. The non-contiguous park is essentially an NPS-authored interpretive overlay on three major sites – some still active, all remaining under DOE management – where work on the world's first atomic bomb took place: Los Alamos, New Mexico; Oak Ridge, Tennessee; and Hanford, Washington.[16] While the interpretive details are still evolving, a commission of scholars met in late 2015 to lay out potential themes, such as the political and ethical issues surrounding the decision to use the bomb; tensions between open scientific inquiry and national security; historical personalities and daily life in the secret cities; and the pivotal position of the nuclear weapons effort within the history of science and technology.[17] Event reflections also identified present-day environmental legacies and relationships with smaller atomic and Cold War museums to be priorities that had not yet received adequate attention.

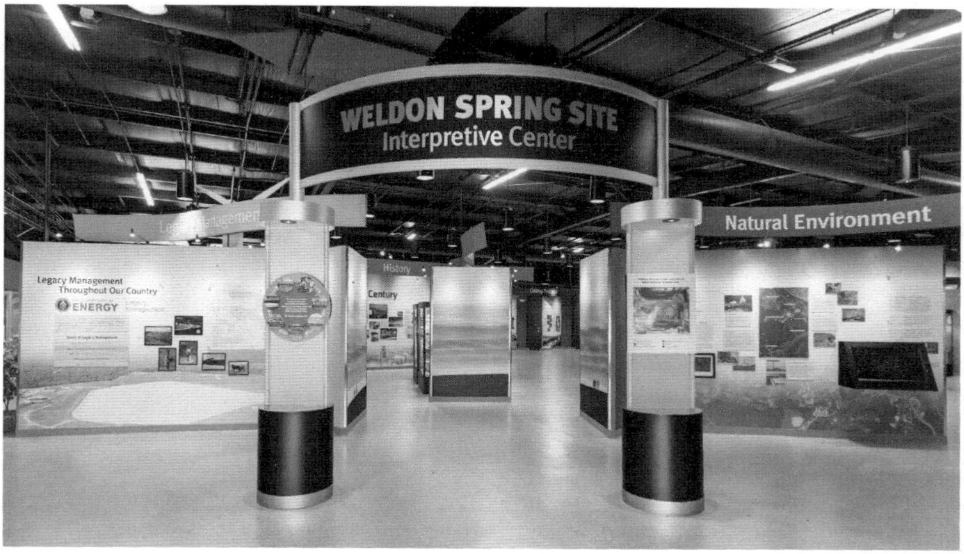

Above:
The interpretive center at the Department of Energy's Weldon Spring uranium feed materials site devotes equal time to the plant's history and the engineering details of the on-site disposal cell.

Alongside the long development period of the Manhattan Project National Historical Park, a number of atomic museums and tours have been established, expanded, or proposed. After many years in Old Town Albuquerque's museum district, for example, the National Museum of Nuclear Science and History moved to a significantly larger new facility, with outdoor exhibition space, in southeast Albuquerque in 2009. The Atomic Testing Museum in Las Vegas achieved national status in 2012, becoming part of a small group of privately operated museums affiliated with the Smithsonian network. The museum's signature exhibition is the sensationalistic "Ground Zero Theater," a bunker-inspired screening room that purports to simulate an atmospheric bomb test, complete with shaking seats and blasts of hot air.[18] On the more academic side, the Wende Museum moved to the armory building in the Los Angeles community of Culver City in 2017 to better present and interpret Cold War history and culture from the Soviet bloc.[19] Finally, numerous site-specific interpretive centers tell local stories of involvement in the nuclear weapons complex. Often funded as part of remediation agreements between municipalities, the federal government, and corporate contractors, these interpretive centers generally emphasize heroic worker stories and highlight the success of environmental clean-up operations, as is the case at Weldon Spring, Missouri, and the Fernald Reserve in Ohio. When continuing funding is not provided, these site-specific museums often struggle to achieve sustainability, as in the case of the former Rocky Flats plutonium production facility in suburban Denver, Colorado.

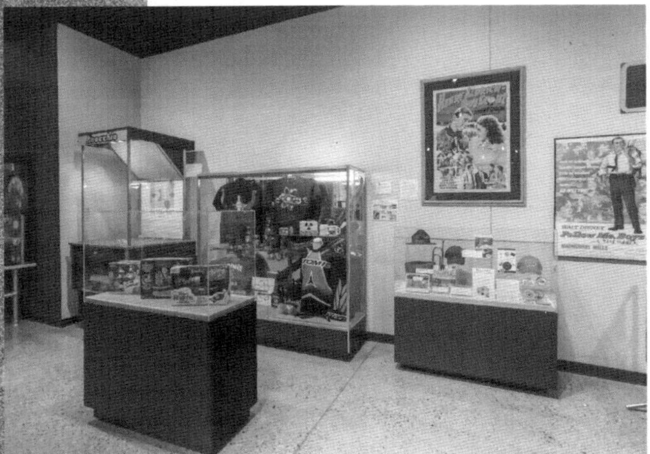

Atomic pop culture memorabilia on display at the National Museum of Nuclear Science and History in Albuquerque, New Mexico.

Displays of atomic pop culture are common features in many of the divergent sites that make up the ad-hoc nuclear public memory industry. While presumably designed to humanize the Cold War period and provide a sense of everyday life in the shadow of the bomb, they tend to dress the era in retro-glam and obscure the profound anxiety within which these cultural objects circulated. Furthermore, the Cold War is indistinguishable from the first period of mass televisual spectacle, and itself has become subject to commemoration via the popular moving image. With stunning cinematography and a score by members of Sigur Rós, the short-lived cable series *Manhattan* presented a highly fictionalized account of the motivations, political machinations, and complicated lives of the scientists sequestered in Los Alamos to build the first atomic bomb. Other, more fanciful references to nuclear weapons crop up in the post-apocalyptic video game series *Fallout* and *The Americans*, a popular spy television serial focusing on the personal lives of KGB agents in deep cover as a 1980s suburban couple.[20] The Cold War memorial complex itself has become part of this fiction/nonfiction cross-referencing system: Albuquerque's National Museum of Nuclear Science and History had its television moment in season two of *Breaking Bad*, when it served as the secret meeting point between a street dealer network and the fledgling meth kingpin who dubbed himself "Heisenberg" after the architect of the failed Nazi atomic bomb program.[21]

Our brief survey of Cold War memory has revealed certain common threads. These include the progressivist conceptions of science; tantalizing glimpses into forbidden worlds of science, espionage, or crime; aesthetic and pop cultural nostalgia; the nuclear sublime and its dystopic mirror image; and tales of resilience, remediation, and recovery. These rhetorical themes are neither exhaustive nor exclusive; they overlap, sometimes draw on each other, and can be analyzed as operating within a more general Cold War commemorative refer-

When did you first become radical? I remember graduate school conversations about "political awakening." I did not really feel authorized to speak. The inquiry usually implied a division between critical intellectuals suspicious of the state and those who would serve as its technocratic limbs. I was born into government administration: My father worked for the DOE and my mother taught at large state public institutions. The higher education I received was supported by federal fellowships related to earlier wartime investments, and I received my PhD from a public university with deep ties to the nuclear complex.[22] I now teach in a School of Foreign Service with a direct pipeline to government departments and their clandestine counterparts. My embeddedness in such institutions has meant that ideological purity and avant-garde distance are neither accessible nor desirable to me.

Bureaucracy and what Michel Foucault calls the "administrative grotesque" have been my "proving ground," directing my attention to the politics of form – a way of seeing radical opportunities even within the most monolithic and lethargic institutions.[23] The usual procedure of analysis in the service of politics is defining one's object, targeting it, and opposing it. Instead, the politics of form comes with the tacit responsibility to question one's own legibility, participation, and investments in governance and subjectivity. This involves a restless exploration of knowledge as a social practice, that enacts alternative institutional arrangements and productions of space. There is nowhere outside the pedagogical and the political.

The National TLC Service was imagined, in part, as a collaborative experiment to carry out art and research as a social practice of knowledge production. In doing this project, both of us challenge conventions within our respective fields of geography and visual arts in academia, by troubling distinctions between scholarship, creative work, and public engagement. The TLC Service pedagogically agitates around democratizing bureaucratic/technocratic expertise in a political era that too often focuses "on "stakeholders," relies on the "non-profit industrial complex," and depends on "big data" for developing policies."[24] The agency treats knowledge as fundamentally a political, not a technical, problem, and considers creative pedagogies to be necessary for demilitarizing society. We see this as an ethical imperative to counteract an ever-expanding global military-industrial complex through process-oriented methodologies that cut across scholarship and public fields. (Shiloh)

encing system. While further analysis of these commemorative modes is required, our preliminary investigation revealed that many attempt to foreclose active interpretation to minimize and contain controversy. When ethical quandaries may be explored, the race to achieve the atomic bomb is nonetheless presented as *the* pinnacle of human scientific achievement, thickened with a dash of unrecognized heroism. In others, nostalgic period reconstructions, easily digestible spectacle, and personality-driven storylines dominate the scene. Even in sites of unquestionable environmental devastation, narratives of scientific progress – now in pursuit of ecological restoration and waste containment – attempt to reassure potentially anxious visitors. The National TLC Service is concerned that dominant Cold War memorial practices distance the era and its atomic and chemical weapons from present and future embodied environmental geographies of war.

PHASE 3: Developing Programs

In a reparative spirit, the National TLC Service considers popular interest in the Cold War to be an opportunity to address the ongoing political, ecological, cultural, and ethical implications of U.S. military projects. In contrast to the spectacular forms of atomic memory and kitsch in our inventory, the vast physical and economic infrastructure of the American military state remains invisible. Between the scales of local lore and national databases, the specificity and reach of the nuclear complex continues to be largely unexplored. The Cold War remains hidden in plain sight due to legacies of secrecy, the domesticating effects of consumer technologies, the invisibility of radiological contamination, and the comparative banality of its landscape. While critical accounts of the Cold War do exist – virtual museums, community-based toxic tours, experimental monuments – they are largely self-funded, temporary and/or conceptual, and usually isolated from the better-known museums, national laboratory visitor centers, and historic sites. Therefore, in our third phase of operations, the National TLC Service initiated a series of public programs for a proposed National Cold War Monuments and Environmental Heritage Trail (NCWMEHT) that would investigate local legacies from the Cold War and provide a platform for communities, artists, and critics to coordinate with each other at a larger scale.

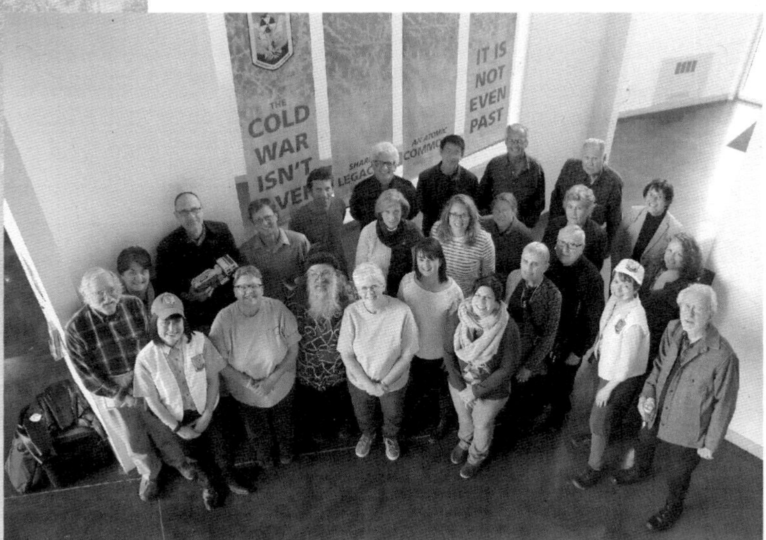

National TLC Service Agents Shiloh Krupar and Sarah Kanouse pose with participants in the National Cold War Monuments and Environmental Heritage Trail design charrette at the National TLC Service temporary field office in Colorado Springs, Colorado.

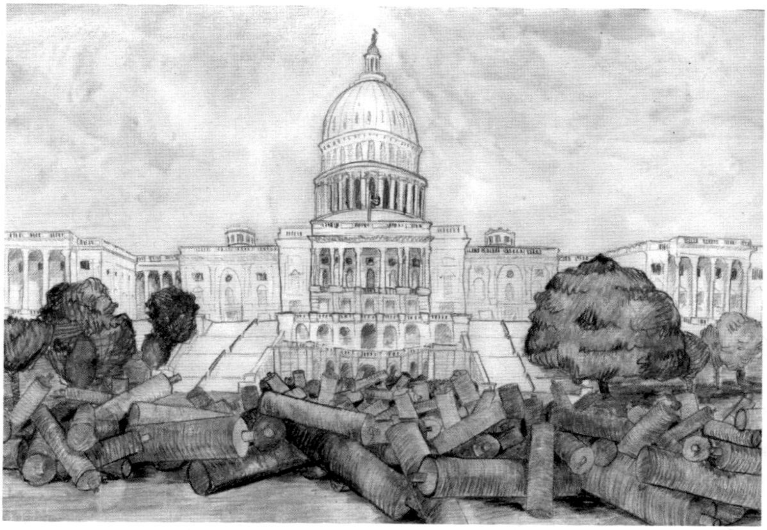

Artist's sketch of the spill monument at the center of the proposed National Cold War Monuments and Environmental Heritage Trail.

NATIONAL COLD WAR MONUMENTS AND ENVIRONMENTAL HERITAGE TRAIL

The National Cold War Monuments and Environmental Heritage Trail (NCWMEHT) is a proposed coast-to-coast route linking both famous and obscure sites involved in the production, testing, and storage of nuclear and chemical weapons since World War II. The goal is to encompass but also extend beyond existing commemorative efforts, in order to create a narrative that is both more comprehensive and responsive to the particularities of local experience. Recognizing that there is no coherent national narrative of the nuclear weapons complex, the National TLC Service has established a process to work with various publics to map a "people's geography" of the Cold War's ongoing, differential impacts. Our trail will incorporate both well-known and unmarked sites, as well as mundane places that rarely register in popular atomic geographies, from training facilities to disposal cells. Given the nuclear weapons complex's vast extent – temporal, geographic, and economic – and the atmosphere of secrecy and/or obfuscation that surrounds it, the National TLC Service recommends that identifying and enrolling sites for the Trail continue through the United States tricentennial in 2076.

At the symbolic center of the NCWMEHT is a monument consisting of a pile of 17,287 bronze forms – cast from smaller nuclear waste storage casks – on the lawn of the U.S. Capitol. The number of cask forms is based on the estimated size of the U.S. nuclear arsenal at the fall of the Soviet Union in 1991. After a year of public display, cask forms will be removed from the pile and installed as site markers along the National Cold War Monuments and Environmental Heritage Trail until the number of casks in the pile equals the number of weapons in the U.S. arsenal (currently ~3,700).[25] The central monument is at all times indexed to the size of the current U.S. nuclear arsenal: When the government, as planned, begins producing new, smaller, tactical weapons, new casks will be added to the monument.[26]

Recognizing the (often intentional) incompleteness of government records, sites for inclusion in the NCWMEHT are selected not only through historical research but also the collective knowledge of affected communities. Many citizens' groups, tribal communities, and grassroots organizations have been conducting their own research, environmental monitoring, and commemorative activities for decades. The National Cold War Monument and Environmental Heritage Trail is an opportunity to gather together and amplify these efforts. Additionally, individuals somewhat more distant from these legacies have been galvanized by National TLC Service programming and asked to participate in the process of developing the Trail. To that end, we have developed two mechanisms of public participation: Site nominations and design charrettes. A third program, the TLC Service Transnatural Park Posters and Stamp Series, fosters public awareness of the NCWMEHT and highlights especially forgotten sites on a state-by-state basis.

The National Cold War Monument echoes post-minimalist and land artworks through its irregular "pile" or "spill" form and the use of fluctuation and potential disappearance of the work as a gesture equal parts poetic and conceptual. Most directly, the piece references Joseph Beuys's participatory social sculpture project "7,000 Eichen/7,000 Oaks," consisting of an initial pile of 7,000 basalt columns removed one at a time to accompany oak trees planted by the artist and community members around the city of Kassel, Germany. In a more personal vein, Félix González-Torres's 1991 ***Untitled (Portrait of Ross in LA)*** turns the disappearing pile into elegy: A pile of candy is figured as the artist's dead lover, his body carried off piece by piece by the viewers and replenished daily by the museum.

The indeterminacy and disintegration of these works align them with what James Young identified in the early 1990s as "countermonuments," charting "metamorphosis of the monument from the heroic, self-aggrandizing figurative icons of the late 19th century … to the antiheroic, often ironic and self-effacing conceptual installations that mark the national ambivalence and uncertainty of late 20th-century postmodernism."[27] Noting that audience activation is more central to the operations of these artworks than a critique of the monument tradition, Mechtild Widrich has recently described as "performative monuments" those that hinge on "the temporal interaction with an audience that itself is no eternal public, but a succession of interacting subjects."[28]

Through this lens, the TLC design charrettes themselves emerge as monuments, the architectural design process eclipsing its product in its capacity to catalyze collaborative citizenship projects that model intergenerational ecological responsibility and find ways of living with contamination and loss. In both their form and substance, the charrettes re-align modernist polarities between monument and countermonument, nature and waste, wildlife and human, community and contamination. Furthermore, the National Cold War Monuments and Environmental Heritage Trail confounds the problematic tendency to privilege Washington, D.C. as the locus of American power and public memory. Our unbuildable proposal for a bronze spill of fake nuclear waste casks on the Capitol grounds is delightful symbolism, but the real work on the monuments and the charrettes is done by small groups of people, in overlooked or marginal places, on an ongoing basis, and over a long period of time.

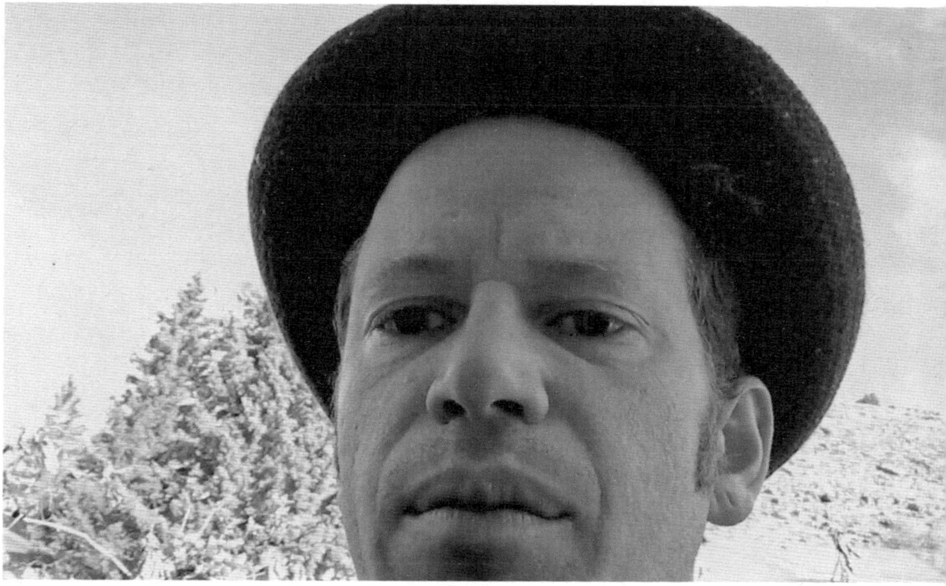

In a video submission, National TLC Service Field Agent Josh McDonald nominated the airspace above the Western United States for inclusion in the National Cold War Monuments and Environmental Heritage Trail.

PUBLIC MARKER SYSTEM AND SITE NOMINATIONS

The National TLC Service website features an open-ended public call for submissions of sites for inclusion in the Trail. Nominators produce a brief video (using a webcam or a cell phone) to describe the site and explain why it should be included in the Trail. Participants are then inducted as TLC Service Field Agents, receiving a miniature replica of the trail marker and an agency patch in exchange for committing to interpret the site actively for others. A number of the inventive and intelligent amateur-produced nominating videos are archived on our website and have been featured in temporary field offices and other exhibition contexts.

Nominations have ranged from familiar locations, like Hanford and the Nevada Test Site, to the creation of non-territorial sites, like the airspace over the entire western United States. Whether nominators ratify an acknowledged site of historical significance or propose an unexpected new stop along the Trail, they become active in a commemorative process that is based less on official markers and tour guides than on a living, communal process of storytelling and continual re-interpretation. Videos often include anecdotes describing personal or family connections to the places described. Through a continual nomination process, the TLC Service hopes to make the Cold War an ongoing and active site of public living memory through mapping and storytelling rituals.

PUBLIC CHARRETTES / WORKSHOPS

The National TLC Service also draws on community design methods in developing the Trail. In 2013, we launched a series of regional *charrettes*, or intensive design workshops, to plan the route of the Trail and to propose a series of monuments to reflect the local character of the nuclear weapons complex. Two regional events have been held so far: Illinois in 2013 and Colorado in 2016. Catalogs of outcomes from both charrettes are available on our website.

The design charrette is a well-established methodology in architecture and urban planning that brings together bureaucratically defined "stakeholders" to collectively gather information, identify needs, and brainstorm design solutions. Its community orientation is seen as a corrective to architectural and public art projects that ignore neighbors' needs and desires. However, the process often unfolds less-than democratically, devolving into a mechanism for consultation with bureaucratically legible "community" representatives. Too often, project control remains in the hands of experts, and the consultation process homogenizes different positions and experiences as one input.

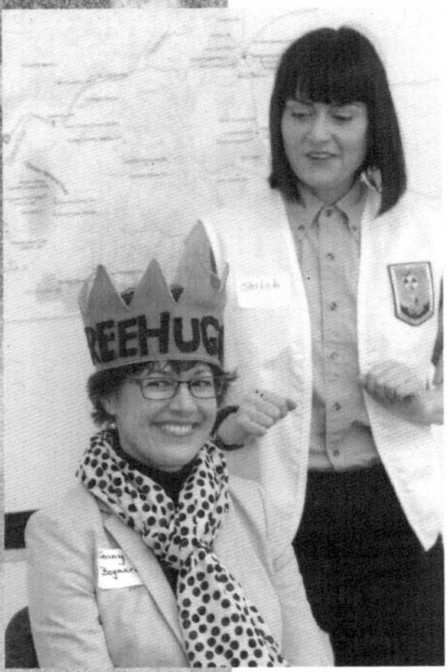

National TLC Service Agent Krupar crowns participant Conny Bogaard "Treehugger" in an icebreaker exercise at the Colorado design charrette for the National Cold War Monuments and Environmental Heritage Trail in March 2016.

Mapping conventions reflect legacies of imperial exploration, resource extraction, colonization, and state control. These strategic activities have required the collection of more precise data and objective scientific procedures, resulting in what seem to be value-neutral mirrorings of space rather than interest-laden renderings of the surface of the planet. But while maps reinforce dynamics of coercion and manipulation, maps can be – and are – used as tools to support social movements and social justice projects.[29] The National TLC Service is interested in the role of maps and contemporary mapping in political struggles – specifically, how maps are political and how mapping can be a political act. Mapping techniques and technologies can serve as applied methodologies to resist and/or supplement dominant knowledge. Counter-mapping projects, such as Indigenous mapping networks and radical atlases and cartographic practices, challenge official maps while also using map conventions to lay claim to resources and land or to render alternative worlds.[30] Mapping appears in TLC work in a number of ways: As a form of protest and geopolitical critique; as social commentary, including alternative history trails, collective walks, and site nomination systems that feature marginalized experiences and critique the status quo; and as a method of community organizing, wherein mapping functions as a form of human communication and process-based participatory learning.[31] We consider map literacy to be an educational tool for envisioning and advocating social change.

By contrast, National TLC Service charrettes invite participants from generationally, politically, and geographically fragmented communities affected by the Cold War, in order to recognize and develop the capacity of non-designers to intervene critically in landscape and the built environment. We emphasize that everyone present offers different kinds of expertise and encourage the self-organization of small groups with complementary skills in order to develop solutions that no one could devise on their own. In the Illinois charrette, for example, an activist fighting Manhattan Project radiological contamination in her neighborhood worked with an ethicist and a visual artist to propose a functional monument consisting of a mobile health facility and oral history center connecting personal stories of potential exposure to citizen-science data collection efforts.

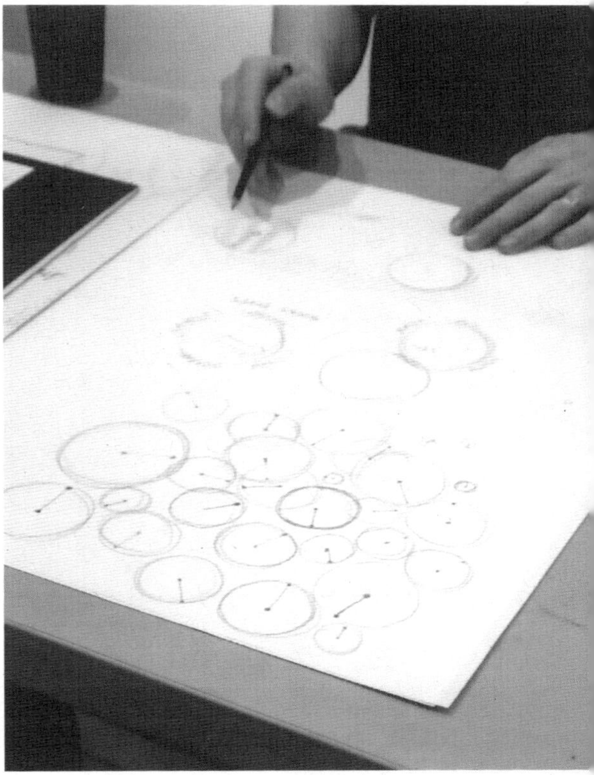

A participant sketches a monument proposal at the Illinois design charrette for the National Cold War Monuments and Environmental Heritage Trail in November 2013.

The charrettes use post-consensus processes that maintain polyphony and the potential for internal dissent. The events become opportunities to publicly work through – though not necessarily overcome – ongoing antagonisms surrounding the nuclear weapons complex. The charrettes begin with a roleplaying exercise designed to help participants recognize and openly discuss their internalized assumptions around various positions commonly associated with the Cold War. These include stereotypes like "The Treehugger" and the "The Patriot," as well as professional roles such as "The Regulator." While this activity fosters critical distance from one's prior assumptions and encourages an appreciation of the contributions made possible by others' perspectives, differences between participants can flare in uncomfortable ways. Rather than glossing over these moments of discomfort, charrette facilitators strive to provide respectful space for the mutual recognition and constructive elaboration of that difference. During a tense moment in the Colorado charrette, an environmental activist who had lost family members to diseases associated with uranium mining expressed disgust at a pro-mining community's celebratory eating of a yellow layer cake, symbolizing raw uranium, at their annual picnic. The entire group listened as the organizer of the picnic expressed regret at the pain the gesture inadvertently caused and explained that it was intended as a light-hearted recognition of the unity and resilience of a now-dispersed community. Simply noting the courage required to raise and respond to conflict establishes that no single perspective has primacy in the commemorative process and models a way to make deep divisions more visible (through intimate proximity to each other) and more open to meaningful reflection.

At the charrettes, National TLC Service facilitators emphasize the wishful nature of the NCWMEHT. The knowledge that the monuments proposed during the workshop will probably never be realized seems to license a wide range of creative responses, from the poetic to the educational to the satiric. The iterative nature of the workshops allows the TLC Service to highlight multiple speculative monuments and accommodate many specific, conflicted, and inter-related positions. The objective of the NCWMEHT is not to rewrite existing historical narratives and rhetorics of nuclear militarism, but to supplement, contextualize, and contest. Liberated from the burdens of pragmatism and the imperative to produce a comprehensive narrative, the charrettes are a provocation for social and political engagement with Cold War landscapes through architectural play. They align documentation with speculation in a common project that articulates diverse positions, landscapes, and conditions of exposure to allow an uneven atomic commons to materialize. Ultimately, the NCWMEHT is an emergent geography, one whose process of creation is as much a part of its "path" as any route traced on a map.

TLC SERVICE TRANSNATURAL POSTER AND STAMP SERIES

Concurrently with the development of the NCWMEHT, the National TLC Service proposes releasing a series of commemorative posters and postage stamps to highlight significant toxic sites in each state and across North America. These posters and stamps echo the well-known, early twentieth-century series of National Park Service posters that rendered the conservation of beautiful public lands as an inextricable part of the U.S.'s national identity. Utilizing these graphic conventions, the National TLC Service posters and stamps play with the aesthetics of nature and waste to galvanize awareness of environmental hazards and the domestic footprint of war but also, simultaneously, to inspire public appreciation of the nation's contaminated but no-less stunning public lands.

The poster series extends the recognition, also found in our logo, that waste and nature coexist and inflect one another. Our charrettes allow publics to develop a broader suite of possibilities for addressing the painful realities of contamination and loss than those afforded by understandings of nature as either pure ***or*** unnatural and "ecocidal." The poster and stamp series fosters what our institutional acronym popularly stands for – "TLC" – that is, ***care of*** and ***accountability for*** the monstrous and uncanny by-products, collateral damage, and multiple bodies affected by the nuclear weapons complex – bodies of land or water, animal bodies, human bodies, administrative bodies, national bodies.

Responses to toxic crisis typically evoke some combination of natural, bodily, or national sovereignty that impede creative demilitarizing efforts. The National TLC Service seeks to cultivate queer understandings of nature instead of resorting to conservative preservation efforts that rely on notions of national exceptionalism, untainted nature, or the pure body. The agency enlists the term "transnatural" to signal both the indeterminacy of and ongoing attachments to nature or the natural. Instead of claiming the privilege of distance from waste or portraying nature as passive victim in a nostalgic realm of lost purity, the agency tactically considers nature to be ongoing social-ecological composition, qualified by the unevenly embodied realities of toxicity and contamination. Our mission applies a queer ecofeminist perspective to environmental education and ethics, in order to reevaluate, even dramatize, different kinds of bodies as sites of material connection and parodic inhabitance capable of converting the fearful into the livable.[32] Our brand of administration is indebted to the rich inventory of social practices by LGBTQIA2S+ communities that survive – flourish – within often unsupportive structures.[33] Catriona Sandilands captures the spirit of our approach, "to queer nature is to 'put out of order' our understandings, so our 'eccentricities' can be produced more forcefully."[34] The agency advocates a more diversified ecological politics that combines concerns about equity of environmental justice efforts with investigations of deviant material agencies and the impurities and inter-corporeal effects of militarism. We see this as a corrective or counter-practice to the militarized resilient life only claimed by those with the means to preserve themselves amidst widespread conditions of insecurity and risk. Our queer environmental programming rethinks relations with waste to challenge the status quo and cultivate irreverence toward the traditional boundary making of classic environmentalism and the bodily security of liberal (white heteronormative) environmental security.

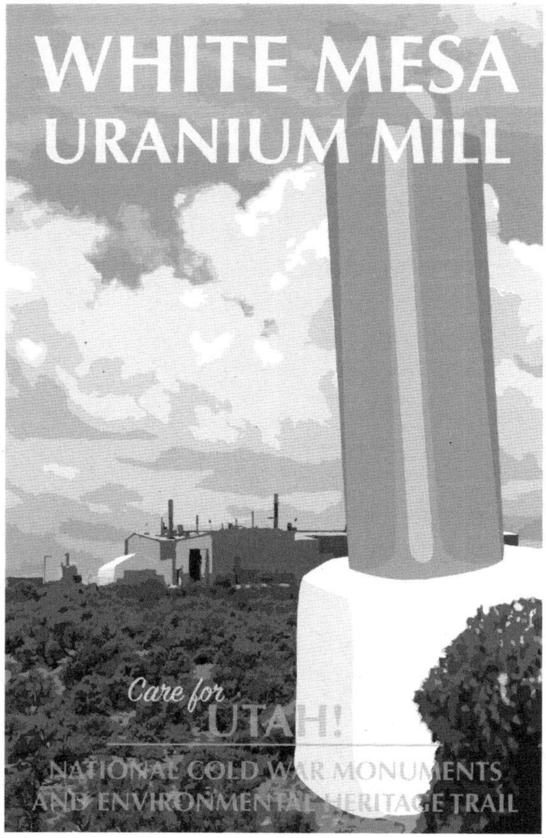

National TLC Service Transnatural Poster for Utah featuring the controversial White Mesa Uranium Mill.

TLC VALUES: Our Atomic Commons

In closing, we – the National TLC Service – believe the National Cold War Monuments and Environmental Heritage Trail is urgently needed:

- To counter the history of misinformation and ignorance regarding the material effects of the Cold War
- To keep uncertainty open with respect to the material transformations initiated by militarization
- To make accountability an ongoing public and government relation

While American mythologies hold that war has not profoundly shaped our "homeland," the existence of sovereign Indigenous nations within the territorial boundaries of the U.S. and nearly 800 U.S. military bases in over 70 countries demonstrate the primacy of force in our history and cast into uncertainty exactly where the nation begins and ends. Material geographies of the Cold War are always already transnational. The National TLC Service, therefore, will not remain tethered to U.S. territory; rather, the Trail and TLC agency "pop-ups" will expansively address sites – and companies, markets, bodies, and other material flows – involved in nuclear testing, waste, and fallout. Furthermore, the inclusive public process is a provocation to explore how we should respond to our contemporary Cold War condition – the subtle forms of exposure, abandonment, and "slow violence" that operate below the threshold of catastrophe to devastate bodies and hollow out landscapes.[35] In doing so, the National TLC Service endeavors to coordinate an "apology" that takes the form of a ***transnatural*** and ***transnational*** ecological politics.

Both the uneventfulness of the Cold War and the profoundly alarming reality of its residues make it difficult – seemingly impossible – to commemorate. Moreover, the fact that the U.S. and Soviet governments subjected the world's population to a vast, uncontrolled radiological experiment threatens to delegitimize any national or international attempts to address its effects. The National TLC Service sees this not as reason for despair and inaction but instead as an occasion for experimentation with a variety of techniques and aesthetics that challenge the status quo, care for the residual, and encourage more exuberant politics. In an era in which the "left hand" of the state is in retreat across the industrialized world, the National TLC Service is a platform on which to rehearse alternative social organization, embodied ecological politics, and open-ended governmental administration.

It has been an exciting decade of operations. We have laid a foundation for how government might operate differently: By embracing controversy and complexity, by facilitating open-ended dialogues, and by responding to the physical, environmental, and cultural impacts of the American nuclear state. In everything we do, we are motivated by our desire to serve the downwind and downstream populations of the Cold War, which is to say, all of us. We conclude this report by formally stepping down as interim co-directors of the National TLC service, with an open call to *decolonize public agencies.*

NOTES

1 University of Iowa, University of Illinois, and University of Maryland; Cooper Union, Georgetown University, Northeastern University, and Colorado College.

2 Historic stockpile figures are from Norris and Kristensen, "Global Nuclear Weapons Inventories." Current figures include both deployed and stockpiled warheads. Refer to Kristensen, Korda, and Norris, "Status of World Nuclear Forces"; and Arms Control Association, "Nuclear Weapons."

3 Kristensen, Korda, and Norris, "Status of World Nuclear Forces."

4 Mehta, "America's Nuclear Weapons."

5 Democracy Now, "U.S. Led 2020 Nuclear Weapons Spending."

6 Schwartz, *Atomic Audit*; US Nuclear Weapons Cost Study Project, "50 Facts about U.S. Nuclear Weapons."
7 See, for instance, the Humanities Without Walls consortium and its Mellon-funded "Global Midwest" research initiative, based at Agent Kanouse's alma mater.
8 Lambert-Beatty, "Make-Believe," 54. See also Sholette, *Dark Matter*, 153.
9 Kester, "FIELD Editorial 4."
10 Full text of the resolution is available at the National TLC Service website.
11 See, for example, Pred, *The Past Is Not Dead.*
12 Democracy Now!, "U.S. Led 2020 Nuclear Weapons Spending."
13 For a discussion of "nuclearity" in Africa during the historic Cold War and the contemporary boom, see Hecht, "An Elemental Force."
14 National TLC Service.
15 Gilmore, "In the Shadow of the Shadow State."
16 National Park Service, "Memorandum of Agreement."
17 National Park Service, "Scholar's Forum Report."
18 See tourism-related reviews at "National Atomic Testing Museum."
19 Refer to the Wende Museum website: https://www.wendemuseum.org/about-us.
20 For fan-sourced information on the *Fallout* series, see "Welcome to Nukapedia."
21 Alcala, dir., "Negro y Azul," *Breaking Bad.*
22 The University of California–Berkeley, where Agent Krupar received her PhD in geography, is part of the UC system, which operated Los Alamos National Laboratory for sixty years and continues to oversee the Lawrence Berkeley and Lawrence Livermore laboratories. Los Alamos National Laboratory conducted classified work toward the design of nuclear weapons.
23 Foucault, *Abnormal*, 11–14.
24 CUNY Graduate Center, "The Detroit Geographical Expedition."
25 Kristensen, Korda, and Norris, "Status of World Nuclear Forces."
26 Despite his ambition to rid the world of nuclear weapons, former President Barack Obama presided over the initiation of a nuclear weapons "revitalization" program estimated to cost $1 trillion over thirty years. Donald Trump sought to further expand the U.S. atomic arsenal, raising the spectre of a multi-player arms race involving China and Russia.
27 Young, "Memory and Countermemory."
28 Widrich, *Performative Monuments*, 8.
29 Krupar, "Map Power and Map Methodologies."
30 See, for example, Lize Mogel and Alexis Bhagat, eds., *An Atlas of Radical Cartography.*
31 Kanouse, "Critical Day Trips." Also refer to Counter Cartographies Collective, Craig Dalton, and Liz Mason-Deese, "Counter (Mapping) Actions."
32 Krupar, *Hot Spotter's Report*, 215–51.
33 Sedgwick, *Touching Feeling*, 149–51.
34 Sandilands, "Lavender's Green?" 22.
35 Nixon, *Slow Violence*; Stoler, "Imperial Debris"; Berlant, "Slow Death."

BIBLIOGRAPHY

Alcala, Felix, dir. "Negro y Azul." *Breaking Bad*, season 2, episode 7. Written by Vince Gilligan and John Shiban. AMC Network, 19 April 2009.

The Americans. Created by Joseph Weisberg. FX Network, 2013–18.

Arms Control Association. "Nuclear Weapons: Who Has What at a Glance." https://www.armscontrol.org/factsheets/Nuclearweaponswhohaswhat.

Berlant, Lauren. "Slow Death (Sovereignty, Obesity, Lateral Agency)." *Critical Inquiry* 33, no. 4 (2007): 754–80.

Counter Cartographies Collective, Craig Dalton, and Liz Mason-Deese. "Counter (Mapping) Actions: Mapping as Militant Research." *ACME: An International E-Journal for Critical Geographies* 11, no. 3 (2012): 439–66.

CUNY Graduate Center. "The Detroit Geographical Expedition: What Is Its Relevance Now?" http://www.gc.cuny.edu/All-GC-Events/Calendar/Detail?id=27573.

Democracy Now! "U.S. Led 2020 Nuclear Weapons Spending; Now Biden Going 'Full Steam Ahead' on Trump's Nuclear Plans." 10 June 2021. https://www.democracynow.org/2021/6/10/biden_putin_nuclear_weapons.

Foucault, Michel. *Abnormal: Lectures at the Collège de France 1974–1975*. Edited by Valerio Marchetti and Antonella Salomoni. Translated by Graham Burchell. New York: Picador, 1999.

Gilmore, Ruth Wilson. "In the Shadow of the Shadow State." In *The Revolution Will Not Be Funded: Beyond the Nonprofit Industrial Complex*, edited by INCITE! Women of Color Against Violence, 41–52. Cambridge, MA: South End Press, 2009.

Hecht, Gabrielle. "An Elemental Force: Uranium Production in Africa, and What It Means to Be Nuclear." *Bulletin of the Atomic Scientists*, 1 March 2012. https://doi.org/10.1177/0096340212440352.

Humanities Without Walls. http://www.humanitieswithoutwalls.illinois.edu.

Kanouse, Sarah. "Critical Day Trips: Tourism and Land-Based Practices." In *Critical Landscapes: Art, Space, Politics*, edited by Emily Eliza Scott and Kirsten Swenson, 91–130. Berkeley: University of California Press, 2015.

Kester, Grant. "FIELD Editorial 4." *FIELD: A Journal of Socially Engaged Art Criticism* 4 (Spring 2016). http://field-journal.com/editorial/field-editorial-4.

Kristensen, Hans M., Matt Korda, and Robert Norris. "Status of World Nuclear Forces." Federation of American Scientists, October 2021. https://fas.org/issues/nuclear-weapons/status-world-nuclear-forces.

Krupar, Shiloh. *Hot Spotter's Report: Military Fables of Toxic Waste*. Minneapolis, MN: University of Minnesota, 2013.

– "Map Power and Map Methodologies for Social Justice." *Georgetown Journal of International Affairs* 16, no. 2 (2015): 91–101.

Lambert-Beatty, Carrie. "Make-Believe: Plausibility and Parafiction." *October* 129 (Summer 2009): 51–84.

Manhattan. Created by Sam Shaw. WGN Network, 2014–15.

Mehta, Aaron. "America's Nuclear Weapons Will Cost $1.2 Trillion over the Next 30 Years." *Defense News*, 31 October 2017. https://www.defensenews.com/breaking-news/2017/10/31/americas-nuclear-weapons-will-cost-12-trillion-over-the-next-30-years.

Mogel, Lize, and Alexis Bhagat, eds. *An Atlas of Radical Cartography*. Los Angeles: Journal of Aesthetics and Protest Press, 2008.

"National Atomic Testing Museum." *Vegas.com*. https://www.vegas.com/attractions/off-the-strip/atomic-testing-museum.

National Park Service. "Memorandum of Agreement between the United States Department of the Interior and the United States Department of Energy for the Manhattan Project National Historical Park." 10 November 2015. https://parkplanning.nps.gov/document.cfm?parkID=482&projectID=57561&documentID=69127.

– "Scholar's Forum Report November 9–10, 2015." https://www.nps.gov/mapr/upload/MAPR_Scholars_Forum_Report-2-2.pdf.

National Toxic Land/Labor Conservation Service. http://www.nationaltlcservice.us/2011/05/national-tlc-service-established.

Nixon, Rob. *Slow Violence and the Environmentalism of the Poor*. Cambridge, MA: Harvard University Press, 2011.

Norris, Robert S., and Hans M. Kristensen. "Global Nuclear Weapons Inventories, 1945–2010." *Bulletin of the Atomic Scientists* 66, no. 4 (2010): 77–83.

Pred, Allan. *The Past Is Not Dead: Facts, Fictions, and Enduring Racial Stereotypes*. Minneapolis: University of Minnesota Press, 2004.

Sandilands, Catriona. "Lavender's Green? Some Thoughts on Queer(y)ing Environmental Politics." *UnderCurrents: Critical Environmental Studies* 6 (1994): 20–4.

Schwartz, Stephen I. *Atomic Audit: The Costs and Consequences of U.S. Nuclear Weapons since 1940*. Washington, DC: Brookings Institution Press, 1998.

Sedgwick, Eve Kosofsky. *Touching Feeling: Affect, Pedagogy, Performativity*. Durham: Duke University Press, 2003.

Sholette, Gregory. *Dark Matter: Art and Politics in the Age of Enterprise Culture*. New York: Pluto Press, 2011.

Stoler, Ann Laura. "Imperial Debris: Reflections on Ruins and Ruination." *Cultural Anthropology* 23, no. 2 (2008): 191–201.

US Nuclear Weapons Cost Study Project. "50 Facts about U.S. Nuclear Weapons." Washington, DC: Brookings Institution, 1998. https://www.brookings.edu/50-facts-about-u-s-nuclear-weapons.

"Welcome to Nukapedia: *Fallout* wiki." http://fallout.fandom.com/wiki/Fallout_Wiki.

Wende Museum. http://www.wendemuseum.org/about-us.

Widrich, Mechtild. *Performative Monuments*. Manchester: University of Manchester Press, 2014.

Young, James E. "Memory and Countermemory." *Harvard Design Magazine* 9 (1999). http://www.harvarddesignmagazine.org/issues/9/memory-and-counter-memory.

8

Sounding Out the Nuclear: An Atomic Opera

JULIET PALMER, JULIE SALVERSON,
AND PETER C. VAN WYCK

Juliet

The sound of a plane arriving, its propellers slicing through the air, engines throbbing, marks both the beginning and the turning point of the opera, *Shelter*. The first plane carries Lise Meitner to the US in 1946, where she is greeted by a bombardment of camera flashbulbs and questions from journalists. Who is this Jewish "Mother of the Bomb" who refused to work on the Manhattan Project? The second plane carries the Pilot as he soars above the landscape – "we could be anywhere ... a desert, a frozen lake, or a wasteland."[1] Both planes transport us into the imaginary realm of history intermingled with allegory. The sound itself is make-believe: a fictional airplane, conjured by lip trills and tremolo, oscillations both instrumental and vocal veering toward the ridiculous. A cartoon plane reminding us that this is not real – but that in singing the plane into being, we understand that any one of us can become the plane and that none of us are bystanders.

As the sound of the propeller engines subsides, the Pilot descends from the skies, climbs out of his plane, and follows the desert road. In his hand is the musical instrument of the post-atomic age: the Geiger counter. Mounted on the dashboard, the instrument plays a starring role in videos of illicit drives through the Fukushima dead zone. It authenticates danger for tourists wandering the abundant greenery of Chernobyl. It verifies history in the desert of Alamogordo. Geiger counter in hand, the Pilot listens for a spasm of clicking. Like a cicada's mating call, the instrument unleashes a flurry of rat-a-tats as the Pilot approaches

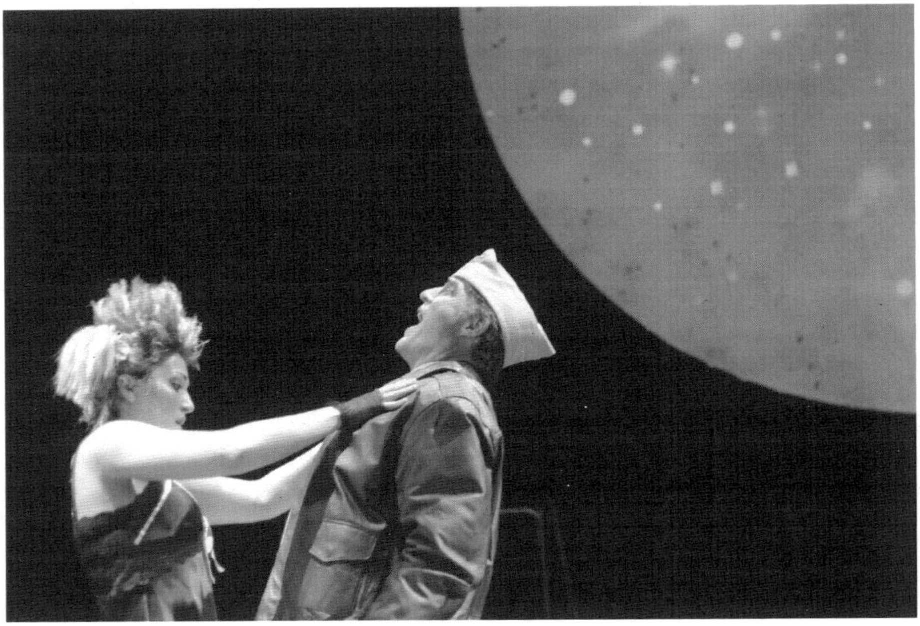

8.1 Hope and Pilot. Teiya Kasahara and Keith Klassen.

the nuclear family home. Locked inside, Hope's breathing quickens, her words fracture, syllables flutter: the Pilot slips in through the window of her bedroom. This is a love scene.

These are all, in fact, love scenes.

Julie

When I was twenty I started going to seminars about how to survive a nuclear war.[2] It was the early seventies, the peace movement was just getting going in North America and the West, the one my generation thought would finally change the world. I remember sneaking out late at night to the field behind our house, staring at the stars to see if a plane overhead would drop something big. I lived in perpetual anticipation of an unexpected explosion. I wanted to be sure that when the world blew up, there would be an escape route, a door with an Exit sign. I clipped an advertisement from a Toronto newspaper and found my way to a convention hall downtown. I spent an August afternoon looking at exhibits about how to keep food for long periods of time and making notes about how to keep warm underground in a Canadian nuclear winter.

In a letter to Gershom Scholem about Kafka, Walter Benjamin wrote: "This much Kafka was absolutely sure of: First, that someone must be a fool if he is to help; second, that only a fool's help is real help. The only uncertain thing is, can such help do a human being any good?"[3] Then, the famous phrase: "There is an infinite amount of hope, but not for us."[4] This suggests risk, engagement, failure. The context is the inevitable pain of ongoing loss in the midst of an almost ridiculous insistence upon life and a hope that is different from messianic hope that does not offer reassurance or certainties. What potential could be offered by a fool's help, a foolish witness? And could a clown opera bear witness to violent histories and act as an intervention in sentimental and melancholic approaches to testimony?

Juliet

In my mind's eye a mushroom cloud blooms in slow motion, its mesmerizing expansion billowing upwards and outwards in silence. The image is black and white, sometimes with a lurid burst of red, blue, or orange, but the image is always silent. If I strain to listen to this inner image, I convince myself that I hear a faint crackle and the hiss of static pops – perhaps the residue of old-time film and tape recording? No. The image remains resolutely silent. Are the countless images of atomic bomb explosions so antique that there is no sound element? Or has my memory stripped away the soundtrack to leave the silence of unspeakable horror?

My first atomic sound memory comes from a much more recent time. It is an explosive boom reverberating through the unlikely modern metropolis of Auckland, New Zealand. The boom is not that of an atomic blast, but rather the crack and tearing of explosives as they rip through the hull of the Greenpeace vessel the *Rainbow Warrior*. Moored in Auckland's Viaduct Basin, the ship was about to head into the controversial Pacific nuclear testing zone of Mururoa. But the French government had dispatched a team of undercover agents to thwart the mission, destroying the boat and killing one crew member. The year of the sinking of the *Rainbow Warrior* was 1985, my first as a student at Auckland University. This astounding act of state violence was a vivid reminder that, although New Zealand seemed to exist at the fringe of global politics, the Pacific Ocean had been host to hundreds of nuclear tests by the governments of Britain, the US, and France. It was a location chosen as the somewhere in the middle of nowhere – *aqua nullius*? Hardly. Shockwaves from the blasts killed schools of fish, which washed up dead on neighbouring islands. Radiation leaked from underground test sites through fissures in fragile atolls. Communities were evacuated from devastated island nations, never to return.

Julie

In 2001 Peter van Wyck, who teaches communications at Concordia University, phoned me up. Did I know that the uranium in the bombs over Hiroshima and Nagasaki came, primarily, from northern Canada? *The Highway of the Atom* was the name given to the Canadian trade route over which ore was transported from Deline, Northwest Territories, to be refined in Port Hope, Ontario. Many of the Dene in the area (Sahtúgot'ine Bear Lake People) were employed to mine and transport the sacks of ore and have since died of cancer. Peter Blow's 1999 documentary calls the place the Village of Widows. When, in the 1990s, the Dene discovered that "their uranium" had been used to build the bombs, they went to Japan to apologize to the survivors, following a path of their own.

In Jewish tradition, there is the path (*haggadah*) and the law (*halakhah*). *Halakhah* is from the root *halakh* – to walk, to go – it is about what is binding, commanded, permitted, forbidden. The *haggadah* is from the root *higgid* – to say, tell, narrate – it is about stories, legends, witticisms. The path is not something followed, it is something transmitted, passed on, listened for. It is dangerous when the requirement to pass on is forgotten, lost.

Juliet

During the 1980s, a group of committed and talented musicians gave voice to the frustrations and outrage of nuclear protesters. The music group From Scratch, led by musician and artist Phil Dadson, performed *Pacific 3, 2, 1, zero,* at the 1984 Paris Biennial. To my ears, the sound of the atomic age is as much the sound of that bomb blast destroying the *Rainbow Warrior,* as it is the visceral drumming and chanting of From Scratch. I became a fan. I bought the LP. I went to live performances. I even became a student of Dadson's. Revisiting *Pacific 3, 2, 1, zero* through the exquisite documentary film by Gregor Nichols, I am struck by the simplicity and clarity of its form. Circles abound: from the formal layout of banks of instruments to the whirling of resonating tubes overhead to the tubes themselves, circular resonators of air and voices. Echoing circular shockwaves of a bomb blast expanding ever outwards. The piece functions as a kind of *aide-mémoire,* inscribing the names of those tiny islands blasted into oblivion by nuclear tests into my ear's memory. Rhythmic bursts of drumming on empty oil cans alternate with the voices of the musicians shouting out the names of the lost and destroyed in scatter-shot syllables: E-ne-we-tak! Mu-ru-ro-a! Bi-ki-ni! Fan-ga-tau-fa!

Peter

A couple of things we know.[5]

First. Until quite recently the Dene of Great Bear Lake, and many others along the Highway, knew nothing of radioactivity. Why would they? How would one even translate such a concept? In the Inuktitut language, I am told, the concept of half-life translates as half-human. In Slavey, the language of the Dene, there just are no words to translate this word. Today – that is, by now – many other translations have been invented, some conceptual and linguistic and some decidedly material and corporeal. This, of course, is only one sense of *translate*.

Second. In August 1998, to coincide with the commemorative events held on the fifty-third anniversary of the nuclear bombing of that city on 6 August 1945, a delegation of ten Dene went to Hiroshima. They went to the end of the circuit, the end of the route that we will call the Highway of the Atom, to convey their apologies for having been involved, and to acknowledge their responsibility for their role: their labour, their complicity, unknowing as it was, and foremost, their land – the territorial archive that was now indelibly stained with the record of their collusion. What had not been registered as traumatic in the first instance – or at least no more or less traumatic than any other of their historical and ongoing contacts with Europeans and southerners – was in a way even less so in the second.

But more than all of the tragic details that came to light – the stories of white and Dene children playing in sandboxes filled with uranium mine tailings, discarded ore sacks used to repair tents and clothing, contaminated materials, and the accidents and spills – more than all of this, the apology on the part of the Dene is incomprehensible, unfathomable. Taking responsibility *for* an other. A singular work of cultural care. Of mourning.

Juliet

I left New Zealand in 1990 for New York. My inner Geiger counter switched off for a number of years. Yes, it was slightly menacing to know that Three-Mile Island wasn't that far away from Brooklyn, but it felt as though the Cold War was over and that the imminent threat from the bomb had subsided. Skip ahead ten years and I became a citizen of Canada, a country whose atomic history is in murky opposition to that of New Zealand and the USA. While the US proudly acknowledges its role in the development of the bomb and the destruction of Hiroshima and Nagasaki, and New Zealand flaunts its nuclear-free status, Canada's relationship to the bomb is more furtive and contradictory. Scientific research and raw materials from Canada helped build the bombs, while Indigenous workers in Canada's northern Dene community paid the price through cancer and early deaths. CANDU reactors have been exported around the world. Closer to home, the lights in my house are powered by nuclear power and my neighbourhood in Toronto hosts a uranium fuel pellet processing plant. At night I lie in bed listening to the haunting sound of train whistles and wonder if another shipment of uranium has arrived from the west.

Did I really hear that explosion on 10 July 1985? Perhaps the answer is irrelevant. The sound and the meaning of the sound had a profound impact on my relationship to atomic science and the politics of the nuclear industry. That sound – imagined or real – continued to resonate in my unconscious, waiting to resurface decades later. The occasion was Tapestry Opera's Lib-Lab, a chance for composers and writers to creatively engage in a ten-day-long dance of serial monogamy. When I met writer Julie Salverson at the composer-librettist laboratory in 2002, our shared fascination with the story of the atomic bomb led us on a decade-long journey. Our arrival point in 2012: the premiere of the chamber opera *Shelter*, charting the course of "a nuclear family adrift in the atomic age."[6] From the blast in Auckland's harbour I reached back to unearth other musical connections to nuclear history, finding the sound world for the opera and its cast of real and imaginary characters. What follows is a sketch of how musical and historical research, along with a slippery quantity of intuition, informed the composition of the opera and the development of its musical language.

Julie

In 2003, Juliet and I wrote a fifteen-minute opera commissioned for the opening of Tapestry New Opera and Nightwood Theatre's studio theatre in Toronto. *Over the Japanese Sea* is about a businessman meeting an office cleaner, who turns out to have been the pilot of the plane that dropped the bomb. The young man is losing his lover, and asks the older man for advice when he overhears the cleaner reminiscing about a lost love:

> Did you love her? Utterly. What was her name? Storm. She took you for a ride? I flew the plane, we carved up the sky. Over the Japanese sea. She cut a fine figure, exquisite curves, arching curves of light, blossoms of light. Did she break your heart? No, it wasn't that kind of murder.[7]

The response of audiences to this piece – staged partly in the Tapestry offices with the audience on their feet among the singers, and partly in the studio with the audience seated – encouraged us. Instead of sitting back immobilized by the idea of the bomb, people chatted about their connections to the atomic story: a cousin who was an engineer, an uncle who was a miner, a neighbour who worked in a physics lab. We wanted this engagement from people, this energy.

In the fall of 2004, Tapestry brought together a professional team to experiment. Steve Hill of Leaky Heaven Circus came from Vancouver, and three actors, three singers, Wayne Strongman, Steve, Juliet, and myself walked into the studio with scraps of text and storyline from me, several years of research notes and conversations in Juliet's and my heads, and absolutely no idea what would happen. *Shelter* is the story of a nuclear family that implodes, giving birth to a radioactive daughter and trying to keep her a secret from the world. What happens when the pilot arrives to take her up into the sky? Weather warnings, a storm is coming, and angels are blowing up heaven.

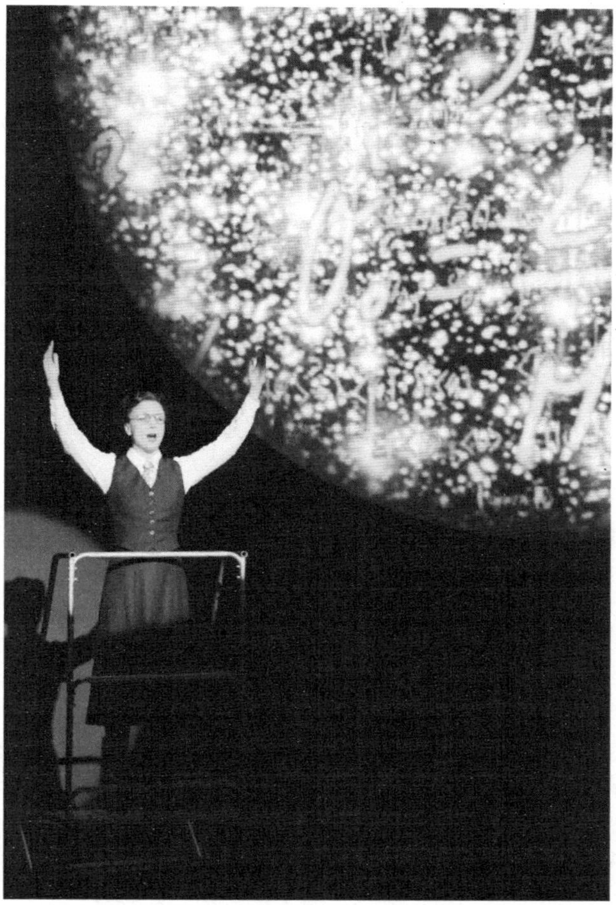

8.2 Lise Meitner with Equations. Andrea Ludwig.

We came away with a blur of fantastic images and sounds, and above all, a sense of energy and involvement. Juliet improvising every sound imaginable from a piano and its strings while everyone played with the characters and text. Clown Martha Ross as the mother, Claire, desperately cleaning her house while tenor Ian Funk, as Thomas the father, paints their new radioactive daughter's face with concealer make-up and croons her a lullaby. Soprano Tammy Hummel as the radioactive daughter Hope making passionate love to bouffon Adam Lazarus – as the stranger, the pilot – on top of a car in a thunderstorm, with panicked parents Martha and Ian arguing inside same car. Adam falling in love with Hope, then pacing alone late at night, listening with his Geiger counter to the "music," the "history" of the house. I went away and wrote the first of many drafts of the full libretto.

Juliet

Chemistry isn't enough.
You need form, elegance, physics
To release this fire, fire,
This hidden energy.
To shatter, to split,
You need physics.[8]

The opera's central real-world character is the physicist Lise Meitner. As a Jewish Austrian citizen, Meitner was forced into exile in Sweden in 1938 following the Anschluss. Her research on nuclear fission in Berlin with Otto Hahn continued through intermediaries and via correspondence. In spite of their thirty-year-long partnership, however, by the summer of 1945 her long-time collaborator was unwilling to acknowledge Meitner's role in the development of fission. This distancing of Meitner in the shadow of the Third Reich was a far cry from their earlier years working side by side in a laboratory at the Kaiser Wilhelm Institute of Chemistry:

> When our work was going well we sang duets, mostly Brahms Lieder, which I could only hum, while Hahn had a very good singing voice … If he was in an especially good mood he would whistle large sections of the Beethoven violin concerto, sometimes purposely changing the rhythm of the last movement just so he could laugh at my protests.[9]

I was struck by this image of two scientists working together, about to change the course of history, all the while singing the music of the nineteenth-century Romantic Brahms. Meitner was ardently committed to her scientific endeavours, and Brahms's love songs began to resonate in my imagination as expressions of her passion for discovery. Much of Meitner's musical language as a character in the opera inhabits this world. Not knowing which Brahms *lieder* Hahn and Meitner sang together, I chose three whose lyrics offer a peculiarly prescient insight into both her character and the imminent atomic age.

"Nein es is nicht auszukommen mit den Leuten" ("No, there is no getting on with other people") from Brahms's *Liebeslieder-Walzer* op. 52, no. 11, sets lyrics from Georg Friedrich Daumer's "Polydora" for vocal quartet and piano duet. A reworking of this furiously fast and defiant waltz becomes the musical anchor for Lise's lone voice as she steadfastly refuses to join the Manhattan Project. Later in the opera, as she struggles to protect the innocence of science from the corruption of military "necessity," she sings, "if you cross this thin line, nothing will remain of beauty. Do not cross me."[10] The piano strides along with Lise, underscoring her fierce resolve.

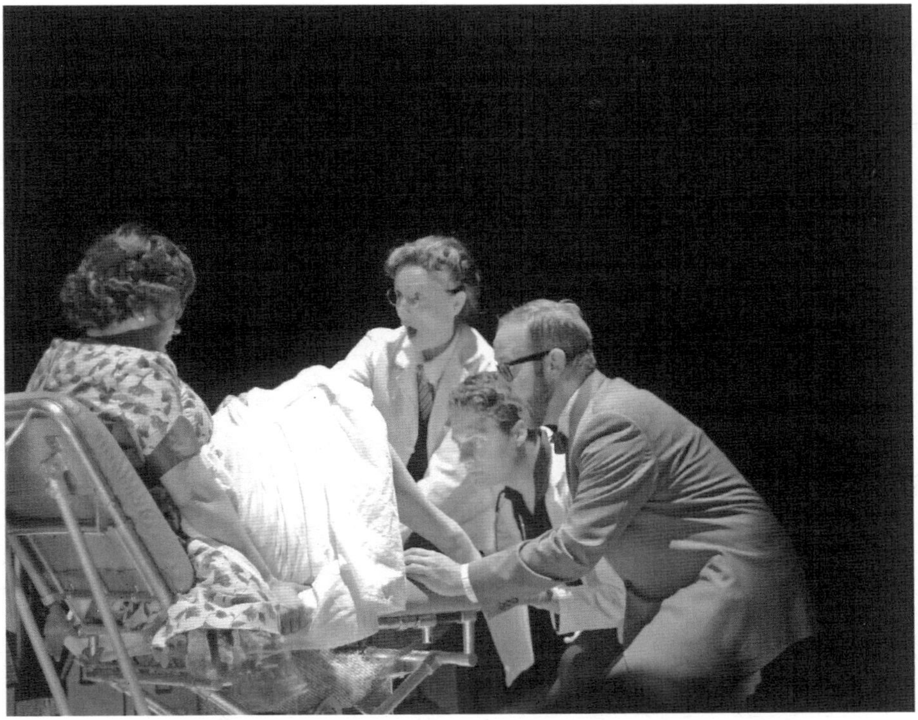

8.3 The Birth. Christine Duncan, Andrea Ludwig, Andrew Love, and Connor Lafarga.

At the heart of Meitner's principled resolve is her love of science. Brahms's *Neue Liebeslieder* no. 14, "Flammenauge," to Georg Daumer's free translation of a Russian poem, speaks to the wonder of scientific discovery but also to the possibility of a reversal of Nature's order:

Flaming eyes, dark hair,
Sweet and audacious boy....

Can the sun's fire make ice,
Or turn day into night?

Is the field so full of light
That the flowers stand in darkness?[11]

Traces of "Flammenauge" surface throughout the opera: this is the music of Lise falling in love with science. It's also the music of the opera's "cartoon" characters Claire and Thomas falling in love with each other. Love as innocent and unknowing, unaware that it contains the seeds of its own demise.

The myth of Prometheus stealing fire from the gods reminds us that knowledge so often outstrips wisdom. Just as scientific passion helped birth the atomic bomb, our cartoon parents Thomas and Claire bring to life the strangely glowing child Hope. As Lise cradles the baby, these two worlds of science and the imagination collapse and collide, Lise acknowledging the pain of the new as she sings a transfigured version of Brahms's "Sind es Schmerzen, sind es Freuden?":

Are they sorrows or are they joys
Which tug at my breast?
All the old desires leave;
A thousand new flowers bloom.[12]

In the penultimate scene of the opera, Hope follows her heart and leaves with the Pilot, Lise singing together with Thomas as black rain falls and the house burns. "Sind es Schmerzen" returns in a filigree re-imagining of Brahms's love song:

Through the dusk of tears
I see shining, two suns[13]

Peter

We have just sat through a screening of Peter Blow's *Village of Widows* in Edmonton – neither of us have seen it in a few years, and never on a big screen. Over a hundred people in the audience. We are here for the premiere of Julie and Juliet's opera, *Shelter.* It is being mounted by the Edmonton Opera Company, and we have also been invited by Edmonton's Festival of Ideas to give a talk before the opening. Danny Gaudet and Walter Bayha, community members from Deline on Great Bear Lake, have agreed to say a few words after the screening. The room is full of mostly white Edmontonians. Danny and Walter take turns speaking. They seem taken aback. Whether this is because they have just watched an intimate film about their community that they may not have seen for a long while, or because of the startling fact of a house full of serious-looking white people, is not clear. The audience is quiet, and moved. Danny and Walter speak with eloquence and care and seem to be saying things they had not planned on saying. They tell us that there are only a handful of elders left in the community – there has just been no way to heal. The elders, they tell us, cannot read, and they cannot write. They have only words for those with ears with which to hear them. And if you can only speak, they say, your speech must be trustworthy. If you're caught in a lie, you're in serious trouble … It's not just something you wrote – speech can always repair the errant writing. But if all you have is speech, there is no damage control –

you cannot switch to another channel to make reparation. And, of course, this all relates in a poignant and tragic manner to the catch-22 of testimony for government relations around remediation and reparation: oral testimony is not valid, and writing can only be understood as the transcription of undependable speech. And this, in its way, is the very predicament that *Shelter* seeks to address; that is, to read and write an atomic narrative for the twenty-first century – sited in the *oikos*, the family, the home – through the craft of song and story.

They tell us about George Blondin, an important Dene writer and activist who died in 2008 – the difference between George's life and his, says Walter, is that he went to school. Everything they tell us is anchored in the present. Not a whiff of *we're beyond all of this, we need to move on*, a position we've heard a lot in the last decade. They emphasize that the cultural transmission of history – even family history – is completely broken in the community. Walter said something about having *just* (in the last few years) begun to gather together his own immediate family history (i.e., his grandfathers'). And then someone asks Walter a question about Blondin's "we are all brothers" speech in the film. This is a very moving scene in which the Dene delegation are visiting a hospital for Korean *hibakusha* in Hiroshima. Blondin's words gather together the Koreans and the Dene as survivors and victims – "we even look alike," he says – and affirms the Dene's sense of the brotherhood of all humanity. Walter's response is completely unexpected: he basically says that all this was just *a generational thing*; they no longer feel this same sense of the meaning of being Dene. Abstractly, yes, but "this is an older sense of being an Indian," he says (a word that Blondin still used, he notes). The channel between elders and children is damaged – the intergenerational pedagogy, based on land, and practice and story, no longer functions in the way it did. Now it seems they just have the stories of the prophet, Ayah, spun into a kind of cultural salvage on the one hand, and an apocalyptics on the other. All of this seems to set the stage for the performance.

But what is this opera, *Shelter*, about? It's not *about* Great Bear Lake, or the highway of the atom, is it? How is it that stories are *about* things? I think back to speaking with elders in the community; my foolish questions always seemed to provoke the stories they deserved. Here, I have no idea what to expect – I have seen only a very early version workshopped several years ago – so, *voilà*. It's a kind of atomic cadenza! Gathering together a thick field of reference for an improvisational reworking of a story. From PJ Harvey to Walter Benjamin by way of Brahms and Ellington, *Shelter* takes and reworks all of the pieces, the stories, the monuments and archives, ruins and photographs that we have assembled and heard and lived with since we began this research. It delivers an allegorical work that is deeply moored to all of these things but also referentially constellated into a beautiful set of figures. That is, it's a story about how to think this history. After the opening, I see it – it actually follows the route, the highway of the atom, but differently.

I remember a caution I wrote to myself as I struggled to pry the concept of a route away from a simple line on the ground.

In following a route such as the highway of the atom, one must take care.[14] "History," Maurice Halbwachs wrote, "may leave us passengers on a boat. As the riverbanks pass by, everything he sees is neatly fitted into the total landscape. But suppose he loses himself in thought … Later on he will be able to remember where he travelled but few details of the landscape … He *will be able to trace his route on a map* [but] he has not really been in contact with the country through which he passed"[15] (emphasis added).

Good advice for a distracted traveller (and for us too); to be lost in thought is surely an occupational hazard. But how to really be in contact with the country? This is the question.

Yet we suffer, here as elsewhere, from too few routes into the past. And perhaps into the present, too. From the point of view of this route – this highway of the atom – what is the power of the past, apart from our efforts to reconstruct it? Is it, as Peirce put it of memory, *a gentle compulsiveness*?[16] It doesn't seem very gentle …

The image that comes to my mind is musical – a kind of contortion of Bruce Chatwin's realization that "music is a kind of memory bank for finding one's way about the world" – in other words, musical phrase is a "kind of map reference."[17]

The route, showing my own European roots, might thus be thought of as contrapuntal – that is, *punctus contra punctum*. We might then shift this thought from the note against note of polyphony, to a moment against moment, or point against moment. Then our route would not be simply a horizontal movement, a line, a vector, a spreading path, an extension, the lay of the land. Rather, we would think of it, simultaneously, in its vertical development – as with bodies of all kinds – in a presupposition that is both generative *and* controlling. Extension *and* intensity. Paradigm *and* expression.

The extension of the route is also a thickness of intensity, a palimpsest. Really, there's no point in making a new portage when there's one already there.

Shelter performs this operation. Yes, it is a kind of map reference, an *aide-mémoire* for finding one's way about the world, but it derives its sonic and narrative force precisely from a kind of contrapuntal development of citation, and place and figure, and time. It is a critical topography – with an itinerary of invisiblities – of the unstable atom.

Hope (scene 6):
My bones hold secrets, blind alleys,
A jumble of paths
From the mine to the lake through burial grounds
Thick with bones to the desert.
Men carried me.

Science handled me.
My bones hold secrets from the mine to the lake
To the desert's burning heat, heat.[18]

Juliet

Witnesses to the awesome power of the first atomic bomb test in the desert of New Mexico spoke of a blinding white light, of colours filling the sky, and of the unexpected crack of sound as it arrived moments after the blast. Of seeing two suns. In one account, scientists listened to big band music on the radio as they waited for the weather to clear in the early hours before the detonation. My curiosity was piqued. What was the music they listened to on that morning of 16 July 1945? As I searched again for a creative touchstone, a warped sense of humour inspired me to choose one of the year's top songs, Duke Ellington's "I'm Beginning to See the Light." The song topped the charts in three different versions in 1945, so it's not inconceivable that it found its way onto the airwaves at that crucial moment. Don George's lyrics were much praised for their virtuosic stringing together of imagery of light and heat:

I never cared much for moonlit skies,
I never winked back at fireflies,
but now that the stars are in your eyes,
I'm beginning to see the light.[19]

This music comes from the cusp of the rock'n'roll era, from a time of tremendous upheaval and rupture. Slivers of the Ellington song insinuate themselves into the opera as the atomic world starts to bleed into the present. Thomas glimpses Claire and begins to fall in love; Lise is on the verge of discovering fission. With Ellington's song, a flickering shadow in the ensemble, Thomas sings of the angel of history and the seductive inevitability of progress:

A storm just blew in from Paradise[20]

The storm promises destruction, but we stand awe-struck by its beauty, unable to stop its progress.

As the opera heads toward its climax and the dropping of the bomb on Hiroshima, I found inspiration in the harsh thrash of Japanese punk. One band in particular caught my ear as I shaped the voice of Hope – Japan's all-female rock group the 5,6,7,8's. Lead singer and guitarist Yoshiko "Ronnie" Fujiyama describes their music: "I like Chuck Berry and we wanted that sound, but we wanted to deconstruct that rock'n'roll into punk music by using distortion and noise and screaming."[21]

If Brahms was the music of Romantic rapturous love and Ellington that of seduction, then this is the music of reckless love. The kind of love that takes you to the brink, to the edge of destruction, and maybe even right over. Hope's vocalizations begin from deconstructed speech and morph into a cartoon rock/punk hybrid. Hers is the voice of youth, of a new era, of the thrill of the bomb, but also of the bombardment which followed that of the bombs on Hiroshima and Nagasaki. This is the cheerful bombardment of American culture dropped onto Japan's postwar: of baseball and rock'n'roll, and beehives, and of the propaganda of the "peaceful atom." Hope sings to us from the atomic future and her visions.

I won't pretend that there is anything historically accurate about the musical choices I made in composing *Shelter.* This grab bag of disparate musical sources was fodder for a creative process. A wanton rummaging in history's closet. An attempt to give voice to five characters along the Highway of the Atom.

Julie

When Peter, Juliet, and I heard about the Dene apology to the Japanese, we asked ourselves, "What does this mean, this taking of responsibility for something that was not their fault?" In his discussion of an ethic that is "infinitely demanding," Simon Critchley addresses the problem of accepting an infinite responsibility without extinguishing oneself as a subject. He suggests that humour offers a human terrain of self-knowledge that "recalls us to the modesty and limitedness of the human condition, a limitedness that calls not for heroic, tragic affirmation, but comic acknowledgement; not Promethean authenticity, but laughable inauthenticity" at the heart of subjective experience.[22] This engagement of the subject through a mature humour "does not paralyze the ego, but rather drives the ethical activity of the subject."[23] It is absurd, even ridiculous, to risk answering the call of another. It is absurd to think that my availability might contribute anything in the face of another's violation, yet I step forward all the same. It is frightening, there is a nakedness in this kind of clownish contact. But it offers another way through the victim/hero polarities, another way through the tragic appropriations of pain. The clown is not a hero, but is heroic in courage, in being available to the possible, no matter how absurd and unlikely.

In *The Clown and the Crocodile*, Joseph C. McLelland writes that comedy depends not on happy endings, but on

> a different sense of the real … Comedy does not agree with the ancient rule according to which Western philosophy and theology have been constructed, namely that "what-is must be intelligible." It renounces this rational decision (a circular argument) to count as real only what is understandable, and bows

before a presence whose name it cannot pronounce. It forces a new question: *Is life worth celebrating, despite its contradictions?*[24]

Peter

Field Note: Tsiigehtchic (Arctic Red River), 6 August 2005

> The Norweta, the Mackenzie river boat I have been travelling on, arrives mid-day.[25] Fourteen eagles sunning themselves at the mouth of the river. I think it was fourteen. I can find no fish being smoked. No bales tied and waiting for transport to Aklavik. It is hot. There is an inviting pond on the large delta area at the confluence of the Red and Mackenzie rivers. High above stands the village. I stop and speak with an old woman who is standing near the church. I ask her about swimming in this pond. "No," she says, and then tells me a long story. After one of the innumerable battles between the Gwitch'in and Inuit, scores of Inuit were thrown into the lake, the wounded to drown, and all the bodies were just left in the lake. Ever since this time, she said, the eagles wait for the bodies to rise, but all that emerge are ghosts. The lake is full of ghosts, too full, so there is just no room to swim. She uses a Slavey word to describe when this happened; I think it means the really long-ago time. Just too many ghosts.

Julie

My father is almost ninety. His most vivid memory of the atomic bomb, he tells me, is of my mother coming into their kitchen with a magazine in her hands. "I was cooking, a pot was simmering, I was stirring. She just stood in the doorway, shaking, in a yellow sundress. *Life* magazine photos, I think. Thin-skinned, your mother." He smiles, and reaches a thin spotted arm to pat me on the knee. "She was missing the tough layer, the one that keeps you safe. A strong woman on the outside. Inside fragile. Things could break."

South of the Arctic Circle on the shores of Great Bear Lake, the surviving elders of Deline say that a very long time ago, an old man, an elder they call the Prophet, warned them. These are the people whose families hauled the raw uranium ore over the waters, whose women sewed tents with used uranium sacks. Who still have no word for radiation.

Juliet

As I drift off to sleep, a screeching tracery of lines and circles fills the deep night sky: a squadron of fighter jets write unseen words, punctuated by the sonic booms of a broken sound barrier. Five years old, I understand why my father does not tuck me into bed this night. He flies above – circling in a lullaby of Cold War anxiety. My father the pilot was never far from my mind as I searched for the voice of the opera's Pilot. Seduced by the thrill of speed and flight to a task compelling in its inevitability, the Pilot stumbles into a state of wonder. Suspended in mid-flight, the music of the final scene rises endlessly upward toward the moment of release. Hovering among clouds "brilliant blue, red, gold": a constellation of amazement and dread, of innocence and horror. "Heaven and earth falling over the Japanese Sea."

NOTES

1 Salverson, "Shelter," 2012.
2 Versions of this passage appear in "Shameless Acts of Foolish Witness," *Comedy Begins with Our Simplest Gestures: Levinas, Ethics and Humor*; and *Lines of Flight: An Atomic Memoir.*
3 Handelman, *Fragments of Redemption.*
4 Ibid.
5 Revised from van Wyck, *The Highway of the Atom*, §2.01, 45; §3.09, 129; and "An Archive of Threat."
6 Salverson, "Shelter," 2012.
7 Salverson, *Over the Japanese Sea*, 2003.
8 Salverson, "Shelter," 2012.
9 Sime, *Lise Meitner*, 35.
10 Salverson, "Shelter," 2012.
11 Brahms and Daumer, "Flammenauge."
12 Brahms and Tieck, "Sind es Schmerzen, sind es Freuden."
13 The original lyric reads: "Through the dusk of tears, I see suns standing in the distance."
14 Versions of this passage appear in van Wyck, "Highway of the Atom," and *The Highway of the Atom.*
15 Halbwachs, *The Collective Memory*, 52.
16 Peirce, *Collected Papers*, 7:390.
17 Chatwin, *Songlines*, 108.
18 Salverson, "Shelter," 2012.
19 Don George, 1945.
20 From Walter Benjamin's description of Paul Klee's painting *Angelus Novus.*
21 Simon Bartz, "The New House Band chez Tarantino."
22 Critchley, *Infinitely Demanding*, 82.
23 Ibid., 84.
24 McLelland, *The Clown and the Crocodile*, 83.
25 Revised from van Wyck, *The Highway of the Atom*, §4.18, 202.

BIBLIOGRAPHY

Bartz, Simon. "The New House Band chez Tarantino." *Japan Times*, 16 November 2003. https://www.japantimes.co.jp/culture/2003/11/16/music/the-new-house-band-chez-tarantino.

Benjamin, Walter. "On the Concept of History." *Selected Writings: 1938–1940*, vol. 4, edited by M.P. Bullock et al., 389–400. Cambridge, MA: Belknap Press, 2003.

Brahms, Johannes, and Georg Daumer. "Flammenauge." Translated by Juliet Palmer. *Neue Liebeslieder*, op. 65, no. 14. Berlin: Simrock, 1875.

Brahms, Johannes, and Johann Ludwig Tieck. "Sind es Schmerzen, sind es Freuden." Translated by Juliet Palmer. *15 Romanzen aus L. Tiecks Magelone*, op. 33, no. 3. Winterthur and Leipzig: Rieter-Biedermann, 1865.

Chatwin, Bruce. *The Songlines*. London: Penguin Books, 1987.

Critchley, Simon. *Infinitely Demanding: Ethics of Commitment, Politics of Resistance*. London: Verso, 2007.

George, Don, lyricist. "I'm Beginning to See the Light" by Duke Ellington. RCA Victor Records, 1945.

Halbwachs, Maurice. *The Collective Memory*. Translated by Lewis A. Coser. New York: Harper Colophon Books, 1980.

Handelman, Susan A. *Fragments of Redemption: Jewish Thought and Literary Theory in Benjamin, Scholem, and Levinas*. Bloomington: Indian University Press, 1991.

McLelland, Joseph C. *The Clown and the Crocodile*. Louisville: John Knox Press, 1970.

Peirce, C.S. *Collected Papers of Charles Sanders Peirce*, vol. v. 7. Cambridge, MA: Belknap Press of Harvard University Press, 1974.

Salverson, Julie. *Lines of Flight: An Atomic Memoir*. Hamilton: Wolsak and Wynn, 2016.

– *Over the Japanese Sea*. Composer, Juliet Palmer. Unpublished libretto, Tapestry New Opera, 2003. (The last sentence of the text, "Heaven and earth falling over the Japanese sea," is a quotation from this unpublished libretto.)

– "Shameless Acts of Foolish Witness." In *Comedy Begins with Our Simplest Gestures*, edited by Brian Bergen-Aurand. Pittsburgh: Duquesne University Press, 2017.

– "Shelter." In *When Words Sing: Seven Canadian Libretti*, edited by Julie Salverson. Toronto: Playwrights Canada Press, 2021.

Sime, Ruth Lewin. *Lise Meitner: A Life in Physics*. Berkeley: University of California Press, 1996.

van Wyck, Peter C. "An Archive of Threat." *Future Anterior* 9, no. 2 (2013): 53–80.

– *The Highway of the Atom*. Montreal and Kingston: McGill-Queen's University Press, 2010.

– "Highway of the Atom: Recollections along a Route." *Topia* 7 (2002): 99–115.

9

Poetry and Anti-Nuclearism: *Τεχνη* and the "Fundamental Project"

JIM KRAUS

To de-animalize human mentality, to purge it of obsolete
evolutionary characteristics, in particular of death,
which foreknowledge terrorizes the contents of skulls with,
is the fundamental project of technology.

– GALWAY KINNELL

"Scattered by the Wind"

What use is poetry in the struggle to understand "the nuclear" and its ominous "ism"? How can an understanding of poetic practice be extended so as to redirect and perhaps oppose the complex of ideas, ideologies, and policies generalized as nuclearism? To the extent that one agrees with Ezra Pound's famous statement, "Literature is news that STAYS news,"[1] one can also agree with his delineation of the three elements that animate poetry: *phanopoeia*, *melopoeia*, and *logopoeia*: "You use a word to throw a visual image on to the reader's imagination, or you charge it by sound, or you use groups of words to do this."[2] Upon understanding that poetry is the merging of image-making with sound and with the logic of language, one can begin to see how poetry might have an effect on the news of nuclearism. Through its capacity to evoke image, music, and various types of logic, poetry prompts its readers to renew their interpretive potentials, thus renewing sensibilities about the immanence of nuclearism and its toxic consequences. In response to the idea of technology's "fundamental project" as a death-seeking juggernaut, one can assert an idea of Pound's that centres on the Greek etymon Τεχνη (*technē or* TEXNE) and a shifting of emphasis from death to creativity.[3] In this way, Τεχνη, meaning skill and more importantly habit of mind,[4] stands in opposition to the technology of nuclearism.

In "The Speed of Darkness," Muriel Rukeyser writes: "The universe is made of stories, / Not of atoms."[5] While the destructive power of the atomic bomb seems to ensure the ascendency of science, poets like Rukeyser strive to assert a moral authority that scientists lack. It is in this assertion that one can begin to show the function of poetry in the context of nuclearism. And in this way one can examine poetry's own fundamental project (not the project described in Galway Kinnell's poem) and consequently ask how the elemental particles of poems begin to take form and how is it that poetry can make a claim to asserting moral authority.

It was in the 1960s that the term nuclearism began to be used to describe advocacy for nuclear weapons.[6] While the imagination transforms the idea of nuclearism into fear and dread – the consequence of ever-expanding arsenals of nuclear weaponry and the ever-accumulating presence of toxic nuclear materials in the environment – poetic language renders fear and dread into a form of abjection capable of evoking a moral sense that influences human action. Since the bombings of Hiroshima and Nagasaki, the size of arsenals has grown to include ever-larger nuclear bombs as well as smaller "tactical" nuclear weapons and even smaller "depleted" uranium bullets (which are already in use). While the destructive potential of nuclear arsenals is well-known, the threat of their use crosses over into the realm of actual terror, if not horror. This is the experience of what Julia Kristeva calls the abject: "a frontier, a repulsive gift … that sublime alienation, a forfeited existence." She further explains, "victims of the abject are its fascinated victims – if not its submissive and willing ones."[7]

Kristeva's use of the term "sublime alienation" suggests consideration of the traditional aesthetic category of the sublime in relationship to nuclearism. In her 1984 article "The Nuclear Sublime," Frances Ferguson shows how the sublime as an aesthetic construct is an experience of utter horror in the proximity of death. In 1989, Rob Wilson extends Ferguson's work by developing a nuanced theoretical framework within which he examines a wide range of poetic responses to the fear of nuclear disaster. Wilson questions whether an anti-nuclear position exists as part of the history of the Cold War, "except in the discursive exile of theory," and advocates a kind of "radical negativity" described by Theodor Adorno. Wilson's examples range from Wallace Stevens to Allen Ginsberg. And of course, the idea of annihilation is in itself a kind of radical negativity. John Gery's *Nuclear Annihilation and Contemporary American Poetry: Ways of Nothingness* also explores this with studies of works by John Ashbery, Amiri Baraka, Gwendolyn Brooks, James Merrill, Adrienne Rich, and others.

It is the point at which radical negativity becomes empathetic and elegiac that it reaches its full potential. One example of this is the song "I Come and Stand at

Every Door," composed by Pete Seeger and based on a poem by Turkish writer Nâzım Hikmet.[8] In the poem, the ghost of a seven-year-old girl who died during the atomic bombing of Hiroshima describes her death and pleads that children be spared from such horror: "My hair was scorched by swirling flame / My eyes grew dim, my eyes grew blind."[9] In a particularly moving version, by the group This Mortal Coil, the song is rendered as a dirge, a slow-moving processional.[10] While the ghost of the dead Hiroshima child makes her door-to-door visitations, she pleads with the world to awake from its complacency. The spectre of the seven-year-old girl might function as a guide, like Dante's Virgil, or even like Beatrice herself, but the emotional force of the song is limited to asking that her listener fight for peace. William Heyen in "A Poetics of Hiroshima" examines the problem of proxy, or proximity, by attempting to drive his poem directly into the epicentre of the experiences of victim and perpetrator: "into the syntax of Hiroshima, 'Little Boy' plunges."[11] In this way, Heyen is able to personify the fractured consciousness of nuclearism itself, seeking to reconcile that which cannot be reconciled; and thus rendering the nuclear sublime at its most problematic: "I have no faith except in the half-life of poetry. I seek radiation's rhythmic sublime. I have no faith except in atrocity. I seek the nebulous ends of time. This is the aria those cities have made of me. I hope my centered lines retain their integrity. I have no faith except in beauty."[12]

In the decades since the bombings of Hiroshima and Nagasaki, nuclearism has become normalized as the world has grown perilously, if not absurdly, confident that the imagined horror of mutually assured destruction has led to a functional type of deterrence that prevents an actual nuclear war from occurring. Whether this is the case, only the future will show. Nonetheless, other forms of nuclearism related to the manufacturing of radioactive materials for nuclear weaponry, nuclear power, as well as nuclear science more broadly, bring unsolved, and perhaps unsolvable, environmental problems deriving from the production, accumulation, and disposal of toxic nuclear waste.

Compost versus Radioactive Decay

In the poetic imagination, images of waste and its disposal, as Kristeva explains, are related to abjection, the intractable confrontation between subject and object. As we experience abjection, terror and horror are the consequence; it is in the experience of abjection that images of excrement and vomit seem to prevail: "Excrement and its equivalents (decay, infection, disease, corpse, etc.) stand for the danger to identity that comes from without: the ego threatened by the non-ego, society threatened by its outside, life by death."[13] Toxic nuclear waste is particularly well-aligned to abjection as it cannot be composted in a time-bound process on a human or biological scale. Instead, nuclear decomposition is measured in the half-life. For the physicist, this is a measure of the instability of a radioactive isotope, the amount of time it takes for a particular isotope to decay, to transform into something else, often another radioisotope. One of the best-known forms of toxic nuclear pollution is the creation of radon gas from the decay of radium. The half-life of radium, once used to create "glow-in-the-dark" watches, is approximately 1,600 years. Whereas the half-life of radon is approximately 92 hours. Thus, radon (a gas) is extremely unstable, highly toxic, and a well-established carrier of lung cancer risk.[14]

It is perhaps an attribute of the Anthropocene that time itself seems to be altered. One has only to remember Zeno's dichotomy paradox to understand the half-lives of the radioactive materials that make up toxic nuclear waste and how they place such waste relatively outside time as it is experienced on a human or biological scale. It is as if radioisotopes are from an alternative universe, interwoven with the human one, forever toxic and forever the source of horror and dread. This is one reason we find concepts of nuclearism nested within the experience of abjection. At its worst, the experience of this sort of abjection is itself outside time, as if it has no end, a seeming eternity of half-lives. It is within the experience of abjection that poetic responses to nuclearism take form. Artistic transformation, as Kristeva explains, "takes from ritual space what theology conceals: trans-symbolic jouissance," a counter to "threatening the unity of the social realm and the subject."[15]

An intense sensibility about the biologically time-bound transformations of toxicity and decay is central to Walt Whitman's poem "This Compost," written before chemists and physicists began bringing us into the nuclear age. It might well be seen as introducing us to the problem of how toxicity is related to infectious diseases and how it seems to be consumed and transformed by the Earth:

> Now I am terrified at the Earth! it is that calm and patient,
> It grows such sweet things out of such corruptions,
> It turns harmless and stainless on its axis, with such endless
>   successions of diseas'd corpses,
> It distills such exquisite winds out of such infused fetor,
> It renews with such unwitting looks its prodigal, annual, sumptuous crops,
> It gives such divine materials to men, and accepts such leavings
> from them at last.[16]

As Whitman celebrates Earth's capacity for distilling "exquisite winds" out of the "infused fetor" of decomposing bodies, he sees that the process "is terrible," showing again an example of how a form of sublime abjection is able to propel the poem forward.

Allen Ginsberg, who famously asserted himself as the literary descendant of Whitman, wrote the poem "Plutonian Ode" on the occasion of an anti-nuclear protest at a plutonium bomb trigger facility run by Rockwell Corporation near Denver, Colorado.[17] In 1981, San Francisco's City Lights Books published *Plutonian Ode and Other Poems, 1977–1980*, which opens with a photograph of Ginsberg and friends from the anti-nuclear group Rocky Flats Truth Force sitting in meditation on railway tracks outside the Rockwell facility. They had stopped a train carrying toxic nuclear waste. The poem, like much of Ginsberg's work, has the feel of a Whitman poem – long lines, often called "breath lines" – with sixty-five lines in three sections of differing tonalities. In the opening lines, Ginsberg addresses the spirit of Whitman directly: "At last inquisitive Whitman a modern epic, detonative, Scientific theme / First penned unmindful by Doctor Seaborg with poisonous hand, named for Death's planet through the sea beyond Uranus." Seaborg is Dr Glenn Seaborg, the chemist who in 1940, along with a group of other scientists, discovered plutonium. Seaborg was also part of the Manhattan Project, which designed and manufactured the bombs dropped on Hiroshima and Nagasaki. Plutonium is the radioactive isotope used in the bomb that was dropped on Nagasaki.[18]

"Plutonium Ode" proceeds with the poet forcing and reinforcing a connotative association between plutonium and the mythological underworld. Ginsberg calls plutonium a "chthonic ore" and asserts that its meaning derives from the deepest known origins of beliefs about hell, fire, and time itself:

> Radioactive Nemesis were you there at the beginning black Dumb
> tongueless unsmelling blast of Disillusion?
> I manifest your Baptismal Word after four billion years
> I guess your birthday in Earthling Night, I salute your dreadful
> presence lasting majestic as the Gods,
> Sabaot, Jehova, Astapheus, Adonaeus, Elohim, Iao, Ialdabaoth,
> Aeon from Aeon born ignorant in an Abyss of Light,
> Sophia's reflections glittering thoughtful galaxies, whirlpools of
> starspume silver-thin as hairs of Einstein!
> Father Whitman I celebrate a matter that renders Self oblivion!

Ginsberg then drives the poem forward as an incantation, evoking the primary meaning of "ode": a poem meant to be sung. In the years since its original composition, "Plutonium Ode" has been set and performed by a range of musicians, actors, and composers, most notably by Philip Glass, whose Symphony no. 6 encompasses an operatic setting of the entire poem.[19] Also, Owen Plotkin created a video setting of the poem consisting of Ginsberg's own reading set against archival images of nuclear devastation.[20]

The poet progressively calls to the spirits of ancient knowledge and mythology in order to merge the pre-Christian, pagan mythos, nominally, with the spirit of contemporary scientific knowledge, denominating places where "nuclear reactors create a new Thing under the Sun, where Rockwell war-plants fabricate this death stuff trigger in nitrogen baths." The poem continues, expanding within its strophic crescendo: "I roar your Lion Roar with mortal mouth," then after a full stop, but without stanza break, shifts to an evocation of the awful minutiae of radioactive materiality: "One microgram inspired to one lung, ten pounds of heavy metal dust adrift grey Alps / the breadth of the planet, how long before your radiance speeds blight and death to sentient beings?" He then merges his consciousness with the plutonium itself as he chants with increasing intensity from his new perspective: "I chant your absolute Vanity. Yeah monster of Anger birthed in fear O most / Ignorant matter ever created unnatural to Earth! Delusion to metal empires! / Destroyer of lying Scientists." This last allusion is likely to the deaths of Manhattan Project scientists from radiation poisoning.[21]

Once the chant acquires the full force of its subject, the true purpose of the ode is revealed to assume the power of a curse: "Canker-Hex on multitudes learned or illiterate!" This section of the poem is unambiguous, as the force of the chanting is directed at and through the various connections that together constitute both the plutonium itself and the "practitioners in Black Arts" in order to "embody your ultimate powers!" Then the poet casts the full force of the chant: "my breath near deathless ever at your / side / to Spell your destiny, I set this verse prophetic on your / mausoleum walls to seal you up Eternally with / Diamond Truth! O doomed Plutonium."

The ode shifts to the physical reality of place, purpose, and time, with a description of the specific part of Colorado near Denver, as if having awakened from an intense dream to describe the particulars of the morning: "as sparrows waked whistling through Marine Street's summer green leafed trees." In the third, final section, the poet addresses his listeners, enjoining them to "enrich this Plutonian Ode" and to "destroy this mountain of Plutonium with ordinary mind and body speech," closing the poem by enjoining his listeners to likewise "empower this Mind-guard spirit gone out, gone out, gone beyond, gone beyond me …"

In this way, the human imagination is able to assert a creative force that strives to actualize what is antithetical to war and to its associated nuclearisms. Closely related to the idea of imagination as creative force is the idea of conscience, which of course is the imagination as moral force. While Whitman's poem reflects a type of sublimity associated with Romanticism, Ginsberg portrays a crisis of conscience, calling out the moral force of religious tradition.

The "Auto-Toxic" Planet

In her prose collection *What Is Found There: Notebook on Poetry and Politics*, Adrienne Rich writes:

> Someone is writing a poem. Words are being set down in a force field. It's as if the words themselves have magnetic charges; they veer together or in polarity, they swerve against each other. Part of the force field, the charge, is the working history of the words themselves, how someone has known them, used them, doubted and relied on them in a life. Part of the movement among the words belongs to sound – the guttural, the liquid, the choppy, the drawn-out, the breathy, the visceral, the downlight. The theater of any poem is a collection of decisions about space and time – how are these words to lie on the page, with what pauses, what headlong motion, what phrasing, how can they meet the breath of the someone who comes along to read them?

Rich applies this to a poem by Lynn Emanuel, "The Planet Krypton," which depicts mother and daughter in Nevada watching an atomic bomb test on TV. Rich identifies the poem's "political core," its essential meaning: "In the suave, brilliant wattage of the bomb, we were / not poor." As Rich explains, the poem acquires its energy from the bomb's "spectacle of power," its false promise, along with the destruction of Hiroshima and Nagasaki and the toxic radioactive pollution of certain islands in the Pacific Ocean. The speaker of the poem describes her mother watching the event on TV. In evoking the Planet Krypton, Emanuel shows how, in Rich's words, "Earth has become its own Planet Krypton – auto-toxic."[22] The poem shows how poetry takes its creative power from the recognition of "the old primary appetites for destruction and creation."[23]

Now, in the midst of the Anthropocene, we Earthlings find ourselves much conflicted about our role in Earth's auto-toxicity, whether from radioactive fallout, leakage from the inefficiency of radioactive waste disposal and storage facilities, or massive environmental contamination from nuclear catastrophes like the ones at Fukushima, Chernobyl, and Three Mile Island. Technology's capacity for mining and enriching uranium and the various parts of its decay chain have served to concentrate and accelerate various auto-toxic effects.

"Some stood up"

One of the most interesting poets in this excavation of twentieth-century antinuclearisms is the Jesuit priest Daniel Berrigan. Having endured his own father's fascination with British Romantic poets as a young man, Berrigan explains finding himself in "a world that included Hopkins and Eliot and Pound and Frost and others who, as he came to understand, had formed the sensibility of his generation."[24] By 1957, Berrigan's poetry had been discovered by Marianne Moore, and by 1959 he had won the prestigious Lamont Poetry Prize for his book *Time without Number.* The book was reviewed in *Poetry* magazine by Galway Kinnell, who particularly praised Berrigan's poem "Some Young God."[25] One sees in Berrigan's early poems, in the emerging social commentary in his work, aspects of his life as priest, poet, and activist that distinguish him from other poets of his generation.

The key to understanding how Berrigan's activism – in particular his anti-nuclear activism – relates to his work as a poet is to understand that Berrigan himself did not seem to make a distinction at all. In a prose piece titled "Not Feeling Poetic," included in a 1988 anthology of his poetry, drama, and prose, he examines this essential connection. He begins by writing, "Not feeling poetic these days. Where's it gone, that state of the art called human?" Then he turns ironic; given the state of things, "Who in the world can feel poetic?" Under the circumstances, he reasons, poetry is basically a kind of "non prose … non conformity … prelude and preparation …" He understands that poetry entails an essentially creative act, so primal in its nature that it takes place in the dark that precedes creation itself: "that darkest blink of night which seems for a horrid moment, like the blinding of the eye of day." This is the way, he explains, that through poetry, "we keep intact the code of the nearly lost." Sounding a bit like Walt Whitman, he concludes: "I hear this unquenchable poetry of survival. I hear it, it prevails, even on the winds of a firestorm. / It will decompose, into the sweet compost of song, even the file prose of hell."[26]

To read Berrigan's poetry is to witness an inner life as a place of devotion to spirituality and to an ongoing inquiry into the linkage between humanity and divinity. Berrigan's spiritually disciplined habit of mind is intent on the creation of hope out of experience. Running through his poetry and his creative living is the persistent insertion of his own life into the harsh reality of injustice. He was arrested numerous times for actions he took in the spirit of Gandhian non-violence, and he helped found the influential Ploughshares anti-nuclear movement. In this way, he, and many others, used the symbolism intrinsic to Christianity to enact public protests projecting the moral force of Christian teaching and scripture against the forces of destruction. Berrigan's poem "Some" addresses the sense of futility that often haunts people who are engaged in public protest. It was written after the death of his friend Mitch Snyder, who for many years had worked with homeless people, but who grew despondent because of the government's unresponsiveness and took his own life:

Some stood up once, and sat down.
Some walked a mile, and walked away.

Some stood up twice, then sat down.
"It's too much," they cried.
Some walked two miles, then walked away.
"I've had it," they cried.

Some stood and stood and stood.
They were taken for fools,
they were taken for being taken in.

Some walked and walked and walked –
they walked the earth,
they walked the waters,
they walked the air.

"Why do you stand?" they were asked, and
"Why do you walk?"

"Because of the children," they said, and
"Because of the heart," and
"Because of the bread,"

"Because the cause is
the heart's beat, and
the children born, and
the risen bread"[27]

Like other religiously grounded activists, Berrigan has much to teach about the importance of being able to move between resistance, that is, active public protest, and contemplation. Contemplation creates the moral fuel for protest, and the writing of poetry is essentially contemplative practice and as such is a discipline or a habit of mind.

In November 1964, Berrigan participated in the three-day Gethsemani Abbey Peacemakers Retreat, organized by Thomas Merton, poet and one of the most highly respected Catholic writers of the twentieth century. It was there that Berrigan and Merton both studied the work of French sociologist Jacques Ellul, whose book *The Technological Society* had recently been translated into English. Influenced by Ellul's ideas, they came to believe that technology often serves to sever the relationship between the individual person and the technology itself, such that any moral sense of responsibility for consequences is either muted or completely diminished.[28]

"The Fundamental Project"

Galway Kinnell writes in "The Fundamental Project of Technology" about the significance of the moment "when a white flash sparkled."[29] That phrase "a white flash sparkled" is continuous with the poem's epigraph, by Hiroshima survivor

Tatsuichiro Akizuki: "A flash! A white flash sparkled!" Some people may feel that a white flash is a euphemism, but in the realm of the poetic imagination, the white flash signifies a complex of meaning, both symbolic and metaphoric. It is symbolic, of course, of the heat and light given off by a nuclear explosion, but it also metaphorically evokes the much-discussed white light seen just prior to death.

Kinnell's poem begins by describing a visit to the scene of devastation, perhaps Nagasaki, many years later, looking at artifacts:

> Under glass: glass dishes which changed
> in color; pieces of transformed beer bottles;
> a household iron; bundles of wire become solid
> lumps of iron; a pair of pliers; a ring of skull-
> bone fused to the inside of a helmet; a pair of eyeglasses
> taken off the eyes of an eyewitness, without glass,
> which vanished, when a white flash sparkled.[30]

Then, "An old man, possibly a soldier back then, / now reduced down to one who soon will die." It becomes clear that the poem is transforming the past into the present by way of pairing the immediate scene with the imagined scene from the past. The poem describes a group of schoolchildren lining up for a photo; they "grin at a flash-pop" and then they run around saying hello and goodbye to strangers. To the poet, these are the remnant ghost words of the dead, who died when "the white flash sparkled."

In this way, the poem finds its vehicle for examining what is at the end of the long arc of consequences that emanates not only from the nuclear annihilation of Hiroshima and Nagasaki, but prior to that from what is near the centre of technology's fundamental "project." But to approach this problem, or project, or perhaps projection, the poem expands on the irony that would have us hear the speechlessness of the dead. One realizes that the children have gone away; they are ghosts who quickly retreat back into the ashes, so that when the day "flashed" no one is there to say, "A flash! A white flash sparkled!" The poem rushes toward its fully realized projection: humanity seems to have lost its most fundamental of attributes in a convolution of ignorance and knowledge:

> To de-animalize human mentality, to purge it of obsolete
> evolutionary characteristics, in particular death,
> which foreknowledge terrorizes the contents of skulls with,
> is the fundamental project of technology; however,
> the mechanisms of *pseudologica fantastica* require:
> to establish deathlessness it is necessary to eliminate
> those who die; a task attempted, when a white flash sparkled.

This dark concept of technology is rooted not only in the cataclysmic and immoral destruction of Hiroshima and Nagasaki, but also in the ideas of Ellul, Merton, and Berrigan, ideas that formed an important part of the anti-nuclear movement. In the early years of anti-nuclearism, Bertrand Russell exposed what Kinnell's poem calls the "*pseudologica fantastica*," the pathologically false logic of those who would claim the validity of a technological society that creates the tools of mass destruction while at the same time promising justice and freedom, and one would presume, morality.

In 1959, Russell writes,

> But the spread of power without wisdom is utterly terrifying, and I cannot much blame those whom it reduces to despair.
>
> But despair is not wise. Men are capable, not only of fear and hate, but also of hope and benevolence. If the populations of the world can be brought to see and to realize in imagination the hell to which hate and fear must condemn them on the one hand, and, on the other, the comparative heaven which hope and benevolence can create by means of new skills, the choice should not be difficult, and our self-tormented species should allow itself a life of joy such as the past has never known.[31]

Russell believed, however, that because the situation of geopolitics fails logical analysis and because of the improbability of nuclear disarmament, a more generalized and popular approach to the problem was necessary. And it is largely out of this that the mode of massive anti-nuclear protests began. Protests continue today, but seldom on the scale of the 1960 Aldermaston March, which poet W.S. Merwin reported on for *Nation* magazine: some 20,000 marchers, filling the highway for over seven miles.[32]

Much of the poetry under discussion here has to do with memory and remembering, for it is in this way that poetry creates images and then makes them into new forms of consciousness. After all, as Kinnell suggests, it is foreknowledge we seek, and in this seeking we seem to apprehend the end, death itself. Imagine sleepwalking and being haunted by both the past and the future. One meets the victims of the two nuclear bombings, one meets the protesters, but there's an absence, an empty space where the physicists, engineers, bomber pilots, and blind politicians would be. So it comes to this. What would one do? What would one say to this congress of ghosts who occupy both the past and the future? Kinnell writes:

> The children go away. By nature they do. And by memory,
> in scorched uniforms, holding tiny crushed lunch tins.
> All the ecstasy-groans of each night call them back, satori
> their ghostliness back into the ashes, in the momentary shrines,

the thankfulness of arms, from which they will go
again and again, until the day flashes and no one lives
to look back and say, a flash, a white flash sparkled.[33]

Embracing Abjection

Today, the most important young poet addressing anti-nuclearism is the Marshallese poet Kathy Jetnil-Kijiner. Her first collection of poems (and the first by a Marshallese author), *Iep Jaltok: Poems from a Marshallese Daughter*, draws readers into the toxic legacy of nuclear testing in the Marshall Islands. Her poem "History Project" depicts a fifteen-year-old Marshallese girl researching the history and ongoing consequences of radioactive pollution. Two of the poet's grandparents died of cancer before she was born, and she lost a young cousin to leukemia. In the poem, the fifteen-year-old speaker knows how sixty-seven times between 1946 and 1957 the US military exploded nuclear weapons in the Marshall Islands, at Enewetok and Bikini Atolls, subjecting the Marshallese people to forced relocations and radioactive poisoning from fallout.

In the poem, a general says, "90,000 people are out there. / Who / cares?" The fifteen-year-old responds by saying, "I'm not mad at all / really / I already knew all of this." It's teenage irony, though, perhaps a specialized type of abjection. What immediately follows is a painful description:

I glance at a photograph
of a boy, peeled skin arms legs suspended
a puppet
next to a lab coat, lost
in his clipboard
I read first hand accounts of what we call
jelly babies
tiny beings with no bones
skin – red tomatoes
the miscarriages gone unspoken
the broken translations
    I never told my husband
    I thought it was my fault
    I thought
                there must be something   wrong
    inside me
I flip through snapshots [...]

The poem recalls US military officers persuading islanders to leave their homes by saying that it was "for the good of mankind" and "God will thank you." This drives the poem though a section that makes concrete the pervasive abjection that links vomit to sin and to a visceral rendering of a specifically Christian form of torture:

> like God just been
> waiting
> for my people
> to vomit
> vomit
> vomit
> all of humanity's sins
> onto impeccable white shores gleaming
> like the cross
> burned
> into our
> open
> scarred palms

Reminiscent of Ginsberg's calling forth the power of plutonium in order to redirect that power back onto itself, "History Project" presents the anger ("I'm not mad at all") of what the fifteen-year-old transformed into:

> I want radioactive energy megatons of tnt a fancy degree
> anything and everything I could ever need to send ripples of death
> through a people who put goats
> before human beings
> so their skin can shrivel beneath the glare of hospital room lights
> three generations later
> as they watch their grandfather their aunty their cousin's life drip
> across that same black screen knots of knuckles tied to steel beds
> cold and absent of any breath[34]

To finish her history project, the fifteen-year-old graphs cancer deaths and US trust fund payments, quotes from ancestors, then spray paints "FOR THE GOOD OF MANKIND" at the top of her poster. Parents and teacher are proud of the project, but when "the three balding white judges" look at it, one of them says, "*but it wasn't really / for the good of mankind, though / was it?*" And the fifteen-year-old concludes, "And I lost."

In 2014, in commemoration of the sixtieth anniversary of the Castle Bravo nuclear bomb test at Bikini Atoll, the Marshallese hosted Nuclear Victims Remembrance Day. At the event, the *hibakusha*, the Japanese survivors of the Hiroshima and Nagasaki bombings, were also honoured.[35]

▪ ▪ ▪ ▪ ▪

Ultimately, poetry is a type of intelligence that blossoms in the act of creating, in making and doing. As such, the practice of poetry is the habit of mind that is continuously conscious of the function of what is being created. In this context, the practice of poetry, painting, music, dance, and architecture all help redirect civilization to creatively confront the forces pushing humanity toward death, war, and nuclear annihilation.

Thomas Merton, speaking to a congress of poets in Mexico City, said:

> We are not persuaders. We are the children of the Unknown. We are the ministers of silence that is needed to cure all victims of absurdity who lie dying of a contrived joy. Let us then recognize ourselves for who we are: dervishes mad with secret therapeutic love which cannot be bought or sold, and which the politician fears more than violent revolution, for violence changes nothing. But love changes everything.
>
> We are stronger than the bomb.
>
> Let us then say "yes" to our own nobility by embracing the insecurity and abjection that a dervish existence entails.
>
> In the Republic of Plato there was already no place for poets and musicians, still less for dervishes and monks. As for the technological Platos who think they now run the world we live in they imagine they can tempt us with banalities and abstractions. But we can elude them merely by stepping into the Heraklitean river which is never crossed twice.
>
> When the poet puts his foot in that ever-moving river, poetry itself is born out of the flashing water. In that unique instant, the truth is manifest to all who are able to receive it.
>
> No one can come near the river unless he walks on his own feet. He cannot come there carried in a vehicle.
>
> No one can enter the river wearing the garments of public and collective ideas. He must feel the water on his skin. He must know that immediacy is for naked minds only, and for the innocent.
>
> Come, dervishes: here is the water of life. Dance in it.[36]

NOTES

1 Pound, *ABC of Reading*, 32.
2 Ibid., 37.
3 Pound, *The Cantos of Ezra Pound*, 566, 570.
4 Pound, *Machine Art and Other Writings*, 23.
5 Rukeyser, *The Speed of Darkness*, 111.
6 "Nuclearism," *OED Online*.
7 Kristeva, *Powers of Horror*, 9.
8 Petrus and Cohen, *Folk City*, 240.
9 Hikmet Ran, "I Come and Stand at Every Door."
10 Ibid.
11 Heyen, *A Poetics of Hiroshima & Other Poems*, 500.
12 Ibid., 500–2.
13 Kristeva, *Powers of Horror*, 71.
14 Jorgensen, *Strange Glow*, 278.
15 Kristeva, *Revolution in Poetic Language*, 80.
16 Whitman, *Leaves of Grass*, 1867.
17 Ginsberg, *Collected Poems, 1947–1980*, 710–13.
18 Jorgensen, *Strange Glow*, 159.
19 Glass et al., *Symphony No. 6*.
20 Plotkin, *Plutionian Ode*.
21 Jorgensen, *Strange Glow*, 110–11.
22 Rich, *What Is Found There*, 88.
23 Ibid., 89.
24 Berrigan, *To Dwell in Peace*, 96.
25 Kinnell, "Four First Volumes," 183; Berrigan, *Time without Number*, 21.
26 Berrigan, *Daniel Berrigan: Poetry, Drama, Prose*, 343–4.
27 Berrigan published several different versions of this poem. Fr John Dear of the Daniel Berrigan Literary Trust has requested that this version be used here.
28 Oyer, *Pursuing the Spiritual Roots of Protest*, 54, 59–60.
29 Kinnell, *The Past*, 47–51.
30 Ibid., 47–50.
31 Russell, *Common Sense and Nuclear Warfare*, 20.
32 Merwin, "Letter from Aldermaston," 466.

33 Kinnell, "The Past," 47.
34 Jetnil-Kijiner, *Iep Jaltok*, 40.
35 Zak, "A Ground Zero Forgotten."
36 Merton, *Raids on the Unspeakable*, 160–1.

BIBLIOGRAPHY

Berrigan, Daniel. *And the Risen Bread: Selected Poems, 1957–1997*. New York: Fordham University Press, 1998.

– *Daniel Berrigan: Poetry, Drama, Prose*, edited by Michael True. Maryknoll: Orbis Books, 1988.

– *Time without Number*. New York: Macmillan, 1957.

– *To Dwell in Peace: An Autobiography*. San Francisco: Harper and Row, 1987.

Ferguson, Frances. "The Nuclear Sublime." *Diacritics: A Review of Contemporary Criticism* 14, no. 2 (1984): 4–10.

Gery, John. *Nuclear Annihilation and Contemporary American Poetry: Ways of Nothingness*. Gainesville: University Press of Florida, 1996.

Ginsberg, Allen. "Plutonian Ode." In *Collected Poems, 1947–1980*. New York: Harper and Row, 1984.

Glass, Philip, Allen Ginsberg, Lauren Flanigan, Dennis Russell Davies, and Bruckner Orchester Linz. *Symphony No. 6: Plutonian Ode*. Sound recording. New York: Orange Mountain Music, 2005.

Heyen, William. *A Poetics of Hiroshima & Other Poems*. Wilkes-Barre: Etruscan Press, 2008.

Hikmet Ran, Nâzim. "I Come and Stand at Every Door." *The Art of Marxism: poetry*. https://www.marxists.org/subject/art/literature/nazim/icomeandstand.html.

Jetnil-Kijiner, Kathy. *Iep Jaltok: Poems from a Marshallese Daughter*. Tuscon: Univeristy of Arizona Press, 2017.

Jorgensen, Timothy J. *Strange Glow: The Story of Radiation*. Princeton: Princeton University Press, 2016.

Kinnell, Galway. "Four First Volumes." *Poetry*, June 1958, 178–83.

– *The Past*. Boston: Houghton Mifflin, 1985.

Kristeva, Julia. *Powers of Horror: An Essay on Abjection*. Translated by Leon S. Roudiez. Reprint edition. New York: Columbia University Press, 1982.

– *Revolution in Poetic Language*. New York: Columbia University Press, 1984.

Merton, Thomas. *Raids on the Unspeakable*. New York: New Directions, 1966.

Merwin, W.S. "Letter from Aldermaston." *Nation* 190, no. 19 (1960): 408–10.

"Nuclearism, N." *OED Online*. Oxford University Press, http://www.oed.com.ezproxy.chaminade.edu/view/Entry/257104.

Oyer, Gordon. *Pursuing the Spiritual Roots of Protest: Merton, Berrigan, Yoder, and Muste at the Gethsemani Abbey Peacemakers Retreat*. Eugene: Cascade Books, 2014.

Petrus, Stephen, and Ronald D. Cohen. *Folk City: New York and the American Folk Music Revival*. Oxford: Oxford University Press, 2015.

Plotkin, Owen. *Plutonium Ode …* YouTube, 2007, https://www.youtube.com/watch?v=1HYEnjnu5RU.

Pound, Ezra. *ABC of Reading.* New Haven: Yale University Press, 1934.

– *The Cantos of Ezra Pound.* New York: New Directions, 1996.

– *Machine Art and Other Writings: The Lost Thought of the Italian Years: Essays.* Durham: Duke University Press, 1996.

Rich, Adrienne. *What Is Found There: Notebooks on Poetry and Politics.* New York: W.W. Norton, 1993.

Rukeyser, Muriel. *The Speed of Darkness.* New York: Random House, 1968.

Russell, Bertrand. *Common Sense and Nuclear Warfare.* New York: Simon and Shuster, 1959.

This Mortal Coil. *I Come and Stand at Every Door.* MP3. Vol. Blood (remastered). 4AD, 1991. Apple Music.

Whitman, Walt. *Leaves of Grass.* New York, 1867. https://whitmanarchive.org/published/LG/figures/ppp.00707.295.

Wilson, Rob. "Towards the Nuclear Sublime: Representations of Technological Vastness in Postmodern American Poetry." *Prospects, 14:* 407–39.

Zak, Dan. "A Ground Zero Forgotten." *Washington Post,* 27 November 2015, http://www.washingtonpost.com/sf/national/2015/11/27/a-ground-zero-forgotten.

Afterword to Part Two

Fears and the (Nuclear) Apocalypse: Who Is Afraid of What?

KARENA KALMBACH

One topic in particular links the five texts of this part of the book: memory. All authors are concerned with the question of what kind of memory individuals and collectives form of the nuclear age. Memories are passed on through different media. As the texts exemplify, these media range from oral narratives to performative acts and crafted commemorative artifacts. These memories evoke emotions, both on the side of the carrier of the memory and on the side of the recipient. These emotions are manifold, they include hope, fear, horror, joy, helplessness, and need for action.

The central role that emotions play in nuclear discourse is brought to the fore, for instance, by the opera *Shelter*, which Juliet Palmer, Julie Salverson and Peter van Wyck present in their text: one of the central characters of this opera, the "radioactive daughter," is called *Hope*.[1] The "give peace a dance" posters that Joseph Masco analyzes in his text also articulate the central role of emotions in nuclear discourse: the public staging of joy to counter nuclear doom.[2]

Despite the broad spectrum of emotions at stake in nuclear discourse, historical research on the nuclear age has focused on only one emotion: fear. What is more, this research has not considered the "multifaceted relationship of fear and technology,"[3] which includes among other things technology *by* fear and fear *for* technology. Instead, it has focused only on one very specific fear/technology interrelation: fear *of* technology. The concentration on this specific fear/technology interrelation is so powerful – both in research and in public discourse – that the term "nuclear fears" has adopted a very one-directional focus. It always means: the fear *of* radiation, the fear *of* the atomic state, the fear *of* the nuclear apocalypse. What is more, on most occasions, "fear" stands here for "exaggerated fear" or "irrational behaviour" and thus distances the "layperson (female) fearful discourse" from the "expert (male) rational discourse."

This dynamic is fuelled by the fact that researchers rarely consider other fear/technology interrelations, such as the pro-active role that various fears and anxieties played in the creation and implementation of nuclear technologies. Instances of such "proactive emotions" include Western Europe's fear of being dependent on Arab oil – a major factor in the large-scale building of nuclear power plants in the 1970s and 1980s – and policy-makers' fear of the political opponent, which became a driving force in the stockpiling of nuclear weapons.[4] Another example: researchers barely pay attention to the fears of losing knowledge of nuclear technologies that many engineers and policy-makers have been talking and writing about – fears that result in keeping alive study programs in nuclear engineering despite students' decreasing interest in this topic. Bryan Taylor, in his text, phrases this fear *for* technology in eloquent terms when he writes about "the undesirable possibility that, in the absence of explosive testing, nuclear weapons design knowledge might decay, gradually and irrevocably, to produce their *de facto* 'uninvention.'"[5]

The proactive and productive role of fear *of* nuclear technologies has often been addressed in scholarship. Jim Kraus's text offers an instructive instance of the productive aspect of fear of nuclear technologies, when he writes that "while the imagination transforms the idea of nuclearism into fear and dread … poetic language renders fear and dread into a form of abjection capable of evoking a moral sense that influences human action."[6] However, I would argue that historians of technology should finally start to examine the proactive and productive role of fear in developing and maintaining nuclear technologies and complexes, rather than focus primarily and obsessively on people's fear *of* nuclear technology.

Joseph Masco alludes to the complex interrelationship between fear and nuclear technology when he writes: "The nervous system of the nuclear state has also always been quite nervous – a quivering universe of experts building systems for, and imaginatively rehearsing over and over again, apocalyptic scenarios."[7] What Masco shows here is of central importance: the apocalypse was not invented by the anti-nuclear campaigners, who were often criticized and ridiculed for their "fear-mongering" narratives. Their "nuclear apocalypse" was a reaction to other apocalyptic scenarios – apocalyptic scenarios that led to further developments and implementations of both military and civilian nuclear technologies, which then sparked the imagination of nuclear apocalypses in the opponents of these technological developments. Thus, it is important to stress that the "nuclear apocalypse" was not the only apocalyptic discourse shaping the nuclear age. The imaginary of the nuclear apocalypse was the result of pro-nuclear actors inventing and implementing nuclear technologies to confront (their fear of) their own apocalyptic scenarios – apocalyptic scenarios related to political and economic future imaginaries. Fears have been omnipresent in the nuclear age, on the pro-nuclear as on the anti-nuclear side.

But even one and the same nuclear fear – that of the nuclear apocalypse – has been considered very differently depending on who reacted upon it. The fearful action of the political opponent formed the very linchpin of the nuclear deterrence strategy. As Bryan Taylor puts it: "adversaries are discouraged from making or acting on threats, for fear of nuclear retaliation."[8] Fear thus gained a central role in Cold War communication strategies in international relations. To again cite Brian Taylor's eloquent argument: "nuclear deterrence has sublimated the physical potential of nuclear weapons in favour of their symbolic function as a messaging apparatus, whose only useful function is to be manipulated for communicative purposes."[9] But while the imaginary of nuclear apocalypse had the clear aim of evoking fears in the political opponent and compelling state leaders to act in rational-choice ways, the fears that the very same imaginary of the nuclear apocalypse sparked in the broader public were framed as irrational, emotional behaviour.

It is of central importance to unpack this multifaceted entanglement of nuclear technology and fear. Only in doing so can we grasp the full "symbolic dimension"[10] of the nuclear enterprise and understand why some people's (e.g., nuclear scientists', politicians', and military planners') acting upon fear was framed as legitimate rational behaviour, while other people's fear (e.g., that of anti-nuclear movements and grassroots activists) was framed as illegitimate or irrational behaviour.

By shedding light on the discursive dynamics that produced the narrative of the fear-driven irrational behaviour of laypeople, and by unpacking the power dynamics that are negotiated within the fear discourse, historians may be able to contribute to the breaking of the "apathetic silence and inarticulateness surrounding the persistence of mutually assured destruction" (Taylor).[11] But they won't be able to do this by means of new research questions alone. Where nuclear dangers once "invited everyone to think about the qualities of life, and the nature of politics and power, and to articulate their commitments to one another and the future" (Masco),[12] they have fully lost centre-stage attention, particularly in younger people.

How could we lose sight of these dangers, in an era in which terms such as the "Anthropocene" have entered the register of everyday language, and in which young people around the world connect in the Fridays for Future movement – that is, in a time that could be considered perfect ground for debating nuclear dangers? How can it be that in a time when historical research on the nuclear age is flourishing as never before, nuclear dangers have become increasingly depoliticized? This will be an interesting question for future historians. For the time being, and for the here and now, historians need to develop new forms of scholarship – scholarship that is not hidden behind expensive pay walls for books and journals or high tuition fees.

What can academics do? They could engage in forms of transdisciplinary collaborative cooperation and writing as nicely exemplified by Palmer, Salverson, and van Wyck's essay in this volume.[13] Or they could follow creative avenues like the one proposed by Sarah Kanouse and Shiloh Krupar.[14] Kanouse and Krupar's exuberant, at once satirical and serious contribution on the "National TLC Service" shows how much we gain when we deviate from the dry and often complicated script of academic articles, how we can reach various audiences without compromising on academic quality. Their work is the opposite of "apathetic silence and inarticulateness" evoked by Bryan Taylor. Their piece deserves admiration and should be an inspiration, particularly in a context in which "impact" and "outreach" have been turned into purely quantitative, administrative evaluation criteria by the persistent logics of the neoliberal, entrepreneurial university. Kanouse and Krupar's work is a story of joyful empowerment: an open-minded, engaged, merry counter-story to the narrow-minded fears that drive the nuclear enterprise.

NOTES

1 Palmer, Salverson, and van Wyck, "Sounding Out the Nuclear," in this volume.
2 Masco, "Shaking, Trembling, Rattling, Shouting," in this volume.
3 Gall et al., "Tech-Fear."
4 Kalmbach et al., "Crises and Technological Futures."
5 Taylor, "What Is the Matter," in this volume.
6 Kraus, "Poetry and Anti-Nuclearism," in this volume.
7 Masco, "Shaking, Trembling, Rattling, Shouting," in this volume.
8 Taylor, "What Is the Matter," in this volume.
9 Ibid.
10 Ibid.
11 Ibid.
12 Masco, "Shaking, Trembling, Rattling, Shouting," in this volume.
13 Palmer, Salverson, and van Wyck, "Sounding Out the Nuclear," in this volume.
14 Kanouse and Krupar, "National Toxic Land," in this volume.

BIBLIOGRAPHY

Gall, A., et al., eds. "Tech-Fear: Histories of a Multifaceted Relationship." Special Issue of *Technikgeschichte* 86, no. 3 (2019).

Kalmbach, Karena, Andreas Marklund, and Anna Åberg. "Crises and Technological Futures: Experiences, Emotion and Action." *Technology and Culture* 61, no. 1 (2020): 272–81.

THREE

Archaeologies and Heritages

10

Emergency/Salvage Archaeology: Excavating Media and Uranium in the Glen Canyon

THOMAS PATRICK PRINGLE

Historicizing Environmental Materiality and Developing an Interdisciplinary Perspective on a Mineral

When Michel Foucault introduced his archaeological method in 1966, he produced a historical perspective for the analysis of the background rules that bound epistemic fields such as linguistics, psychology, philosophy, and economics.[1] The emergence of "media archaeology," a media studies methodology that emphasizes the historical "rummag[ing of] textual, visual, and auditory archives as well as collections of artifacts, emphasizing both the discursive and the material manifestations of culture," is indebted to Foucault's approach to history, language, and epistemology.[2] Foucault's archaeology of knowledge involved the scholarly excavation of what *could be said and thought* in a given time or place, in other words, "the limits and forms of the sayable."[3] Then, archaeology is the study of "historical *a priori* which … served as an almost self-evident ground for our thought."[4] Within media studies, this theoretical disposition toward excavating knowledge formations within discontinuous ruptures of history inspired a vein of much-debated technological determinist thought, epitomized by Friedrich Kittler's advocacy for the media archaeology of "technological *a priori*."[5]

Media archaeology relies on a definition of media as historical, or *excavatable*: media are both situated communication discourses and historically contingent material technologies. Importantly, within this view, media also have an *operative* status in that they facilitate exchanges between discourse and materiality as though a *hinge*. In Jussi Parikka's media archaeological response to recent environmental crises, *A Geology of Media* (2015), he terms the historical strata of such mediated interchanges between discourse and materiality "medianatures": "a concept that crystallizes the 'double bind' of media and nature as co-constituting spheres, where the ties are intensively connected in material nonhuman realities as much as in relations of power, economy, and work."[6] Parikka's development of archaeology within media studies complicates Foucault's original emphasis on the semiotic definition of *epistemes* as linguistically delineated fields of nonlinear historical *a priori*. For Parikka, media archaeology instead reveals how historicized cultural designations, such as what counts as "nature" and the according valuation of what is "natural" in a designated society, are embedded in the storage and transmission capacities of technological media. At once, media archaeology surfaces how those same cultures produce material media technologies in ways consonant with the periodized knowledge preserved in media. Thus, media history indexes human attitudes toward nature across both the form and the content of the media being historicized.

Where Foucault stressed discontinuity in the historical *a priori* of knowledge, media archaeology, as Alexander Monea and Jeremy Packer elaborate, instead underscores how media technologies "allow for the extension of culture across time, for culture's duration and endurance. As such, [media] have a priori stakes in the realms of the political, the ethical, and the epistemological. Media collect, store, process, and transmit data that are variously used to rate, coordinate, create, obfuscate, obliterate, translate, demonstrate, and even create virtuality, materiality, and reality itself."[7] In this chapter, I interpret media archaeology as a method for discerning transformations in the social and economic valuation of "nature" through the historical analysis of a series of media-technological enunciations describing the biophysical reality of a selected place. These media record the same geographical space and act as storage units conveying different cultural attitudes as they originate in different times. I use these disparate media to reconstruct how, when, and for whom the space counted as nature, by isolating the historiographical inquiry to a delimited place. Holding these heterogeneous and discontinuous attitudes toward a specified region together as recorded across the variegated media history of the Glen Canyon, I build on the theorization of "medianatural" relationships that I have argued elsewhere bind the material properties of uranium to the cultural practice of photography, generously conceived as "the general barometry of light."[8]

Parikka writes that "medianatures [are] regime[s] constituted as much by the work of micro-organisms, chemical components, minerals, and metals as by the work of underpaid laborers in mines or in high-tech entertainment device component production factories, or people in Pakistan and China sacrificing their health for scraps of leftover electronics."[9] In Parikka's project, media archaeology turns toward the intersection of technology and historical definitions of "nature," which inform the politics of knowledge that shape the human use-valuation of environmental materiality as resources. Mineral resources play a particularly influential role as inorganic components in media history. Where the mining and refinement of aluminum facilitated "new forms of speed and transport" with the "material affordances of lightness," historical discourses in the twentieth and twenty-first centuries defined rare earth metals as "materials of production" for the media fabricated from mechanical and digital technological advancement.[10] By following how a selected mineral is valued by media as a useful material for human society, and paying attention to the historical specificity of how its use-value is positioned or oriented for the interest of specific groups of people, media archaeology becomes a political methodology for analyzing socio-environmental problems through a "media history of matter."[11]

In this chapter, I build on Parikka's suggestion to study media in terms of historical attachment to minerals, and the media archaeological method more broadly, by drawing on a theoretical conversation in environmental studies that unpacks how periodized epistemological accounts of what is considered "nature" benefits from thinking about conceptions of "the natural" as a co-constitution of social knowledge and biophysical reality. With less focus than Parikka on the literary and mediated qualities of non-human environments, environmental studies conceives the agencies of the natural as an anthropogenic outcome of regional histories of socio-environmental interaction. The natural is not afforded a blanket or uninterrogated status, nor is it cleanly distinguished in binary terms from culture. Instead, the actions of the natural are determined through an asserted anthropomorphic epistemological frame assembled from historical exchanges between social values, scientific knowledge, and landscapes undergoing human-initiated change. This chapter employs an interdisciplinary theoretical lens to delineate local knowledge of the historical conditions of "nature" in a specific region – here, the Glen Canyon in southern Utah – as these conditions came under repeated political contestation in the twentieth century. Excavating, tracing, and reanimating how this landscape, and its corresponding social valuation, were framed and reframed through photographic practices demonstrates how media archaeology can contribute to reconstructing the political history of the canyon's environmental materiality that has been relatively and strategically understood as a resource.

Paralleling debates on the new materialist turn in the humanities, environmental studies asks how, and with what limitations, non-human and material agency needs to be figured into theories of human politics. Of particular use is the debate regarding the shifting epistemological status of biophysical reality under the bracketing term of *the natural*. For example, William Freudenberg, Scott Frickel, and Robert Gramling outline a historical methodology with reference to a specific geographic case study undergoing historical changes by attending to the serial associated social perceptions of what societies valued as natural in the site over a period. Their proposal involves "a greater awareness of the extent to which (what we take to be) the physical is influenced by (what we take to be) the social, and vice versa," or, a "conjoint constitution" that accounts for how "human beings who have thought about 'environment' and 'society' have often been hindered as well as helped by the hidden assumptions that are often built into the ways in which the two concepts have been distinguished in the past."[12] Their method expresses a historical perspective that accepts the discrete categories of "environment" and "society," albeit as fluid, dynamic, interactive, and contradistinctive definitions that are politically activated and operationalized by historical actors to achieve certain gains. Focusing on the shifting epistemological appraisal and according use-value endowed to the category of "natural resources," they argue that "while the term is often seen as referring to physical characteristics, however, what are called 'natural' resources are in fact social *as well as* natural; they are products of biogeochemical processes."[13]

There are complications in taking up environmental political perspectives that advocate the political potential of material agencies (organic and non-organic) unto themselves, as the will or action attributed to non-humans will inevitably derive from a historically determined epistemic position. In other words, the advances proposed by the new materialist turn inescapably reflect contemporaneous formations of technological, social, and scientific knowledge that seek to define what is "natural" about an environment at a given time: "whether the environment is seen as 'acting' or as 'sitting,' however, it is seen as doing so by human beings."[14] In anticipation of the new materialisms, the method proposed by Freudenberg and colleagues requires the assessment of how natural knowledge is historically conditioned and situated, which they term "the naturalization of 'nature'": "without attention to the nature of the interplay of the social and the physical, there is a significant risk that what in fact are social as well as physical properties *will* be ignored. Such oversights, in turn, can cause us to develop unrealistically constrained analyses of socially significant questions and problems."[15]

In recent years, critics have underscored Foucault's blind spot in addressing the enduring legacies and state violence of colonialism. As a methodological response, scholars have, for instance, drawn together postcolonial theory with media archaeological approaches to show how the production of supercomputing microchips relied on racialized representations of Indigenous women labourers,[16] or how ideas about ethnicity and bodily difference have endured in mechanized media from early automata such as orientalized chess-playing Turks through to recent AI.[17] As a part of this recuperative theoretical gesture instilling a political position critical of settler-colonialism within media historiography, Kyla Schuller's essay, "The Fossil and the Photograph," most productively elaborates upon Parikka's argument about environmental materiality. As she argues, "expanding our notions of media into the geological requires careful attention in turn to what has been deemed to matter, to the finely tuned political process of denoting entities as lifeless, nearly extinct, or artifacts of the past."[18] Inspired by this literature, this chapter adopts a post- and de-colonial theoretical position on media archaeologies of the nuclear, seeking to account for how the valuation of uranium as a natural resource contributed to the settler-colonial project of Indigenous dispossession in the United States.

Then, the task for this chapter is to demonstrate how the media archaeological perspective offers a technological-epistemic framework within which to think about the history of environmental materiality, or the shifting status of what counts as "natural," whereas the environmental studies perspective makes an analytical commitment to emphasizing the historical and geographical specificity with which "natural resources" are conjointly defined by biophysical reality and corresponding conditions of social knowledge. Filtered through a serialized historical narrative of the medianatural values ascribed to the Glen Canyon, this chapter approaches the medium of photography as a discursive operation that both reveals and imparts social values as a process of naturalizing space. In this case, photography as an epistemic project complemented the history of uranium extraction and economy in the region. Photography, here, is examined as a media practice that attributes use-value to represented environments in discursive correspondence with knowledge produced by historical scientific and extractive projects. In addition to traditional photographic stills, I look to fictional cultural productions reliant on photography, such as Hollywood films, that participate in the socio-cultural valuation of this region over time. As such, I highlight how photography, as a media technique that indexes and stores an ecology of light for duration across time, played a role in framing and reframing what American settler-colonial society valued as natural in the anthropogenic history of the Glen Canyon's environment.

The Glen Canyon through the Lens of the Natural

In 1956, as authorized during the 84th Congress under President Dwight Eisenhower, the Colorado River Storage Project (CRSP) greenlit the construction of the Glen Canyon Dam as part of a "billion-dollar project promis[ing] to convert the river's Upper Basin – Colorado, Wyoming, Utah, New Mexico – into a land of prosperity."[19] The resulting Lake Powell that filled the canyon had a shoreline of more than 1,800 miles and is the second-largest artificial reservoir in the United States. The scale of the 710-foot dam's impact, physically and in terms of its ecosystemic, cultural, and economic effects, has been immense. The dam created an artificial lake more than 300 kilometres in length, which currently attracts more than 3 million tourists annually. It has also submerged an untold number of Colorado River tributaries, streams, and fragile desert ecosystems.

The ethos behind this anthropogenic process is most clearly illustrated by the Bureau of Reclamation's 1947 manifesto espousing the ideology of "cash register" infrastructure through a document reviewing the Hoover Dam waterworks, transparently titled *The Colorado River: "A Natural Menace Becomes a National Resource."*[20]

The document's focus – treating the "natural" qualities of the Colorado River under the distinctly utilitarian idiom of its capacity to be a "resource" – signals more than a shift toward an interior settler mentality promising greater habitation on allegedly unlivable territory: "[The river's] terrifying energy will be harnessed completely to do an even bigger job building bulwarks for peace."[21] For the Bureau of Reclamation, the river is both a "national" resource and a "natural" one, framed in the geopolitical language of "energy" and "peace" as convenient euphemisms for the development of the region's nuclear minerals that accompanied the Cold War arms race and, in turn, defined postwar environmental destruction in the American Southwest.

In addition to the familiar narratives of atomic energy expansion and ecocide during this period in American history, this chapter charts the knowledge politics of discrete medianatural knowledge formations illustrating social change in the natural valuation of Utah's Glen Canyon. Beginning with the incursion of the uranium industry in the late 1940s, I show how the social values of a regional extractive economy gave way to a reconceptualization of the area in correspondence with knowledge produced by an archaeological effort endeavoured by dozens of scientists from various disciplinary backgrounds, whose mandate included categorically documenting the canyon before flooding began in 1963. As outlined above, the canyon became an engineered energy reserve in the eyes of the federal government in the 1950s and 1960s. During the floods, the canyon was briefly a site of cultural production for three Hollywood films shot in the Glen Canyon, dramatizing the rising water between 1962 and 1965.

As the intended contribution of this chapter is to initiate an interdisciplinary perspective synthesizing media archaeology and environmental studies, my focus is on establishing a theoretical line through these historicized knowledge formations to delineate how each subsequent perspective articulates a certain vision of "nature" and how these shifting perspectives weigh on future environmental politics for the region. Then I focus on how the Glen Canyon, as a specific biophysical site, is a location that exemplifies successive medianatural narratives that record the discursive status of what media forms index as natural in *this* environment. As the archaeologists I study rely heavily on photographic techniques for their method of documentation, and the photographic production of the Hollywood presence in the area reflects a similar audiovisual index to the regional geography, I deploy the category of the natural through the metaphor of the lens to indicate the term's work in registering a shifting social, cultural, and epistemic heuristic characterized by techniques of focus, framing, and serialization. I articulate the natural in medianatures as primarily a construction of *human* knowledge co-constituted with environmental change, wherein each successive state of "nature" in this region's historical series of definitions indicates a situated perception of reality and not a monolithic or ahistorical agential force attributed to the area's non-human environmental composition.

From Natural Resource to Nature Preserve: Uranium Excavation in the Glen Canyon

The American military production and testing of atomic weapons involved a vast apparatus of legal, civic, and economic institutions concentrated in the Four Corners area of the United States. The uranium mining and refinement boom in the late 1940s and early 1950s has been widely critiqued by social scientists and geographers as an extended act of "nuclear colonialism" directed toward Indigenous populations.[22] As Valerie Kuletz has argued, this term works to situate "the emergent nuclear landscape in the arid regions of the American Southwest within a larger history of U.S. internal colonialism."[23] Prominent in this history are the bio-horror narratives of the Navajo, the Southern Paiute, and the Western Shoshone peoples, as well as the Mormon "downwinders" who lived in eastern Nevada and southwestern Utah. In the 1950s – with full awareness of the health hazards of radioactivity – the Atomic Energy Commission (AEC) deliberately planned for the fallout from the test site in Nevada to blow over targeted areas so as to avoid Los Angeles and Las Vegas.[24] The AEC described the areas as "virtually uninhabited," leading to the cynical moniker adopted by those activists and victims suffering from forced displacement and radiation-related illness for years after: "the virtual uninhabitants."[25] In an astonishing AEC internal memo from 1951, these citizens were evaluated in terms of their civic utility-value: "a minority of low education, skill, low-use segment of the American population."[26] Infamously, in the words of Michael Carricato, the Pentagon's former top environmental official, the desert in the Southwest became the "national sacrifice zone."[27]

In the environmental history of the resulting toxic landscape, the legacy of uranium extraction in the southwestern US emerges alongside stories of state-sanctioned ecocide and colonial techniques of displacement. This includes the 1979 Church Rock, New Mexico, spill that released more than 1,000 tons of mill waste into the groundwater of the Navajo nation, where cancer rates remain higher than national averages.[28] As Gabrielle Hecht explains, American nuclear colonialism relied on the state and corporate administration of epistemic conditions coding the ontology of radioactive natural resources, wherein knowledge of the material risk of uranium was strategically managed so as to amount to the organized abandonment of Indigenous communities living near extraction and refinement of the mineral.

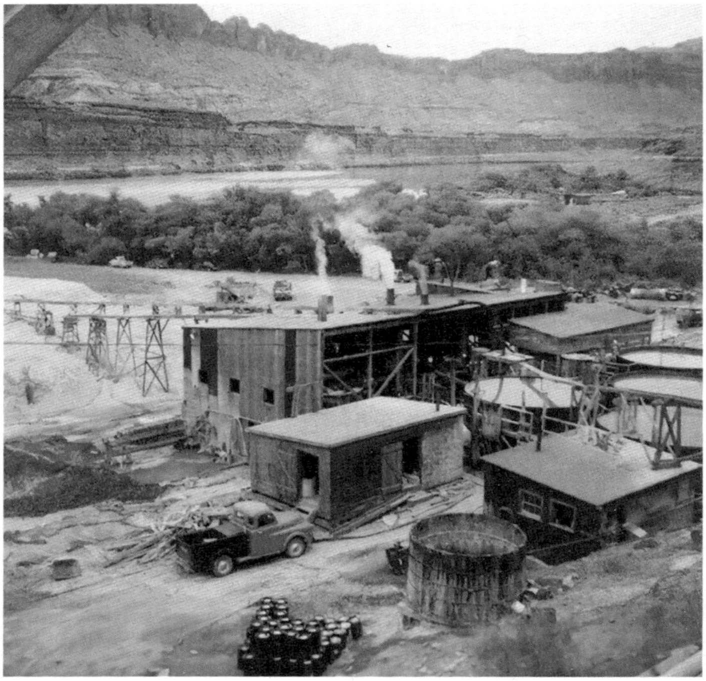

10.1 "Vanadium Corporation of America's White Canyon Processing Plant located in Hite, Utah," 1958.

Families of poorly paid Navajo miners "often camped by mine entrances and drank mine water. Their children played in mine galleries. As with Anglo, Hispano, and Mormon mineworkers, who died in significant numbers, no one told the Navajo about the dangers. Many succumbed to lung cancer before they could figure out why."[29] In recognition of the multigenerational inheritance of genetic defects and the circulation of these stories in southwestern localities, Stephanie Malin recounts how "there is a widespread community belief that the region was a national sacrifice zone."[30] Yet there remain endemic ideological positions motivated by the potential economic benefits of uranium due to the mineral's enduring signification as a natural resource and not a national menace, instilling regional beliefs that "contemporary environmental health outcomes in uranium communities are *not* the result of governmental ignorance in the 1940s and 1950s."[31]

Malin's findings testify to a geographical distribution of both environmental risk amnesia and pro-uranium ideology present in recent understandings of regional history. In the case of the Glen Canyon's nuclear history, this is best illustrated by Tom McCourt's 2003 narration of the uranium boom town of Hite, Utah, which was ultimately submerged by the damming of the Colorado River and the creation of Lake Powell:

> People in the uranium industry were on the cutting-edge of technology and morality. They were trailblazers and pioneers on a new frontier, and they were leading us into a brave new world. They were soldiers, out to save the world from the horrors of war by making explosive devices too terrible to ever be used. They were conservationists, offering atomic power as a way to save other, less valuable natural resources.[32]

For those living in and around Hite, the area's social hub and home to several hundred people living and working in the ill-fated uranium-rich bed of the White Canyon extension of the Glen Canyon, their American citizenry included a sense of Cold War duty to the hazardous mineral that built their boomtown economy.[33] Following the uranium economy's crash, the resource then dismantled and buried their municipality.

In 1946, President Harry S. Truman passed the Atomic Energy Act designating the AEC the sole organization legally authorized to purchase uranium ore.[34] The government thus gained control over the "source material after its removal from its place of deposit in nature."[35] This meant that unmined and recently mined uranium stayed in private hands until delivered to the government for processing. The ambiguity of these laws meant that private companies bore virtually no responsibility for the mining and milling the dangerous mineral. In 1949, the Vanadium Corporation of America (VCA) established a uranium mill in the basin of the Glen Canyon system near Hite and a series of independent mines several miles away.[36] The mill operated until 1953, when it was largely dismantled.[37] Then, unbelievably, the corporation impounded an estimated 26,000 tons of tailings pond materials into the canyon's landscape before evacuating the area.[38]

In 1963, the US government stalled the Colorado River at the bottom of Glen Canyon, creating a 48-million-acre-foot reservoir over the course of twenty years of flooding with the completion of the massive dam. The engineers were aware of the tailings pond enclosed in the canyon, and "the Nuclear Regulatory Commission ordered VCA to remove an undisclosed amount of high-grade ore material from the tailings pile ... The remainder of the tailings, an unrecorded amount, were covered by more than one hundred feet of silt as well as Lake Powell water."[39] The risk taken by the planners of the dam largely paid off: the rising silt over the canyon bed proved an effective containment mechanism for what could have been a disaster.[40] Two studies of large-mouth bass in Lake Powell report dangerously high selenium concentrations, indicating heavy metals in the ecosystem, but beyond this, the danger is buried as a poisonous time capsule below the surface.[41]

Burying and submerging toxic waste as the outcome of mining a natural resource is precisely the layered social valuation of geography that makes this region of scholarly interest; contemporary environmentalists' efforts to reclaim the canyon are threatened in advance, for the uranium at the bottom of Lake

Powell presents a significant roadblock to reclamation efforts. As Tom Dansie writes: "Mitigating contamination from the pile presents a complex problem. Currently there is a likely water contamination problem, but the tailings are fairly stabilized. Any attempt to mitigate the pollution with the site submerged will disturb the tailings and cause extensive spreading of the contaminants ... Thus the 26,000-ton mill tailings pile at White Canyon presents certain present pollution and contamination issues in the reservoir."[42] Dansie's impact report on the buried and submerged tailings pond concludes that remediation is not an impossible task, citing encouraging results from several successful containment projects of larger exposed tailings ponds under way in the region.[43] Nonetheless, proposals to return the Glen Canyon to its former state remain unaware of the danger or do not refer to the submerged toxic waste at all.[44]

Today, the tailings pond site at the bottom of Lake Powell is located within a protected "National Recreation Area" and there is a boat jetty several hundred feet from the submerged tailings pond.[45] These layers of environmental knowledge recoding the "nature" of the site accord to a temporality of deferred toxicity presented by abandoned industrial sites. Philosopher of science Michel Serres best theorizes this rationale of use, waste, and territorial settlement: "Pollution comes from measurable residues of the work and transformations related to energy, but fundamentally it emanates from our will to appropriate, our desire to conquer and expand the space of our properties. He who creates viscous and poisoned lakes or garish posters is making sure no one will take away the spaces he has occupied, now or after he is gone."[46] In this sense, the lost Glen Canyon adheres to a settler mentality that carves out territory through postponed toxicity; the canyon's status as a "recreation area" is, in part, held in place by a rain-checked potential ecological disaster, which is an outcome of a period of environmental managerialism designating the physical landscape as a resource. As a Utah instance of "the cold-war reinvention of the Nevada Desert as an empty, isolate space, sealed against culture and memory," the disappeared space beneath Lake Powell exemplifies the techno-political historical transformation that turned vast amounts of the Southwest into target ranges, chemical laboratories, and, eventually, nature preserves.[47]

Natural History: Cultural Excavation in a State of Emergency

One of the first uranium claims in the Glen Canyon system near Hite involved a group of archaeologists exploring a cave while looking for Indigenous artifacts. This was in 1898 during a lunch break, as Jana Mellis recounts of the anthropological-become-prospecting expedition:

> There were no Indian ruins in the cave ... However, once inside the cave, [the archaeologist] saw something which momentarily made him forget the heat. In the cave was a petrified tree, surrounded by yellow stains, and, just outside the cave entrance was another petrified log whose carbonaceous material had been almost entirely replaced by vivid yellow, blue, black and green minerals ... Between the two flat rocks near the base of the monument, he placed a piece of paper on which he had written his claim to the minerals located there.[48]

That a nineteenth-century archaeologist would make a uranium claim while searching for Indigenous artifacts is an unintended fortune worth closer examination. As Schuller writes, "natural history practice in the US nineteenth century suggested the prominence of something we might call the 'geological machine,'" an impulse to taxonomically survey and categorize material environments in support of settler-colonial projects.[49] This process unfolded through a program reminiscent of Giorgio Agamben's notion of the "'anthropological machine,' which secures the notion of 'man' through the elaborate categorization of groups whose humanity can be denied."[50] Anthropologists and archaeologists at the time sought to categorize Indigenous peoples, their culture, land, remains, and instruments as though they were prehistoric fossils, extinct relics that "belonged to the earth, not to a people."[51] Relegating Indigenous livelihood to natural history illustrates how the territorialization of settler-colonialism works in tandem with scientific research: "settler colonialism depends on dispossession in perpetuity."[52] By finding uranium ore in place of expected relics, the archaeological practice of natural history in Glen Canyon unearthed an economic argument for settlement while resignifying the canyon system from a place of Indigenous habitation to a site for extraction. The uranium claim, and the technological modernity it promised, helped delineate the canyon's period of Indigenous use as natural history in contradistinction to a novel view of the region as a resource.

The postwar decision to flood the Glen Canyon generated at least some pushback from environmental activists. However, it was particularly controversial for a team of scientists from the University of Utah who were working on a number of sizable Ancestral Puebloan archaeological sites and cliff dwellings dating from AD 500 to 1300.[53] Having been given eight years' notice by the US Congress, a huge team of researchers from all scientific backgrounds descended on the canyon to document not just the perceived anthropological crisis but all manner of biological, geological, and ecological knowledge. These frantic excavations are detailed in a comprehensive report published by lead researcher Jesse Jennings, drolly titled "Glen Canyon: A Summary."

There is a palpable frustration in Jennings's writing as he recounts the endless amount of information and artifacts the team felt responsible for recovering before the area was flooded. Out of a discernable fear that the canyon and their labour could be forgotten, he writes:

> With the closing of the gates of Glen Canyon Dam in 1963 ... the field aspects of the emergency ended. As [of] this writing (May, 1966) Lake Powell is one-third filled and has become a recreational resource for the nation, entirely obscuring the canyons where researchers worked in heat, dust and hardship. During 1965, Powell attracted boaters, fishermen, campers, waterskiers, and photographers to a total exceeding 250,000 persons.[54]

While priming the area for its eventuality as a "recreational resource for the nation," Jennings and the team were the pioneers of a novel anthropological technique they termed "emergency/salvage archaeology."[55]

Reviewing prior archaeological expeditions in the canyon, Jennings points out that, due to a dearth of research on the artifacts, "so little was known of this sparsely settled region, but rumors were so rife, that the salvage operation here reviewed was necessary before the vast region could be certified as being understood well enough that inundation posed no threat of significant loss to science."[56] Not only was the salvage operation, in a sense, a targeting operation that produced scientific knowledge in pre-emption of the landscape's destruction and total resignification, but it also spoke to a long-standing academic problem in archaeology: Was it better to preserve the heritage site as is, or destroy it for the prospect of preserving history? Jennings, clearly, preferred liquidation for the purpose of knowledge production:

> Emergency programs have been equally valuable in the freeing of technique from rote or ritual procedures and devices which have often hampered archaeological field work. One shibboleth is the injunction on young men that the "aim of the record is that one be able, if it is desired, to restore the site exactly as it was" ... As a summary, I suggest that in virtually any detail, and certainly in overall results, emergency salvage archaeology is superior to most other work done in America.[57]

The consequence of his statement is that the violent historical loss entailed by this specific mode of scientific knowledge production is, also, a state-sanctioned destruction of the site under study. They produced their total catalogue of the landscape with full knowledge that the space was slated for demolition and renovation. Implied, at least discursively, is that the archaeological project acted in collusion with the statist goal of transforming the Glen Canyon's "natural" formations, and its Indigenous cultural heritage, first into a cash register energy reserve and then into a recreational resource.

The photographic archive from the expedition is wide-ranging, covering geological, ecological, and archaeological features. Reflecting on the anthropocentrism of their work, Jennings insightfully wrote: "In some ways this account *is* natural history because human behavior is not only tied to the resources – physiographic, biotic and climatic – of a land, but all these are in some degree modified (or modifiable) by human behavior, and man is a vital force in nature."[58] As a result, the photographs range from landscapes to images of relatively recent mining practices, oil wells, plant life, the social lives of the researchers, and river rafters, as well as petroglyphs, burial grounds, and cliff dwellings.

This constellation of photos, and Jennings's conceit that natural history involves humans as a "vital force in nature," co-constitutionally apprehend what is natural about the canyon as a historicized social encounter with an anthropogenic biophysical reality. For the archaeologists, their photographs were "natural" in the sense of natural history: the knowledge they produced yielded a desire for total legibility amid the staged human purposes of occupation, destruction, and transfiguration of the landscape's valuation. The emergency/salvage operation was a politicized natural history endeavour that both recorded and destroyed, easing the movement from natural resource to nature preserve in the accumulation of scientific records.

Natural history, here, adopts the technological mechanism of the lens by pulling successive waves of human relationships to landscape into and out of focus. With a team of scientists seeking the impossible by trying to document the entirety of a soon-to-be-submerged area within a distinctly American-national and colonial universe of reference, their action itself became a photographic metaphor for the impulse to capture how the history of a landscape contains human relationships to socially and culturally valued environmental materiality. (In this case, the archaeological photograph of "natural" knowledge production *targets* for the purpose of preservation anticipating destruction, which entailed sacrificing Indigenous history and cultural relations to the geography that shared the same space.) Jennings and company primed the area by including the geographic site within a dominant system of knowledge, effectively clearing the epistemological ground for the facsimile of landscape: the national recreation area.

A series of photographs featuring one of the archaeologists inspecting a series of "steps" carved into a near-vertical canyon face stand out from the rest. Jennings, speaking in an anthropological voice, makes an interpretation: "In terms of understanding the year-in year-out use of the canyon, the great number and lengths of the aboriginal trails, including the hand and toe paths up the sheer cliffs, offer unexpected insights. These trails create a network of easy communication between canyon and highland all along the Glen."[59] More than

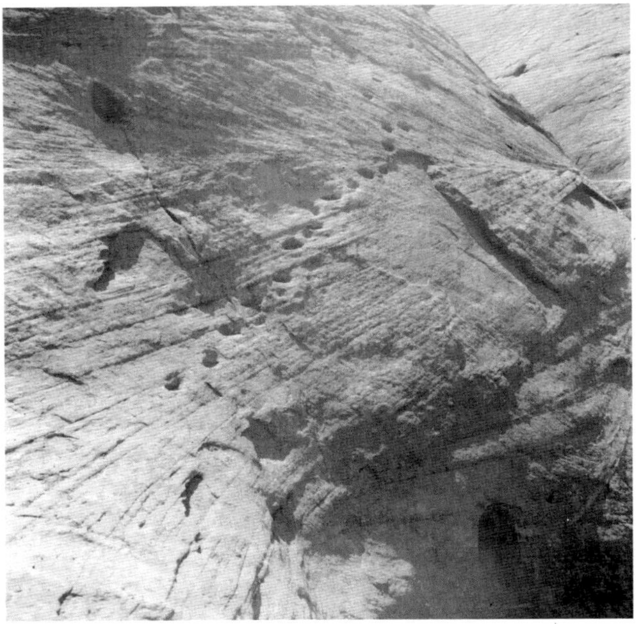

10.2 "Closer view of prehistoric steps cut into slickrock." Jesse D. Jennings, *Glen Canyon: A Summary.*

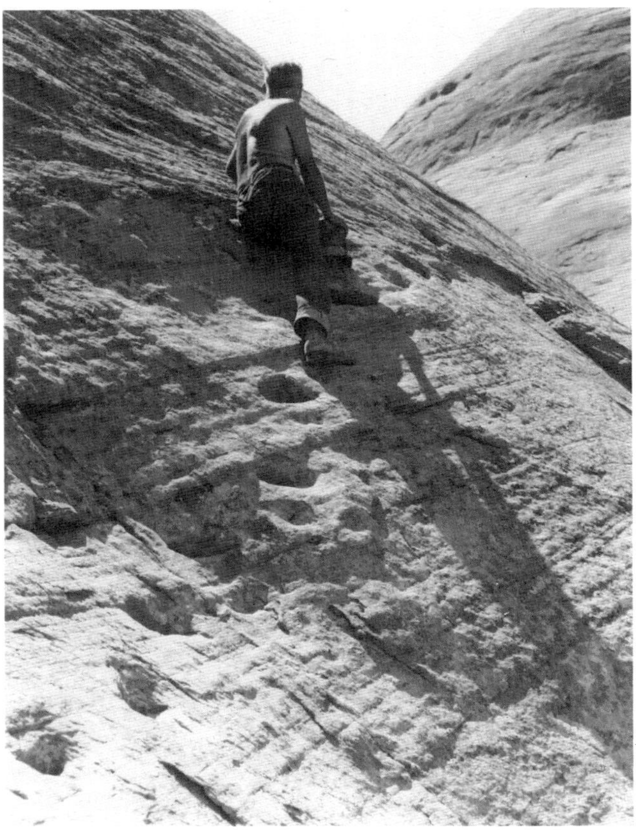

10.3 "Close-up of same prehistoric steps." Jesse D. Jennings, *Glen Canyon: A Summary.*

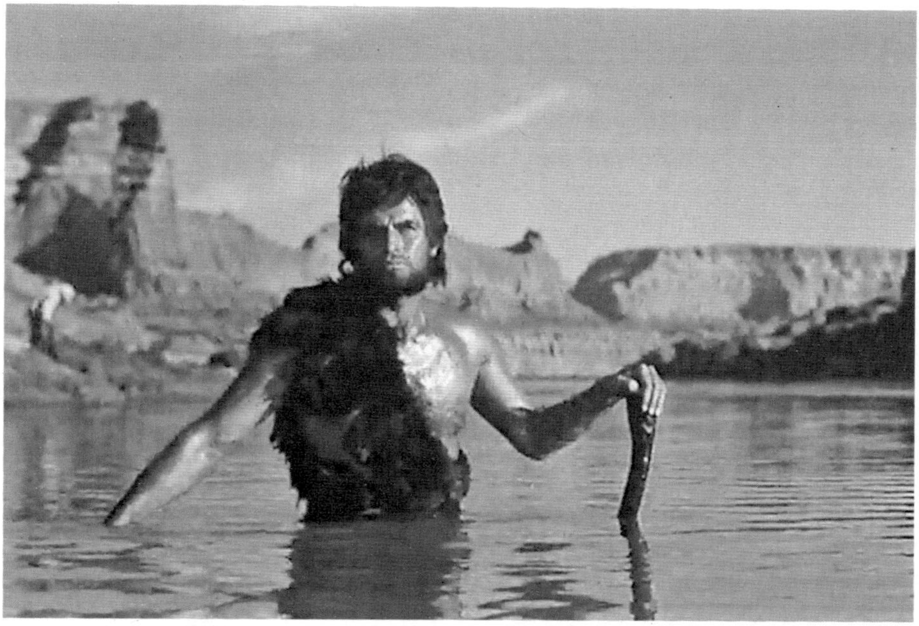

10.4 Charlton Heston as John the Baptist in the flooding Glen Canyon, *The Greatest Story Ever Told* (MGM, 1966).

thirty of these structures were documented within the canyon, as "historic trails up the naked rock [consist] of flat steps hewn with steel tools, for the passage of animals – sheep, goats, tools, and horses, as well as man."[60] The study renders the image legible under a human-scientific and anthropological rubric, but the targeted people are gone and only the photographic shadows of their steps remain. These shadows, easily filled by the traversing researcher and surveyor, provide an apt allegory for the contested politics of socio-environmental knowledge in the Glen Canyon. Jennings's survey, operating in anticipation of destruction, is a case study in how the lens of the natural was repeatedly deployed and focused to extract, exhaust, survey, recode, and revalue a landscape.

Environmental Media Archaeology and Excavating Uranium's Afterimage

Jared Farmer, in his introduction to the history of the Glen Canyon, recounts how George Stevens filmed the biblical epic *The Greatest Story Ever Told* while depicting the rising waters of Lake Powell as the Jordan River. Farmer quotes Charlton Heston – who played John the Baptist – as stating in an interview: "Not scenery like that of the Holy Land … but more as the Holy Land should be, still with the fingerprints of God on it" (xi). Diegetically, the character's lines are even more revealing of the postmodern reticulation of social and physical knowledge that marks this particular site: "Listen to the voice of one crying in the wilderness. Repent! The anger of the lord is upon all nations." The film's cultural representation, even as fiction, assisted in the epistemic project of recoding a landscape in flux. Heston's voice "crying in the wilderness" both performed a narrative baptism and served the Glen Canyon its last rites as ravine geological formations gave way to the anthropogenic Lake Powell.

The status of Heston-the-Baptist's "wilderness" is contradictory, growing only more specious with Hollywood's enlistment of the shifting landscape into a narrative about religious discovery. *The Greatest Story Ever Told* at once perverts and extends the American settler-colonial project that Fredrick Jackson Turner sought to impose through a vision of "wilderness" that was, as William Cronon influentially argues, always already under human sway: "Seen in this way, wild country became a place not just of religious redemption but of national renewal, the quintessential location for experiencing what it meant to be an American."[61] The canyon, by then a nuclear waste repository as well as a source for energy, anthropological knowledge, and recreation, became articulated as a fictional setting of "wilderness" for the redemption story of Jesus Christ and his followers.

With Stevens's film, the historical contradictions embedded in the landscape accumulate to the point of absurdity. It is useful, then, to further recall Cronon's prescription that the historically contested status of a society's relationship to the natural can

> teach us to recognize the wildness we did not see in the tree we planted in our own backyard. By seeing the otherness in that which is most unfamiliar, we can learn to see it too in that which at first seemed merely ordinary. If wilderness can do this – if it can help us perceive and respect a nature we had forgotten to recognize as natural – then it will become part of the solution to our environmental dilemmas rather than part of the problem.[62]

Likewise, a film's presentation of a fictional setting and narrative diegesis atop a real place and time serves as a co-constitution of biophysical reality and social values. Reading the merely ordinary *The Greatest Story Ever Told* for its otherness and unfamiliarity provides lessons for how the image passes as quotidian, as natural. Heston, waist-deep in conflicting definitions of the space surrounding him, paints a clear picture of the Gordian status of what the natural means in the media history of the Glen Canyon. Regarding the contradictory strata of medianatures, including the infrastructural history of the dam, the manufactured status of Lake Powell, and the radioactive pile buried beneath it, this chapter's archaeology emphasizes that the layered recognition of what counts as natural in a contested geography is relevant to current debates in environmental politics.

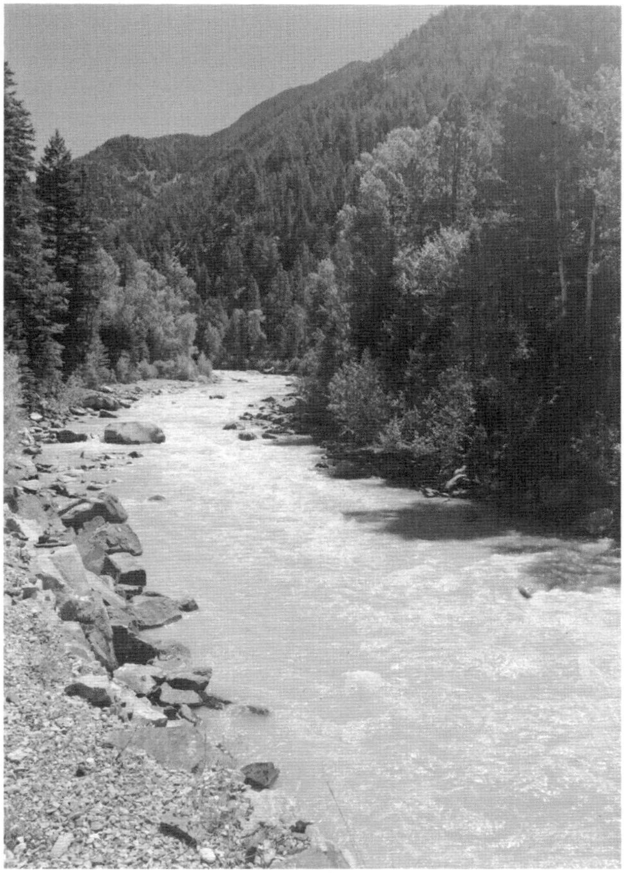

10.5 The Animas River between Silverton and Durango in Colorado, USA, within twenty-four hours of the 2015 Gold King Mine wastewater spill.

For instance, the Environmental Protection Agency's spectacular 2015 accidental release of 3 million gallons of gold mine wastewater into the Animas river – a tributary to Lake Powell – presents a reminder of the risks characterizing the sealed-away socio-environmental problems of the Southwest. As several news outlets pointed out while responding to the disaster, there are tens of thousands of such abandoned sites in the region and the corporate parties responsible for their construction and maintenance are rarely known or still surviving.[63] Is the EPA responsible for the remediation of the landscape *and* the reconstruction of the landscape's history? With what methodological tools are these lost environmental histories resurfaced and analyzed? Excavating the layers of social/physical strata defining the archaeology of what counts as natural today may only lead to a terminal revelation of the legacy of colonial violence, but the possibility of compiling media archives to articulate the contradictory and historical "natures" that lay the groundwork for subsequent environmental political problems remains an important pursuit.

Sean Cubitt reflects on how the disaster occupied the same geographical space as an iconic scene shot for *Butch Cassidy and the Sundance Kid* (1969), in which characters played by Paul Newman and Robert Redford dive into a remote river to escape pursuers. Cubitt uses the film's depiction of the landscape as natural and uninhabited wilderness to comment on the politics of environmental historical research surrounding media production, noting the semiotic shift in value belonging to Colorado gold extraction, which changed from an industry for currency to a latent threat as wastewater pollution for the Navajo nation. Most recently, Colorado gold is finding use in the commercial production of electronics: "Has the spill communicated to us a demand to revalue not only today's gold but the mutual responsibility of metal and men for the permanence of gold's release into the world?"[64] His inquiry, in other words, asks: In recognizing the history of social relations to a mineral, how can it be reinterpreted in the future?

Cubitt's reading suggests that archaeological inquiries into historical strata of medianatures can begin with familiar depictions of natural space, such as the desolate canyon in *Butch Cassidy and the Sundance Kid* that would become a spectacular yellow flood in 2015. Much like the Glen Canyon's impounded uranium, the Animas River environmental disaster had origins in a knowledge regime of natural resource signification, becoming a by-product of what Schuller identifies as settler-colonialism's "geological machine."[65] Cronon encouraged readers to begin environmental historical inquiry by looking closely and recognizing what enables us to find nature in one's own backyard; Cubitt shows how the media archaeology of photographic cultural objects can realize this reading strategy by focusing on how the history of a place surfaces in familiar cultural depictions, such as a well-known film. Constellating the series of enduring media depictions produced in one location allows the media archaeologist to denaturalize how communities view the area in the present. As *Butch Cassidy and the Sundance Kid* bore witness to one layer of medianatural strata signifying nature, the film becomes an invitation to reconstruct a place's layers of meaning as made and recorded through media.

Glen Canyon provides one final example: Monte Hellman's *The Shooting* (1966), produced in southern Utah alongside *Ride in the Whirlwind* (1966) in 1965. Hellman did not know it at the time, but his cameras caught the final days of the Glen Canyon bed before the dammed water rendered the space inaccessible.[66] The film's unsettling aesthetics and treatment of the disappearing landscape – photographed with sparseness, distance, and an urgency coextensive with a landscape undergoing rapid transformation – communicate a desire to index an area slated to become primarily knowable as natural history. The emergency conditions that drove Jennings and colleagues' study to pioneer the method of salvage archaeology surfaces in Helmann's film form too. *The Shooting* treats geography as though its own act of recording both archives and destroys history, preserving landscape under knowledge of the coming inundation.

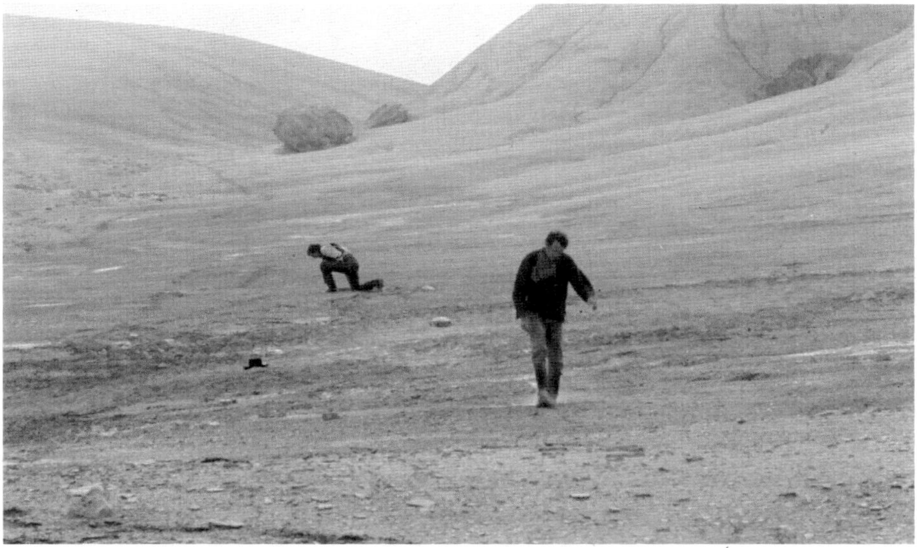

10.6 Glen Canyon in *The Shooting*.

Similarly, for Jennings and company, their interdisciplinary study – literal archaeology – corresponded with a broader discursive project that destroyed and transformed biophysical reality for a different and instrumental use. *The Shooting* marks a cultural formation analogous to salvage archaeology: the fiction film presents an audiovisual archive that recorded and therefore participated in producing the Glen Canyon as natural history. I interpret cultural projects like *The Greatest Story Ever Told* and *The Shooting* as representations of the epistemological project of settlement that helped prime national consciousness for what would become cash register infrastructure, a national recreation area, and a looming environmental disaster.

In Schuller's methodological argument about adopting a post- or decolonial position in media archaeological inquiry, she writes that viewing media forms that capture the past "enables us to see how ecological information about … just who and what is relegated to prehistory – plays a central role in settler colonial politics."[67] This is a crucial theoretical development for environmental media archaeologists, for "excavating materiality and temporality as hinges of settler colonialism exposes the sedimentary layers of biopower that complicate Foucauldian-inspired binary schematics of the power that fosters life or forces death, a framework infamously inattentive to colonialism."[68] While Schuller's account works to show and underscore the agency of Indigenous communities that experienced erasure while being recorded through the mediatized inscription of the American natural historical record, this chapter illustrates how "excavating materiality and temporality as hinges of settler colonialism" can assist in assembling a photographic archive so to show how media index site-specific cultural attitudes about waste, energy, and the historical forces holding the settlement of territory in place.

At the time of writing, the Glen Canyon Dam remains in place, fixed by a buried pile of uranium tailings and the social need for fresh water and energy in desert cities. Hellman's films are among the last images of the canyon bed, now irrevocably altered by the geohistorical processes of settlement. Media archaeology, when synthesized with approaches to space developed by theorists in environmental studies, becomes an important tool for understanding how settler-colonialism is a perpetual act of violence and dispossession. It also shows that settler-colonialism is an epistemic project that engineers geographical space across time, giving rise to the nature experienced today through cultural valuations of terrain such as the national recreation area.

As these varying photographic media forms record different attitudes toward what is considered natural about the Glen Canyon – and thereby support the diverse practical resource agendas of settlement – they at once record just how open to resignification a landscape is. While remediation appears an impossibility, the enduring violence of settler-colonialism emphasizes the need for a media aesthetics of reparation and repatriation for future definitions of nuclear landscapes. However, Jennings's destructive natural historical method might ask questions to media archaeology in return. In assembling discontinuous and heterogenous media histories, what is destroyed? And do the emergency conditions of recent environmental crises support exigency to accomplish historical research as motivated by present crises? In the least, this chapter argues for methods tuned to the recursive epistemologies presented by media environmental histories of settler-colonialism, and the toxic landscapes they leave behind.

NOTES

1 Foucault, *The Order of Things*.
2 Huhtamo and Parikka, *Media Archaeology*, 3.
3 Foucault, "Politics and the Study of Discourse," 59.
4 Foucault, *Order of Things*, 375. See also Foucault's definition of "historical *a priori*" in *The Archaeology of Knowledge*: "The fact that discourse has not only a meaning or a truth, but a history, and a specific history that does not refer it back to the laws of an alien development … This *a priori* does not elude historicity: it does not constitute, above events and in an unmoving heaven, an atemporal structure," 127.
5 Kittler, *Gramophone*, 117.
6 Parikka, *A Geology*, 14.
7 Monea and Packer, "Media Genealogy," 3152.
8 Pringle, "Photographed by the Earth," 136.

9 Parikka, *A Geology*, 14.
10 Ibid., 26.
11 Ibid., 25.
12 Freudenburg, Frickel, and Gramling, "Beyond the Nature/Society Divide," 366–7.
13 Ibid., 369.
14 Ibid., 367.
15 Ibid., 369, 388.
16 Nakamura, "Indigenous Circuits."
17 Geoghegan, "Orientalism and Informatics."
18 Schuller, "The Fossil and the Photograph," 231.
19 Farmer, *Glen Canyon Dammed*, xii.
20 Ibid., 147.
21 Ibid., 134.
22 Kuletz, *The Tainted Desert*, 7.
23 Ibid., 7.
24 See Davis, "Dead West," 49–73.
25 Nixon, *Slow Violence*, 153.
26 Gallagher, *American Ground Zero*, xxiii.
27 Davis, "Dead West," 49.
28 See Pasternak, *Yellow Dirt*, 133–64.
29 Hecht, *Being Nuclear*, 177.
30 Malin, *The Price of Nuclear Power*, 72.
31 Ibid., 74.
32 McCourt, *White Canyon*, 55.
33 This is supported in Darroll P. Young's ideologically suffused "Economic Impact of the Uranium Industry in San Juan County," which depicts the resource as locally comprehended as a divine gift endowing the earth with both endless energy and world peace: "Proper use of the natural resources of the earth is not evil in nature. Let us look forward, in great anticipation, for the time to come when the intelligence of mankind will overrule its foolishness. The human inclinations toward greed, selfishness, foolishness, and ignorance of right and wrong, good or bad, destroys our understanding that the earth is the Lord's. Let us also understand that we have been commanded to take good care of it, to subdue it, and use the resources thereof for the proper benefit of mankind," 40. Throughout the volume, much of the local historical reflection – while thorough in recording interviews, photographs, and written accounts – is replete with uncritical nostalgia for the days of the uranium boom, often employing messianic rhetoric to describe the mineral and its economic benefits. The portion of Jana Mellis's dissertation enclosed in the issue, "White Canyon," seeks a more balanced historical account.
34 Buck, "The Atomic Energy Commission."
35 Mazuzan and Walker, *Controlling the Atom*, 307.
36 McCourt writes that the majority of this uranium came from the nearby Happy Jack mine: "The mill processed 26,258 tons of raw ore between April 1949 and the end of December 1953. 128,145 pounds (64 tons) of uranium sodium diuranate was

extracted and sold to the Atomic Energy Commission during the life of the mill." *White Canyon*, 44.

37 Chenoweth, *The Geology*, 12.

38 Dansie, "A Study," 7.

39 Ibid. In Dansie's follow-up report one year later, he revises "undisclosed" to be "remove a *small amount.*" *"Uranium Mill Tailings," n.p.*

40 As Dansie comments: "Due to the silt cover over the tailings at White Canyon there is little potential for mechanical transportation of the sediments away from the site and into the reservoir. Also due to the silt cover there is little danger of radioactive decay emissions or radon gas emissions from the pile. However, as effective as the silt cover is in trapping the release of these contaminants it is ineffective in isolating the tailings from groundwater," "Uranium Mill Tailings," n.p.

41 Dansie continues: "Selenium transported from the submerged tailings piles is a likely contributor to the high concentration … Concentrated in mill tailings piles are a number of heavy metals including arsenic, barium, cadmium, lead, vanadium and selenium. In addition to these contaminants mill tailings piles contain radioactive materials not removed in the production process. In fact, 85% of the radioactive material in ore remains after the milling process. Radionuclides concentrated in tailings piles include Thorium-230, Radium-226 and Radon-222 … These contaminants in tailings piles are introduced to human contact through a number of pathways. Continued radioactive decay through alpha and gamma particle emissions, inhalation of windblown particles, and inhalation of radon gas, a daughter product of Radon-222, are all potential contaminant exposure pathways," "Uranium Mill Tailings," n.p.

42 Ibid.

43 Regarding reclamation, Dansie writes: "Recently nine major tailings piles in the Colorado River Basin which posed threats similar to those anticipated at the Hite pile (the main concerns with exposed tailings are contamination of nearby water sources, eolian distribution of tailings, inhalation of radon gas and gamma radiation) have been relocated to suitable permanent and appropriate disposal sites." "A Study," 10.

44 Reviewing the literature on proposals involving the decommission of the Glen Canyon Dam, Carothers and House do not mention the tailings pond at the bottom of the lake ("Environmental Restoration"), while the Glen Canyon Institute's "Citizens' Environmental Assessment (CEA) on Decommissioning of the Glen Canyon Dam" states: "Water percolating through tailings becom[ing] contaminated with radioactivity from Thorium-230, Radium-226 and Radon-22 left behind in the milling process" and "Uranium mill tailings contain high concentrations of heavy metals including arsenic, barium, cadmium, lead, selenium and vanadium. These materials pose a significant health risk." "Decommissioning," 11.

45 In Dansie, "Uranium Mill Tailings"; and McCourt, *White Canyon*, 221–2.

46 Serres, *Malfeasance*, 42.

47 Nixon, *Slow Violence*, 152–3.

48 Mellis, "White Canyon," 65.

49 Schuller, "The Fossil and the Photograph," 238.

50 Ibid.
51 Ibid.
52 Ibid.
53 Documented in Farmer, *Glen Canyon Dammed*, 158–62.
54 Jennings, *Glen Canyon*, 5.
55 While the Jennings expedition was comprehensive in its interdisciplinary scientific record, its methodological contribution was constrained to archaeology. For another theoretical perspective on salvage science in an environmental humanities context, Serena Stein has developed the concept of "salvage ecology" to account for research practices of rescue, documentation, and conservation during the climate emergency, with reference to endangered species in threatened ecologies. Stein, "Remnant Rainforest."
56 Jennings, *Glen Canyon*, 6.
57 Ibid., 7.
58 Ibid., 1. Author's emphasis.
59 Ibid., 45.
60 Ibid., 45.
61 Cronon, "The Trouble with Wilderness," 76.
62 Ibid., 88.
63 See Handwerk, "Why Tens of Thousands"; and Loomis, "Toxic Frontiers."
64 Cubitt, *Finite Media*, 179.
65 Schuller, "The Fossil and the Photograph," 238.
66 Hellman states as much in the audio commentary for the Criterion DVD release, but this is also supported by the distributor's website: https://www.criterion.com/current/posts/3360-the-shooting-and-ride-in-the-whirlwind-we-can-bring-a-good-bit-of-rope.
67 Schuller, "The Fossil and the Photograph," 260.
68 Ibid., 232.

BIBLIOGRAPHY

Buck, Alice. "The Atomic Energy Commission." US Department of Energy, 1983. https://energy.gov/sites/prod/files/AEC%20History.pdf.

Carothers, S.W., and D.A. House. "Decommissioning Glen Canyon Dam: The Key to Colorado River Ecosystem Restoration and Recovery of Endangered Species?" *Arizona Law Review* 42 (2000): 215–38.

Chenoweth, William L. *The Geology and Production History of the Uranium Deposits in the White Canyon Mining District, San Juan County, Utah*. Salt Lake City: Utah Geological Survey, 2000.

Cronon, William. "The Trouble with Wilderness; or, Getting Back to the Wrong Nature." *Uncommon Ground: Rethinking the Human Place in Nature*, edited by William Cronon. New York: W.W. Norton, 2000.

Cubitt, Sean. *Finite Media: Environmental Implications of Digital Technologies*. Durham: Duke University Press, 2000.

Dansie, Tom. "A Study of the White Canyon Mill Tailings at Hite, Utah." *Glen Canyon Institute*, September 1999.

– "Uranium Mill Tailings in the Colorado River Basin." *Hidden Passage: The Journal of the Glen Canyon Institute* 5 (Spring 2000).

Davis, Mike. "Dead West: Ecocide in Marlboro Country." *New Left Review* 200 (1993): 49–73.

Farmer, Jared. *Glen Canyon Dammed: Inventing Lake Powell and the Canyon Country.* Tucson: University of Arizona Press, 1999.

Foucault, Michel. *The Archaeology of Knowledge* [1969]. New York: Routledge Classics, 1989.

– *The Order of Things: An Archaeology of the Human Sciences* [1966]. *New York: Routledge, 1989.*

– "Politics and the Study of Discourse." In *The Foucault Effect: Studies in Governmentality,* edited by Gordon Graham Burchell and Peter Miller. Chicago: University of Chicago Press, 1991.

Freudenburg, William, Scott Frickel, and Robert Gramling. "Beyond the Nature/Society Divide: Learning to Think about a Mountain." *Sociological Forum 10, no. 3 (1995): 361–92.*

Gallagher, Carole. *American Ground Zero: The Secret Nuclear War.* Cambridge, MA: MIT Press, 1993.

Geoghegan, Bernard Dionysius. "Orientalism and Informatics: Alterity from the Chess-Playing Turk to Amazon's Mechanical Turk." Ex-position 43 (2020): 45–90.

Glen Canyon Institute. *Citizens' Environmental Assessment (CEA) on the Decommissioning of Glen Canyon Dam: Report on Initial Studies.* Flagstaff: Glen Canyon Institute, 2000.

Handwerk, Brian. "Why Tens of Thousands of Toxic Mines Litter the U.S. West." *Smithsonian.com*, 13 August 2015. http://www.smithsonianmag.com/science-nature/why-tens-thousands-toxic-mines-litter-us-west-180956265/?no-ist.

Hecht, Gabrielle. *Being Nuclear: Africans and the Global Uranium Trade.* Cambridge, MA: MIT Press, 2012.

Hellman, Monte, Millie Perkins, Cameron Mitchell, and Jack Nicholson. *The Shooting; Ride in the Whirlwind.* Irvington: Criterion Collection, 2014.

Huhtamo, Erkki, and Jussi Parikka. *Media Archaeology: Approaches, Applications, and Implications.* Berkeley: University of California Press, 2011.

Jennings, Jesse D. *Glen Canyon: A Summary.* Department of Anthropology, University of Utah, Salt Lake City, 1966.

Kittler, Friedrich. *Gramophone, Film, Typewriter.* Stanford: Stanford University Press, 1999.

Kuletz, Valerie. *The Tainted Desert: Environmental Ruin in the American West.* New York: Routledge, 1998.

Loomis, Erik. "Toxic Frontiers." *Dissento* 15 September 2015. https://www.dissentmagazine.org/online_articles/toxic-legacy-animas-river-spill-mining-companies.

Malin, Stephanie A. *The Price of Nuclear Power: Uranium Communities and Environmental Justice.* New Brunswick: Rutgers University Press, 2015.

Mazuzan, George T., and J. Samuel Walker. *Controlling the Atom: The Beginnings of Nuclear Regulation, 1946–1962.* Berkeley: University of California Press, 1985.

Mellis, Jana. "White Canyon." *Blue Mountain Shadows: The Magazine of San Juan County History* 16 (1995–6): 62–78.

McCourt, Tom. *White Canyon: Remembering the Little Town at the Bottom of Lake Powell.* Price: Southpaw, 2003.

Monea, Alexander, and Jeremy Packer. "Media Genealogy and the Politics of Archaeology." International Journal of Communication 10 (2016): 3141–59.

Nakamura, Lisa. "Indigenous Circuits: Navajo Women and the Racialization of Early Electronic Manufacture." American Quarterly 66, no. 4 (2015): 919–41.

Nixon, Rob. *Slow Violence and the Environmentalism of the Poor.* Cambridge, MA: Harvard University Press, 2011.

Parikka, Jussi. *A Geology of Media.* Minneapolis: University of Minnesota Press, 2015.

Pasternak, Judy. *Yellow Dirt: An American Story of a Poisoned Land and a People Betrayed.* New York: Free Press, 2010.

Pringle, Thomas Patrick. "Photographed by the Earth: War and Media in Light of Nuclear Events." NECSUS European Journal of Media Studies 3, no. 2 (2014): 131–54.

Schuller, Kyla. "The Fossil and the Photograph: Red Cloud, Prehistoric Media, and Dispossession in Perpetuity." Configurations 24, no. 2 (2016): 229–61.

Serres, Michel. *Malfeasance: Appropriation through Pollution?* Stanford: Stanford University Press, 2011.

Stein, Serena. "Remnant Rainforest: Salvage Ecologies and the Politics of Leftovers in Mozambique." Workshop Presentation for Hinterlands: A Project in the Rural, Literary, and Environmental Humanities, 2021.

Stevens, George. *The Greatest Story Ever Told.* Santa Monica: MGM Home Entertainment, 2001.

Young, Darroll P. "Economic Impact of the Uranium Industry in San Juan County." *Blue Mountain Shadows: The Magazine of San Juan County History* 16 (1995–6): 2–40.

11

Nuclear Waste as Critical Heritage

CORNELIUS HOLTORF AND ANDERS HÖGBERG

Within the nuclear environmental humanities, it falls especially to the disciplines of archaeology and heritage studies to develop approaches and perspectives about remembering and planning extending across long periods of time. Archaeology and heritage studies are versed in paying attention to both the tangible and the intangible realms and to varying perceptions and complex interrelations between past, present, and future. Archaeology commonly adopts long-term perspectives in research. Archaeologists working in the heritage sector commonly manage tangible and intangible human legacies, the cultural heritage, with present and future needs in mind.

In this chapter we offer two lines of thought related to the emerging field of nuclear environmental humanities: first, an archaeological understanding of time in the context of historical consciousness, and indeed, as we will argue, future consciousness, and how it can inspire critical thinking about the nuclear domain across disciplines; and second, the notion of nuclear waste as cultural heritage, which offers particular critical insights that have the potential to challenge not only current thinking in the nuclear waste sector but also current thinking in the humanities themselves.

We are both researchers in archaeology and heritage studies. For more than ten years now we have been investigating the permanent preservation of material items, long-term memory-keeping, and knowledge transfer to future generations, to the extent that these are shared concerns in managing cultural heritage and nuclear waste. Long-standing collaboration with the Swedish Nuclear Fuel and Waste Management Company (SKB), as well as participation in a number of national and international conferences, workshops, and seminars about the challenges of long-term communication regarding final repositories of nuclear waste, have left us firmly convinced that although the starting points for both archaeology/heritage studies and nuclear waste management could hardly be more different, the potential for mutual learning and knowledge transfer is enormous.[1]

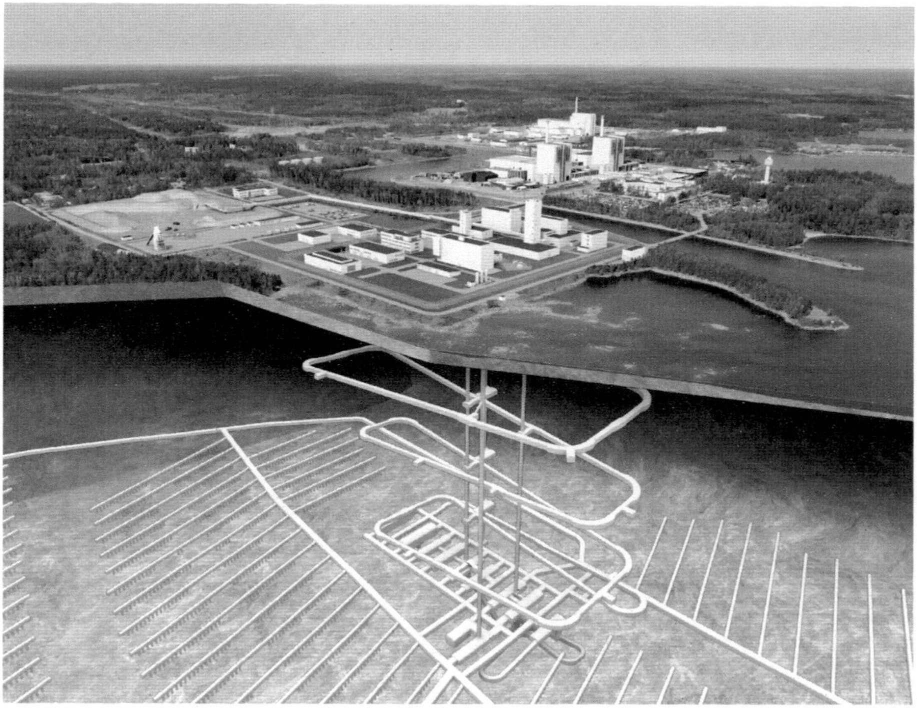

11.1 Image of planned final nuclear waste repository at Forsmark in Sweden. Presently, this site is planned to be built over the coming decades. In operation, it is estimated that it will take more than eighty years to fill up before it is finally sealed.

Nuclear Waste Management and Long-term Communication

The principal challenges in nuclear waste management may or may not be greater than those facing archaeology/heritage studies, but the stakes are usually considered higher. More than 200,000 tonnes of high-level radioactive nuclear waste currently exist on Earth.[2] This waste originates mainly from the world's more than 600 nuclear power plants (at the time of writing, operable: 448; permanent shutdown: 160; long-term shutdown: 2; under construction: 60).[3] It is radioactive and must be kept safely away from humans and nature for thousands of years before contaminated areas can be declared safe. Safe in this case is when radiation dose levels are equal to or lower than natural background radiation.[4] To facilitate safe, long-term storage, deep geological final repositories are planned in many parts of the world (see Figure 11.1).[5] It is thought that these will protect the waste from humans and nature, and humans and nature from the waste, for thousands of years to come. The length of time prescribed for the storage varies from country to country, depending on differences in how national decision-makers have responded to civic movement protests and critical discussions.[6] In

Sweden, for example, regulations state 100,000 years, in the US, 10,000 years, and in Germany, 1 million years.[7]

Once the waste is buried and the repository is sealed, there is a need to communicate the implications of drilling and exploring at this location for a very long time span. Knowledge of location, context, and quality of the waste therefore needs to be transmitted to the future so that coming generations can make informed decisions on how to interact with these sites and their contents. Consequently, we are faced with the task of creating long-lasting media and messages that are able to inform and provide knowledge and memory about the final repositories for nuclear waste for thousands of years, or even hundreds of thousands of years.[8] Yet reliably preserving relevant knowledge about nuclear waste over very long time periods is, while necessary, also impossible.

Many thousands of years is a long time. Too long a time for anybody to understand unless it is divided into sections. The nuclear waste sector commonly employs four timescales regarding the preservation of record, knowledge, and memory at nuclear waste repositories. These are *very short term*, *short term*, *medium term*, and *long term*.[9]

- *Very short term* refers to a time span that is consistent with staff stability in terms of roles, cycles of organizational change, and the regulatory expectations of periodic safety reviews. Typical timescales are ten to twenty years.
- *Short term* refers to the period that ends with the repository's closure. This period includes both the preoperational and the operational phases of the repository. Timescales are in the order of hundreds of years. Consequently, already the short term extends beyond what is a normal lifespan for a human being.
- *Medium term* refers to the period of indirect oversight activities that would follow the repository's closure. Timescales are in the order of a few hundred years.
- *Long term* refers to the time span with no repository oversight.[10] This period extends over the time of concern in the safety regulations, typically tens or hundreds of thousands of years in the case of high-level waste.[11]

Compared to the timescales common in heritage studies and archaeology, we can see that the *medium term* deals with a time span that equals historic time and the *long term* with deep history and evolutionary time.[12] In other words, repositories for long-term storage of nuclear waste are physical and social systems locked into a very distant future.

Future Consciousness

The ways humans make sense of pasts and futures are important for how they situate themselves in the present. Arguably, it is the ability to understand the present as a result of change over time and the ways we plan for the future that separate us from other species.[13] However, the ways that individuals embody this ability vary. This variation depends, among other things, on variable skills and knowledge about how the past, the present, and the future are interlinked and how these time dimensions are manoeuvred in the present.[14]

In this section we introduce the concept of future consciousness as a development of historical consciousness, which is an established concept in the humanities. Historical consciousness refers to the underlying thought structures that generate meaning when a particular historical perspective is given significance, implying certain consequences for the way we see the present and the future.[15] Similar thought structures can also be given significance by particular perspectives on the future, with consequences for how we see the past and the present. We argue, therefore, that when we shift attention from the past in the present to the future in the present, historical consciousness turns into future consciousness. Just as it is possible to analyze historical consciousness from the way it manifests itself in uses of the past, future consciousness can be analyzed from the way it manifests itself in perceptions and uses of the future. We argue that the notion of future consciousness can inspire critical thinking in the nuclear humanities.

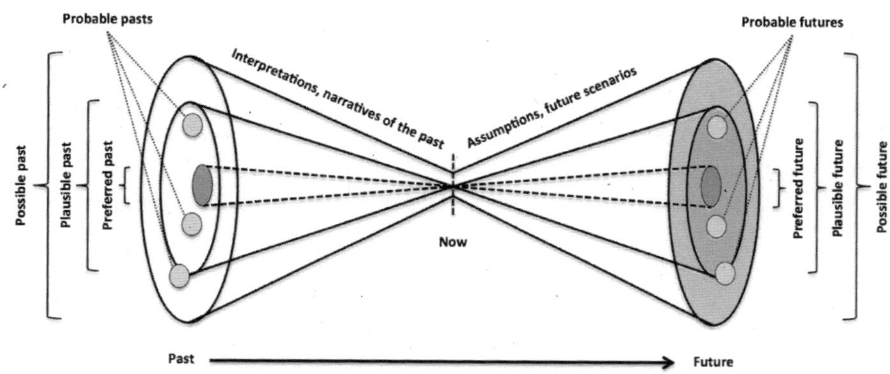

11.2 Future consciousness: schematic illustration of how human interpretations and narratives of the past are transformed through the needle's eye of Now into assumptions and future scenarios. Crucially, this is a "rolling now" constantly moving along the axis of time as the future becomes present and the present becomes past. The grey tone in the future part of the diagram marks the fact that even though past and future are conceptually closely linked, there are significant differences in how they are perceived.

Table 11.1

Different pasts and futures as illustrated in future consciousness schematic.

Preferred past	what someone would like to have happened	Preferred future	what someone would like to happen
Probable past	what presumably has happened	Probable future	what presumably will happen
Plausible past	what arguably might have happened	Plausible future	what arguably might happen
Possible past	what possibly could have happened	Possible future	what possibly could happen

Figure 11.2 illustrates how we can understand the processing of past and future in the present. We live in the *Now*. This *Now* is the present moment in which past and future are perceived based on certain assumptions about pasts and futures.

Human beings are capable of imagining a great many events and processes that may have happened in the past. After research and reasoning about the available evidence, we find, however, that some events and processes offer more plausible or credible indications of having occurred in the past. Among such plausible pasts, some are more *probable* than others in a given historical scenario. For example, it is certainly possible that extraterrestrials visited Earth in the past, but it is not particularly plausible, given the lack of direct evidence; for most people it is even less probable, given the existence of more convincing accounts of what presumably happened in the human past. Finally, in any given present context there may be one particular past that one or more people, for example politicians, would like to have happened for some reason. This *preferred past* will be more convincing to the extent that it is also considered possible, plausible, and even probable.

By the same token, people may have a *preferred future*. But even when they would like something to happen, this does not mean that it presumably will happen and is therefore *probable*. Indeed, there are arguably many different futures that are *plausible* enough to happen, within a wide spectrum of *possible* future events and processes that can be imagined by people in any *Now*. To stay with the same example, we can certainly imagine today that extraterrestrial life forms will one day try to make contact with humans on Earth, but there is not much evidence that this might plausibly happen; except in some movies, there are no scenarios of probable futures where such contact presumably will happen, no matter how much anybody might prefer it to happen.

These pasts and futures all have to be processed in our present, the *Now* in Figure 11.2. Just as sand in an hourglass has to flow from one container to another through a narrow passage, time is constantly flowing through an ever-changing present. By passing through that narrow passage – the needle's eye – comprising our present, pasts, and futures, time constantly modifies or reformats our perception of these temporal assemblages. Consequently, through time, the spectrum of preferred, probable, plausible, and possible pasts and futures will change, but the shape of the figure will stay the same. Every past and every future present has specific repercussions as to how people understand their *Now* in relation to how they perceive past and future. Even though we speak here about a singular *Now*, there may be various *Nows* and multiple presents, but the point is that Figure 11.2 applies to each of these.[16]

There is however a crucial difference between our perceptions of past and future. Regardless of how we perceive the past, we have to live with its consequences. Yet how we perceive the future is partly shaped by our own actions. This realization brings to life particular human emotions linked to what we desire and what we hate, what we fear and what we hope for, what we dare to do and what we choose not to risk, informing our actions that will intentionally shape the future. At the same time, we accept a particular moral responsibility for our actions, especially when they will impinge upon many people's lives in the long-term future. Memory and future thinking are closely related, and imagining futures and pasts rest on the same cognitive and neural processes.[17] As MacLeod suggests, there is however a key difference: the "uncertainty inherent in future-thinking implies a greater role for semantic memory in how people think about the future compared to how they remember the past."[18] Consequently, thinking about the past and the future activates the same parts of the human mind, but in different ways. These affective dynamics of future consciousness may explain why we spend more time worrying about the future than about the past.

This discussion of a rolling now, associated perceptions of past and future, and the affective dynamics of future consciousness has one main implication that applies equally to heritage conservation and the preservation of record, knowledge, and memory as it concerns final repositories of nuclear waste: future societies will presumably have their own perceptions of the past and the future, and they will in all likelihood be just as committed as we are to acting on the basis of these perceptions so as to shape their own futures.

Reinterpretation and indeed a change of perspective is thus not a problem to be eliminated from our own strategies of preservation and conservation; rather, it is a basic condition of human development that we need to understand and adapt to in our long-term plans. The question is not, therefore, how to create a long-term strategy based on our own perspectives and perceptions of the challenges ahead; rather, it is how to create a long-term strategy that will allow future *Nows* to create their own perspectives and perceptions to act upon so that future human beings will act as creatively and innovatively as we do, generating perspectives and perceptions that we ourselves cannot imagine today.

Allowing for future societies to have their own views of the past and the future leads to a new way for us to conceive of the nuclear domain and its long-term challenges. The evolving *Now* changes the nature of planning from the domain of making futures to the domain of accommodating future presents that make their own pasts, presents, and futures.

UNIMAGINED FUTURES

Does our argument thus far imply that the future cannot reliably be predicted and therefore cannot be planned for at all? No, but it does suggest that what we can say about the future with some confidence lies in the mutability of perspectives and perceptions that inform what will happen, rather than in the accuracy of predicting what will happen. Planning, therefore, requires a different approach as well.

A common way to plan futures in the present is to build trend analyses of what happened in the past and from that infer what may happen in the future.[19] When we take this track, the future is often expressed in terms of anticipations or prognostic trends, to be discerned from analyzing past changes and emerging patterns over time and extrapolating into the future. In theory, this type of forecasting domesticates time by imposing present knowledge on the future. As such, it creates a future that cannot develop into anything more than an extension of what is already known. There is, however, an alternative way of imagining the future: as *unimagined futures*. UNESCO's futurologist Riel[20] has argued that we do not have to project the present onto the future. The unimagined future is framed by a future consciousness that encourages us to investigate what we need to imagine, know, and change in order to create futures that will be desirable but that we cannot realize at the moment.[21] In this sense, the unimagined future lies beyond the possible future in Figure 11.2. It implies finding new ways to understand the present in order to strengthen our capacity for innovative thinking beyond what we currently know. The unimagined future also requires creativity and calls upon us to "stretch our ontological bounds."[22] In such processes, it may be helpful to explore unimagined pasts as well.

In relation to nuclear waste, we arguably need an understanding of how unimagined futures can provide us with creative tools for making the kinds of futures needed in our present to keep knowledge alive in the short and medium term. Such thinking aims at taking advantage of an unknown future rather than becoming its victim. It is a form of future consciousness that helps us question dominant perceptions in the present of what is right and proper. This strategy, then, is not about predicting the future, but about creating capacity in order to deal with a continuously changing world, by allowing us to use *the unimagined* as a resource in the work we conduct to shape the future.

Nuclear Waste as Cultural Heritage

In the second part of our discussion, we explore *the unimagined* by elaborating on the previously unthought idea of nuclear waste as cultural heritage. We return to our initial observation that cultural heritage management and nuclear waste management share important concerns related to the permanent preservation of material items, long-term memory-keeping, and knowledge transfer to future generations. We argue that nuclear waste is a very particular kind of cultural heritage. This argument will offer a number of critical insights that challenge established approaches not only in the nuclear waste sector but also in the humanities. In that sense, nuclear waste is not only cultural heritage but also critical heritage.

Bringing together nuclear waste and cultural heritage has a critical potential in at least two complementary ways. On the one hand, it facilitates encounters of the nuclear waste sector with existing insights and knowledge about cultural heritage, in particular about the mutability and politics of heritage. This might improve the judgments made when designing geological repositories for nuclear waste and associated messages for the long-term future.[23] On the other hand, it prompts the humanities to consider specific futures and to debate more thoroughly the long-term trajectories of memory and preservation. Ultimately, both realms will be able to make better-informed judgments and thus better meet their responsibilities in the present.

Arguably, deep geological repositories and the nuclear wastes they hold, in tandem with carefully designed accompanying messages in particularly long-living media, are a very particular kind of historical legacy of our atomic culture.[24] This legacy is particularly durable and will likely remind future generations of the twentieth and twenty-first centuries, thus becoming part of the cultural heritage of the future.[25] Indeed, a short look at history demonstrates that nuclear waste is already now part of our modern heritage.

A SHORT HISTORY OF NUCLEAR WASTE

Transforming uranium into energy in nuclear power plants was regarded in the 1950s and 1960s as a virtually unlimited approach to generating power that would advance humanity's mastering of nature, science, and technology and facilitate energy supply for an infinite future.[26] This optimism and confidence affected how nuclear waste was understood.

The knowledge existed that nuclear power produced highly radioactive waste to which direct exposure was lethal. The reactors of the time, however, were thought to be just the first in a series of technological developments that would lead to more efficient ones. When these came into operation, the amount of waste would be reduced to virtually nil. In this way an undeniably modern belief in progress[27] influenced how society generally thought of nuclear waste. People knew it was dangerous and that it needed to be looked after, but since the waste would be processed as technology evolved, the highly radioactive nuclear waste was viewed more as a resource than as waste.[28]

In the 1960s, the insight that nuclear waste might become a problem was increasingly foregrounded. A radical increase in volume, new knowledge about the long half-life of radioactivity, and difficulties in reprocessing the waste caused a swing in perception, from the earlier optimism to caution and uncertainty. In 1977, US president Jimmy Carter announced that reprocessing of nuclear waste was to be stopped. The purpose was to limit access to plutonium for making nuclear weapons and thus prevent their proliferation.[29] This had consequences for how society regarded nuclear waste. Gradually, awareness developed that the nuclear waste that had been produced, and that was still being produced, would need to be placed in storage or in repositories safely away from humans and nature for a very long time.[30]

In the early 1990s, the discussion of new technologies gained momentum once again. Transmutation technology seemed capable of reducing the storage time. In simplified terms, this technology involves bombarding highly radioactive nuclear waste with powerful neutron flows, thus generating new energy. The residual waste is small in quantity and has a decay time of about a thousand years.[31] This development was viewed positively by many, and the technology was discussed as a first step toward the idea of recycling in the nuclear energy industry. It soon became clear, however, that transmutation technology was far from fully developed. It was also uncertain when, and indeed whether, the technology could ever come into use.[32]

Today, it is generally accepted that nuclear waste will have to be stored for a very long time so as to prevent its radioactivity from harming humans, ecosystems, and the environment. Although there is ongoing research on alternative ways to handle nuclear waste, the major part of today's work with nuclear waste is geared to final storage in deep geological repositories.[33]

The lesson to be drawn here is that technoscientific and popular conceptions of nuclear waste have varied historically. This implies that nuclear waste, notwithstanding its dangerous material properties, is socially and culturally negotiable.[34] The main properties of this particular type of nuclear material are conditioned by both zeitgeist and physics. Accordingly, the notion of nuclear waste may be viewed as part of modern history and its physical legacy as part of the tangible cultural heritage of our age. Much as archaeological remains, ruins, and monuments point to vanished cultures and histories and function as lures for decoding and interpretative activities, nuclear waste reminds us – and will likely remind future generations – of the multiple histories and legacies of the nuclear age.

THE POLITICS OF ARCHAEOLOGICAL HERITAGE

Previous applications of archaeology to questions about the long-term preservation of knowledge and meaning have often attempted to draw lessons from what has been preserved from ancient monuments to the present day. For example, the Pyramids of Giza and the stone circle of Stonehenge are often mentioned as surviving monuments from the past containing ancient information and meaning for us to decode today.[35] These and similar ancient monuments have inspired discussions about how permanent markers at nuclear waste repositories could be designed to mark out the sites and warn future generations against exploring them.[36] In our view, however, the possibility of maintaining or recovering knowledge and meaning over long time spans needs to be approached with caution. The example of European megalithic tombs shows how people's interpretations of their significance have changed drastically over the 6,000 or so years of their existence.[37] Indeed, many have been completely destroyed. What has been preserved or recovered is, at best, the assumption that these monuments stemmed from a distant past beyond human memory and that somebody may have been buried in them. But these assumptions hardly constitute the kind of maintenance or recovery of relatively complex information, knowledge, and meaning that is relevant in the present context of nuclear waste repositories. The lesson to be drawn from this example is that information, knowledge, and meaning are created in every present. Historical developments, including histories of interpretations, are not predictable and may be subject to dramatic fluctuations.

In each present, the meaning of heritage is determined by its value for different stakeholders and communities. That value is variable and may be often contested.[38] There is a politics of heritage.

The insight that nuclear waste can be viewed as a form of present and future cultural heritage has critical implications. We know already that nuclear waste is deeply implicated in the politics of the present.[39] A focus on nuclear waste as cultural heritage suggests that the legacy of nuclear power in the twentieth and twenty-first centuries will be enmeshed in the politics of the future. Drawing on our discussion of nuclear waste as heritage and the politics of archaeological heritage, an important conclusion can be drawn: what we today think and know about nuclear waste is not going to be the same as what future generations will think and know about it. This is a perspective not yet fully developed in the nuclear waste management sector.[40]

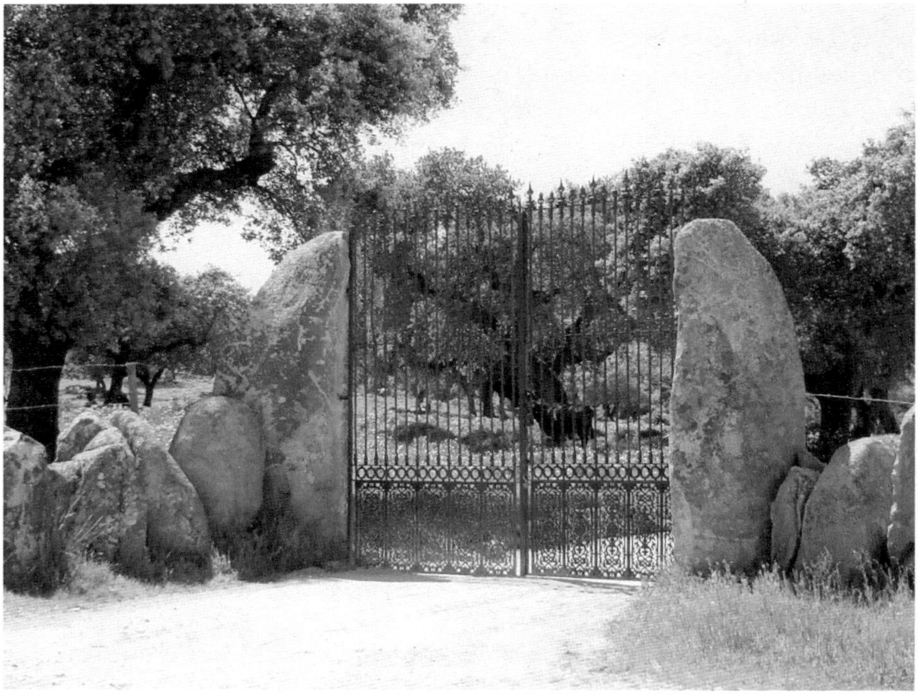

11.3 A megalithic tomb that has changed its meaning and significance over time.

The future meaning of nuclear waste, just like that of any other type of heritage, will be determined by heritage values that not only will be interpreted differently and be contested by different communities according to their own values in each (future) present but also will keep changing over time. Any long-term strategy for preserving records, knowledge, and memory will therefore have to allow for continuing transformations in relevant stakeholders' and communities' understanding of what we call nuclear waste and of the latter's highly political and contested nature. This means, among other things, that the perceived risks and properties of nuclear waste as part of the human socio-cultural heritage are subject to change and contestation as this heritage is transmitted to future generations. For this reason, we cannot assume that our toxic waste is also the future's toxic waste. This is not because we wish to relativize physics, but because we understand the mutability and political nature of heritage.

THE FUTURE IN HERITAGE PRACTICES

If nuclear waste will last many thousands of years and if this material, as we have argued, can be understood as cultural heritage, it is relevant to consider how archaeologists and other managers of cultural heritage normally perceive the future in their work.

A substantial part of the work carried out within the heritage sector is motivated by a present-day desire to preserve past objects and knowledge about the past for the benefit of future generations.[41] The underlying preservation paradigm relies on a strong conservation ethos, based on the assumption that future generations will in one way or another value what we leave for them. Policy documents justifying the underlying thinking and motivations of this conservation ethos typically include phrases like "hand on to future generations" or "stewardship for tomorrow's generations."[42] This conservation philosophy dominates the heritage sector and enjoys wide public support, yet it has rarely been critically discussed,[43] and few alternatives to its perspective have been proposed.[44]

In a recent study we analyzed how the future is thought about, conceptualized, and linked to practice in the heritage sector.[45] Sixty-six experienced professionals, often in senior positions, working in the heritage sector in Sweden, England, and globally were interviewed. Our aim was to understand how individual heritage managers and experts think about the future in their professional role and how the future informs their work.

Our findings reveal a paradox. Archaeologists and heritage managers are familiar with the politics of heritage and the variability of heritage values, yet they are struggling to come to terms with the future. They focus on the future as a fundamental premise of their work and motivate this focus with the conservation ethos, yet this future is not explicitly formulated. The future tends to remain implicit in daily heritage practice that operates in the continuing present. Several of those interviewed offered the insight that the future may not value what we preserve, as we value it today. This understanding, however, was not translated into practice-related ideas about how to change the day-to-day work of heritage management. Frequently used phrases such as "only time will tell" and "history will judge" emphasize a perception of backward-looking future judgments of present decisions over a forward-looking present practice of realizing possibilities for a meaningful future. In heritage practices, as it were, the future does not stretch forward from the present but instead is envisioned as preceding the latter.

We were surprised to find that professionals working in the heritage sector perceive the future not from a professional point of view based on trained skills and institutionalized knowledge, but as any other person in society would. For example, when being explicit about time frames, and not just referring to an indefinite future, heritage professionals tend to think of a future that lies, at the very most, three generations ahead. Studies have found that three generations (about seventy-five years) is a common interval in relation to how many societies understand time.[46] Three generations are what many today are familiar with from personal experience. It is the time scale that includes our parents and grandparents or children and grandchildren. Time spans beyond four generations are difficult to grasp for anybody not trained in thinking in time (such as members of cultures with orally transmitted genealogies that can extend over several centuries).

Beyond our grasp of the generations we can meet, the future becomes a blur, an eternity. Hence, our results show that there is no difference between how professionals working in the heritage sector perceive the future and how any other person in society would.

Another instance of heritage sector professionals' lay or non-professional understanding of the future relates to the assumption held by many in the field that the future will be a continuation of the present. We have not come across substantial efforts to understand how the future will differ from today and may therefore require decisions that differ from what would be best for our own society now. Having studied a random selection of individuals, Quoidbach and colleagues found that the future is commonly understood as a continuation of the present.[47] When individuals who over the previous ten years had experienced dramatic changes in their lives – for example, falling seriously ill, moving house with the family, changing jobs, being involved in an accident, or losing a family member – were asked how they saw the coming ten years of their life, the vast majority replied that they perceived the future as the continuation of a non-changing present. Despite having experienced major changes in their lives, when talking about the future people generally tend to assume that what they have in the present will be what they have in the future. So even in this respect, those who work in the heritage sector do not perceive the future from a professional point of view, but rather tend to view it as an extension of the present, like any ordinary person would.

A close encounter with the world of nuclear waste can potentially make a critical difference for those working in the fields of archaeology and the humanities insofar as it prompts them to consider the future explicitly and in their professional roles. For example, when the nuclear reactor at Dounreay, Scotland, was being decommissioned, nuclear safety and heritage values were pitted against each other.[48] After the reactor was shut down and the decommissioning of the site had started, issues of preservation were raised. One controversy in relation to the site's cultural legacy concerned the famous spherical Dounreay Dome, completed in 1958 and often highlighted as an emblem of the atomic age. The discussion was about whether the dome could and should be preserved. Did its cultural significance warrant the costs of cleaning a radioactively contaminated structure? Can radioactive waste be of cultural significance for the future? In the end, the dome was not preserved. The nuclear waste, on the other hand, will need to be taken care of for an almost unthinkably long period of time. In this debate, heritage managers were called upon to assess the cultural significance of a nuclear site, and in engaging with the processes of nuclear decommissioning they were also prompted to consider more professionally which futures they were actually working for and how the needs of those futures might be best addressed in present-day decisions about heritage preservation and management.

Conclusion: Critical Dimensions of Nuclear Heritage

Cultural heritage management and nuclear waste management share concerns regarding the permanent preservation of material items, long-term memory-keeping, and knowledge transfer to future generations. However, the ways in which these concerns are negotiated differ between the two sectors. In this chapter, we have approached nuclear waste as culture heritage. We have also proposed new perspectives on future consciousness and on how to perceive memory and planning over long periods of time. Our work has led us to important critical reflections. Professionals in the nuclear waste sector can think about the future but seem to have difficulties dealing with contested values and the changing politics of heritage. Experts in cultural heritage management, by contrast, understand the politics of heritage but do not appear to come to terms with addressing the long-term future in their work.

Discussing nuclear waste in relation to future consciousness and in terms of cultural heritage fosters new thinking in both domains. We have argued that planning for the future in order to conserve the heritage or preserve records, knowledge, and memory concerning final repositories of nuclear waste requires a long-term strategy that allows for future presents in which human beings will be creating their own perspectives and perceptions on the basis of which they will act as creatively and decisively as we do. This future cannot be limited to predictions based on what we can infer from the past, but will depend, in part, on what is unimagined in the present.

In the second part of the chapter, we argued that the future meaning of nuclear waste repositories and their contents, just as with other forms of heritage, will be determined by future heritage values, which will be contested and variable in space and time. This is a previously unimagined idea. What we usually call nuclear waste is in fact a form of cultural heritage; this means we cannot be certain that it will always be viewed predominantly as hazardous material posing a threat to humanity, or that its radioactivity or other physical properties will always be its most significant properties.

Although radioactive substances are very dangerous – for example, when they enter the food chain or are used in dirty bombs – the repositories we create to house those substances will not forever be seen as areas of deadly threats; they may over time be transformed into altogether different things. This is not to trivialize or deny the real dangers posed by radioactive materials for future generations, but rather to look at those dangers in the present from a different perspective as to their possible future meaning.

Nuclear states, nuclear industries, and national and international regulatory institutions need to think seriously about how to build long-lasting, durable, deep geological repositories for nuclear waste in light of the understanding that these sites will become a contested and mutable heritage whose meaning will be transformed in accordance with as yet unimagined future presents. Arguably, the best chance to transmit knowledge in the long-term under changing circumstances may be to keep records, knowledge, and memory alive, inviting future generations to interpret and use these sites in their own way. This critical insight opens a new door for discussions about how best to design final repositories of nuclear waste and also for the nuclear environmental humanities in how to engage with long-term responsibilities.

At the same time, nuclear waste is critical heritage in the sense that it can act as a catalyst for heritage managers to better address the long-term future implied in their work. Here, too, the potency of the nuclear environmental humanities becomes apparent when very different areas of expertise are brought together for the benefit of both. The encounter of the two fields will foster critical discussion on how the management of cultural heritage expresses future consciousness in the present and also on how the heritage may actually benefit future societies.

NOTES

A different version of this chapter was published as "What Lies Ahead? Nuclear Waste as Cultural Heritage of the Future," in *Cultural Heritage and the Future*, edited by Cornelius Holtorf and Anders Högberg (New York and London: Routledge, 2021).

1 Holtorf and Högberg, "Nuclear Waste," "Communicating with Future Generations," "Archaeology and the Future," "The Contemporary Archaeology," "Contemporary Heritage and the Future," and "Heritage Futures and the Future of Heritage."
2 Feiveson et al., *Spent Fuel.*
3 World Nuclear Association, "Reactor Database," https://www.world-nuclear.org/information-library/facts-and-figures/reactor-database.aspx.
4 Chapman and McCombie, *Principles and Standards.*
5 Elfwing et al., *SFL Concept Study*, 13–14.
6 Alexander and McKinley, "Deep Geological Disposal."
7 Jensen, *Conservation and Retrieval*; Trauth, Hora, and Guzowski, *Expert Judgement*; Herrmann and Röthemeyer, *Langfristig sichere Deponien*; Elfwing et al., *SFL Concept Study.*
8 Schröder, *Preservation of Records*; see also van Wyck, *Signs of Danger*; and Moisey, "Considering the Desire."
9 *NEA/RWM 14/REV4.*
10 Ibid.

11 Elfwing et al., *SFL Concept Study*, 13–14.
12 Small, *On Deep History.*
13 Suddendorf and Corballis, "The Evolution of Foresight."
14 Karlsson, "Historia."
15 Rüsen, "Historical Consciousness."
16 Högberg and Holtorf, "Contemporary Archaeology."
17 Schacter and Madore, "Remembering the Past."
18 MacLeod, "Prospection," 266.
19 Bell, *Foundations of Future Studies*; Urry, *What Is the Future?*
20 Miller, "Opinion"; Miller, Poli, and Rossel, "The Discipline"; Miller, "Transforming the Future."
21 Miller, Poli, and Rossel, "The Discipline."
22 Harvey and Perry, "Heritage and Climate Change," 3.
23 Holtorf and Högberg, "Nuclear Waste"; Holtorf and Högberg, "Archaeology and the Future."
24 Zeman and Amundson, *Atomic Culture.*
25 Holtorf and Högberg, "Contemporary Archaeology."
26 Storm, *Post-Industrial Landscape Scars.*
27 Bauman, *Modernity and Ambivalence.*
28 Anshelm, *Bergsäkert eller våghalsigt?*
29 Carter, "Nuclear Power Policy Statement."
30 Anshelm, *Bergsäkert.*
31 Nakajima, *Nuclear Back-End.*
32 Anshelm, *Bergsäkert eller våghalsigt?*
33 Nakajima, *Nuclear Back-End.*
34 Sundqvist and Elam, "Sociologin."
35 Kaplan and Adams, "Using the Past"; Joyce, "Future."
36 Moisey, "Considering the Desire."
37 Holtorf, *Monumental Past.*
38 Clark, "Preserving What Matters"; Joyce, "Future."
39 Andrén, *Nuclear Waste Management.*
40 Moisey, "Considering the Desire."
41 Holtorf, "Preservation Paradigm."
42 Spennemann, "The Futurist Stance"; Holtorf and Högberg, "Communicating with Future Generations."
43 Rüsch, "Vergangenheitsfalle"; Spennemann, "The Futurist Stance" and "Futurist Rhetoric"; Holtorf and Ortman, "Endangerment"; Vidal and Dias, "The Endangerment Sensibility."
44 Holtorf and Högberg, "Communicating with Future Generations"; Harrison, "Beyond 'Natural'"; Lavau, "Climate Change"; DeSilvey, *Curated Decay.*
45 Högberg et al., "No Future."
46 Irvine, "Deep Time."
47 Quoidbach, "The End of History Illusion."
48 Holtorf and Högberg, "Communicating with Future Generations."

BIBLIOGRAPHY

Alexander, W. R., and Linda McKinley. "Deep Geological Disposal of Radioactive Waste." *Radioactivity in the Environment* 9 (2007): 1–273.

Andrén, Mats. *Nuclear Waste Management and Legitimacy: Nihilism and Responsibility.* New York: Routledge, 2012.

Anshelm, Jonas. *Bergsäkert eller våghalsigt? Frågan om kärnavfallets hantering i det offentliga samtalet i Sverige 1950–2002.* Lund: Arkiv förlag, 2006.

Bauman, Zygmunt. *Modernity and Ambivalence.* Cambridge: Polity Press, 1991.

Bell, Wendell. *Foundations of Future Studies: History, Purposes, and Knowledge. Human Science for a New Era*, vol. 1. London: Transaction, 2009.

Carter, Jimmy. 1977. "Nuclear Power Policy Statement on Decisions Reached Following a Review." 7 April 1977. Online by Gerhard Peters and John T. Woolley, *The American Presidency Project.* http://www.presidency.ucsb.edu/ws/?pid=7316.

Chapman, Neil A., and Charles McCombie. *Principles and Standards for the Disposal of Long-lived Radioactive Wastes.* Waste Management Series no. 3. Amsterdam: Pergamon, 2003.

Clark, Kate. "Preserving What Matters: Value-Led Planning for Cultural Heritage Sites." *Getty Conservation Institute Newsletter* 16, no. 3 (2001). http://www.getty.edu/conservation/publications_resources/newsletters/16_3/feature.html.

DeSilvey, Caitlin. *Curated Decay: Heritage beyond saving.* Minneapolis: University of Minnesota Press, 2017.

Dunne, Anthony, and Raby Fiona. *Speculative Everything: Design, Fiction, and Social Dreaming.* Cambridge, MA: MIT Press, 2013.

Elfwing, Mattias, Lena Z. Evins, Mikael Gontier, Pär Grahm, Per Mårtensson, and Sofie Tunbrant. *SFL Concept Study: Main Report.* Svensk Kärnbränslehantering AB. Technical Report TR-13-14, 2013.

Feiveson, Harold, Zia Mian, M.V. Ramana, and Frank von Hippel, eds. *Spent Fuel from Nuclear Power Reactors: An Overview of a New Study: International Panel on Fissile Materials*, 2011. http://fissilematerials.org/library/ipfm-spent-fuel-overview-june-2011.pdf.

Harrison, Rodney. "Beyond 'Natural' and 'Cultural' Heritage: Toward an Ontological Politics of Heritage in the Age of Anthropocene." *Heritage and Society* 8, no. 1 (2015): 24–42.

Harvey, David C., and Jim Perry. "Heritage and Climate Change: The Future Is Not the Past." In *The Future of Heritage as Climate Change: Loss, Adaptation, and Creativity*, edited by David C. Harvey and Jim Perry, 3–21. London and New York: Routledge, 2015.

Herrmann, Albert G., and Helmut Röthemeyer. *Langfristig sichere Deponien. Situation, Grundlagen, Realisierung.* Berlin: Springer, 1998.

Holtorf, Cornelius. *Monumental Past: The Life-Histories of Megalithic Monuments in Mecklenburg-Vorpommern (Germany).* Electronic monograph. Toronto: University of Toronto, Centre for Instructional Technology Development, 2000–8. http://hdl.handle.net/1807/245.

– "Preservation Paradigm in Heritage Management." In *The Encyclopedia of Global Archaeology*, edited by C. Smith, 6128–31. New York: Springer, 2014.

Holtorf, Cornelius, and Anders Högberg. "Archaeology and the Future: Managing Nuclear Waste as a Living Heritage." *Radioactive Waste Management and Constructing Memory for Future Generations*. Nuclear Energy Agency. Paris: OECD. no. 7259 (2015): 97–101.

– "Communicating with Future Generations: What Are the Benefits of Preserving Cultural Heritage? Nuclear Power and Beyond." *European Journal of Post-Classical Archaeologies* 4 (2014): 315–30.

– "The Contemporary Archaeology of Nuclear Waste. Communicating with the Future." *Arkæologisk Forum* 35 (2016): 31–7.

– "Contemporary Heritage and the Future." In *The Palgrave Handbook of Contemporary Heritage Research*, edited by Emma Waterton and Steve Watson, 509–23. New York: Palgrave Macmillan, 2015.

– "Heritage Futures and the Future of Heritage." In *Counterpoint: Essays in Archaeology and Heritage Studies in Honour of Professor Kristian Kristiansen*, edited by Sophie Bergerbrant and Sabatini Serena, 739–46. BAR International Series 25. 08. Archaeopress: Oxford, 2013.

– "Nuclear Waste as Cultural Heritage of the Future." WM2014 Conference Proceedings, Atlanta, 2014. https://www.diva-portal.org/smash/get/diva2:718845/FULLTEXT01.pdf.

Holtorf, Cornelius, and Oscar Ortman. "Endangerment and Conservation Ethos in Natural and Cultural Heritage: The Case of Zoos and Archaeological Sites." *International Journal of Heritage Studies* 14, no. 1 (2008): 74–90.

Högberg, Anders, Cornelius Holtorf, Sarah May, and Gustaf Wollentz. "No Future in Archaeological Heritage Management?" *World Archaeology* 49, no. 5 (2017): 639–47.

Irvine, R.D.G. "Deep Time: An Anthropological Problem." *Social Anthropology* 22, no. 2 (2014): 157–72.

Jensen, Mikael. *Conservation and Retrieval of Information: Elements of a Strategy to Inform Future Societies about Nuclear Waste Repositories*. Nordic Committee for Nuclear Safety Research, Nordic Council of Ministers, 1993. http://www.iaea.org/inis/collection/NCLCollectionStore/_Public/28/038/28038113.pdf.

Joyce, Rosemary. *The Future of Nuclear Waste: What Art and Archaeology Can Tell Us about Securing the World's Most Hazardous Material*. Oxford and New York: Oxford University Press, 2020.

Kaplan, Maureen F., and Mel Adams. "Using the Past to Protect the Future: Marking Nuclear Waste Disposal Sites." *Archaeology* 39, no. 5 (1986): 107–12.

Karlsson, Klas-Göran. "Historia, historiedidaktik och historiekultur – teori och perspektiv." In *Historien* är *närvarande. Historiedidaktik som teori och tillämpning*, edited by Klas-Göran Karlsson and Ulf Zander, 13–89. Lund: Studentlitteratur, 2014.

Lavau, Stephanie. "Climate Change and the Changing Nature of Conservation." In *The Future of Heritage as Climates Change: Loss, Adaptation, and Creativity*, edited by D.C. Harvey and J. Perry, 111–29. London and New York: Routledge, 2015.

MacLeod, Andrew. "Prospection, Well-Being, and Memory." *Memory Studies* 9, no. 3 (2016): 266–74.

Miller, Riel. "Opinion: Futures Literacy – Embracing Complexity and Using the Future." *Ethos* 10 (2011): 23–38. https://www.cscollege.gov.sg/knowledge/ethos/issue%2010%20oct%202011/pages/Opinion%20Futures%20Literacy.aspx.

Miller, Riel, Roberto Poli, and Pierre Rossel. "The Discipline of Anticipation: Exploring Key Issues. Scoping Global/Local Anticipatory Capacities," Working paper no. 1. Unesco Foresight. 2013. https://www.academia.edu/3523348/The_Discipline_of_Anticipation_Miller_Poli_Rossel_-_DRAFT.

Miller, Riel, ed. *Transforming the Future. Anticipation in the 21st Century.* Paris: UNESCO; and London and New York: Routledge, 2018.

Moisey, Andrew. "Considering the Desire to Mark Our Buried Nuclear Waste: Into Eternity and the Waste Isolation Pilot Plant." *Qui Parle: Critical Humanities and Social Sciences* 20, no. 2 (2012): 101–25.

Nakajima, Ken, ed. *Nuclear Back-End and Transmutation Technology for Waste Disposal: Beyond the Fukushima Accident.* New York and London: Springer Open, 2015.

NEA/RWM(2011)14/REV4. Glossary of Terms. Preservation of Records, Knowledge and Memory (RK&M) Across Generations. OECD Nuclear Energy Agency.

Radioactive Waste Management Committee. https://www.oecd-nea.org/rwm/docs/2011/rwm2011-14-rev4.pdf.

Quoidbach, Jordi, Daniel T. Gilbert, and Timothy D. Wilson. "The End of History Illusion." *Science* 339 (2013): 96–8.

Rüsch, Eckart. 2004. "Vergangenheitsfalle oder Zukunftsentsorgung? Folgen einer Denkmalpflege ohne Gegenwartsbewusstsein." *Kunsttexte.de* 1 (2004). http://edoc.hu-berlin.de/kunsttexte/download/denk/sym3-ruesch-v.pdf.

Rüsen, Jörn. "Historical Consciousness: Narrative Structure, Moral Function, and Ontogenetic Development." In *Theorizing Historical Consciousness*, edited by P. Seixas, 63–85. Toronto: University of Toronto Press, 2004.

Schacter, Daniel L., and Kevin P. Madore. "Remembering the Past and Imagining the Future: Identifying and Enhancing the Contribution of Episodic Memory." *Memory Studies* 9, no. 3 (2016): 245–55.

Schröder, Jantine. *Preservation of Records, Knowledge, and Memory (RK&M) across Generations: Final Report of the RK&M Initiative.* Paris: OECD, Nuclear Energy Agency, 2019.

Small, Daniel L. *On Deep History and the Brain.* Berkeley: University of California Press, 2008.

Spennemann, Dirk H.R. "Futurist Rhetoric in US Historic Preservation: A Review of Current Practice." *International Review of Public and Non-Profit Marketing* 4, nos. 1–2 (2007): 91–9.

– "The Futurist Stance of Historical Societies: An Analysis of Position Statements." *International Journal of Arts Management* 9, no. 2 (2007): 4–15.

Storm, Anna. *Post-Industrial Landscape Scars.* New York: Palgrave Macmillan, 2014.

Suddendorf, Thomas, and Michael C. Corballis. "The Evolution of Foresight: What Is Mental Time Travel, and Is It Unique to Humans?" *Behavioral and Brain Sciences* 30, no. 3 (2007): 299–313.

Sundqvist, Göran, and Mark Elam. "Sociologin, hybriderna och den sociala verkligheten - exemplet kärnavfall." *Sociologisk forskning* 46, no. 2 (2009): 4–25.

Trauth, Kathleen M., Stephen C. Hora, and Robert V. Guzowski. *Expert Judgement on Markers to Deter Inadvertent Human Intrusion into the Waste Isolation Pilot Plant.* Sandia Report SAND92–1282, UC–721. Albuquerque, 1993.

Urry, John. *What Is the Future?* Cambridge: Polity Press, 2016.

Vidal, Fernando, and Nélia Dias. "The Endangerment Sensibility." In *Endangerment, Biodiversity, and Culture*, edited by Fernando Vidal and Nélia Dias, 1–38. London and New York: Routledge, 2015.

van Wyck, Peter C. *Signs of Danger: Waste, Trauma, and Nuclear Threat.* Minneapolis: University of Minnesota Press, 2005.

Zeman, Scott C., and Michael Al Amundson. *Atomic Culture: How We Learned to Stop Worrying and Love the Bomb.* Boulder: University Press of Colorado, 2004.

Afterword to Part Three

Lingering Radiation: On Violent Pasts and Open-Ended Futures

RUBY DE VOS

In *Being Nuclear* (2012), Gabrielle Hecht pinpoints a crucial distinction between radiation and what she calls "nuclearity." She writes: "Radiation is a physical phenomenon that exists independently of how it is detected or politicized. *Nuclearity* is a *technopolitical* phenomenon that emerges from political and cultural configurations of technical and scientific things, from the social relations where knowledge is produced."[1] In other words: while radiation persists out in the world, our specific encounters with it are deeply shaped by material-semiotic configurations and practices, in each of which radiation functions differently. Thus it is possible, Hecht suggests, that uranium mining is kept outside of the realm of "the nuclear" despite uranium's importance to the nuclear industry, or that radiation can simultaneously be considered exceptional as well as banal.[2] Hecht makes these observations in a very particular context, one that aims to understand the workings of nuclearity as one category that subsumes the phenomenon of radiation. However, the general thrust of her argument – that our understanding of radiation is both mutable and deeply political – resonates with the essays of Thomas Patrick Pringle and Cornelius Holtorf and Anders Högberg in Part 3: "Archaeologies and Heritages." Although Pringle looks to the past, while Holtorf and Högberg turn to the future, both texts for me prompt the question: what are the consequences of radioactive material's shifting meaning?

Focusing on the infamous question of how underground repositories for high-level radioactive waste can be marked for the future, Holtorf and Högberg identify the central paradox that shapes this wicked problem: "reliably preserving relevant knowledge about nuclear waste over very long time periods is, while necessary, also impossible."[3] The challenge of marking nuclear waste repositories has always focused on navigating the instability of communication and perceptions over extended periods of time. Rather than eliminate this instability, however, we need to take it into account, Holtorf and Högberg argue, as instability is the one thing we can be certain about. If perceptions of nuclear waste shifted almost every decade during the twentieth century, it is only to be assumed that this "socially negotiable" quality of the waste extends also to the future.[4] Thus the future need not be an extension of what is already here, Holtorf and Högberg suggest – it can be *unimagined* still. They evoke a radically different perspective on the often-assumed stasis of nuclear waste's deep future underground: the radioactivity of nuclear waste may become something we cannot yet even begin to envision.

At the other end of the spectrum, Pringle's chapter offers a historical counterpart to radiation's slippery signification by turning to uranium mining, the backbone of the nuclear industry. His text zooms in on the Glen Canyon and Lake Powell in the US, following uranium around in order to trace how "the shifting status of what counts as 'natural'" came to shape the geographical region.[5] He demonstrates that the frame of uranium as a "natural resource" was used to intrude on and pollute Indigenous lands, letting Navajo people die unnatural deaths, after which the same extractive logic could be used to reinstate the land as a National Recreation Area. The mutability of the meaning attributed to radiation here aided (and continues to aid) in obscuring the deep roots of imperial and settler-colonial violence that have enabled the nuclear industry. Pringle's excavation of "the layers of socio-physical strata defining the archaeology of 'the natural'" sheds light on the manifold ways that the material practice of uranium mining can be folded into the discourse of the natural, inviting us to consider how meanings of radioactive materials were shifted in the past, as well as who benefited and who suffered from this.[6]

Both chapters achieve much more than I have the space to convey here, especially when it comes to the methodological tools they develop for their respective fields – notably Holtorf and Högberg's redefining of nuclear waste as cultural heritage, and Pringle's convincing intervention in the depoliticized field of new materialism through media archaeology. What intrigued me, however, was the dialogue that emerged between the two texts on the implications of radiation's unstable meaning over time. For while Holtorf and Högberg convincingly show that engagement with nuclear waste demands a grappling with the fact that our perception and understanding of it will inevitably change in the future, Pringle's text reads like a warning of what happens when we forget the violence the nuclear industry has perpetrated by mobilizing radiation's unstable meanings. Thinking of nuclear waste as a heritage or a legacy (also outside of the more specialized definitions used in heritage studies) thus always needs to be a reckoning with the past as well as with the future. How can we hold that open-endedness of radiation together with the need to take care of it in a way that is also always attentive to its histories of violence?

While there may be various ways to answer this, I propose that art and literature form one such site where these multiple temporal orientations of radiation's

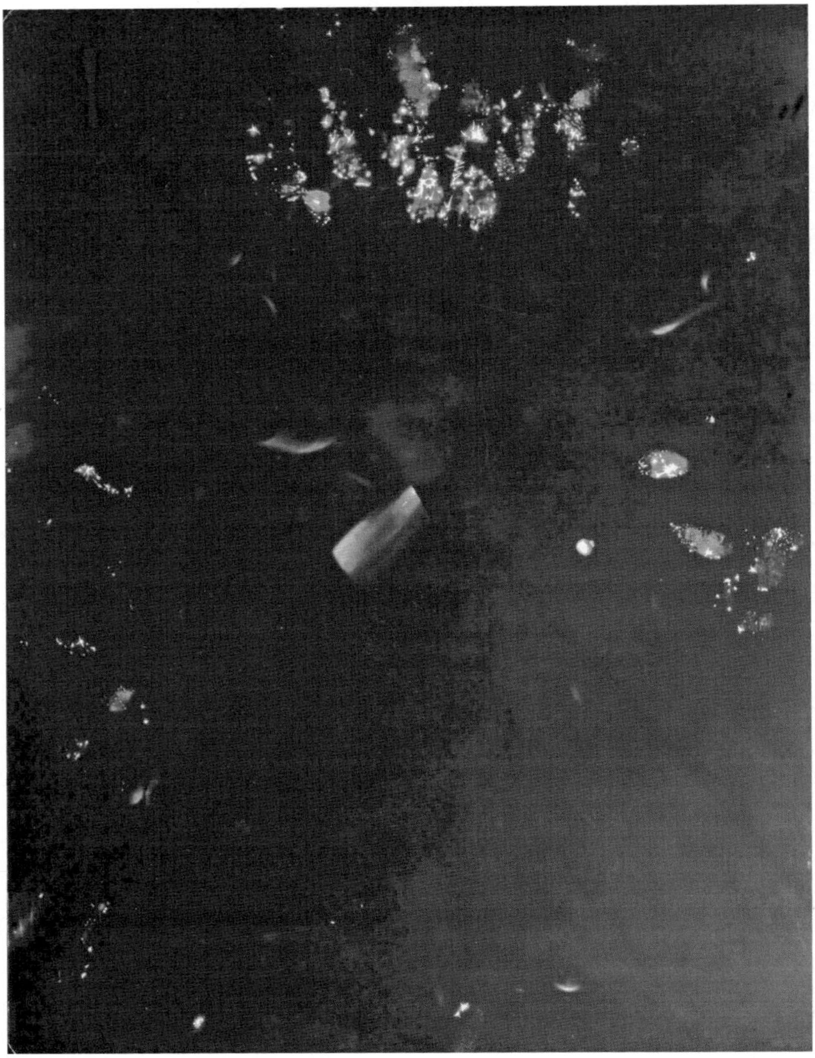

A3.1 elin O'Hara slavick, Lingering Radiation, 2008. Autoradiograph, contact print of X-ray exposed from A-bombed tree stump fragment.

meanings and effects can be held together and thought through. As an exemplary case study I turn here to a work by US artist elin o'Hara slavick with the fitting name *Lingering Radiation* (2008), which I will read with particular attention to the artist's comments on the creation process. *Lingering Radiation* shows a black background with brightly illuminated sparks clustered in petal-like shapes toward the top and the sides of the print. slavick made the work with objects collected in the archive of the Hiroshima Peace Memorial Museum (HPMM), which was built to commemorate the US bombing of the city in 1945. The large archive contains many objects that have been brought in over the years, often broken or melted by the impact and the heat of the bomb. The size of the archive speaks of the immensity of loss produced by the bomb, yet *Lingering Radiation* draws our attention to something that may have remained: radiation.

Lingering Radiation is an autoradiograph. An image is produced as film is exposed by light emanating from radioactive material, without the interference of a camera. In a dark room in the HPMM, slavick placed several objects from the archive onto X-ray film, including fragments from a Hiroshima tree destroyed by the bombing. She then put the film into light-tight bags and left them for ten days.[7] Once developed, the film with the tree fragments stood out: light was singed into the paper like brightly lit embers, crackling. If the pieces of the tree were initially collected in the archive because they were touched by the bomb, then *Lingering Radiation* suggests that this touch remains felt until this very day. Susan Schuppli writes about autoradiographs: "the radiological contact print is immanent to and continuous with the event."[8] Similarly, the traces of radiation on slavick's work read both as evidence of and as testament to the violence of 6 August 1945, persisting over time, a haunting reminder of the irrevocable violence of the atomic bomb.

But once we know more, the work gently pushes against such a one-sided interpretation. slavick explained in an interview that she did not know in advance whether the film would actually pick up any radiation.[9] Out of the six plates she prepared, *Lingering Radiation* showed the most visual registration, which one could read as evidence of the bark's remaining radiation. Yet slavick holds open the space for another source of origin: "I have to say this – it was not really scientific. It could have been some of the oils from my fingers that made some of those marks. There could have been a light leak. It could be background radiation."[10] Knowing this, the petal-like forms may suddenly suggest the possibility of fingerprints, while the cracks of light become a record of a world that has become increasingly radioactive throughout the twentieth century. Similarly, slavick, as an American making work about the bombing of Japan, works from "within conflict," as she has put it herself, yet the work also inevitably escapes that context somewhat if the imprint of the radiation could also have another origin.[11] In this moment, the meaning of radiation in *Lingering Radiation* becomes slippery and manifold, and it begins to resonate with a plurality of stories.

Literary studies, art history, and cultural analysis have shown over the past decades that art allows for ongoing reinterpretation and for changing signification. This same openness became a problem in the design of nuclear waste markers for the future. In an essay on the design of the markers for the US Waste Isolation Power Plant, art historian Julia Bryan-Wilson points out that while the designers for the WIPP did explicitly draw on art to develop the markers – including Munch's *The Scream* and land art – they nevertheless rejected the open-endedness of meaning that characterizes art. The "interpretative ambiguity" of Munch's screaming face had to be flattened into singularity because it was only supposed to communicate a singular meaning: danger.[12] As Peter C. van Wyck writes in his extensive study of the WIPP: "The designs are not about the present ... The future witness to the monument is called upon to understand only the site itself – not the reason that the wastes are there."[13] Consequently, these markers attest neither to previous forms of violence nor to shifting understandings of what radioactive waste might mean: the instrumentalization of art for a single purpose serves to strip it of its interpretative ambiguity and closes off its openness of meaning in relation to past and future.

My point here is not so much to question whether this is or is not the right approach to a nuclear marker. Rather, I offer the markers as a counter-example to suggest that, because *Lingering Radiation* does not just function as a mark of danger to come or as evidence of the violence of the past, it is able to hold radiation within a space of shifting meaning. This point is made all the more forcefully because the autoradiograph has its roots in scientific research in the late nineteenth century, and because it was used as a form of data production during the Cold War.[14] *Lingering Radiation* approximates those registers of the scientific and the evidentiary but never inhabits them fully. Of course the uncertainty of radiation's origin has often been used to political advantage by those who incite violence – Holtorf and Högberg's and Pringle's texts each speak to that. I would suggest, however, that this history itself is materially evoked through *Lingering Radiation*'s indeterminate imprint. Who gets to decide the meaning of radiation? The point of slavick's work is not to resolve uncertainty: the very possibility for dispute is contained within the work, as it allows these various interpretations to coexist uneasily.

So perhaps the registered radiation came from somewhere else; perhaps future scholars will interpret the work in a way I cannot even begin to conceive. And yet – the plate also was touched by the bark for ten days, in the dark. Even if it was not radioactive, there is a poignant connection to touch, always bringing us back to Hiroshima as well. If *Lingering Radiation* does not obtain an indelible memory of the bomb, perhaps it gives us something else: its very materiality already speaks to the diffusion of radiation, the difficulty of fully grasping its workings and meanings in the past, present, and future. Lingering right there might prove a productive model for nuclear studies/nuclear environmental humanities, too.

NOTES

1 Hecht, *Being Nuclear*, 15.
2 Ibid., 8.
3 Holtorf and Högberg, "Nuclear Waste as Critical Heritage," in this volume.
4 Ibid.

5 Pringle, "Emergency/Salvage Archaeology," in this volume.
6 Ibid.
7 slavick, "After Hiroshima," in *After Hiroshima*, 28.
8 Schuppli, "Radical Contact Prints," 280.
9 slavick, "After Hiroshima," *The Asia-Pacific Journal*, 4.
10 Ibid. I have expanded on the imperfection of this photographic process of *Lingering Radiation* in "Exposed." See Vos, "Exposed."
11 slavick, "After Hiroshima," in *After Hiroshima*, 27.
12 Bryan-Wilson, "Building a Marker," 188.
13 Van Wyck, *Signs of Danger*, 79.
14 Schuppli, 281, 285. For more on radioactive traces as evidence, see also Schuppli, *Material Witness*.

BIBLIOGRAPHY

Bryan Wilson, Julia. "Building a Marker of Nuclear Warning." In *Monuments and Memory, Made and Unmade*, edited by Robert S. Nelson and Margaret Olin, 183–204. Chicago: University of Chicago Press, 2003.

Hecht, Gabrielle. *Being Nuclear: Africans and the Global Uranium Trade*. Cambridge, MA: MIT Press, 2012.

Schuppli, Susan. *Material Witness: Media, Forensics, Evidence*. Cambridge, MA: MIT Press, 2020.

– "Radical Contact Prints." In *Camera Atomica*, edited by John O'Brian, 277–91. London: Black Dog, 2015.

slavick, elin o'Hara. "After Hiroshima." In *After Hiroshima*, 26–31. Chapel Hill: Daylight Books, 2013.

– "After Hiroshima." *Asia-Pacific Journal* 11, no. 3 (2013): 1–16.

Van Wyck, Peter C. *Signs of Danger: Waste, Trauma, and Nuclear Threat*. Minneapolis: University of Minnesota Press, 2005.

Vos, Ruby de. "Exposed: Radiation's Imperfect Traces." *Sublime Imperfections*, 27 November 2018. http://sublimeimperfections.org/ruby-de-vos.

FOUR

Nuclear Aesthetics: Contemporary Art, Nuclear Colonialism, and the Transformation of Life and the Environment

12

Atomic Aborigines: Appropriation and Colonization of Indigenous Australia during British Nuclear Testing

MICK BRODERICK

> Overall, the attempts to ensure Aboriginal safety during the [nuclear] series demonstrate ignorance, incompetence and cynicism on the part of those responsible for that safety. The inescapable conclusion is that if Aborigines were not injured or killed as a result of the explosions, this was a matter of luck rather than adequate organization, management and resources allocated to ensuring safety.[1]

Eighteen years after Captain James Cook's "discovery" of eastern Australia in 1770 – territory claimed in the name of King George III – the "first fleet" arrived at Sydney Cove in 1788. The convoy was led by Commodore Arthur Phillip, comprising British naval officers with detachments of marines, several hundred convicts (male, female, and children) and more than eighty free settlers. Within two years, Aboriginal tribes began resisting the encroachment of the European invaders. Bediagal man Pemulwuy led scores of Aboriginal warriors to attack the settlers, often as "payback" for atrocities conducted against their kin. He was shot and killed in 1802, his head cut off, placed in "spirits," and dispatched to Sir Joseph Banks in England.[2]

By 1803–04 Mathew Flinders had circumnavigated the island continent, naming the land "terra Australis" or "Australia" on his charts and maps. Claimed as a British territorial possession, Australia was no longer considered "terra incognitus" or "Hollandia Nova" (New Holland), named after the seventeenth-century Dutch sailors who skirted the western, northern, and southern coastlines (including "Tasmania"), although early British maps continued to assign the western portion as New Holland.[3]

Within a generation after the arrival of Europeans at Sydney Cove in 1788, the Crown surveyors' discourse had begun to gradually shift from naming new colonial locations after British royalty, military leaders, the political elite, and English geographical toponyms – such as the new colonies of "New South Wales" and "Victoria" – toward Aboriginal names, including Woolloomooloo, Illawarra, and Parramatta (the upriver site of the first Government House). By the time the colonial states and territories merged into the Federation of the Commonwealth of Australia in 1901, a new prime minister was keen to allocate an Aboriginal name to the freshly designated national capital, which became known as Canberra, ostensibly meaning meeting place.[4]

The tendency to (re)designate sites continues into the twenty-first century, with contemporary planning authorities controversially naming or renaming sites of significance using Indigenous terms. For instance, former Australian prime minister Paul Keating railed against the arbitrary branding of a landmark waterfront section of Sydney Harbour "Barangaroo," ostensibly after an Aboriginal woman: "I regard these completely unassociated Aboriginal names as a form of Aboriginal kitsch."[5] Sam Furphy has noted the historical role of appropriating First Nation language for European settler colonization, asserting that the use of "Aboriginal place names both at an official and popular level, has very rarely been the result of sensitive and meaningful cultural interchange. In most cases, words were appropriated with little or no reference to their layers of meaning, history, or linguistic origins. In fact, the use of Aboriginal place names has very rarely had anything to do with Aboriginal people beyond a very superficial level."[6]

Whether or not these appellations resulted from deliberate or casual etymological purloining, they aided and abetted the overarching colonial dispossession of Indigenous cultures. As Furphy suggests, Aboriginal names were deployed "with little or no reference to their linguistic and cultural heritage."[7]

In the late 1940s the Australian and UK governments agreed to create a long-range weapons establishment (LRWE) in South Australia. A new, remote oasis town named Woomera was constructed for scientists, technicians, the military, security personnel, and their families. Hence the colonial toponymic trend was similarly adopted by UK project managers, who grafted the Sydney-area language term *woomera* ("throwing stick" or "spear thrower") onto the newly excised "prohibited area" 1,000 kilometres away in South Australia, which had been assigned for the launching of rockets, jets, and cruise missiles across central and northwestern Australia, even though the local Anangu word for spear thrower is *miru*.[8] Such transposition from colonial to modern Australia lazily ignores the fundamental differences among the two-hundred-plus First Nations language groups throughout the island continent pre-European settlement.[9]

Atomic Australis

A little more than seven months after Britain used the uninhabited Monte Bello Islands off the northwest coast of Australia to demonstrate its first experimental atomic explosive in Operation Hurricane (October 1952), a new trial was designed to be conducted in central Australia at "Emu" field. After vapourizing the metal support tower and incinerating spinifex grass, mulga, and clusters of stout sheoaks and eucalypts, a broiling mushroom cloud sucked up the outback ochre-red dirt to form a conventional, and by then familiar, mushroom cloud.[10] The ominous countdown had passed along with the searing flash and blinding glare of atmospheric ignition. Twelve miles away, the assembled press, military, and scientific boffins turned quickly to view the nuclear achievement. Almost instantly a voice declared loudly, while pointing at the apparition: "Look! Do you see it? A perfect portrait of a myall blackfeller written with atomic dust; the new and the old have come together today."[11]

This anecdote was related by Len Beadell, an Australian Army surveyor who had been seconded to reconnoitre the vast tract of territory excised by the Australian Commonwealth for the LRW trials. A seasoned bushman, raconteur, and cartoonist, upon his official retirement Beadell wrote a series of humorous and beguiling books about his outback experiences, including *Too Long in the Bush* (1964), *Still in the Bush* (1965), and *Blast the Bush* (1967), all recounting his adventures surveying the scrub, zigzagging the continent, and building a network of roads thousands of kilometres long. In relating this tale, Beadell is bemused by the willingness of the press corps to embrace this particular angle to the atomic story: "one by one they all agreed there was no doubt about it."[12]

Consequently, print headlines in Australia and the UK carried the story, many with accompanying photos of the "atomic head." Under the title "Atomic Cloud Forms Head of Aborigine," Adelaide's *The Advertiser* newspaper of 16 October 1953 described how at an elevation of 15,000 feet the "atomic explosion twice formed the gigantic head profile of an aborigine."[13] One scientific observer was quoted saying: "It made me wonder just how an old native witch doctor would explain the weird significance of the greatest smoke signal and the most destructive happening man has ever caused on the Australian mainland."[14]

Apart from the ignorant application of the term "witch doctor" to central Australian Aborigines – popularly designating African medicine men – the quote invokes the importance of the traditional use of smoke to signal or communicate over vast distances. Newspapers such as *The Chronicle* of 22 October 1953 paraphrased the sentiment or simply parroted it without attribution to accompany

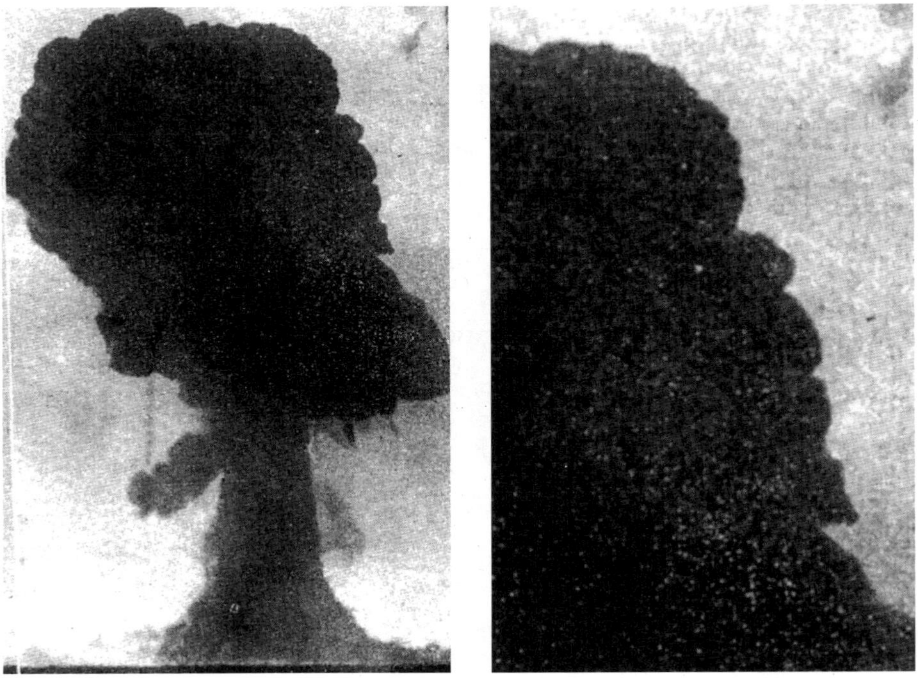

12.1 Detail of press photos of Operation Totem mushroom cloud with close-up outline of "mayall" head.

printed images of the mushroom cloud. One caption accentuates wild-ness, connoting both savagery and the uncultivated: "After mushrooming, the cloud assumed the rough shape of a wild aborigine's profile. The silhouette of the black-fellow's head can be seen at the top right of the picture.[15]

No one seemed to appreciate the irony that this anthropomorphized attribution immediately conflated existing associations with primitivism – a form of atomic atavism. The British had code-named this series of twin nuclear detonations "Totem 1" and "Totem 2." The rhetorical conjuring of domestic indigeneity was further evinced by the naming of the test area "Emu," an imported name for a large flightless bird, after a UK scientist had asked about animal tracks he had found on the large clay pan (*tjintjira*), the flat terrain that was being used as a natural airstrip.[16] They did not seek out or apply the traditional and local Pitjantjatjara terminology for emu, *kalaya*.

This Anglo-Australian gaze and its cultural imaginary correspond to Homi K. Bhabha's 1988 reflection on Frantz Fanon's masterful *Black Skin, White Masks* (1952), where he notes: "The White man's eyes break up the Black man's body and in that act of epistemic violence its own frame of reference is transgressed, its field of vision disturbed."[17] Hence such an attribution of primitivism intrinsically and inevitably echoes the dispossession and subjugation of First Nation peoples. The process of "indigenization," politically unconscious or otherwise, resonates with Terry Goldie's notion of the colonial paradox as connoting the "Aborigine as Other and therefore Alien," while simultaneously, and seemingly contradictorily, recognizing that Aborigines are "indigenous" and hence "cannot be alien."[18] According to Bhabha, Fanon identified the colonial subject as always "historicized" and heterogeneously inscribed in the texts of history, literature, science, myth. The colonial subject is always "overdetermined from without ... It is through image and fantasy – those orders that figure transgressively on the borders of history and the unconscious – that Fanon most profoundly evokes the colonial condition."[19]

For Sam Furphy, while such naming practices incorporated Aboriginal culture into more mainstream Australian (and British) cultural identity, they also glossed over "the realities of imperialism and invasion."[20] First Peoples faced these stark realities up to and throughout the Cold War era. They were still subjected to forcible removal from their traditional lands, they were counted among the nation's fauna for official census purposes, their mixed-race children were abducted by authorities (the Stolen Generation), and they were commonly deemed ineligible to vote in state or federal elections until after a 1967 referendum extended that right to them.[21]

The desire to indigenize these test site locations – which were alien to the visiting Britons and mostly unseen by even the remotest of colonial explorers – signified a concerted mid-twentieth-century Eurocentric domestication of place. The hybrid imprinting of Aboriginal names and terms for place continued with "Maralinga," a northern Australian Aboriginal word for "thunder," given to the site that replaced Emu as a permanent nuclear testing ground. It was at odds with the local Anangu word *tuuni*, which identifies thunder.[22] Nearly 1,500 kilometres away, the down-range missile and warhead instrumentation facility constructed south of Broome was designated "Talgarno" – incongruously adopted from the local Aboriginal description of "lizards running." Similarly, the US space surveillance and early warning facility established near Woomera (1969–99) was named "Nurrungar," purportedly meaning "listen" in the local language.[23]

Some media commentators of the era invoked a clash of civilizations in the central Australian desert, where nomadic peoples were being displaced by an emerging Anglo-Australian military-industrial complex. Newspaper and magazine cartoonists frequently lampooned this development with a "stone age meets the space age" (or "atomic age") trope. Throughout the Cold War, official Australian stationery, envelopes, and first-day covers celebrating successive Woomera technological milestones, or the British atomic tests, were commonly adorned with imagery juxtaposing Aborigines with rocketry and nuclear explosions.[24] Government envelopes, for example, showed Aborigines pointing at distant mushroom clouds; others displayed similar postures of "noble" (Rousseauian) Aborigines making smoke signals from bush campfires.[25]

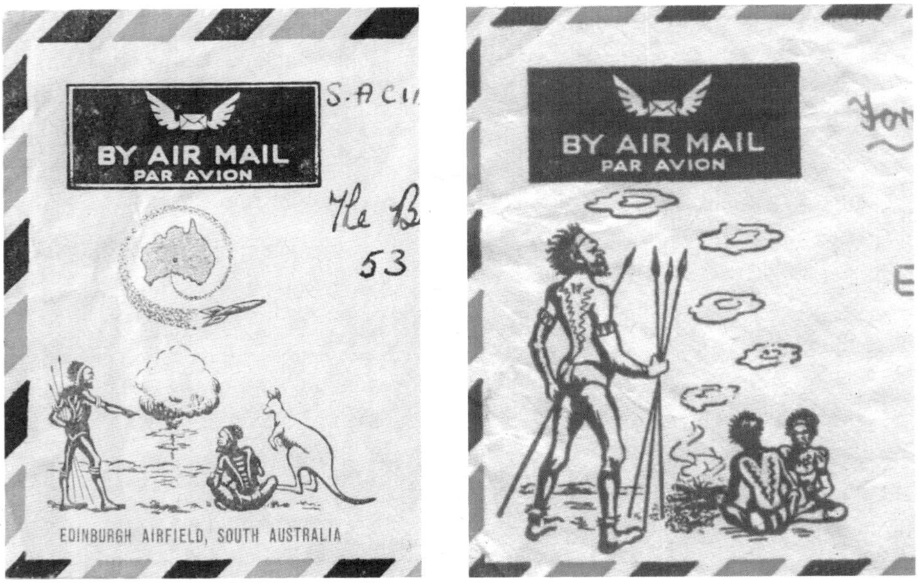

12.2 Australian airmail envelopes with Indigenous figures, ca late 1950s.

However, not all representations of the period were as paternalistic or derogatory. The cover image chosen for a report titled *Adam and Atoms* recounting the 1956 Western Australian Parliamentary Select Committee investigation of "the plight of inland natives" affected by the joint UK–Australian atomic tests and LRW trials juxtaposed the prominent visage of an Aborigine against a backdrop of a stylized mushroom cloud. While a contemporary reader will likely interpret this imagery as kitsch – akin to the mass-produced (usually foreign) Kewpie dolls and tasteless "Abo art" of the era[26] – the report is entirely sympathetic to the cause of Indigenous representation and agency. The book frankly foregrounds its findings: "The facts reveal an adamant refusal by official after official to accept the aborigines as human beings and measure their plight by human standards."[27]

These invented place names or operational designations were standard modes of legitimizing conquest and acquisition in the nuclear age and were correspondingly deployed by US military and scientific staff from the mid-1940s onward in their colonial spheres of influence in the South Pacific and continental US. Native American names such as Wigwam, Cherokee, and Aztec were frequently provided for various atomic and thermonuclear tests in the 1950s and 1960s. As Carol Cohn has demonstrated in relation to gendered and domesticating "technostrategic" parlance: "This language both reflects and shapes the nature of the American nuclear strategic project [and] plays a central role in allowing defense intellectuals to think and act as they do."[28] Such labelling therefore helped reify – and often deify – nuclear technology and its associated operations.[29]

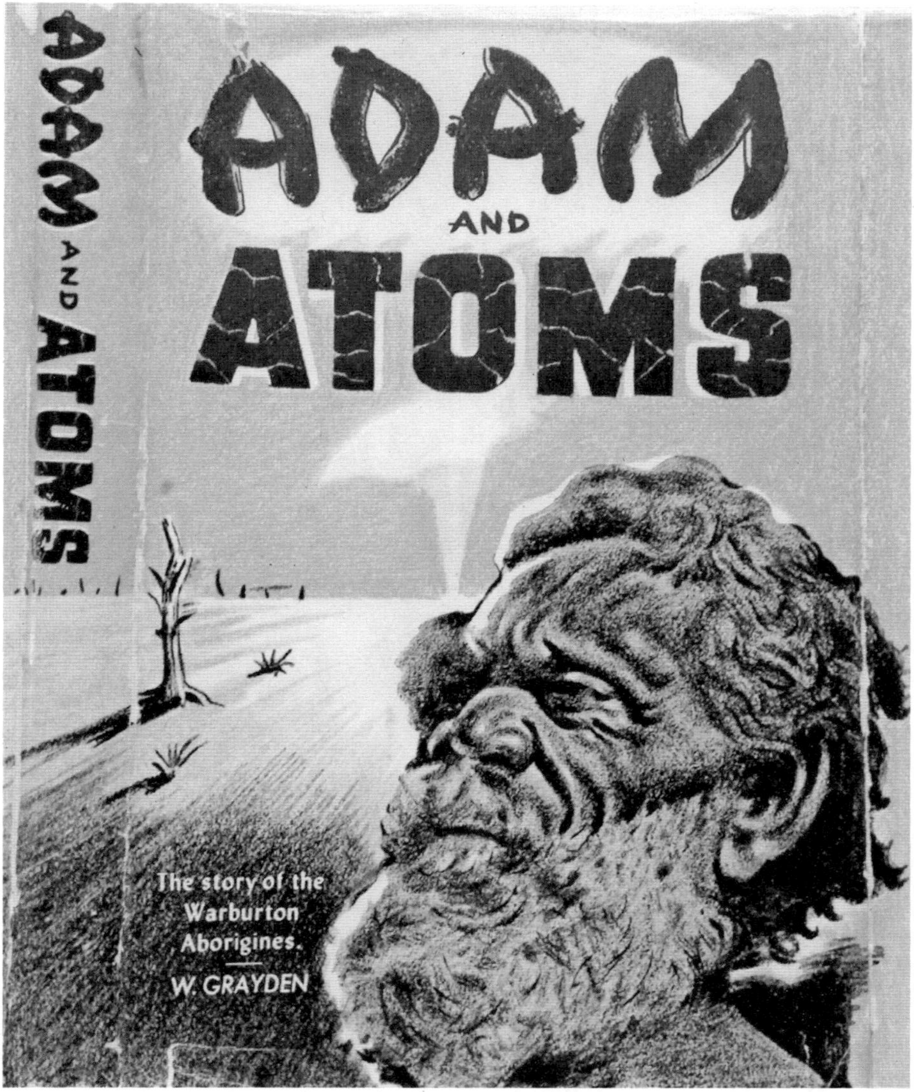

12.3 Cover art of *Adam and Atoms*, 1957.

Post-Nuclear Royal Commission

After a series of exposés by whistle-blowers who defied the punitive Official Secrets Act and spoke publicly in the late 1970s and early 1980s, the Australian government established a Royal Commission to investigate the undertakings and legacy of the British nuclear tests. Australian and British military service personnel, scientists, civilian employees, Indigenous people, and downwinders gave testimony and evidence throughout 1984 and 1985.

The multivolume final report differed substantively from official UK histories of the trials and included a multitude of recommendations regarding the remediation and further clean-up of the affected lands as well as compensation to aggrieved Aboriginal communities. The commission heard of Aborigines subjected to a "black mist" that rolled over the land following the Totem tests that afflicted camps to the northeast, causing sickness, blindness, stillbirths, and deaths. Witnesses detailed soldiers discovering Aboriginal men, women, and children camping next to a ground-burst crater some months after the firing. When discovered, another Indigenous family was ordered to leave the test site and to follow a newly constructed road, a directive that saw them perish as it removed the party from their traditional tracks and waterholes. Reportedly, up to 20 kilograms of plutonium and a large quantity of long-lived and toxic radionuclides continued to pollute the Indigenous lands, some buried in shallow pits, with much of the contaminants scattered about the surface in deliberate dispersal experiments. The commission was incredulous that the weapons safety committee had assigned only one or two Aboriginal Welfare Officers to scout and search an area half the size of Europe. As the commission noted: "The [nuclear] test site was chosen on the false assumption that the area was not used by its traditional owners. Aborigines continued to move around and through the Prohibited Zone and inadequate resources were allocated to locating them and to ensuring safety. The reporting of sightings of Aboriginal people was discouraged and ignored."[30]

Coeval with the 1980s Royal Commission hearings and recommendations was a concerted Indigenous pushback against the ongoing dispossession of traditional lands, with the assertion of native title and governance via self-determination. Like many settler nations, Australia has a long history of hegemonic interventions aimed at assimilating First Nations into the majority culture. Increasingly, state and federal governments have sought to undermine racial discrimination laws and ignore environmental and cultural protection in favour of globalized engineering projects that marginalize Aboriginal communities. These projects include the continued operation of huge uranium mines on Indigenous lands (some adjacent to World Heritage–protected wilderness areas) and the proposed siting of low- to high-level radioactive waste repositories on, or adjacent to, remote traditional lands.[31] An added insult has been the cultural appropriation of increasingly popular Indigenous art: little financial benefit returns to individual artists or Aboriginal communities, while white gallery owners and commissioning agents profit greatly.[32]

Coinciding with the renewed political agency and empowerment sought by Aboriginal peoples (in coalition with various NGOs and white support networks) has been a rise in the production and dissemination of Indigenous artworks that reflect and record these protests at local levels.[33] Over the past twenty years, and likely catalysed by the Royal Commission, a new generation of Indigenous artists and cultural practitioners have presented texts, artworks, videos, music, and performances that recognize and commemorate the long struggle of Aboriginal peoples confronting imposed atomic colonialism.[34]

But what of the Indigenous gaze? Where does this situated, alternative history and representation reside? Anecdotally, it is preserved and circulates within community, spoken and passed down to generations as oral history. In more recent years, artistic renderings of Aboriginal encounters with the colonial bomb-making endeavours have increasingly been recorded in traditional dot-paintings and in publications. Indigenous Australians also lend their considerable artistic talents to social media campaigns to protect remote (native title) lands, raise consciousness, and fight against uranium mining and the further despoiling and contaminating of their country by low-, intermediate-, and high-level nuclear waste storage facilities.[35] Often these campaigns and their collective history-making are conducted in concert with fellow Australian supporters and mentors, who have formed collaborative artistic and/or grassroots political ventures.

One significant outcome has been the storytelling and artmaking workshops facilitated by writer and editor Christobel Mattingley with the elders at Yalata and Oak Valley communities. The Indigenous informants related the harrowing experience of the inexplicable "black mist" that rolled into their campsites following the British nuclear detonations at Emu and Maralinga, as well as the subsequent illness and death that resulted from them. To help ensure that the *tjilpi* (old people) passed on their knowledge and wisdom to the next generations, a children's book *Maralinga: The Anangu Story* (2009) was produced recounting *Tjukurpa* (Dreamtime) stories of origins and spiritual teaching from the ancestors. It also describes the *walypala* (whitefella) invaders who travelled from the east to explore and claim Anangu lands and "gave their own names to many sacred and deeply spiritual places" such as "the Great Victoria Desert, after a whitefella queen that Anangu had never heard of, in a faraway land they did not know existed."[36] This richly illustrated work is full of "sorry business," stolen generations, people forcibly removed from their lands by governments and missionaries, inexplicable sickness and death befalling communities, and Anangu employed to undertake dangerous work in the contaminated areas after the test sites were abandoned. The displaced elders who were removed from their traditional lands thought that the long and loud rumblings in the desert caused by the bombs exploding were *mamu* (evil spirits) coming. It terrified the community, whose members feared for the people still camped out in the bush.[37]

Among the Indigenous artworks assembled for the book are images of mushroom clouds, some foregrounded against the familiar and contemporary dot-style paintings from the central desert. One painting portrays a white, grey, and pink mushroom cloud. Within it is the portrait of a bearded tribal man (or ancestral spirit), with a painted face and wearing a narrow headband, crying blue

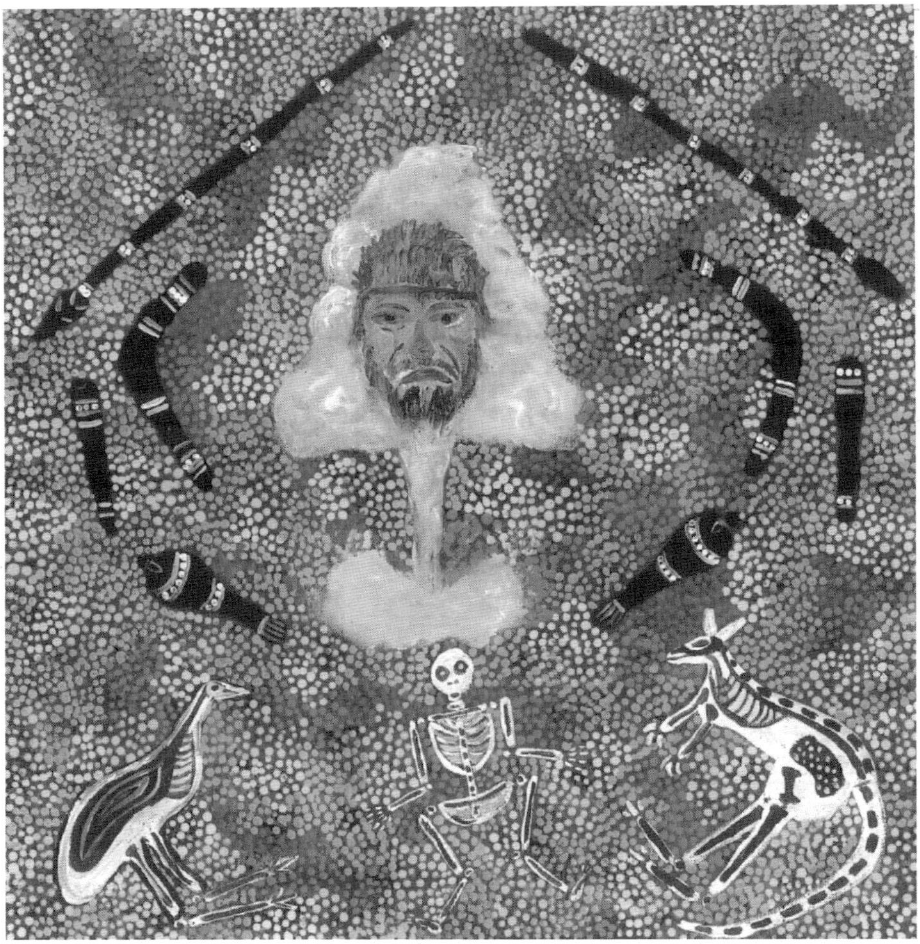

12.4 Maralinga artwork by Yvonne Edwards, from *Maralinga, the Anangu Story*, Yalata and Oak Valley Communities, with Christobel Mattingley (Allen and Unwin, 2009, 2012); and *Maralinga's Long Shadow: Yvonne's Story*, Christobel Mattingley (Allen and Unwin, 2016).

tears from both eyes, resolutely returning the gaze of the viewer. His image is surrounded by geometrically mirrored spears, *kali* (boomerangs), *woomeras*, and clubs overlaying the dot composition of country. Depicted immediately below the mushroom cloud are human skeletal remains, perhaps the bones of the departed spirit-face man rendered in the cloud, with adjacent X-ray imagery of *kalaya* (emu) and *malu* (kangaroo) skeletons, ironically evoking the modern Australian coat of arms.

Mattingley reprised this earlier artistic collaboration with the 2016 publication of *Maralinga's Long Shadow: Yvonne's Story*, which reveals that the above painting was by Yvonne Edwards, titled *Maralinga [1]*. She describes the image as an "elder

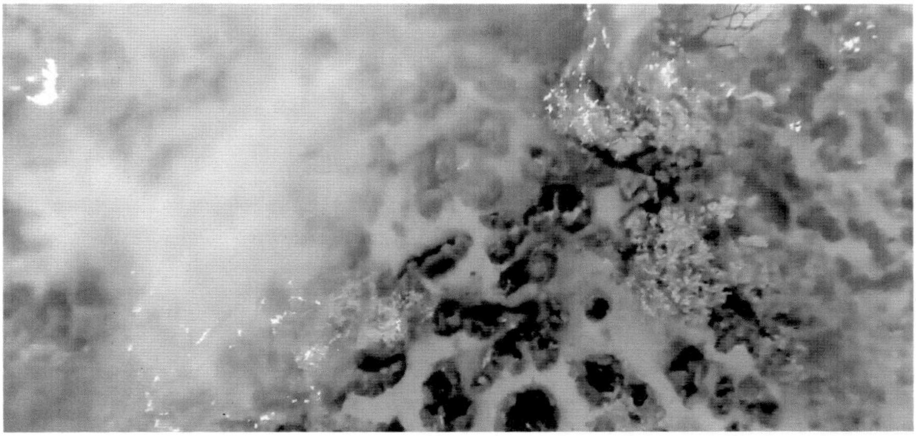

12.5 Virtual reality aerial view of Pilbara bush on fire with smoke rising.

grieving at the desecration of country and the deaths of its people and animals.[38] The author recounts the tragic illnesses and deaths that befell the Edwards family and the profound intergenerational effect on the community.

However, perhaps the zenith of these cross-cultural artistic collaborations to date can be found in the captivating virtual reality (VR) documentary/testimony of Martu elder Nyarri Nyarri Morgan, working with his grandson, Curtis Taylor, and the digital media artist Lynette Wallworth. Morgan's recollection of his encounter with a British nuclear weapon test in the outback desert is steeped in a unique perspective of place, brilliantly acquired through the capacity to experience Morgan's history in a 360-degree, immersive video titled *Collisions* (2015). The technology employed to enliven Morgan's oral testimony is deftly deployed to provide an embedded sense of spirituality, one that evokes an Indigenous connection to place traversed by Dreamtime spirits and ancestors, harmonized to the cyclical movement of the circling cosmos and seasonal patterns of climate and ecology.

A hovering drone overhead permits stunning views of elder Morgan amid the ancient Pilbara landscape as he sets fire to the spinifex and grasses to unlock the natural, regenerative process, an agricultural practice undertaken by his forebears scores of millennia past. The smoke seemingly passes through the VR observer/audience as it rises from the ground with fires crackling below between the gum trees. Participants assume a perspective (in both elevation and motion) that is transcendental in its fluidity and vantage, an animistic or shamanistic perspective of great beauty and omniscience.

The contemporary Martu community is shown going about its daily business, kids at play and family members introducing us to the country. Perched on a

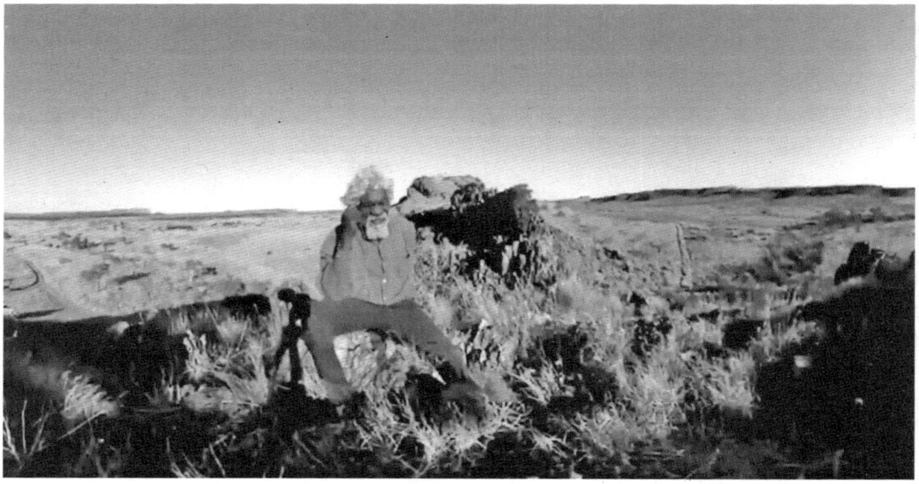

12.6 Virtual reality view of Martu elder Nyarri Morgan singing his welcome to country.

rocky outcrop high above the desert, Morgan sings an ancient song, bridging law and land, as he explains the importance of First People's custodianship of their sacred places from time immemorial, now possibly endangered by the prospect of uranium mining.

Outdoors at night, Morgan and other Indigenous elders sit before videos projected onto a makeshift screen. One clip reveals a sombre and drawn-looking Robert Oppenheimer from 1965 recalling his initial reaction to the Trinity test, invoking an apocalyptic deity: "I remembered the line from the Hindu scripture, the *Bhagavad-Gita*; Vishnu is trying to persuade the Prince that he should do his duty and, to impress him, takes on his multi-armed form and says, 'Now I am become Death, the destroyer of worlds.' I suppose we all thought that, one way or another."[39]

Other clips depict Cold War documentary footage of US troops advancing toward ground zero. At this point, if viewers turn away from the startling images of the towering mushroom cloud recorded from less than ten miles distant, Morgan can be seen on the left looking to his right (at us, the spectator) with finger pointing, speaking inaudibly. He is addressing the (unseen) VR apparatus, announcing that the image on screen is exactly what he saw in the desert decades earlier.[40]

But it is Morgan's experience of witnessing the results of an atomic bomb blast that dominates the VR production. The apparition he unexpectedly encountered in the bush is presented in compelling CGI. In dream-like slow motion a mob of large, panicked kangaroos rapidly leap toward the spectator, who has assumed Nyarri's first-person vantage via VR. A cloud of dirt and dust spreads out across the horizon as the kangaroos desperately attempt to flee the encroaching blast wave. They fall dead all about the viewer as we see the sky turning dark. The ascending

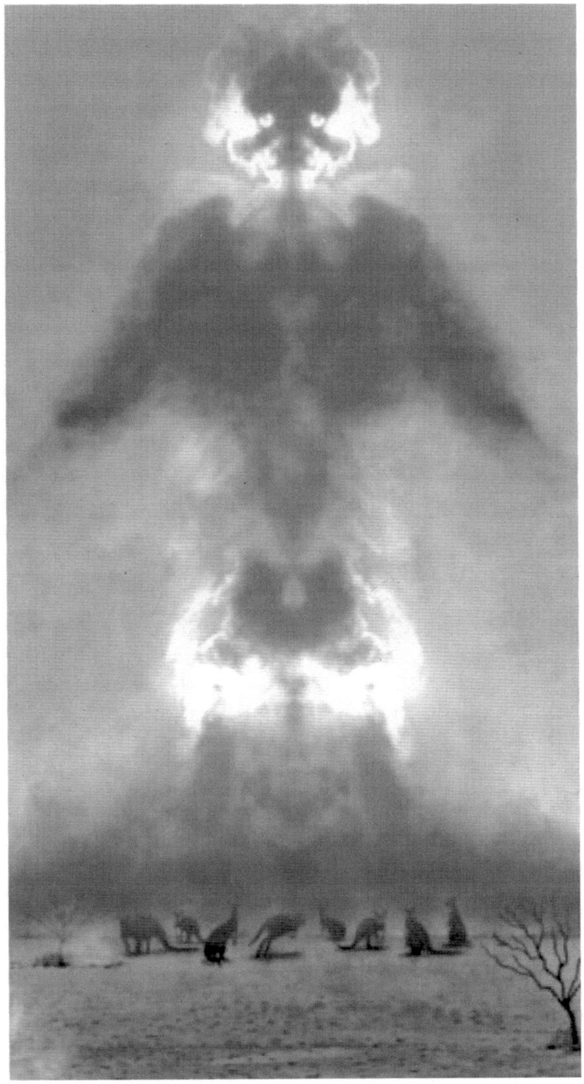

12.7 Virtual reality depicting Nyarri Morgan's vision of the British nuclear test mushroom cloud, becoming god-like. Collisions (2015).

mushroom cloud blots the outback sun, which seemingly morphs into one of two giant eyes. Nyarri relates that he understood this supernatural phenomenon to be the appearance of his ancestral gods, providing a welcome bounty of freshly killed wildlife. The cloud slowly assumes a humanoid figure before dissipating into a haze of dark ash with particles drifting down as fallout across the terrain.

Unlike the paternalistic anthropomorphism that imposed a *faux* indigeneity on the Totem atomic detonation at Emu in 1953, where colonial narrative was literally imprinted on the "myall" profile of the mushroom cloud by British and Australian observers, Nyarri Morgan's stark recollection and vision is intended to

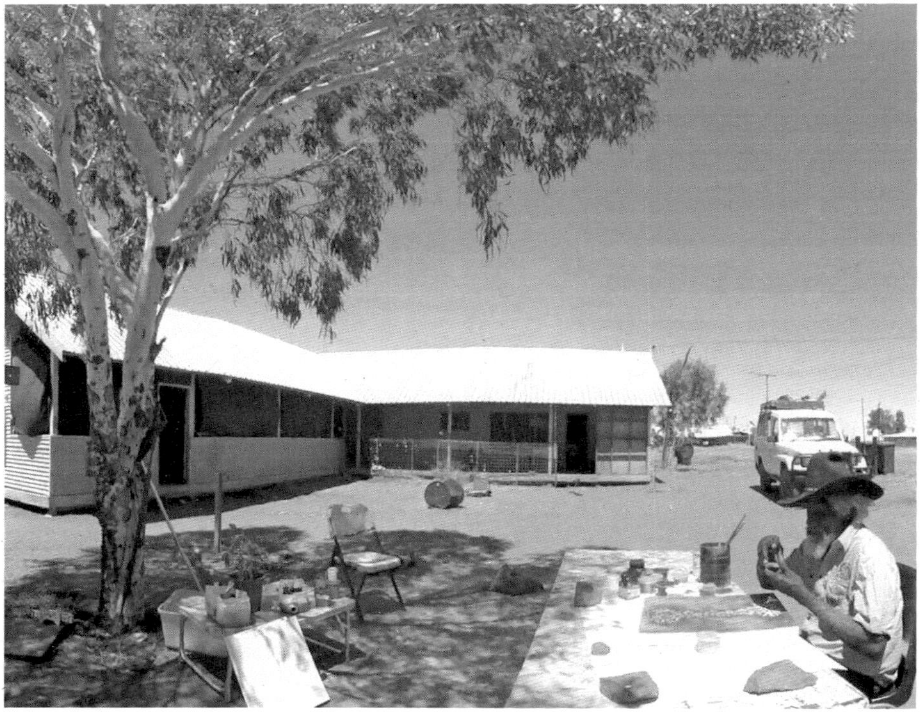

12.8 Virtual reality view of Nyarri Morgan in the act of (dot) painting his vision. Collisions (2015).

be experienced phenomenologically in full 360-degree virtual reality. Immersive digital artmaker Wallworth says that "the camera allows you to feel the experience of being in his home, under that enormous sky ... to place the viewer in relation to this community, to this land and give a sense of place ... The camera allows you to feel like you are in his home."[41] Wallworth recognizes the fundamental importance of using "the newest technology to talk about something which is ancient in this country – Martu's sense of stewardship, how you look after something for a hundred generations."[42]

Collisions also records on screen Nyarri creating a dot-painting of the mushroom cloud that he witnessed first-hand. It corresponds closely with the Maralinga art from the Oak Valley and Yalata communities.[43] However, Nyarri's work is more evidential, in that he was an eyewitness, whereas many other Aboriginal elders' art responds to an inherited, second-hand interpretation drawn from their relatives' stories and experiences.

Through VR technology, Nyarri Morgan's narrated – and importantly embodied – testimony provides a vantage of special significance.[44] While it connotes a cognitive dissonance similar to that of Oppenheimer's reaction at Trinity, and of the speechwriter's text penned for Harry Truman's announcement of the

Hiroshima bombing, in which he evoked "God" and "His" weapon, it is neither a carefully anticipated result of massive wartime experiment nor the second-hand prose reflection of a "civilian" commander-in-chief relaying news to the world of that manifestation. Morgan's vision is cognately one of spiritual cosmology and epistemology: an ancestor god had presented himself in thunder, heat, and smoke to provide the bounty of fallen game. Unlike those present at the Totem 1 atomic test in the central Australian desert in 1953 – who ran with the "myall blackfeller" trope as a press angle to ascribe a parochial inflection to the colonial enterprise – Morgan's interpretation is entirely consistent with his localized Dreamtime engagement.[45] This is not a mystical vision in the desert, or a mirage, but a common (albeit overwhelming) encounter with spirits who share space and time ("everywhen") beyond and outside of Judeo-Christian teleological precepts.[46]

Yet in Morgan's recounting there is also a tinge of apostasy that recalls Oppenheimer's interview, given that the stricken kangaroos and broiling waterholes contained an unknown and invisible poison, as did the particles of ash that fell from the sky onto his mostly naked body. This is an atomic advent *we* can experience, as digitally immersed audience members, granted from Morgan's first-person point of view and artfully realized by his black and white collaborators. Oppenheimer's guilt-laden lament and reflection on US television came twenty years after Trinity. Morgan's revelation has waited much longer, more than six decades, to be told and presented in a manner that sensitively and powerfully gives voice to a unique Indigenous perspective and encounter with the nuclear age.

NOTES

1 Commonwealth of Australia, *Royal Commission*, 15.
2 Kohen, "Pemulwuy (1750–1802)."
3 Cooper, "Flinders, Matthew (1774–1814)."
4 "Canberra" was also the name given to a British jet bomber used in the nuclear test trials in Australia.
5 Pearlman, "Not hungry."
6 Furphy, "Place Names."
7 Ibid.
8 Yalata and Christobel, *Maralinga: The Anangu Story*, 26.
9 An interactive map of these language groups pre-European settlement can be found at https://aiatsis.gov.au/explore/map-indigenous-australia.
10 At the time of Operation Hurricane, the first British nuclear test off Trimouille Island in the Monte Bellos in October 1952, the atomic cloud formed a distinctive and unconventional Z-shape, surprising observers. See Smith, *Clouds of Deceit*; and Broderick, "Filmic Mutation."
11 Beadell, *Blast the Bush, 211.*
12 Ibid.

13 [Anon.], "Atomic Cloud Forms Head of Aborigine."
14 Ibid.
15 [Anon.], *The Chronicle.*
16 Beadell, *Still in the Bush*, 42.
17 Bhabha, "Remembering Fanon," xxv.
18 Goldie, "Signifier Resignified: Aborigines in Australian Literature," 63.
19 Bhabha, "Remembering Fanon: Self, Psyche and the Colonial Condition," xxvi.
20 Furphy, "Place Names," 73.
21 Davenport, Johnson, and Nixon, *Cleared Out.*
22 Yalata and Mattingley, *Maralinga: The Anangu Story*, 37–8.
23 Australian Bases Tracking Stations, "Nurrungar Satellite Tracking Station."
24 Broderick, *Nuclear Movies.*
25 One airmail envelope depicting the mushroom cloud is stamped "Edinburgh Airfield," named after the Duke of Edinburgh, Prince Philip. The official history of the Woomera range notes that "after various Aboriginal names were suggested, it was decided, with the Duke's approval, to call the base 'Edinburgh.'"
26 Healey, *Forgetting Aborigines.*
27 Grayden, *Adam and Atoms, vii.*
28 Cohn, "Sex and Death," 690.
29 Broderick, *Nuclear Movies*, 3.
30 Commonwealth of Australia, *Royal Commission*, 15.
31 Green, "The Nuclear War."
32 Bell, "Bell's Theorem."
33 Mittmann, *Black Mist Burnt Country.*
34 See for example the recent documentary *Maralinga Tjarutja* (2020), directed by the Aboriginal academic and filmmaker Larissa Behrendt, in collaboration with the Maralinga Tjarutja community. See also Broderick, "Sixty Years On."
35 Bell, "Bell's Theorem."
36 Yalata and Mattingley, *Maralinga: The Anangu Story*, 11–12.
37 Ibid., 51.
38 Mattingley, *Maralinga's Long Shadow*, 147.
39 Wellerstein, "Oppenheimer and the Gita."
40 Taylor, interview.
41 Wallworth, "Production Notes."
42 Ibid.
43 See for example the stunning artworks *Destruction I and II (2002) by Jeffrey Queama and Hilda Moodoo, in* Mittmann, *Black Mist Burnt Country*, 31, 48.
44 On phenomenological readings of presence and embodiment immersive digital media, see Bender and Broderick, *Virtual Realities.*
45 Nicholls, "'Dreamtime' and 'The Dreaming.'"
46 Ibid.

BIBLIOGRAPHY

[Anon.]. "Atomic Cloud Forms Head of Aborigine." *The Advertiser, 16 October 1953.*

[Anon.]. "1953." *The Chronicle*, 22 October.

Australian Bases Tracking Stations. "Nurrungar Satellite Tracking Station," n.pag. http://www.thelivingmoon.com/45jack_files/03files/Australia_Nurrungar_Satellite_Tracking_Station.html.

Beadell, Len. *Blast the Bush*. Sydney: New Holland Press, 1967.

– *Still in the Bush*. Sydney: New Holland Press, 1965.

– *Too Long in the Bush*. Sydney: New Holland Press, 1967.

Bell, Richard. "Bell's Theorem: Aboriginal Art – It's a White Thing!," n.pag. http://www.kooriweb.org/foley/great/art/bell.html.

Bender, Stuart Marshall, and Mick Broderick. *Virtual Realities: Case Studies in Immersion and Phenomenology*. New York: Palgrave, 2021.

Bhabha, Homi K. "Remembering Fanon: Self, Psyche, and the Colonial Condition." In Frantz Fanon, *Black Skin, White Masks*, xxi–xxxvii. London: Pluto Press, 1988.

Broderick, Mick. "Atomic Pop." In *Black Mist Burnt Country*, edited by J.D. Mittmann, 66–7. Upwey: Burrinja, 2016.

– "Filmic Mutation: British Nuclear Tests in Australia 1952–63." *Studies in Documentary Film*, in press (2022).

– *Nuclear Movies*. Jefferson: McFarland, 1991.

– "Sixty Years On, Two TV Programs Revisit Australia's Nuclear History at Maralinga." *The Conversation*, 4 June 2020. https://theconversation.com/sixty-years-on-two-tv-programs-revisit-australias-nuclear-history-at-maralinga-139313.

Cohn, Carol. "Sex and Death in the Rational World of Defense Intellectuals." *Signs* 12, no. 4 (1987): 687–718.

Commonwealth of Australia. *Royal Commission into British Nuclear Tests in Australia*, vols. 1–3. Canberra: Australian Government Printing Office, 1985.

Davenport, Sue, Peter Johnson, and Yuwali Nixon. *Cleared Out: First Contact in the Western Desert*. Canberra: Aboriginal Studies Press, 2005.

Cooper, H.M. "Flinders, Matthew (1774–1814)." *Australian Dictionary of Biography*, National Centre of Biography, Australian National University, n.pag. http://adb.anu.edu.au/biography/flinders-matthew-2050/text2541.

Furphy, Sam. "Place Names and the Settler Australian Identity." *Melbourne Historical Journal* 1, no. 29 (2001): 71–8.

Grayden, William. *Adam and Atoms*. Perth: Frank Daniels, 1957.

Green, Jim. "The Nuclear War against Australia's Aboriginal People." *The Ecologist*, 2014. http://www.theecologist.org/News/news_analysis/2476704/the_nuclear_war_against_australias_aboriginal_people.html.

Healy, Chris. *Forgetting Aborigines*. Sydney: University of NSW Press, 2008.

Kohen, J.L. "Pemulwuy (1750–1802)." *Australian Dictionary of Biography*, National Centre of Biography, Australian National University, 2005. http://adb.anu.edu.au/biography/pemulwuy-13147/text23797.

Mattingley, Christobel. *Maralinga's Long Shadow: Yvonne's Story*. Sydney: Allen and Unwin, 2016.

Mittmann, Jan Dirk, ed. *Black Mist Burnt Country: Testing the Bomb – Maralinga and Australian Art*. Upwey: Burrinja, 2016.

Nicholls, Christine Judith. "'Dreamtime' and 'The Dreaming' – an Introduction." *The Conversation*, 2014. http://theconversation.com/dreamtime-and-the-dreaming-an-introduction-20833.

Pearlman, Jonathon. "Not Hungry, Just Aboriginal Kitsch, Says Former PM." *Sydney Morning Herald*, 2006. http://www.smh.com.au/news/national/not-hungry-just-aboriginal-kitsch/2006/10/18/1160850998879.html.

Smith, Joan. *Clouds of Deceit: The Deadly Legacy of Britain's Bomb Tests*. London: Bloomsbury, 2011.

Taylor, Curtis. Telephone interview, 3 March 2017.

Wallworth, Lynette. "Production Notes." Australian Centre for the Moving Image, 2016, https://guides.acmi.net.au/collisions/?_ga=1.168197729.1392034973.1490485815.

Wellerstein, Alex. "Oppenheimer and the Gita." *Restricted Data: The Nuclear Secrecy Blog*, 2014. http://blog.nuclearsecrecy.com/2014/05/23/oppenheimer-gita.

Yalata, Oak Valley Communities, and Christobel Mattingley, eds. *Maralinga: The Anangu Story*. Sydney: Allen and Unwin, 2009.

13

The Antipodean Stance of Pam Debenham's 1980s Screenprints

N.A.J. TAYLOR

Anglo-American and Eurocentric perspectives continue to dominate nuclear discourse and nuclear scholarship. In this chapter I develop further the argument that an Antipodean perspective on the nuclear age offers unique eco-political insights.[1] Such an Antipodean stance situates Australia in Oceania, but also in opposition to the dominant statist and anthropocentric approach – a move that offers a platform for a shared understanding between settler and First Nations approaches to the underlying problem of nuclear harm. In this way, the Antipodean stance is not so much a simple inversion of Eurocentrism, but rather what Paul Giles has described as a "systematic subversion."[2]

This Antipodean approach originates in the oceans and deserts of the Southern Hemisphere, where between 1946 and 1996 the French, the British, and the Americans conducted more than 300 nuclear weapon and radiological tests, including at widely known sites such as Maralinga and the Marshall Islands.[3] Several of the US tests at the Pacific Proving Grounds in the Marshall Islands remain among the largest by yield and the most disastrous ever conducted in terms of radioactive contamination and the effects of the fallout and ionizing radiation on local environments, ecosystems, and human and non-human life. The Marshall Islands site is now at the centre of stratigraphical investigations into the origins of the Anthropocene epoch.[4] Although the consequences – ranging from the death and disfigurement of individuals to forced displacement – remain contested, harms to human health and environmental damage have been well catalogued by the Royal Commission into British Nuclear Tests in Australia.[5]

By the time France concluded its final Pacific nuclear tests at Mururoa in 1996, the negotiations for a comprehensive test ban treaty were well under way. At the time, France was roundly and globally condemned for maintaining that no remediation or monitoring was required at Mururoa on "radiological protection grounds."[6]

Despite the enduring legacy of US, British, and French nuclear colonialism in the Pacific, it is commonplace to frame such nuclear weapons testing as mere "non-use." For instance, on receiving the Nobel Peace Prize, Thomas C. Schelling remarked that it had been "sixty-two and a half years since the second, and last, nuclear weapon exploded in anger."[7] Put this way, the term "use" would appear to be confined to cases where hostilities between enemies provided the context for nuclear events, thus framing the inflicted harm in purely human-centred terms. That is to say, it is to frame nuclear testing anthropocentrically as "non-use." The notion of an Antipodean stance problematizes this notion of "non-use" by exploring the nuclear weapons tests' violations not only of the human body and the biosphere on which all life depends, but also of the very cultures through which meanings between the two are transmitted.

Oceania is particularly interesting as a site for study because it remains home to the oldest continuous cultures in the world's largest ocean. Yet the region and its peoples are often neglected or marginalized in environmental histories and political-ecological studies.[8] As a direct response to its nuclear colonial history, Oceania was one of the first regions in the world to comprehensively denuclearize in 1986. It is also home to the world's first state-level nuclear-free zone, Aotearoa/New Zealand, which declared itself nuclear-free in 1987. Oceania is thus singularly important in that it has forgone both nuclear weapons (through its regional treaty obligations) and nuclear energy (voluntarily and unilaterally). From the perspective of Oceanic peoples, the violence, harm, and injury perpetrated by nuclear weapons tests have reached beyond the human body and the biosphere,[9] shattering the *cultures* through which the interconnectivity and entanglement of the living and non-living are transmitted, as well as the meanings of those relations.[10]

This chapter proceeds as follows. In the first section I introduce the Tin Sheds screenprinting workshop at the University of Sydney, which was a productive source of anti-nuclear political posters in the 1970s and 1980s. In the second section I profile the life and work of Pam Debenham so that we come to understand the personal – both moral and political – motivations that drive her art practice. In so doing I characterize Debenham according to the Antipodean stance. Such a stance situates her in Oceania and also in opposition to statist, corporate-technocratic, and anthropocentric approaches to nuclear history, culture, and geopolitics. Finally, I conduct a close reading of five of Debenham's anti-nuclear posters that to my mind evince an Antipodean perspective. By situating Debenham's screenprints in an Oceanic or Antipodean oppositional framework, another narrative emerges in which artists can be viewed at once as major actors in Antipodean anti-nuclear thinking and as creatives who have made a substantial contribution to the Antipodean/Oceanic local and global anti-nuclear politics that culminated in the 1985 Rarotonga Treaty, which established the South Pacific nuclear-weapon-free zone. Much neglected in nuclear humanities and social science scholarship, the work of artists such as Debenham serves as both a corrective and an alternative to a male-centred Anglo-American nuclear discourse.

The Visual Politics of Posters: The Tin Sheds Screenprinting Workshop

Political posters and printmaking are a rich source of political ferment. Unlike other political activities, such as protests and meetings, posters have an ancillary function as a form of accessible documentation. Thus, posters go some way toward democratizing political discourse through both production and consumption. For this reason, the struggles of various social movements, groups, and causes – such as the feminist, women's, and queer rights movements in the 1960s and 1970s – are relatively well-documented by the medium.[11] As Deborah Clark contends in reviewing the first Australian survey exhibition *The Streets as Art Galleries* at the National Gallery of Australia:

> The medium of screenprinting was attractive for alternative media groups because it was cheap, quick to produce and did not require expensive technology or large premises. In line with the decade's move away from the art in the gallery, poster workshops flung art into the streets and made it part of the urban vernacular. Collective art production and community art were both significant aspects of postermaking, promoting accessibility in the visual arts and an end of the exclusivity of the individual museum object.[12]

One especially prolific and influential site of political poster-making and display in the Australian context was the Tin Sheds Arts Workshop ("Tin Sheds"), a loosely "radical" mixed-media circle at the University of Sydney, established in 1969 and active in various guises through to the 1990s.[13] Although there is some dispute as to how and why the many movements that comprise the Tin Sheds formed, few question that co-founders Marr and Joan Grounds's experience of campus politics at the University of California at Berkeley, in the United States, provided the impetus. The extant scholarly literature on the Tin Sheds is largely preoccupied with its internal politics and that politics' impact on the many social causes to which its members applied themselves.[14] Here I am more concerned with the assertion that the Tin Sheds was a site of "art made in an oppositional spirit." Kenyon's *Under a Hot Tin Roof*, a detailed study of the Tin Sheds' institutional and art-historical importance, captures the tension between overtly political art and the fine arts in the group's activities:

> The Sheds was one of the first places in Australia to recognize the need for a "laboratory" for experimentation around art and the built environment with open access and live-in studio space for artists ... The threads that keep breaking the surface ... are the ongoing struggles for physical survival, funding, and the contest between radical attitudes to artistic practice and traditional skills-based teaching required by faculties. Other strong influences were the counterculture movement, Marxist politics, the anarchism closely allied to post-object art, and the developing feminist principle of "the personal is political." That led to an appreciation of a tradition of women's art and craft, which then led to the idea of community arts, and for the Sheds, a policy of providing access to the local population.[15]

The output and influence of the Tin Sheds movement was sufficiently pronounced that major retrospective exhibitions were recently curated at the University of Sydney Art Gallery as well as at the Verge Gallery, where Tin Sheds once stood.[16] *Girls at the Tin Sheds: Sydney Feminist Posters 1975–1990*, exhibited at the university's premier gallery, displayed the works as posters are on the street – in a salon hang – chronologically so that shifts in the social and political discourse would be easily discernible. Whereas at the former site of the Tin Sheds, for the exhibition *Girls at the Tin Sheds [Duplicated]*, the works were arranged in a colour schematic and presented individually, as is common in art galleries. Decisions were made at both venues, both curatorially and practically, to decontextualize the posters so as to let them speak to their contemporary relevance. At the latter opening, Anne Bickford, a former member of the Tin Sheds, intervened and asked whether decontextualizing the final product from its production and consumption was productive,[17] since in doing so much meaning-making was lost.

Pat Debenham stands out among the Tin Sheds artists because her designs helped construct an anti-nuclear visual politics and popularize the anti-nuclear movement in Australia and internationally.[18] Her anti-nuclear prints display an aesthetic whose appeal is both local, or culturally and geographically specific, and transpacific or global. However, despite several recent major exhibitions of nuclear art in Australia, there have been very few scholarly in-depth studies of Debenham's anti-nuclear works, and close examinations of these prints' underlying ethics and politics are practically non-existent.[19] For instance, in drawing connections between Debenham and Roderick Shaw, a pioneering poster-maker, curator Therese Kenyon characterizes Debenham as being among a core group who received the "'artist as activist' baton" from Shaw.[20] While comments such as Kenyon's call attention to genealogies that have remained understudied, and while Debenham's work is discussed in passing in several art catalogues, her anti-nuclear art has received relatively little attention in recent scholarship on the nuclear arts and culture in Australia and the Pacific region.

The Life and Work of Pam Debenham

Pamela Elizabeth Debenham was born in Launceston, Tasmania, in 1955. She was mentored at high school by the noted artist Joyce Allen and graduated from the Sydney College of the Arts in 1982. In 1980, during her studies, Debenham became a member of the Tin Sheds Art Workshop, becoming its print workshop coordinator in 1982, a position she held until 1989.[21] In a recent interview with the author, she said that her anti-nuclear posters were her way of internalizing the nuclear threat posed by Ronald Reagan's Star Wars project, the escalation of the Cold War with the former Soviet Union, and – closer to home – the strategic importance of Pine Gap in the US nuclear weapons complex. She recalls participating in several demonstrations in Sydney at a time when hundreds of thousands of people across the country were marching in opposition to nuclear weapons, nuclear war, and Australia's nuclear umbrella.[22] She sold her T-shirts and her monotone posters at these demonstrations (see below). Although her art practice encompassed several media, her anti-nuclear oeuvre is confined solely to the medium of screenprinting. Speaking to Debenham's influence on Australian popular culture, Chris McAuliffe contends that her prints were precursors to the iconic Australian streetwear brand, Mambo, which was widely worn among Australian youth in the 1980s and 1990s. However, Mambo clothing was mass-produced, whereas many of Debenham's works had print runs of 200 copies, although this was increased to 400 if there were two colours (e.g., red and black).

Notwithstanding my earlier claims about Debenham's importance in Antipodean nuclear art, there are three main reasons why it is especially difficult to conduct an art-historical study of Debenham. First, the medium of screenprinting does not easily lend itself to formal conventions of documentation and critique. The result is that very little exists by way of artist or curatorial statements on, and critical responses to, Debenham's nuclear works, and what can be gleaned is largely drawn from passing references made by the artist in interviews or by curators and critics in retrospective art catalogues. Because her anti-nuclear posters were meant to be exhibited on the street and in public spaces, they have generally been neglected by art historians and are rarely included in art histories. Existing studies of screenprinting and posters tend to survey the field from a street art or political art perspective – seldom from the vantage point of nuclear aesthetics or nuclear culture. One of the world's largest and most significant collections of social justice and protest posters, housed at the Centre for the Study of Political Graphics in Los Angeles, focuses largely on political texts and iconography. The available documentation of and secondary literature on Debenham's anti-nuclear works is scant. This chapter, then, hopes to expand and reorient the scope and methodology of the growing field of nuclear arts, nuclear aesthetics, and nuclear culture studies, in particular of the feminist-oriented sector within this scholarly field.

Second, the authorship of individual posters is often difficult to ascertain without prior knowledge of the artist's practice. Anarchist principles were variously implemented (or enforced) at the Tin Sheds, and that was one reason why many artists associated with the group did not sign their works. Adding to this, Debenham stated in a recent interview with the author that her works were unsigned, or else marked "Tin Sheds," "P.D.," or "Debenaire," mainly to protect her against possible police action: in the 1980s, it was illegal to put up street posters.[23] The reasons for Tin Sheds artists' mostly anonymous production of screenprints and other artworks are thus many and varied. Interestingly, Louise Mayhew, drawing on recent interviews with contemporary anarchist and feminist printmakers, has pointed to an additional conundrum Debenham and other female printmakers faced that their male counterparts did not: "The use of the collective logos at the very same time that women were attempting to demand access to the label 'artist' and recognition of their artistic creativity generated a difficult situation for female printmakers."[24]

Third – and finally – we must contend with the fact that Debenham's nuclear oeuvre is rather small – fewer than a dozen works. For present purposes I have selected only five of them that in my view express an Antipodean stance. A fuller account might venture deeper into Debenham's entire oeuvre, which includes unpublished sketches of nuclear works in development or abandoned in process. For instance, we know that Debenham's archive contains preliminary drawings for her posters as well as several linocuts that have nuclear imagery. One of these is Debenham's work, *History*, produced for the 1988 Bicentennial travelling exhibition *Right Here Right Now*.[25] This work displays images of American B-52 strategic bombers and the Joint Defence Facility Pine Gap, a US satellite surveillance base jointly operated by Australia and the US. The five works examined in this chapter were selected on the basis of their distinctiveness from other works in Debenham's oeuvre as well as for their capacity to articulate or (re) imagine an Antipodean perspective.

Furthermore, the significance of Debenham's work in global nuclear art and culture stems in part from the location of her art and the publics it addresses – that is, the environment and populations of the Oceanic region. A close reading of Debenham's work may provide new insights into and understandings of various aspects of Oceanic (nuclear) art and culture, although it should be noted that the region displays much diversity and disagreement between and within communities.

Taking seriously Debenham's visual politics is part of a broader concern for visual global politics, which, as Roland Bleiker has argued, has the potential to blur the boundaries between visual images, such as photographs and paintings, and visual artifacts, such as monuments and televised events. Bleiker contends that the significance of the visual for our understanding of politics extends beyond mere representation and that visual images and artifacts are "political forces in themselves":[26]

> Although we live in a visual age, knowledge conventions – both in academia and in the wider realm – are still very much focused on texts and textual analysis. What would a true political appreciation of the visual look like? What would it mean to communicate and think and act in visual ways? How would the media, books, classrooms, and other realms be transformed if we were to treat images not just as illustrations or as representations but as political forces themselves?[27]

Echoing Bleiker's contention that images can be regarded as political forces, Alex Danchev argues that "contrary to popular belief, it is given to artists, not politicians, to create a new world order."[28] Bleiker's approach to visual global politics is especially relevant for the analysis I propose in this chapter, for his perspective enables me to view Debenham's prints as both image *and* artifact. So there is a need here not to perform yet another survey of nuclear art and culture, but rather to gain a better understanding of how Debenham's work evidences the patterning or imprinting of nuclear colonialism in Oceania onto an individual's consciousness. As Bleiker puts it: "Images are political in the most fundamental sense: they delineate what we, as collectives, see and what we don't and thus, by extension, how politics is perceived, sensed, framed, articulated, carried out and legitimized."[29] Bleiker has also argued that an aesthetic reading can provide insights into how events in world politics are "internalized in our minds, our habits and our collective political consciousness."[30] Following Bleiker, the aim here is to "reclaim the political value of the aesthetic, not to replace social science, but to broaden our abilities to comprehend and deal with the key dilemmas of world politics."[31]

Toward these ends, Debenham recognized the qualities of screenprinting that marked it out as a medium of political art: "As a form of social and political expression the screen-printed poster has a long history as an effective medium. It is relatively inexpensive to set up, it can communicate to a wide audience through multiple prints, it has a strong visual impact in terms of image and strength of colour and can subsidize its primary role as public art through print sales."[32]

Debenham's Antipodean Stance

Debenham can be regarded as the quintessential Antipodean nuclear artist since her work very clearly and forcefully evinces the oppositional stance that is characteristic of an Antipodean approach to the problem of nuclear harm. Debenham's iconography is both precise and direct. Where subtlety appears –

such as in the depiction of the non-idyllic nuclear explosion motif on the Hawaiian shirt in *No Nukes in the Pacific* – it is employed both judiciously and satirically to great effect.[33] The iconography and the words are, for Debenham, always both in dialogue and in tension. In *Still Life*, for instance, Debenham's foregrounded subject – the phallic obsession with nuclear militarism (which was later brilliantly critiqued by Carol Cohn) – is placed in tension with a background dominated by the word "NO."[34] Overall, Debenham's work is forcefully oppositional.

Such vehement anti-nuclearism was increasingly common in the mid-1980s when each of the prints was made, but Debenham's posters rendered it in powerful new ways. The word "no" features throughout Debenham's nuclear oeuvre, in fact in four of the five selected posters, and all of the works are "for" some clearly stated anti-nuclear ends. It is through this oppositional stance that Debenham is inherently relational and open to comparative interpretation; in her wordplay and iconography, she is not merely speaking *about* something or someone but *to* something and someone.

It is crucial, however, not to overstate the impact of Debenham's work. Though there is no direct causal relationship between Debenham's works and the founding of regional nuclear non-proliferation and disarmament instruments, her posters do illuminate the social and political ferment of the time. For instance, during one of our recent interview sessions, Debenham recalled that she mailed her *Paradise Lost* Hawaiian shirt to prominent musicians and other public figures such as Midnight Oil front man Peter Garrett, the Torres Strait Islander singer Christine Anu, and the lead singer from Mental As Anything, Andrew "Greedy" Smith, to give it public exposure.[35] Others received Debenham's gesture as a provocation. According to Debenham, then minister for foreign affairs, Gareth Evans, had his office staff send it back. By way of contrast, the Australian Greens leader and co-founder Bob Brown received the shirt at a later date, when he attended an opening of his photographic show at the Helen Maxwell Gallery in Canberra. In 1985, soon after Debenham produced her posters, actions and mass demonstrations organized by the anti-nuclear movement led to New Zealand prime minister David Lange outlawing visits by nuclear-powered and -armed vessels. He also declared New Zealand to be nuclear-free, a position it holds to this day. That same year, Lange won the argument for an anti-nuclear world at the Oxford Union Debate.

Debenham's best-known print, *No Nukes in the Pacific*, appears at first sight to depict a Hawaiian-style shirt replete with swaying palm trees and clear waters.[36] As the curator of a major survey exhibition of Australian political posters, Geoffrey J. Wallis, stated in his catalogue notes: "Hawaiian shirts are associated with an escape from day to day existence, but Debenham associates them with the harsh reality of the arms race and possible nuclear conflict."[37] In reality, however, what Debenham had interspersed in the print were depictions of French and American nuclear weapons tests in the Pacific alongside a lone boat named "Pacific Peacemaker."

Deeply political, Debenham was prone to eschewing the satire and humour favoured by Toni Robertson and other contemporary artists.[38] Such visual politics are evident in the spelling of "Moruroa," unpopular with the French, for whom that spelling has connotations of death, as does the overall colour palette, with its fluorescent orange. Such was the effectiveness of the colours that one reviewer remarked that the poster was "irradiated."[39] The poster was reworked as an item of clothing in 1995, and as Roger Butler – the driving force behind the National Gallery of Australia's print collection – has remarked, the trend toward issuing posters on garments was not simply a strategy to increase both their visibility and longevity and to defray their production costs. It also had symbolic import:

> One of the most successful translations of a poster image to fabric was by Pam Debenham. Her 1984 screenprint *No Nukes in the Pacific* featured a figure wearing a pastiche of an [*sic*] ubiquitous Hawaiian shirt – the palm trees in the design interspersed with nuclear mushroom clouds. This protest against French nuclear tests in the Pacific regained its relevance in 1995 when the French government resumed tests on Mururoa Atoll [*sic*]. Debenham's response was not to reissue the poster but to present it in a new form – producing fabric and making shirts; this was a complete inversion of the usual object-to-image process. When marketed by Greenpeace the shirts had a distribution and impact that far exceed the original poster.[40]

In actual fact, the shirt had been retitled – *Paradise Lost* – to reflect that it had been wholly redesigned from the earlier *No Nukes in the Pacific* poster so that it had a repeat pattern when printed on fabric. At the same time, various of its visual elements had been updated.[41] In addition, Debenham had revised and renamed the boat as if to express solidarity with the plight of Greenpeace's *Rainbow Warrior*, which had been sunk by the French on 10 July 1985, and she had inserted allusions to the most recent nuclear weapons test by the French on Moruroa and Fangataufa Atolls.[42] As Debenham herself remarked at the time in her press release accompanying the shirt, French prime minister Jacques Chirac's resumption of nuclear weapons tests meant that anti-nuclear sentiment "has come to the fore again."[43]

No Nukes in the Pacific grounds Debenham's opposition to the military application of nuclear technology in the Pacific region, while retaining the necessary universal message calling for a moratorium on nuclear weapons. The overarching message is therefore one of peace, linking the environmental and anti-nuclear movements with the broader peace movement in a singular image. The poster was the product of a peace movement that had become aligned with the anti-nuclear movement. Thus, the boat on the shirt has a peace symbol on its sail. As Pam Debenham has recalled:

> The original idea for this poster was inspired by an American friend who collects antique Hawaiian shirts telling me that the rarest shirt from the 1950s was one with a design of a bomb blast. The shirt was produced in *celebration* of the United States doing atmospheric testing on Bikini Atoll. This seemed so contradictory I thought it was worth reinventing the idea in the context of protest. This poster was one I produced in the early 1980s dealing with the continual armaments build-up by the superpowers and nuclear testing in the Pacific. Like many Australians, I feel an affinity for the coastal landscape of our country and the sense of our close proximity to our Pacific Island neighbors. The voice of the individual protester is conveyed through the visual map of the shirt.[44]

However, Debenham is quick to assert the global significance of her message, as opposed to its purely local or regional connotations, as when she states: "The reactions I got to the first poster convinced me it was worth using this image of beauty and fear. The contrast drives home the shock value of the No Tests message. We mustn't allow ourselves to forget what's at stake. It goes beyond the French, and their colonial arrogance – it's a global issue."[45]

In *Still Life*, Debenham uses satire to reimagine a familiar genre. Here the subject that is foregrounded is military technology, not mimetically represented but depicted as children's toys and magazine cut-outs.[46] In an interview with the author in 2019, Debenham recalled that the toys were in fact Lego models made by Daniel Burn, the son of the art critic Ian Burn.[47] The military toys and cut-out bombs are not presented as a collection of isolated and inanimate objects; instead, they are positioned in relation to Earth as the stage onto which military action is unleashed. The spatial arrangement of the military objects in the print suggests there is an epicentre to this activity – the "Fat Man" device that was detonated over the city and people of Nagasaki on 9 August 1945. As if to further animate the scene, Debenham has concocted a background wallpaper decorated with atomic mushroom clouds interspersed with the word "NO." The artist's oppositional, anti-nuclear stance is thus powerfully driven home and strengthened by the tension between the explosive arrangement of military devices and weapons in the foreground and the background's obsessive repetition of both the iconic mushroom cloud motif and a resounding "NO" epitomizing an unequivocal rejection of nuclear weapons tests.

In a 1992 interview, Debenham remarked that her works *No Nukes in the Pacific* and *Still Life* arose from her outrage at the nuclear states' colonial violation of the Pacific:

> [They] were produced out of my concern during the 1980s for the continual armaments build-up by the superpowers and the continued nuclear blasts in the Pacific. The poster *No Nukes in the Pacific* was inspired by the

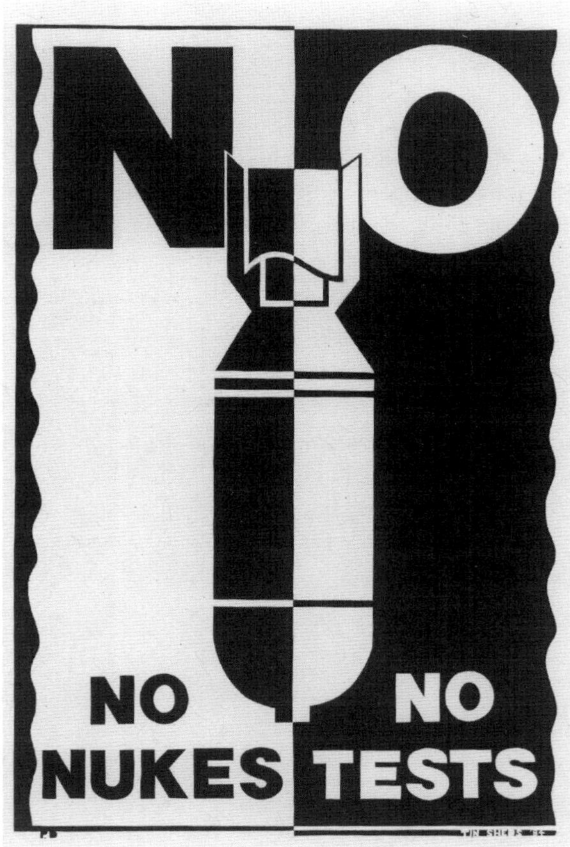

13.1 Pam Debenham, *No Nukes/No Tests*, 1984. Screen print on paper, 76 x 51 cm, National Gallery of Australia.

> knowledge that a traditional Hawaiian shirt with a design of a bomb blast was produced in affirmation of the first tests on the Bikini Islands in the 1950s. Also, like many Australians, I feel an affinity for the coastal landscape of our country and the sense of our close proximity to our Pacific Island neighbours. Having replaced these Pacific resort names by the nuclear test sites with the mushroom clouds punctuating the sky, the Pacific peacemaker boat charts its message between the affected atolls and islands. The voice of the individual protector is conveyed through the visual map of the shirt.[48]

While Debenham's best-known work is *No Nukes in the Pacific* (and the related Hawaiian shirt), *Still Life* is arguably Debenham's strongest work, not necessarily because of its political iconography, or as an instance 1980s anti-nuclear visual culture, but as a distillation of an Antipodean stance. The visual politics of this poster expose the internalizing of nuclear colonialism in the Oceanic region while at the same denouncing the socio-environmental crime perpetrated by that regime.

The poster *No Nukes/No Tests* was printed in two monotype colourways, using either black or red ink, to enable mass production at relatively little cost as well as to heighten its visual message on the streets.[49] The print centres on the silhouette of the "Little Boy" device that was detonated over Hiroshima on 6 August 1945. Reduced to its essence by way of colour and form, Debenham's poster plays with visual contrast horizontally across the print's surface, as if the bomb had been captured at the precise moment of its detonation. The contrast is enhanced through the positioning of a text that boldly calls for both "No Nukes" and "No Tests." Although this abolitionist message was commonplace in the 1980s, the framing of nuclear weapons testing as nuclear attacks or nuclear war resonated strongly with communities affected by the fallout and radiation emanating from the blasts.

No Nukes/No Tests may be read as a critique of the assertion that nuclear knowledge is purely scientific and therefore also universally applicable, devoid of place. Indeed, the notion that there exists a singular nuclear knowledge is widely held by those invested in nuclear science, strategy, and policy. "Expertise" is routinely used to discount the voices of those outside the nuclear establishment; the claim of authority in nuclear matters brings with it a pretense to intellectual hegemony. So when it comes to nuclear discourse, the insights and perspectives of artists, writers, filmmakers, and other "experts" in the creative faculty of the imagination are variously neglected, omitted, or marginalized. Also absent are community voices grounded in local knowledge and experience. Remedying this global tendency takes on additional importance in the Australian context in which Debenham, as a white settler, was working. For pre-nuclear knowledges such as those developed by the Aboriginal peoples on the land they have inhabited for more than 65,000 years must come prior to a singular form of nuclear knowledge derived from a twentieth-century scientific project.[50]

15 More Years Testing in the Pacific? No. is named by curators and institutions in various ways, most notably *No, 15 More Years Testing in the Pacific?*[51] Here, Debenham was responding to the French government's announcement at the time that it would continue testing for a further fifteen years – a declaration to which she was delivering a clear answer. The poster was also available in monotones of red or black on off-white paper.

Unlike the clear visual language of *No Nukes/No Tests*, here Debenham uses the contrast between the ink and the colorations of the off-white paper to great effect to produce an interplay between the text and the iconography. For instance, through a foreshore of palm trees with its impressive canopy one can catch a glimpse of the open ocean, which is punctuated by an atomic mushroom cloud on the horizon where the word "NO" is ironically positioned. The text and iconography therefore are in a dialectic with each other, wherein the wishes of the protagonist fall on deaf ears. Debenham has stated to the author in interview that "the clear and stark use of positive and negative design for the two posters [*No Nukes/No Tests* and

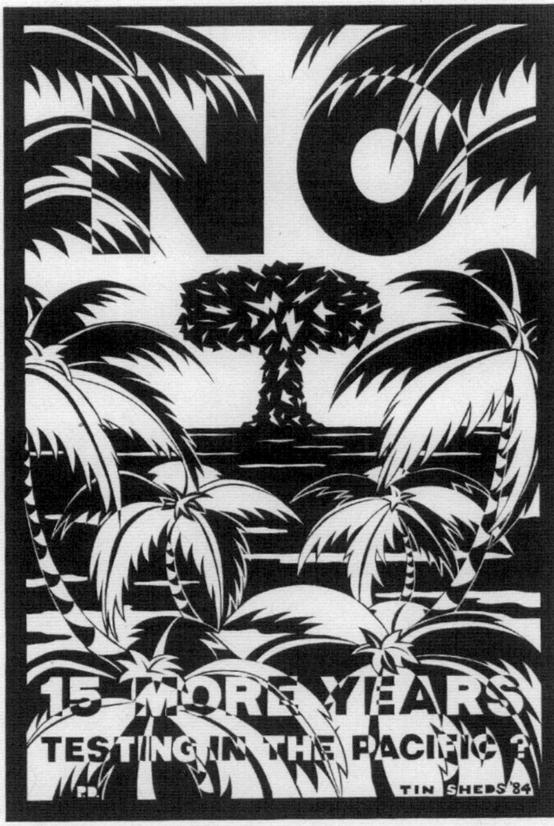

13.2 Pam Debenham, *15 More Years Testing in the Pacific?* No., 1984. Screen print on paper, 71.6 x 49 cm. National Gallery of Australia.

15 More Years Testing in the Pacific? No.] was to create a visually striking image that could be read from a distance. In creating them, I intended for the posters to be pasted on walls on the streets."[52]

Here it is movement, not the nuclear test in the background, that is foregrounded. We imagine the movement as one of violent disruption, and when read alongside other works in Debenham's oeuvre, such as *For a Nuclear Free Pacific* and to a lesser extent *No Nukes in the Pacific*, the stillness and sanctity of the places depicted – a peacefulness that was ruptured and violated by the nuclear weapons tests – is readily apparent.

In many ways Debenham's *For a Nuclear Free Pacific* stands apart from the other works examined in this chapter.[53] During a recent interview with the author, when I probed whether the work was possibly an earlier study for later works such as *No Nukes in the Pacific*,[54] Debenham confirmed that its design and production came about "as a demonstration for students on how to print."[55] Yet it is included here on the basis of its character as an exemplar of Debenham's Antipodean stance.

Depicting the tanned upper torso of a person sporting a Hawaiian shirt with a badge pinned on the lapel calling simply "FOR A NUCLEAR FREE PACIFIC," *For a Nuclear Free Pacific* foreshadows many elements found in several of Debenham's more strident later works. It is the only work that is positioned merely "for" something rather than in opposition to something (nuclear testing and nuclear colonialism) through a declaratory "NO." Indeed, although the phrase "For a Nuclear Free Pacific" does reappear as a secondary message in the footer of *No Nukes in the Pacific*, it is finessed in the latter poster as "FOR A NUCLEAR FREE AND INDEPENDENT PACIFIC." This nuance was arguably a response to New Zealand decision to establish the world's first national nuclear-free zone in 1985. This reformulation may also be viewed as a reflection on the implications of that assertion for New Zealand's disengagement from its former geopolitical alliance with Australia and the US.

A different Antipodean perspective on nuclear weapons and nuclear war is found in Nevil Shute's post-apocalyptic novel *On the Beach*. That book became an international bestseller and was made into a film by Stanley Kramer in 1959. Shute's novel focuses on the notion of nuclear winter, and in particular on the unique plight of Antipodeans, who must wait to experience a Northern War that has already taken place.[56]

Shute had not written anything about the nuclear age before he migrated to Australia to live on farmland in Langwarrin, southeast of Melbourne, in 1950. Yet within several years, his protagonist reasoned as follows:

> "But no wind does blow down right into the Southern Hemisphere from the Northern Hemisphere. If it did we'd all be dead right now."
> "I wish we were," she said bitterly.
> "It's like waiting to be hung."
> "Maybe it is. Or maybe it's a period of grace."[57]

On the Beach articulates what may be regarded as a quintessential Antipodean nuclear perspective.[58] Thus, according to Shute's biographer, Gideon Haigh:

> Shute was British. But no novel could be more explicitly Australian than *On the Beach*, set in his new hometown of Melbourne. Nor could any novel make more provocative creative use of our distance from the rest of the world: as the last habitable continent, Australia is suddenly the most important place on Earth, at the very moment of its greatest impotence and ignorance, awaiting dooming winds from an incomprehensible war in the Northern Hemisphere.[59]

Speaking at the Oxford Union in 1985, New Zealand prime minister David Lange unpacked the moral and (geo)political reasons for New Zealand to remove itself from the powerful Australia–New Zealand–United States Security Treaty (ANZUS), which his government did less than a year later as a consequence of establishing the world's first national nuclear-free zone:

> I want to pass over here the preparations which are constantly being made for the winnable or even survivable nuclear war. I would ignore those and wholeheartedly embrace the logic of the unthinkable war if it could be established that the damage which could result from the collapse of that logic could be confined to nuclear weapon states. Unfortunately, and demonstrably, it would not. We in New Zealand, you know, used to be able to relax a bit, to be able to think that we would sit comfortably while the rest of the world seared, singed, withered. We were enraptured!
>
> And the fact is that we used to have the vision of our being some kind of an antipodean Noah's Ark, which would, from within its quite isolated preserve, spawn a whole new world of realistic humankind. Now, the fact is that we know that that is not achievable. We know that if the nuclear winter comes, we freeze; we join the rest of you. And that means that there is now a total denouement as far as any argument in favour of moral purpose goes. It is a strange, dubious and totally unacceptable moral purpose which holds the whole of the world to ransom.[60]

As a study for her more widely known work, *No Nukes in the Pacific* (which was discussed at the outset of this analysis), Debenham's *For a Nuclear Free Pacific* clearly evinces an Antipodean stance. Thus, whereas in *No Nukes in the Pacific* the Hawaiian shirt motif is a subversive device for delivering her anti-nuclear message, in *For a Nuclear Free Pacific* the shirt's motif with its singular palm leaf earing performs rather a celebration of the sanctity of the Pacific lands and waters.

Conclusion

In this chapter I have made the case that an Antipodean perspective on the nuclear age – a position characterized by a firm grounding in an Oceanic imaginary or in Pacific cultures and traditions, and that is opposed to statist and anthropocentric approaches to nuclear technology – is highly ecologically and politically relevant to how we construct nuclear histories.

By introducing and critically evaluating a selection of anti-nuclear posters by the artist Pam Debenham – posters whose anti-nuclear iconography provided a radical, feminist visual politics for the alliance between the peace, environmental, and anti-nuclear movements in Australia while at the same time also finding an enthusiastic following internationally – I have sought to draw attention to the distinctiveness of an Antipodean stance in nuclear art and culture. From this emerges another narrative in which the region's art is viewed as an essential ingredient both in Antipodean nuclear thinking and in its politics – a powerful anti-nuclear politics that culminated in the establishment in 1985 of the South Pacific nuclear-weapon-free zone through the Treaty of Rarotonga. Much neglected in nuclear scholarship, the Antipodean nuclear art of individuals like Debenham serves as both a corrective and an alternative to the Anglo-American voices that continue to dominate nuclear discourse.

NOTES

1 See Taylor, "Explorations in Antipodean Nuclear Thinking"; "An Ecology of Antipodean Nuclear Art"; "Situated Nuclear Knowledges"; "The Problem of Nuclear Harm"; and "The Visual Politics of Maralinga."

2 J.R. McNeill and Corinna R. Unger's *Environmental Histories of the Cold War*, an edited volume on the relationship between the environment and the Cold War, is one exception and in fact does include a chapter that addresses the question of the "environmental impact" of Pacific nuclear testing: Merlin and Gonzalez, "Environmental Impacts of Nuclear Testing." See also Hamblin, "A Global Contamination Zone"; and Farish, "Creating Cold War Climates."

3 Giles, *Antipodean America*, 28.

4 Whether measured by yield or by quantity, most of the world's nuclear weapons tests have occurred on the sovereign territory of other states, though exclusively by Western states. For instance, while the vast majority of American tests by number were conducted in the Nevada desert, when measured by yield, more than 80 percent took place at the Pacific Proving Grounds, or what we commonly refer to as the Marshall Islands.

5 Although nuclear testing took place in the Pacific between 1946 and 1996, the radiological signature of post–c. 1952 fusion thermonuclear weapons designs is the primary focus of current stratigraphic best practice according to the Anthropocene Working Group and several of its individual participants. For this reason, Colin N. Waters et al., "Can Nuclear Weapons Fallout Mark the Beginning of the Anthropocene Epoch?," 142, have deduced that the Anthropocene's radiological signature is likely "concentrated in the mid-latitudes and highest in the Northern Hemisphere, where most of the testing occurred." See also Zalasiewicz et al., "When Did the Anthropocene Begin?"

6 McLelland, *The Report of the Royal Commission*, vols. 1 and 2.

7 International Advisory Committee, "The Radiological Situation."

8 Schelling, "Prize Lecture."

9 See McNeill, "Of Rats and Men"; Rapaport, *The Pacific Islands*; Garden, *Australia, New Zealand, and the Pacific*; D'Arcy, "Oceania"; and Lobban et al., *Tropical Pacific Island Environments.*

10 See van Wyck, *Signs of Danger*; Masco, *The Nuclear Borderlands*; Hamblin, *Poison in the Well*; Jacobs, *The Dragon's Tail*; van Wyck, *Highway of the Atom*; Brown,

Plutopia; Hamblin, *Arming Mother Nature*; Brown, *Dispatches from Dystopia*; Taylor and Jacobs, *Reimagining Hiroshima and Nagasaki*; Taylor, "The Problem of Nuclear Harm"; Brown, *Manual for Survival*; and Freeman, *This Atom Bomb in Me.*

11 See Roy, "The Politics of Death"; Jacobs, *Filling the Hole in the Nuclear Future*; O'Brian and Borsos, *Atomic Postcards*; Kaur, "The Nuclear Imaginary and Indian Popular Cinema"; O'Brian, *Camera Atomica*; Carpenter, *The Nuclear Culture Source Book*; Sharma, "Nuclear Iconography"; Marder and Tondeur, *The Chernobyl Herbarium*; Taylor, Brown, and Barkley, *Reimagining Maralinga*; and Decamous, *Invisible Colors.*

12 Clark, "Political Picture Post," 464. By way of contrast, Butler in "From Walls to Webs," counters that by the 1990s the situation had markedly changed such that: "health and safety issues, particularly the use of toxic printing inks and solvents and the proliferation of repetition strain injuries, were a factor. Costs of poster-making materials and equipment escalated; upgrading workshops to meet industry requirement was also expensive, and possible only with funding assistance from federal and state sources" (103).

13 See Kenyon, *Under a Hot Tin Roof*; Alam, "The Hothouse." Butler, in "From Walls to Webs," contends that in fact Redback Graphix in Wollongong was "the most influential workshop during the 1980s" (106), in large part due to its professionalism and volume of commissions.

14 For instance, Kenyon's *Under a Hot Tin Roof*, 6. That definitive account "is intended to be read both as an internal record of the Tin Sheds Art Workshop, and as a chronicle of the concerns of the artists, students and academics, the social and political networks, that operated in Sydney over this twenty-five year period."

15 Kenyon, *Under a Hot Tin Roof*, 9.

16 See Mayhew, "Girls at the Tin Sheds (Duplicated)"; and Yuill, *Girls at the Tin Sheds.* See also Mosman, "Girls at the Tin Sheds."

17 See Mosman, "Girls at the Tin Sheds," 183–5.

18 Debenham is featured in several global surveys of political posters and in several major collections, including a widely circulated book by Phaidon. See McQuiston, *Graphic Agitation*, and the archive of the Center for the Study of Political Graphics in California.

19 For instance, both Rod James, "Representation of the Bomb in Australian Art and Culture, 1945–1959," 12, and J.D. Mittmann, "Atomic Testing in Australia," 44, make only passing reference to Debenham in their exhibition catalogues, and only to presage the four works of Debenham's that were curated into their respective exhibitions.

20 Kenyon, "Relationships: The Next Generation," 20–1.

21 Kenyon, *Under a Hot Tin Roof*, 88.

22 Interview with Pam Debenham, 26 May 2019.

23 Ibid. See Kenyon, *Under a Hot Tin Roof.*

24 Mayhew, "Women, Womyn, Wimmin," 19.

25 Debenham, *History.*

26 Bleiker, "Mapping Visual Global Politics," 3.

27 Ibid., 1.

28 Danchev, *On Good and Evil and the Grey Zone*, 91.
29 Bleiker, "Mapping Visual Global Politics," 4.
30 Bleiker, *Aesthetics and World Politics*, 8
31 Ibid., 14.
32 Debenham, "Art Workshop Posters," 74.
33 Toni Robertson's works are also worth mentioning in this regard, although Robertson shows less restraint in employing the satirical register in her posters, and the text and iconography are frequently more parochial in content and tone, as in Robertson's *The Royal Nuclear Show*.
34 Cohn, "Sex and Death."
35 Interview with Pam Debenham, 26 May 2019.
36 Debenham, *No Nukes in the Pacific*.
37 Wallis, "Or Cry?," 92.
38 Robertson, *The Royal Nuclear Show*.
39 Stephen, "From Wilderness to Nature Strip," 7.
40 Butler, "From Walls to Webs," 105.
41 In a recent interview with the author Debenham confessed that the dressmaker did not execute her design as intended, "to save fabric," so the pattern did not align when the shirt was fully buttoned up. One dividend to arise from the autonomous decision by the manufacturer was that Debenham was left with several large samples, which were subsequently collected by the Manly Art Gallery & Museum and the Powerhouse Museum in Sydney. Later, smaller print runs of men's ties were made, and with the excess material two pieces of swimwear were made.
42 Debenham's boat in both prints more closely resembles the smaller sailing vessel "Vega" than it does the trawler "Rainbow Warrior," at least before it was given approval to fit a sailing rig in June 1984.
43 Debenham, "Press Release: Paradise Lost."
44 Debenham, "Tin Sheds Posters."
45 Debenham, "Press Release: Paradise Lost."
46 Debenham, *Still Life*.
47 Debenham, "Interview with Pam Debenham," 6 June 2019.
48 Debenham, "Tin Sheds Posters.'"
49 Debenham, *No Nukes/No Tests*.
50 For evidence that humans arrived on the Australian continent *at least* 65,000 years ago, see Chris Clarkson et al., "Human Occupation of Northern Australia." Following Muir, Rose, and Sullivan, "From the Other Side of the Knowledge Frontier," 264, situated nuclear knowledge has to contend with such encounters in the Australian context "on Aboriginal people's terms and according to Aboriginal people's customs, and not within Western frameworks."
51 Debenham, *No, 15 More Years Testing in the Pacific?*
52 Interview with Pam Debenham, 26 May 2019.
53 Debenham, *For a Nuclear Free Pacific*.
54 Debenham, *No Nukes in the Pacific*.
55 Interview with Pam Debenham, 26 May 2019.

56 See Kramer, *On the Beach.*
57 Shute, *On The Beach*, 100.
58 See Ball, "The Probabilities of *On the Beach*"; Broderick, "Fallout *On the Beach*."
59 Haigh, "Shute the Messenger," 1.
60 Lange, "Nuclear Weapons Are Morally Indefensible."

BIBLIOGRAPHY

Allam, Lorena. "The Hothouse: Art and Politics at the Tin Sheds." ABC Radio National, 24 June 2007. https://www.abc.net.au/radionational/programs/archived/hindsight/the-hothouse-art-and-politics-at-the-tin-sheds/3241670.

Ball, Desmond. "The Probabilities of *On the Beach*: Assessing 'Armageddon Scenarios' in the 21st Century." *Working Paper* no. 401, Canberra Papers on Strategy and Defence Series (May 2006).

Bleiker, Roland. *Aesthetics and World Politics*. Houndmills: Palgrave Macmillan, 2009.

– "Mapping Visual Global Politics." In *Visual Global Politics*, edited by Roland Bleiker, 1–29. Oxford: Routledge, 2018.

Broderick, Mick. "Fallout *On the Beach*." *Screening the Past* 36 (2013). http://www.screeningthepast.com/2013/06/fallout-on-the-beach.

Brown, Kate. *Dispatches from Dystopia: Histories of Places Not Yet Forgotten*. Chicago: University of Chicago Press, 2015.

– *Manual for Survival: A Chernobyl Guide to the Future*. London: Penguin, 2019.

– *Plutopia: Nuclear Families, Atomic Cities, and the Great Soviet and American Plutonium Disasters*. Oxford: Oxford University Press, 2013.

Butler, Roger. "From Walls to Webs: Contemporary Art: Contemporary Poster Art in Australia." *Art and Australia* 37, no. 1 (1999): 102–7.

Carpenter, Ele, ed. *The Nuclear Culture Source Book*. London: Black Dog, 2016.

Clark, Deborah. "Political Picture Post." *Art and Australia* 31, no. 4 (1994): 462–4.

Clarkson, Chris, Zenobia Jacobs, Ben Marwick, Richard Fullagar, Lynley Wallis, Mike Smith, Richard G. Roberts, et al. "Human Occupation of Northern Australia by 65,000 Years Ago." *Nature* 547, no. 7663 (20 July 2017): 306–10.

Cohn, Carol. "Sex and Death in the Rational World of Defense Intellectuals." *Signs* 12, no. 4 (1987): 687–718.

Danchev, Alex. *On Good and Evil and the Grey Zone*. Edinburgh: Edinburgh University Press, 2015.

D'Arcy, Paul. "Oceania: The Environmental History of One-Third of the Globe." In *A Companion to Global Environmental History*, edited by J.R. McNeill and Erin Stewart Mauldin, 196–221. Hoboken: John Wiley and Sons, 2012.

Debenham, Pam. "Art Workshop Posters." In *Pictures from the Collection*, edited by Pamela Bell, 74.University of Sydney, 1988.

– *For a Nuclear Free Pacific*. 1983. Poster. Accessed 10 January 2018. http://artsearch.nga.gov.au/Detail.cfm?IRN=101352.

– *History*. 1987. Edition of 150. https://cs.nga.gov.au/detail.cfm?irn=1121.

– "Interview with Pam Debenham." Interview by N.A.J. Taylor. Email correspondence between the Sunshine Coast, Australia, and Paris, France, 26 May 2019.
– "Interview with Pam Debenham." Interview by N.A.J. Taylor. Face-to-face interview in Paris, France, 6 June 2019.
– *No, 15 More Years Testing in the Pacific?* 1984. Screenprint, printed in colour, from one stencil from Australian Prints + Printmaking. Edition of 200. http://www.printsandprintmaking.gov.au/works/18924.
– *No Nukes in the Pacific.* 1984. Poster. Edition of 200. http://artsearch.nga.gov.au/Detail.cfm?IRN=101399.
– *No Nukes/No Tests.* 1984. http://artsearch.nga.gov.au/Detail.cfm?IRN=101397.
– "Press Release: Paradise Lost." Sydney: Tin Sheds Gallery, 1995.
– *Still Life.* 1984. http://artsearch.nga.gov.au/Detail.cfm?IRN=101398.
– "Tin Sheds Posters." In *My Head Is a Map: A Decade of Australian Prints*, edited by Roger Butler, 10. Canberra: Australian National Gallery, 1992.
Decamous, Gabrielle. *Invisible Colors: The Arts of the Atomic Age.* Cambridge, MA: MIT Press, 2018.
Farish, Matthew. "Creating Cold War Climates: The Laboratories of American Globalism." In *Environmental Histories of the Cold War*, edited by J.R. McNeill and Corinna R. Unger, 51–84. New York: Cambridge University Press, 2010.
Freeman, Lindsey A. *This Atom Bomb in Me.* San Francisco: Redwood Press, 2019.
Garden, Donald Stuart. *Australia, New Zealand, and the Pacific: An Environmental History.* Santa Monica: ABC-CLIO, 2005.
Giles, Paul. *Antipodean America: Australasia and the Constitution of U.S. Literature.* Oxford: Oxford University Press, 2013.
Haigh, Gideon. "Shute the Messenger." *The Monthly*, 6 June 2007. https://www.themonthly.com.au/issue/2007/june/1268876839/gideon-haigh/shute-messenger.
Hamblin, Jacob Darwin. *Arming Mother Nature: The Birth of Catastrophic Environmentalism.* New York: Oxford University Press, 2013.
– "A Global Contamination Zone: Early Cold War Planning for Environmental Warfare." In *Environmental Histories of the Cold War*, edited by J.R. McNeill and Corinna R. Unger, 85–114. New York: Cambridge University Press, 2010.
– *Poison in the Well: Radioactive Waste in the Oceans at the Dawn of the Nuclear Age.* New Brunswick: Rutgers University Press, 2008.
International Advisory Committee. "The Radiological Situation at the Atolls of Mururoa and Fangataufa." https://www.iaea.org/publications/4728/the-radiological-situation-at-the-atolls-of-mururoa-and-fangataufa.
Jacobs, Robert. *Arming Mother Nature: The Birth of Catastrophic Environmentalism.* New York: Oxford University Press, 2013.
– *The Dragon's Tail: Americans Face the Atomic Age.* Amherst: University of Massachusetts Press, 2010.
– *Filling the Hole in the Nuclear Future: Art and Popular Culture Respond to the Bomb.* Lanham: Lexington Books, 2010.
James, Rodney B. "Representation of the Bomb in Australian Art and Culture, 1945–1959." MA thesis, 1990.

Kaur, Raminder. "The Nuclear Imaginary and Indian Popular Cinema." *South Asia: Journal of South Asian Studies* 36, no. 4 (2013): 539–53.

Kenyon, Therese. "Relationships: The Next Generation." In *Artists as Social Commentators and Activists: Tracing the Influence of Artist/Publisher Roderick Shaw (1915–92)*, 20–1. Sydney: Manly Art Gallery, 2007.

– *Under a Hot Tin Roof: Art, Passion, and Politics at the Tin Sheds Art Workshop*. Sydney: State Library of New South Wales Press, 1995.

Kramer, Stanley, dir. *On the Beach*, 1959. http://www.imdb.com/title/tt0053137.

Lange, David. *Nuclear Free: The New Zealand Way*. London: Penguin Books, 1990.

– "Nuclear Weapons Are Morally Indefensible." Debate presented at the Oxford Union Debate, Oxford University, 1 March 1985. https://nzhistory.govt.nz/media/sound/oxford-union-debate.

Lobban, Christopher S., Maria Schefter, Frank Camacho, and John Jocson. *Tropical Pacific Island Environments*. Honolulu: Bess Press, 2014.

Marder, Michael, and Anaïs Tondeur. *The Chernobyl Herbarium: Fragments of an Exploded Consciousness*. London: Open Humanities Press, 2016.

Masco, Joseph. *The Nuclear Borderlands: The Manhattan Project in Post–Cold War New Mexico*. Princeton: Princeton University Press, 2006.

Mayhew, Louise. "Girls at the Tin Sheds (Duplicated)." Sydney: Verge Gallery, 26 March 2015. https://verge-gallery.net/wp-content/uploads/2017/03/room-sheet.pdf.

– "Women, Womyn, Wimmin: Girls in Sydney." In *Girls at the Tin Sheds: Sydney Feminist Posters 1975–90*, edited by Katie Yuill, 17–19. Sydney: University Art Gallery, University of Sydney, 2015.

McLelland, J.R. *The Report of the Royal Commission into British Nuclear Tests in Australia*, vols. 1 and 2. Canberra: Australian Government Publishing Service, 1985.

McNeill, J.R. "Of Rats and Men: A Synoptic Environmental History of the Island Pacific." *Journal of World History* 5, no. 2 (1994): 299–349.

McNeill, J.R., and Corinna R. Unger, eds. *Environmental Histories of the Cold War*. New York: Cambridge University Press, 2010.

McQuiston, Liz. *Graphic Agitation: Social and Political Graphics Since the Sixties*. New York: Phaidon, 1993.

Merlin, Mark D., and Ricardo Gonzalez. "Environmental Impacts of Nuclear Testing in Remote Oceania: 1946–1996." In *Environmental Histories of the Cold War*, edited by J.R. McNeill and Corinna R. Unger, 167–202. New York: Cambridge University Press, 2010.

Mittmann, J.D. "Atomic Testing in Australia." In *Black Mist Burnt Country*, 37–65. Upwey: Burrinja (Dandenong Ranges Cultural Centre), 2016.

Mosman, Petra. "Girls at the Tin Sheds: Sydney Feminist Posters 1975–1990." *Museum Worlds* 3 (2015): 182–6.

Muir, Cameron, Deborah Rose, and Phillip Sullivan. "From the Other Side of the Knowledge Frontier: Indigenous Knowledge, Social-Ecological Relationships, and New Perspectives." *Rangeland Journal* 32, no. 3 (2013): 259.

O'Brian, John, ed. *Camera Atomica: Photographing the Nuclear World*. London: Black Dog, 2014.

O'Brian, John, and Jeremy Borsos. *Atomic Postcards: Radioactive Messages from the Cold War.* Bristol: Intellect Books, 2011.

Rapaport, Moshe, ed. *The Pacific Islands: Environment and Society.* Honolulu: Bess Press, 1999.

Robertson, Toni. *The Royal Nuclear Show*, 1981. http://artsearch.nga.gov.au/Detail.cfm?IRN=87846.

Roy, Srirupa. "The Politics of Death: The Antinuclear Imaginary in India." In *South Asian Cultures of the Bomb: Atomic Publics and the State in India and Pakistan*, edited by Itty Abraham. Bloomington: Indiana University Press, 2009.

Schelling, Thomas C. "Prize Lecture: An Astonishing Sixty Years: The Legacy of Hiroshima." Stockholm: Nobel Media, 2005. http://www.nobelprize.org/nobel_prizes/economic-sciences/laureates/2005/schelling-lecture.html.

Sharma, Manu. "Nuclear Iconography in Indian Popular Culture." Master of Philosophy thesis, Diplomacy and Disarmament Division, Centre for International Politics, Organisation, and Disarmament, School of International Studies, Jawaharlal Nehru University, 2016.

Shute, Nevil. *On The Beach.* New York: W. Morrow, 1957.

Stephen, Ann. "From Wilderness to Nature Strip." In *The Lie of the Land* catalogue, 7. Sydney: Powerhouse Museum, 1992.

Taylor, N.A.J. "An Ecology of Antipodean Nuclear Art," 26. University of Montreal, Montreal, Canada, 2017.

– "An Ecology of Antipodean Nuclear Art." In *Cultural Enquiry Research Group*, 25. Ballarat: Federation University, 2017.

– "Explorations in Antipodean Nuclear Thinking: Val Routley/Plumwood and Richard Routley/Sylvan." In *The Seed Box: Environmental Humanities Collaboratory and The Posthumanities Hub*, 13. Linköping Linköping University, 2015.

– "The Problem of Nuclear Harm: An Ethical Ecology." PhD diss., University of Queensland, 2018.

– "Situated Nuclear Knowledges: An Ecology of Antipodean Nuclear Art." *Unlikely: Journal for Creative Arts* 5 (2018): 13.

– "The Visual Politics of Maralinga: (Re)presentations, Vulnerabilities, and Experiences." *Journal of the History of Biology* 54, no. 1 (2021): 95–106.

Taylor, N.A.J., Paul Brown, and Ellise Barkley, eds. "Reimagining Maralinga." *Unlikely: Journal for Creative Arts*, Special Issue, 2018.

Taylor, N.A.J., and Robert Jacobs, eds. *Reimagining Hiroshima and Nagasaki: Nuclear Humanities in the Post–Cold War: War, Politics and Experience.* London: Routledge, 2017.

van Wyck, Peter C.. *Highway of the Atom.* Montreal and Kingston: McGill-Queen's University Press, 2010.

– *Signs of Danger: Waste, Trauma, and Nuclear Threat.* Minneapolis: University of Minnesota Press, 2005.

Wallis, Geoffrey J. "Or Cry?" In *Got the Message? 50 Years of Political Posters*, 92. Ballarat: Ballarat Art Gallery, 2013.

Waters, Colin N., James P.M. Syvitski, Agnieszka Gałuszka, Gary J. Hancock, Jan Zalasiewicz, Alejandro Cearreta, Jacques Grinevald, et al. "Can Nuclear Weapons

Fallout Mark the Beginning of the Anthropocene Epoch?" *Bulletin of the Atomic Scientists* 71, no. 3 (January 2015): 46–57.

Yalata and Oak Valley Communities, Christobel, and Christobel Mattingley. *Maralinga: The Anangu Story*. Crows Nest: Allen and Unwin, 2009.

Yuill, Katie, ed. *Girls at the Tin Sheds: Sydney Feminist Posters 1975–90*. Sydney: University Art Gallery, The University of Sydney, 2015.

Zalasiewicz, Jan, Colin N. Waters, Mark Williams, Anthony D. Barnosky, Alejandro Cearreta, Paul Crutzen, Erle Ellis, et al. "When Did the Anthropocene Begin? A Mid-Twentieth Century Boundary Level Is Stratigraphically Optimal." *Quaternary International: The Quaternary System, and Its Formal Subdivision*, 383 (5 October 2015): 196–203.

14

The Immanation-Image: Immanent Experience and Kazakhstan's Socialist and Postsocialist Modernity in Almagul Menlibayeva's Video Installation *Transformation* (2016)

LIVIA MONNET

Renowned Kazakhstani artist Almagul Menlibayeva's monumental video installation, *Transformation*, was shown at the Salon d'honneur of the Grand Palais in Paris from 17 December 2016 to 2 January 2017. The installation consisted of three multiscreen videos, two of which were made for this show, as well as a huge floor carpet designed by the artist. Through visually arresting compositions, a stunning electronic soundtrack produced by her long-time collaborator OMFO (German Popov), and an intricate non-linear narrative, Menlibayeva's experimental work critically explores Kazakhstan's Soviet and post-socialist history and culture.

This chapter argues that *Transformation's* reimagining and re-enactment of several dramatic episodes in Kazakhstan's socialist and postsocialist modernity – episodes such as the Soviet nuclear tests at the Semipalatinsk Test Site – is predicated on the immanent perspective and experience of human and non-human subject formations situated in the northeastern part of the country as well as on a more globalized imaginary. The human and non-human subject formations and configurations whose immanent experience is highlighted in the installation include witness-survivors of the Soviet nuclear tests, nuclear and medical scientists, geologic or geographic configurations such as the steppe, uranium and other materials used in the production of nuclear weapons and nuclear energy, and toxic ecologies produced by the nuclear and petrochemical industries. As will become clear in my discussion in the central sections of this chapter, Menlibayeva situates these figures both as nodal points and concepts and as revealing material-discursive configurations in various assemblages of performative image-events I will call *immanation-images*. I will also show that the feminist decolonial aesthetic of these assemblages reveals differential variations of immanent experience (e.g., alterlife, toxic sovereignty, subaltern endurance, technoscientific embeddedness) as well as imaginaries that were critical to the functioning of Soviet industrial-colonial modernity, or that have played a central role in post-Soviet Kazakhstan's transition to and gradual consolidation of a neoliberal capitalist economy.

The next section of the chapter briefly describes Menlibayeva's installation and situates her work in the context of contemporary Central Asian art. Subsequent sections respectively offer a survey of Kazakhstan's modern history; close readings of the artist's reimagining of her native country's experience of Soviet socialism and the Cold War, and of its transition to neoliberal capitalism; and, finally, *Transformation*'s probing of nuclear technoscience as well as its vision of a(n) (im)possible decolonial future. These readings are enabled and mediated by the experimental assemblages of the immanation-image evoked above.

Almagul Menlibayeva, *Transformation*, and the Emergence of Post-Soviet Central Asian Art

A multimedia artist working in video installation, film, photography, and painting, Almagul Menlibayeva (b. 1969) was educated at the Academy of Art and Theatre in Almaty, Kazakhstan. From 1988 to 1995 she was a member of the underground artists' group Green Triangle in Almaty. A recipient of several awards, including the French Ministry of Culture Chevalier de l'Ordre des Arts et des Lettres (Knight of the Order of Arts and Letters) and the 2013 Main Award of the International Film Festival Kino der Kunst in Munich, Germany, Menlibayeva divides her time between Germany and Kazakhstan. Her work has been exhibited in numerous solo and group shows in Germany, France, Kazakhstan, Russia, Austria, Turkey, the United States, the United Arab Emirates, South Korea, and Taiwan.[1]

In the form in which it was shown at the Salon d'honneur of the Grand Palais in Paris, *Transformation* consisted of three interrelated installations: *Kurchatov 22* (2013), an eight-channel video installation; *Tokamak* (2016), also an eight-channel video installation; and *Astana* (2016), a three-channel video installation. A floor carpet designed by the artist that evoked the traditional felt rugs of Central Asian nomadic tribes completed the installation. *Kurchatov 22* and *Tokamak* were projected in succession in a loop on five screens at the front of the Salon d'honneur and on the walls of the vaults overlooking the screens (one above and two on each side of the five screens). The three large screens on which the *Astana* video was projected were suspended from the ceiling. *Kurchatov 22* is a 2013 installation to which two new videos were added; *Tokamak* and *Astana* were created for the

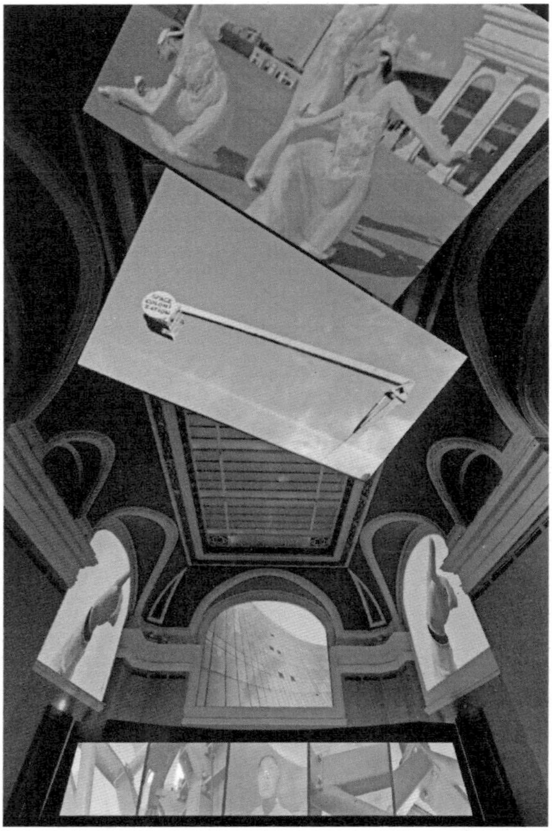

14.1 Almagul Menlibayeva, *Transformation*, installation in the Salon d'honneur, Grand Palais, Paris, 17 December 2016–2 January 2017.

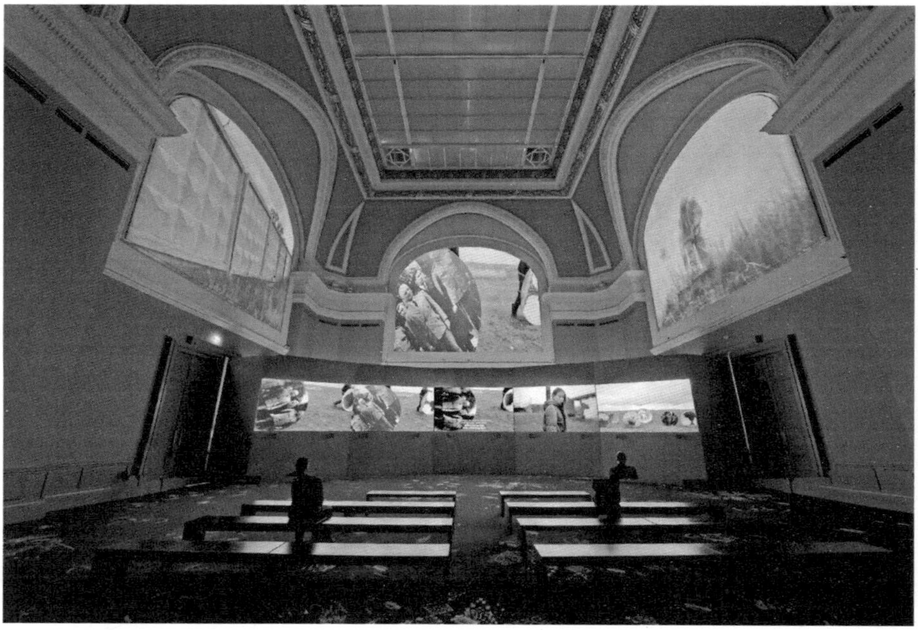

Grand Palais solo show. *Kurchatov 22* revisits the legacy of the Soviet nuclear tests at the Semipalatinsk Test Site. *Tokamak* uses the tropes of science fiction to question the research on nuclear fusion currently being conducted at the National Nuclear Centre (NNC) in Kurchatov. *Astana* highlights the construction of Nur Alem, Kazakhstan's national pavilion at *Expo 2017: Future Energy*, which was held in Kazakhstan's capital, Nur-Sultan (formerly Astana), from 10 June 2017 to 10 September 2017.

Menlibayeva shares her interest in reimagining Kazakhstan's and Central Asia's (pre)modern history and culture, and in questioning the nation-building policies and state capitalism of the region's authoritarian post-Soviet regimes, with a host of local independent artists and diasporic artists hailing from the region. Despite the growing number of large group exhibitions and solo shows in recent years featuring the work of contemporary Central Asian artists, the post-1991 art of the former Soviet republics of Kazakhstan, Uzbekistan, Tajikistan, Kyrgyzstan, and Turkmenistan remains little-known. As Diana T. Kudaibergenova, Madina Tlostanova, Georgy Mamedov, and other critics have noted, contemporary independent art in Central Asia represents a forum for critical thinking and alternative views on a variety of contemporary issues: post-Soviet nationalist agendas and regime-legitimating multiculturalism and globalism; widespread corruption, negligence, and shaky infrastructure in the public sector; the censorship and repression of oppositional groups; the tension between state-sponsored, propagandistic art and independent artists' ironic or fantastic reimaginings of historical traditions such as nomadic pastoralism, shamanism, animism, and Tengrism; globalized pop and consumer culture; the devastating environmental legacies of Soviet-era nuclear testing and industrial megaprojects and of contemporary extractive and manufacturing industries; and local perspectives on gender, the body, and ethnicity.[2]

Kazakhstan's Modern History and Nuclear Experience

HIGHLIGHTS OF KAZAKHSTAN'S MODERN HISTORY

Traditionally a pastoral, nomadic population, the Kazakhs came to dominate the Central Asian steppes in the sixteenth century. In the nineteenth century the three main Kazakh hordes (*zhus*) were incorporated into the Russian empire, and their territories were gradually colonized by Russian European and Ukrainian peasants. The idea of a Kazakh nation and a modern Kazakh state emerged among the Russian-educated Kazakh intelligentsia in the late nineteenth and early twentieth centuries. In 1936 Kazakhstan formally became a Soviet republic. A range of brutal policies inflicted by the central government in Moscow upon nomadic and semi-nomadic livestock herders in the Kazakh SSR – dispossession of farmsteads, forced collectivization, harshly imposed procurements of meat and grains – brought about the great Kazakh famine (1930–33) and ensuing epidemics and protests, as well as the mass slaughter of starving refugees. It has been estimated that more than a million people died during that famine. As historian Sarah Cameron has documented, these tragic events decimated the Kazakh population and consolidated Soviet Kazakhstan's national identity and its integration into the nation-building and modernization projects of the Soviet Union in the interwar decades.[3] During the Second World War, large numbers of Germans, Muslims, and Crimean Tatars were deported to Kazakhstan from the North Caucasus. In the postwar decades, Kazakhstan assumed a major role in the Soviet Union's military build-up and in its massive programs for boosting agricultural and industrial production. The principal site for the Soviets' nuclear weapons development and testing was established in Kazakhstan; so were large detention and labour camps (gulags), as well as the Baikonur Cosmodrome, which has been the launch site for Soviet and, later, Russian space missions since 1957. The Virgin Lands Campaign, a program initiated by First Communist Party Secretary Nikita Khrushchev with the aim of alleviating food shortages in the Soviet Union through greatly expanded cereal production, was launched on the grasslands of northern Kazakhstan in 1954. The overuse of pesticides and fertilizers and huge irrigation projects undertaken under this program led to the drying up of most of the Aral Sea.

With the collapse of the Soviet Union, Kazakhstan became a major player in Central Asia and declared its independence in December 1991. Under President Nursultan Nazarbayev (1991–2019), Kazakhstan became a major player in Central Asia and a leading actor in regional and global nuclear disarmament and non-proliferation treaties and initiatives. The country's economy has grown rapidly in the twenty-first century, fuelled by huge exports of oil, natural gas, and agricultural and chemical products, as well as by the creation of favourable conditions for foreign investment. In his State of the Nation address on 15 December 2012, Nazarbayev announced the *Kazakhstan 2050 Strategy*, which called for sweeping economic, social, and political reforms as well as policies aimed at positioning Kazakhstan among the world's thirty most developed economies by mid-century. The proposed policies included the development of an advanced knowledge economy, as well as infrastructure development under the Nurly Zhol ("the Path to the Future") program.[4]

LEGACIES OF SOVIET NUCLEAR TESTING AND KAZAKHSTAN'S PLANS FOR NUCLEAR DEVELOPMENT

The story of the Semipalatinsk Nuclear Test Site and its devastating legacies is well-known, but it bears retelling here given its central significance in Menlibayeva's installation. The Soviet Union tested its first atomic bomb at the Semipalatinsk Nuclear Test Site (STS) on 29 August 1949. An area of about 18,000 km^2 in

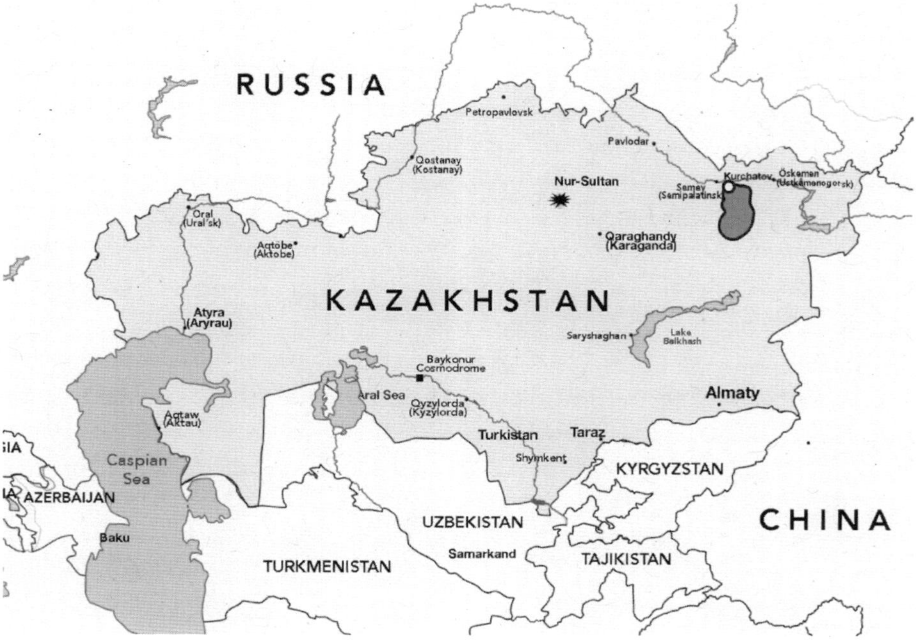

14.2 Map of Kazakhstan with the Semipalatinsk Nuclear Test Site, redrawn by Almagul Menlibayeva.

northeastern Kazakhstan, the STS would be the site for most Soviet nuclear tests during the Cold War.

The headquarters for the research and military operations on the STS was the closed city of Kurchatov (also called Semipalatinsk 21, Moscow 400, and other names). Like all Soviet secret military-scientific cities, it appeared on no maps. Between 1949 and 1989 the Soviets conducted around 456 nuclear tests on the STS, 116 of which were above-ground. After signing the Limited (Partial) Nuclear Test Ban Treaty (LTBT/PTBT) in August 1963, the Soviet Union conducted an estimated 340 underground nuclear tests on what was known as the Polygon. The entire population of the STS, including state farm workers who had been mobilized under the Virgin Lands campaign, were exposed to substantial fallout and radiation from the atmospheric and underground tests. While notifications of upcoming tests by the military and orders for partial evacuations of local villages were not uncommon (especially during the period of above-ground tests), the population was never informed of the real nature of the "military exercises" taking place on the Polygon or about the health hazards of radiation exposure. In 1957 the Soviet government began carrying out secret radiological and medical research on the effects of radiation on human, animal and plant populations in the villages near the Semipalatinsk Test Site. Much of this research was conducted in the city of Semipalatinsk (now Semey) at a secret clinic called Anti-Brucellosis Dispensary no. 4.

Through Mikhail Gorbachev's *perestroika* policies, the people of the Kazakh SSR, including residents of the STS area, learned for the first time about the nuclear tests and their environmental and health consequences. A powerful anti-nuclear movement took shape in February 1989 under the leadership of Olzhas Suleimenov and other activist intellectuals and artists. That movement joined forces with the anti-nuclear protests held at the Nevada Test Site (now the Nevada National Security Site). Together, the Nevada–Semipalatinsk anti-nuclear movement adopted a broad platform that aimed to achieve an immediate halt to nuclear testing, the clean-up and restoration of the test sites, and compensation for the populations affected by nuclear detonations. In December 1990 the Kazakh SSR parliament passed a bill banning all nuclear testing in the republic. In August 1991 Kazakhstan's president Nursultan Nazarbayev officially closed the Semipalatinsk Test Site.[5]

In the three decades since its closing, the STS has become the most thoroughly studied former nuclear test site in the world. Researchers estimate that 1.5 million people in Kazakhstan have been affected by the Soviet nuclear tests. Studies have found that communities on the Polygon and in surrounding areas have higher rates of cancer, genetic mutations, and birth defects than those in other parts of Kazakhstan. However, the relationship between the ill health of a significant part of the population in the East Kazakhstan Region (East Kazakhstan Oblast, where the STS is situated) and radiation exposure during and after the nuclear testing period is a contested issue, and much more long-term, broad-based research remains to be done.

Post-independence Kazakhstan's engagement with nuclear energy has amounted to a rather confused balancing act between nation-building, seeking adequate ways to address the legacies of Soviet-era nuclear testing, active international campaigning for global nuclear disarmament and non-proliferation, and ambitious plans for nuclear development. Post-Soviet Kazakhstan's new national identity has been built not only through the revisiting of narratives and tropes of Kazakhness, Eurasianism, nomadic pastoralism, and pre-Islamic religious beliefs, but also through a strong emphasis on nuclear victimhood (owing to the traumatic experience of the Soviet nuclear tests). The government of Kazakhstan has appealed to international institutions and organizations such as IAEA (International Atomic Energy Association) and WHO (World Health Organization) for assistance in assessing the environmental and health effects of radiation in the Semipalatinsk and East Kazakhstan regions. It has established three concentric radiation exposure zones, as well as four categories of exposed population groups, which are entitled to a comprehensive compensation and benefits program. The implementation of this program was delayed throughout the 1990s, and the compensation amounts and state pensions that have been or continue to be paid to these groups and their descendants are not enough to live on.[6]

In terms of nuclear disarmament and non-proliferation, Kazakhstan has signed the Treaty on the Non-Proliferation of Nuclear Weapons (NPT) as well as the Comprehensive Nuclear Test Ban Treaty (CTBT) and has ratified the Partial Nuclear Test Ban Treaty (PTBT). It has also ratified the Central Asian Nuclear Weapons Free Zone (CANWFZ), which was established through the Treaty of Semipalatinsk (Treaty of Semey), signed at the former Semipalatinsk Test Site on 8 September 2006 by all five post-Soviet Central Asian republics (Kazakhstan, Uzbekistan, Tajikistan, Kyrgyzstan, and Turkmenistan).

Kazakhstan is currently the world's leading uranium producer. It is also the host of an IAEA-sponsored low-enriched uranium (LEU) fuel bank, which can supply 90 tons LEU for nuclear power production to any state in need of uranium fuel if that fuel cannot be accessed from other sources.[7]

By way of establishing a framework for my readings of Menlibayeva's powerful *Transformation* installation, I offer below an outline of the conceptions of immanence, immanation, and immanent experience that fed my understanding of these notions. Building on this philosophical groundwork, I then propose a definition of global neoliberal capitalism's radical ecology of immanence. The concluding part of this section of the chapter engages with the notion of the immanation-image.

Approaches to Immanence, Immanation, and the Immanation-Image

CONTEMPORARY THEORIES AND PHILOSOPHIES OF IMMANENCE

A central notion in Deleuze's philosophy of transcendental empiricism, immanence receives one if its most radical elaborations in "Immanence: A Life." In this short essay, pure or absolute immanence is defined as an indefinite life, a plane of immanence "that is no longer immanent to anything other than itself ... an immanent life carrying with it the events or singularities that are merely actualized in subjects and objects ... the immensity of an empty time where one sees the event yet to come and already happened, in the absolute of an immediate consciousness."[8] The absolute horizon from which thought emerges, immanence in Deleuze's thought is also the virtual plane or process of life's potential, the infinite movement, or duration constituting the events, intensity, becomings and between-times of existence as a problem of life and non-life.

In *99 Theses on the Revaluation of Value*, Brian Massumi weaves a complex, incisive argument on the urgent need to retrieve the concept of value from neoliberal capitalism's deadly quantifications and to articulate a framework for a collective struggle for a post-capitalist future. Massumi argues that globalized neoliberal capitalism is a planetary ecology that has accelerated and aggravated the terraforming impact of twentieth-century capitalist modernity. A speculative machine driven by the excess of capitalist surplus-value, the capitalist system in its globalized neoliberal form has produced the individuating, asubjective subjectivity of human-capital. While the immanence or processual ecology of the neoliberal capitalist system is not absolute or all-encompassing – one is immersed in it but not completely determined by it – there is no position of ethico-political purity outside capitalism from which to critique it. Complicity is an ontological condition in neoliberal capitalism.

Massumi further contends that a major factor in neoliberal capitalism's constant reinvention of itself, and in its unabated advance into all geographies and modes of existence, is its appropriation and conversion of the surplus-value of life – the qualitative experience of life's potential and intensity – into quantitative capitalist surplus-value. Capitalism, warns Massumi, "has its hand on the pulse of life." To recapture the surplus-value of life and lay the groundwork for a non-market-based, post-capitalist alter-economy and a post-capitalist future, techniques and tactics of counter-ontopower such as creative duplicity and creative experimentation with autonomous micro-economies based on alternative systems of value and governance are necessary.[9]

In a recent essay on the relationship between immanence and ecology, Hanjo Berressem argues that a logic of immanence – which, following Deleuze, he defines as the condition of being simultaneously a part of and different from one's milieu – constitutes the foundation of ecological thought, as well as the premise of an ecological attitude (of being adequate to the world). Berressem further contends that immanence should be seen as an ecology, the given to which humans as well as all living and non-living matter are immanent while at the same time trailing this given's givenness behind. This ecology is a gift in the sense of *pharmakon* – a condition that makes life possible while at the same time harbouring great risks for the living. A non-sentimental, non-totalizing immanent ecological attitude must insist on the infinite multiplicity, complexity, and positivity sustaining any ecology of immanence. The infinite multiplicity and interconnectedness of the world, Berressem goes on to argue, may be seen as a given, as the inherent value of the world system. The ethics of an ecology of immanence sees the world, nature, and the cosmos not as a friendly, benevolent milieu, but rather as change, choice, potentiality, and becoming, as the abstract machine that sustains the processes of the world.[10]

Before moving on to my notion of a radical ecology (and ecosophy) of immanence, and the notion of the immanation-image, I would like to briefly introduce Austrian philosopher Arno Bohler's theorization of immanation.

Bohler defines immanation as "a life of immanence within itself." Building on Spinoza, Deleuze, and other thinkers, Bohler argues that all expressions of immanation are immanent to the plane of immanence. Wherever and whenever a body is born or something takes place at a place in space, that body, or the event of taking place, starts to immanate. Immanation is change, becoming, the advent of the new, an object's or a body's experience of or affection or traversal by the plane of immanence. Immanation, or the life of immanence within itself, is indistinguishable from the life, becomings, affects, and relations of and among bodies or populations within the void. These populations express at once the plane of immanence they populate and themselves.[11]

TOWARD A DECOLONIZING ECOSOPHY OF GLOBAL CAPITALISM'S ECOLOGY OF IMMANENCE

While I subscribe to the notions of immanence proposed by Deleuze, Massumi, and Berressem, the understanding of a *radical ecology of immanence* I advocate in this chapter attempts to steer these thinkers' conception of immanence into a riskier, more deeply existential, and experiential zone. I argue that the thinking of a *radical ecology of immanence* – which I envision as the inscription on the earth systems and the biosphere of a planetary ecology developed and gradually globalized by European colonialism and industrial and postindustrial capitalist modernity, with all its systems of governance, despoilment, and ravagement (including slavery, genocide, racism, wars, biopower, and necropower) – in the current context of climate emergency and runaway neoliberal capitalism needs itself to be radical. Deploying a visionary, post-capitalist immanent critique of local infrastructures and material-discursive articulations of global extractive and finance capitalism, *a radical ecosophy or counter-theory and counter-practice of the latter system's radical ecology of immanence* must also develop a decolonial or decolonizing critique of the very notions of immanence, life, and ecology. It must acknowledge neoliberal capitalism's geo-engineering of the earth systems as well as its systematic depredation and depletion of living ecosystems, which make global catastrophe and mass extinction unavoidable. The ethics, politics, and aesthetics of a radical ecosophy of immanence should also remain alert, as Massumi has warned, to global capitalism's mechanisms for capturing and converting the qualitative experience of life's potential into quantitative capitalist surplus-value. A decolonial ecosophical pragmatics that simultaneously reveals and subverts global capitalism's radical ecology of immanence also needs to find ways to account for and resist this ravenous system's biopolitics of disposability.[12]

As Massumi has suggested in his *99 Theses*, collective experiments in building post-capitalist alter-economies need to adopt affirmative, experimental techniques and practices of creative duplicity. A decolonial, post-capitalist tactics of creative duplicity will need to rethink *sovereignty as immanent sovereignty, life as alterlife, and resistance as endurance or duplicitous resistance.* A radical decolonial theory, practice, and struggle for building a decolonial, post-capitalist future and alter-economy will not only honour the infinite multiplicity and interconnectedness of life and non-life/matter, of the planet and the cosmos, but also remain cognizant of the earth's and exoplanetary systems' *radical alterity*: the fact that these systems did not emerge and evolve for us and will long outlast humans and most living species; the fundamental reality of life's non-universal, localized nature; and, finally, the fact that life, or survival at all costs, may not be the best option for (and may not be possible in) a post-capitalist future due to yet another ecology of immanence – toxicity. Anthropogenic toxicities – ionizing radiation from nuclear weapons testing and nuclear energy production, CO_2 emissions from petrochemical industries, chemical environmental pollution from industrial wastes and pesticides, microplastics – constitute an ecology of immanence that is inextricably enmeshed with any experimental practice, ethics, and struggle for laying the groundwork for a decolonial, post-capitalist future and its radical ecology of (immanent) practices. The harmful effects of this toxic ecology of immanence – its slow violence and the varied ways in which its toxic processes and relations transform, sicken, and over time kill bodies, ecosystems, and the environment – may be described as its immanation, its (deadly) immanent "life" or becoming.

AN AESTHETICS OF THE IMMANATION-IMAGE

In view of the above observations, the notion of a radical ecosophy that attempts to resist and subvert global capitalism's radical ecology of immanence might be envisioned as an immanent, decolonizing theory and critique doubled by a feminist, anarcho-ecosocialist practice for imagining and building a (perhaps impossible, compromised, or contaminated) decolonial postcapitalist future. The notion of the *immanation-image* I propose in this chapter may be defined as a performative image-event that attempts to capture or make perceptible living and non-living existents' immanent experience of the radical ecology of immanence of global capitalism and colonial modernity as well as these beings' immanation in the virtual ecology of thought, time, chaos, and the infinite multiplicity and interconnectedness of the world. *An aesthetics of the immanation-image* in the context of the radical ecosophical practice evoked above may be theorized as an assemblage of immanent, performative image-events expressing both sensuously and non-sensously (through thought and affect), virtual, ontogenetic, experiential, and ecosocial "lives": the lived experience of a militarized and racialized, globalized (neo)colonial capitalist system; the immanence, duration or half-life, and transformative effects of ionizing radiation and chemical pollution; the plane of immanence of specific geologic formations or climate patterns; the embodied intensity, pain, and mutant forms of alterlife (i.e., life already altered or open to alteration). Encompassing a range of forms and articulations such as the *radiation/radioactive immanation-image* and *the re-enactment immanation-image*, these performative image-events, as I will show in my discussion of the three video installations constituting *Transformation*, may also feature a character or figure embodying the function of a fractal formation or abstract machine I shall call the *immanation strange attractor*. In chaos theory the strange attractor is a hidden fractal structure or shape in a chaotic non-linear system "that reflects all possible states within which the system can be found, regardless of the time sequence … A continuous process of compression and folding of the phase space," the strange attractor occupies an apparently circumscribed area of that space, yet it cannot be specified or delimited because it unfolds endlessly within the latter. The invisible fractal structure of the strange attractor in a chaotic system reveals a kind of paradoxical stability within the system. It makes possible the discerning and mapping of the potentiality developing within a chaotic system and its differentiation from random noise.[13]

The *immanation strange attractors* featured in the three videos constituting *Transformation* are embodied both by actual individuals interviewed by Menlibayeva and by actors performing roles and functions extracted from the imaginaries and histories of Soviet socialist and neoliberal capitalist modernity. Whether in the form of the *subaltern nuclear witness-survivor/hibakusha endurant* as embodied by actual survivors of the Soviet nuclear tests in *Kurchatov 22*, or as the figure and social function of the *embedded scientist, the independent artist* or the *baksy shaman* as respectively enacted in several performances in *Tokamak* and *Astana* (see below), the virtual yet real immanation strange attractors in these videos are visualized as intensive morphological expressions of the immanent potentialities, differences, fractures, and crises that have emerged in the chaotic system of global industrial capitalism/modernity. They render visible, both through carefully orchestrated retrospective re-enactments and through speculative scenarios, the thought experiments, legacies, and differential immanent experience of twentieth- and twenty-first-century industrial modernity as it played out in Kazakhstan (and in the larger geographies and territorializations of global capitalism).

The following sections examine a sample of the forms of immanent experience, as well as the strange attractor figures and assemblages of immanation-images embodying this experience, that appear in *Kurchatov 22*, *Tokamak*, and *Astana*.

The Memory and Immanent Experience of Soviet Modernity in *Kurchatov 22*

TESTIMONIAL NARRATIVES AND THE SUBALTERN SURVIVOR-WITNESS (HIBAKUSHA) ENDURANT

Among the three videos making up Menlibayeva's *Transformation* installation, it is *Kurchatov 22* that most visibly and insistently engages in reimagining Kazakhstan's Soviet and post-socialist history from the perspective of surviving witnesses of the Soviet nuclear tests at the Semipalatinsk Test Site. While the video features interviews with these survivor-witnesses or *hibakusha*[14] and re-enactments of their experience, these interviews and re-enactments also offer a critical re-envisioning of other dramatic episodes of Kazakhstan's Soviet history such as the deportations of ethnic minorities before, during, and after the Second World War and the Virgin Lands campaign.

Kurchatov 22 was shot in the city of Kurchatov, which during the Soviet period was the secret headquarters of the Soviet nuclear weapons testing program; in the village of Semiyarka, about 50 kilometres north of Kurchatov; in Chagan, a former secret military town; and in other locations in the Semipalatinsk Test Site area. It mixes testimonies by surviving witnesses of the nuclear tests with performances (enacted by young people from Semiyarka and Kurchatov) that re-create the local population's experience of living with above-ground and underground nuclear blasts for forty years. The video also features long stretches of the steppe in the Polygon area, scenes shot among the ruins and rubble of Soviet buildings and structures in the towns of Kurchatov and Chagan, and scenes shot in Kurchatov's National Nuclear Centre.

In a particularly telling sequence, the camera captures the testimony of an aged female witness-survivor of the Soviet nuclear tests from the village of Semiyarka. Sitting next to her disabled grandson in her home, the woman recounts how she was caught by surprise by a nuclear test that was conducted back in 1955, when she was pregnant with her first son. She states that when the bomb was detonated, she felt as if she was traversed by fire, and that the nuclear blasts often destroyed parts of residents' homes as well as infrastructures and public buildings such as the village school and the House of Culture. She also testifies to the inconsistent or contradictory instructions and disinformation that were conveyed to residents of the village of Semiyarka and neighbouring villages in the Polygon area before, during, and after nuclear detonations: the villagers were either told to run out and lie face down, or stay indoors, or rush toward the end of the village – only to be immediately told to rush back toward the village centre.

The old woman's narrative, the awkward, disconnected movements and body language of her grandson, and the disquieting framing of the scenes in which they appear suggest that the young man's disability is a result of disruptions in the genetic material caused by his grandmother's direct exposure to radiation from nuclear testing during pregnancy.[15]

Other testimonies by witness-survivors of the Soviet nuclear tests featured in *Kurchatov 22* also attest to the violence of the nuclear detonations, to the locals' exposure to radiation imposed by the military authorities supervising the tests, and to the secrecy and disinformation surrounding the latter. For instance, in one scene a bus driver recounts how his teacher at the local primary school would instruct his students to leave the classroom and lie down during a particular test, and how he would resume teaching as soon as the test was over. Another witness-

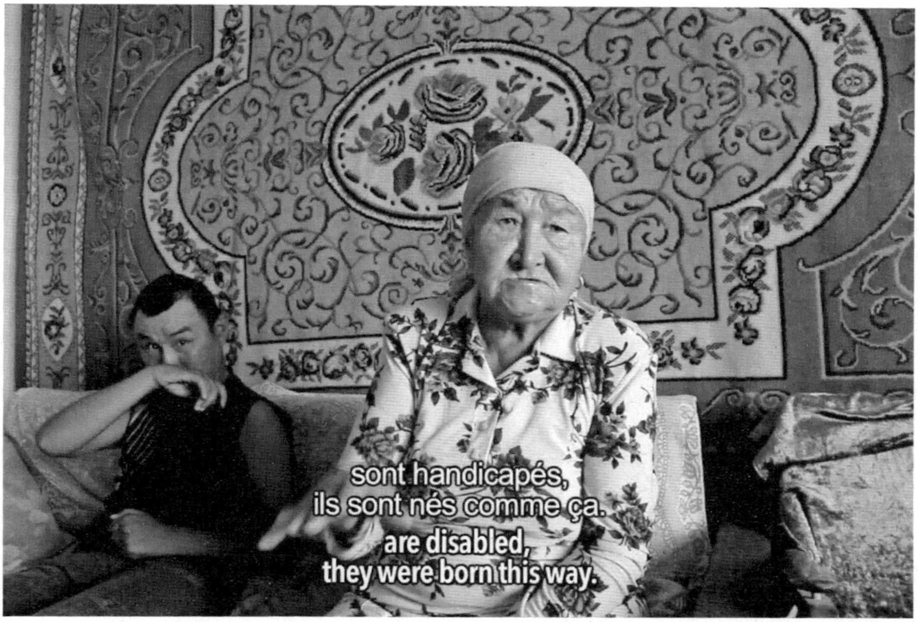

14.3 Old woman testifying, her grandson sitting next to her. Kurchatov 22, *Transformation*, 2016.

survivor from the village of Semiyarka who appears in the video sums up the terrifying, decades-long experience of living with the Soviet nuclear tests in a brief statement accompanied by a deep sigh: "What can you take from those years? It [the testing] made us all disabled."

The sequences featuring testimonies of survivor-witnesses of the Soviet nuclear tests may be described as assemblages of *testimonial immanation-images*. These testimonial image-events attest to the differential yet often painful and disabling reality of *alterlife* – which Michelle Murphy, as we saw in the introduction to this volume, has theorized as a figure of life indexing on the one hand the latter's shaping and alteration by global capitalism and by its unavoidable coexistence with toxic exposures, and on the other hand living beings' capacity to act and resist beyond the individualized body[16] – as experienced by residents of the former Semipalatinsk Test Site area during the four decades of Soviet nuclear tests and in the post-socialist period. The forms of alterlife that are documented or implied in these testimonial assemblages range from illnesses and injuries suffered by the first generation of survivor-witnesses, to the various disabilities and birth defects manifest in the second, third, and fourth generations of the offspring of these *hibakusha*, to genetic malformations and mutations of local animals and plants (including farm cattle, sheep, and goats as well as cereals). *Kurchatov 22*'s assemblages and montages of testimonial narratives of survivor-witnesses of the

Soviet nuclear weapons tests also attest to the subaltern mode of existence of these population groups as *test subjects* during the Cold War nuclear testing and as *rural hibakusha endurants* who have been abandoned both by the state and by local authorities in the post-socialist era. As Magdalena Stawkowski, Susanne Bauer, and other scholars have shown, not only were the rural populations in the Semipalatinsk Test Site area studied as experimental subjects by Soviet military and scientists, but in the post-socialist period residents of most villages in the region have been regarded as backward and unfit to adapt to neoliberal economic policies and capitalist consumerist lifestyles.[17]

The narratives of witness-survivors of the Soviet nuclear tests featured in *Kurchatov 22* also include stories by members of ethnic minorities. One such narrative is recounted by a Chinese-Russian former *sovkhoz* (large state farm) brigade leader, Nikolai Nikolaevich Van See Yu. Another testimony is that of Maria Ivanovna Felik, who as a German Russian was deported to Kazakhstan from her native Volga region. These accounts by two *hibakusha* members of the Chinese and German ethnic minorities respectively allude to the mass deportations of ethnic minorities to Central Asia and Siberia under Stalin's totalitarian regime as well as to the Virgin Lands program (which was itself accompanied by massive transfer of workers from other parts of the Soviet Union). The deportations, forced resettlement, and internment in labour camps of millions of people during Stalin's regime have been described as ethnic cleansing or genocide. Even though it produced spectacular crops in its early years, the Virgin Lands program (1954–1963) resulted in soil erosion and degradation, the disappearance of the Aral Sea, deforestation, the drying up of rivers, desertification, and the pollution and contamination of large tracts of land. In the Semipalatinsk Test Site area this program's "scientific" methods of intensive industrial agriculture left a legacy of long-lasting contamination that has compounded the already devastating effects of the nuclear tests.[18]

The testimonial narratives of Felik and Nikolai Nikolaevich Van See Yu also highlight that the nuclear detonations at the Semipalatinsk Test Site affected not only the Kazakh population but also resettled minorities as well as those who worked on state farms during the Virgin Lands program (who included labour camp inmates). This is an explicit corrective to post-socialist Kazakhstan's dominant narrative of mostly Kazakh nuclear victimhood. In conjunction with other nuclear witness-survivor narratives featured in *Kurchatov 22*, the testimonies of Felik and Nikolai Nikolaevich also call attention to the central role of the immanation-strange-attractor figure of the *subaltern hibakusha endurant* in Menlibayeva's revisionist tale of Soviet modernity. Building on the work of Elizabeth Povinelli – who has described at length the *endurance* and alterlife, or the stubbornly persisting mode of existence, continuance, and world-making, of some Aboriginal groups in abandoned toxic spaces in the Northern Territory of Australia who have resisted capture by and integration with racial neoliberal capitalism – the immanent experience and mode of life of exposed and abandoned rural populations in the former Semipalatinsk Test Site area may be theorized as *toxic endurance* and *duplicitous resistance*. Like the Aboriginal endurant described by Povinelli, the subaltern *hibakusha endurant* featured in *Transformation* is neither allegorical nor heroic, but rather an intensive, both diagnostic and symptomatic decolonial figure epitomizing perseverance in a way of living otherwise or living as resistance in the (mostly polluted and contaminated) cracks and fissures of contemporary Kazakhstan's neoliberal political economy.[19]

RADIOACTIVE CONTAMINATION, RE-ENACTMENT, RUINATION: THE IMMANATION-IMAGE AS THEATRE OF THE NUCLEAR UNCANNY, TOXIC ENDURANCE, TOXIC SOVEREIGNTY, AND SOCIO-ECOLOGICAL DESTRUCTION

In another scene in *Kurchatov 22* we see a young woman in a dotted brown dress running in terror through rows of empty, decaying Soviet apartment blocks while the horizon is illuminated by the blinding white flash of an out-of-frame nuclear explosion.

A similar sequence shows the same young woman sitting among the same post-Soviet ruins and turning repeatedly to look at another simulated nuclear detonation.

Shot in the abandoned military town of Chagan, these scenes perform simultaneously four functions: visualization and (virtual) instantiation of radioactivity; re-enactment of witness-survivors' experience of the Soviet nuclear tests and of lived socio-ecological ruination; immanent subaltern or decolonial critique; and toxic endurance as resistance and toxic sovereignty as resurgence. As I will show in this section, what enables these sequences to re-enact these forms of experience are three types of mutually constitutive immanation-images: the *radiation(or radioactive)-immanation-image*, the *re-enactment immanation-image*, and the *immanation-image of ruination*.

The scenes featuring the young woman in a brown dotted dress running in panic or sitting among the desolate ruins of the former secret military town of Chagan in *Kurchatov 22* clearly attempt to visualize the radioactive contamination of the Semipalatinsk Test Site area (as well as, implicitly, other toxic legacies in which this radiation ecology is enmeshed). Re-creating the embodied experience, memory, and positionality of local *hibakusha* witness-survivors of the Soviet nuclear tests, as well as the subjectivity and lived experience of these survivors' descendants and of other residents, these two scenes articulate a powerful immanent critique of the Soviet nuclear complex's production of a pervasive ecology of ionizing radiation in the STS area and more broadly in northeastern Kazakhstan.[20]

As an *immanent cinematic recreation or re-enactment* both of a Cold War nuclear theatre of operations and of residents' living (and dying) with or because of this decades-long rehearsal for a nuclear war, these two scenes structured by radiation/radioactive immanation-images are significant also for their

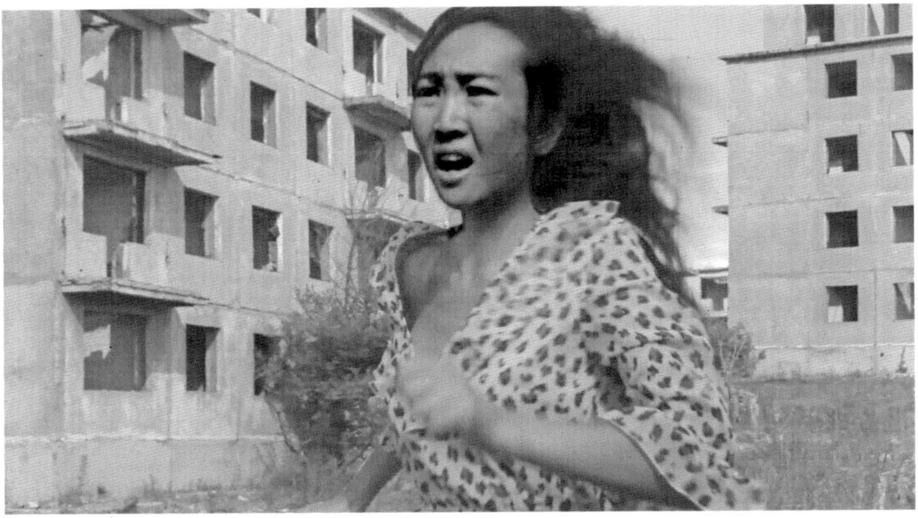

14.4 Woman running through ghost town with simulated nuclear blast in the background. *Kurchatov 22, Transformation*, 2016.

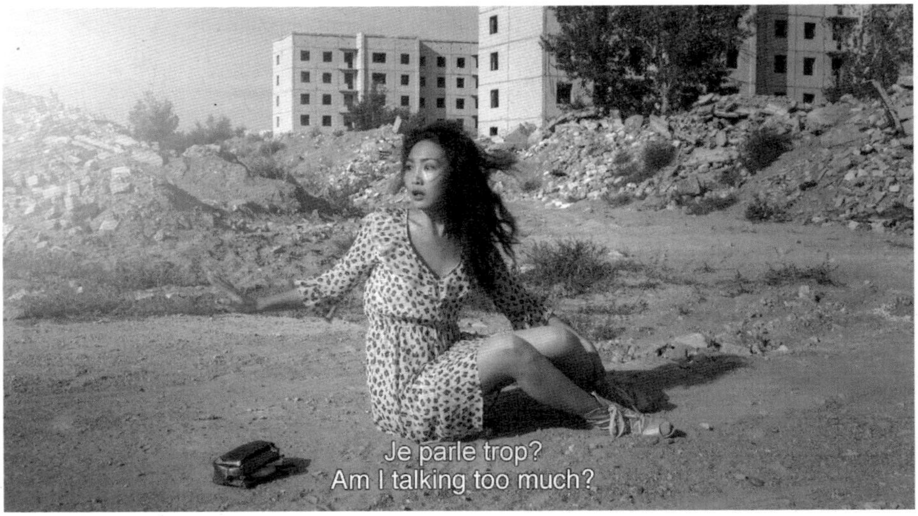

14.5 Young woman turning to look at nuclear detonation. *Kurchatov 22, Transformation*, 2016.

poignant visualization of the local population's immanent experience of the *nuclear uncanny* – everyday life rendered strange or *unheimlich* by military-industrial radiation.[21]

The assemblage of overlapping radioactive and re-enactment immanation-images in the two sequences showing a terrified young woman running through, or sitting amidst, the ruins of the abandoned military town of Chagan also re-creates for contemporary viewers the panic and helplessness of the population exposed to the tests, as well as the calculated way local residents were used as test subjects (through their deliberate positioning in open places, where they absorbed high doses of high-level radiation) during the Soviet nuclear experiments. This is one level, or stratum, of the performative re-enactment or reimagining we witness in these scenes.

The theatrical *mise-en-scène* in these two scenes shot in Chagan also symbolically and metonymically re-enacts the harnessing of the anarchic geochemical de- and re-composition of geologic substratum and atmospheric forces in a nuclear fission– or nuclear fusion–based bomb detonation. The harnessing and governing of nuclear geopower[22] during the Cold War, as has been richly documented by nuclear and environmental historians, went hand in hand with the development of climate science and other environmental sciences, as well as with an ecological vision that was often more invested in the biopolitical control and management of populations and the strategic transformation of environments than in the protection of local or global ecosystems.[23] A second level of re-enactment in the two scenes featuring the young woman in the ghost city of Chagan may thus be detected in these sequences' implicit reimagining or re-enacting of the Soviet state's harnessing and management of nuclear geopower for nuclear weapons tests – military-scientific and geopolitical experiments that were arguably, in terms of the radiation and environmental hazards they unleashed, colonial-sacrificial and immensely destructive.

The same scenes showing the young woman amidst the ruins of the city of Chagan also embody other sets of implicit or virtual re-enactments. Situated on the banks of the Irtysh River in the East Kazakhstan Oblast (East Kazakhstan Region), the former closed military town of Chagan was founded in 1950. It was an extended military base serving the nearby military airfield of Dolon (also known as Chagan or Semipalatinsk Dolon), which was used for long-range heavy bomber aviation. In 1995, the military units stationed in Chagan were withdrawn and sent back to Russia, which led to the depopulation and finally abandonment of the town. The ghost town we see in the scenes under discussion consists mainly of ruins of 4- and 5-storey *khrushchyovkas*, low-cost brick- or concrete-sided apartment buildings whose prototype was developed in the early 1960s, and which continued to be built throughout the Soviet Union until the collapse of the communist state. This empty ghost town testifies to the enormous military power of a vanished socialist empire and its vast network of closed military cities. It is a landscape of imperial debris, attesting to the boom and bust of the Soviet Union's gigantic military apparatus, to the development of infrastructures, industries, and cities catering to the latter, and to the environmental devastation that often accompanied the construction and maintenance of these sites. The scenes featuring the young woman among the ruins of the abandoned town of Chagan may thus be said to testify to a past Soviet history through *ruination immanation-images* – performative image-events whose theatrical *mise-en-scène* attests to a vast landscape of socio-ecological devastation. At the same time, these performative cinematic renderings of radioactive contamination, re-enactment, and ruination also implicitly attest – like the assemblages of testimonial immanation-images examined above – to the resilience and endurance of the population in the Semipalatinsk Test Site area. As Magdalena Stawkowski's ethnographic research in the STS area has documented, in some villages residents have crafted a defiant, self-reliant form of "mutant subjectivity" and endurant resistance, as well as an informal economy that supports their survival and the occasional upward mobility of family members who have moved to the city. Similar informal economies and modes of self-reliant collective existence have been operating in the area surrounding the Chernobyl Exclusion Zone and in the contaminated area in the Northern Territory of Australia, where the Aboriginal Karrabing Film Collective live and make their films. Following the work of Stawkowski and Povinelli on emerging forms of subjectivity (re)construction in nuclear/toxic areas abandoned by the state, the assemblages of radioactive, re-enactment, and ruination immanation-images in *Kurchatov 22* we have examined may also be seen as a *paradoxical cinematic theatre of toxic endurance-as-resistance and toxic resurgent sovereignty.*[24]

The remaining sections of this chapter focus respectively on the *Tokamak* and *Astana* video installations. Whereas the *immanation-image of undone nuclear science* is prevalent in *Tokamak*, the *Astana* video imagines (im)possible feminist decolonial futures through speculative assemblages of ironic spiritual-mythical and/or non-human immanation-images inspired by Central Asian nomadic traditions, folk arts, and religious beliefs.

Tokamak's Envisioning of Nuclear Science as Undone Science

NUCLEAR SCIENCE AS UNDONE SCIENCE

Nuclear science and technologies in both their military and civilian applications have long been overshadowed by undone science. Undone science – which David Hess has defined as "a kind of non-knowledge that is systematically produced through the unequal distribution of power in society," and which often involves the systematic underfunding of scientific knowledge deemed potentially damaging to the interests of industry, the military, political elites, and other groups and organizations representing corporate capitalism[25] – in the field of nuclear weapons research, production, and testing as well as in the nuclear power industry during the Cold War was produced and maintained through a thick wall of legally entrenched, institutionalized, and normalized secrecy, censorship, and classified documentation. This production and upholding of nuclear undone science and ignorance entailed the lack of access to crucial information for researchers, medical workers, media outlets, and the public, including communities and populations that were seriously affected by the fallout from nuclear activities. Nuclear undone science and non-knowledge have also been a product of pro-nuclear governments' and nuclear industries' disinformation, denial, and obfuscation of responsibility, and the downplaying of environmental and health hazards in the wake of major accidents such as Chernobyl and Fukushima, as well as of nuclear complexes' racist and colonial attitude toward Indigenous, poor, and marginalized populations living in or near nuclear testing sites. Undone nuclear science both during and after the Cold War should also be attributed to the power and authority of international bodies such as the IAEA.[26]

The complex realities of post-socialist Kazakhstan – the reinvention of the state and the nation respectively as an authoritarian neo-patrimonial regime and a multicultural society with Kazakh as the primary official language, and the transformation of the economy into a neoliberal, export-oriented capitalist economy based on resource extraction – require that analytic tools such as undone science be coupled with an examination of phenomena such as the state's abandonment of poor rural populations, ongoing legacies of the Soviet and Cold War past, and the tension between the political elites' forward-looking, ambitious modernization projects such as *Kazakhstan 2050* and the country's worsening environmental ruination. A characteristic expression of the latter tension is the deployment, by scientists and experts associated with the Institute for Radiation Safety and Ecology and the National Nuclear Centre in Kurchatov, of a specific form of undone science (which is at the same time a powerful ideological and political tool of the nuclear lobby not only in Kazakhstan, but in all nuclear states): radiophobia. As Magdalena Stawkowski has shown, radiophobia, or the irrational fear of radiation, has in recent years often been invoked by corporate scientists and government bureaucrats to counter arguments by environmental groups and NGOs that the Semipalatinsk Test Site is still polluted by dangerous residual radioactivity from the Soviet-era nuclear testing as well as by a toxic cocktail of chemicals and heavy metals from Soviet and post-Soviet industrial wastes, and that because of this widespread toxicity the site should not be privatized and licensed for industrial and agricultural development.[27]

Transformation articulates a powerful immanent critique of Soviet and post-Soviet nuclear and medical undone science and technology, of the social, political, economic, and ideological conditions that have produced and continue to maintain it as well as of the serious toxicity and pollution affecting the Semipalatinsk Test Site area and other regions in contemporary Kazakhstan. The installation's experimental, decolonizing aesthetic of the immanation-image also critically highlights the Kazakhstani government's ambitious plans for economic development and infrastructural modernization, plans that once again entail the strategic production of non-knowledge and a disastrous disregard for citizens' well-being and for environmental protection.

In the *Tokamak* video Menlibayeva's immanent critique of nuclear science is expressed as an ironic portrayal of the science of fusion energy.

THE HYPERBOLE OF FUSION SCIENCE

The *Tokamak* video focuses on Kazakhstan's participation in the large-scale, international fusion energy development project ITER (International Thermonuclear Experimental Reactor, an acronym that in Latin also means "the way"). This project, which has seven members – China, the EU, the US, Russia, Japan, India, and South Korea – and thirty-five participating nations, is currently under construction at Saint Paul-lès-Durance/Cadarache in southern France. It will be the world's largest tokamak, an experimental magnetic fusion device with a capacity to hold plasmas ten times greater than the largest tokamak currently in operation, the JET (based in the UK). The ITER's main objective is to demonstrate the ability of science and technology to achieve fusion power in a laboratory setting by confining and controlling large hydrogen plasmas in the tokamak's vacuum chamber by means of powerful magnetic coils and producing energy through the fusion of deuterium and tritium (two hydrogen isotopes) within these plasmas.[28]

The first plasma in the ITER tokamak is expected to be achieved in December 2025; the DT (deuterium-tritium) operation is scheduled to begin by 2035. The first experimental DEMO fusion reactor, currently in the conceptual phase, is expected to begin operating in the 2040s. Note that the ITER tokamak is not intended to produce electricity.

Kazakhstan's contribution to the ITER project is the KTM Tokamak, or Kazakhstan Tokamak for Material Testing. The KTM was designed in 2000 to test the plasma-facing materials in the vacuum chamber of the tokamak under conditions of extreme heat and neutron fluence (exposure). Based at the National Nuclear Centre in Kurchatov, the KTM Tokamak completed the first phase of its physical launch in June 2017 with the achievement of its first plasma discharge pulse.

The completion of the first stage of the physical start-up of Kurchatov's KTM machine and the achievement of the first plasma discharge in June 2017 were described in Kazakhstan's media as an important scientific event. The KTM tokamak was also showcased as a mock-up prototype in the national pavilion Nur Alem during *Expo 2017: Future Energy* (10 June–10 September 2017) in Astana/Nur-Sultan. Rather than celebrating Kazakhstan's participation in international fusion science and fusion energy development projects, Menlibayeva's *Tokamak* video stages a playful, ironic performance that raises provocative questions about the science of nuclear fusion and about Kazakhstan's continued commitment to nuclear energy.[29]

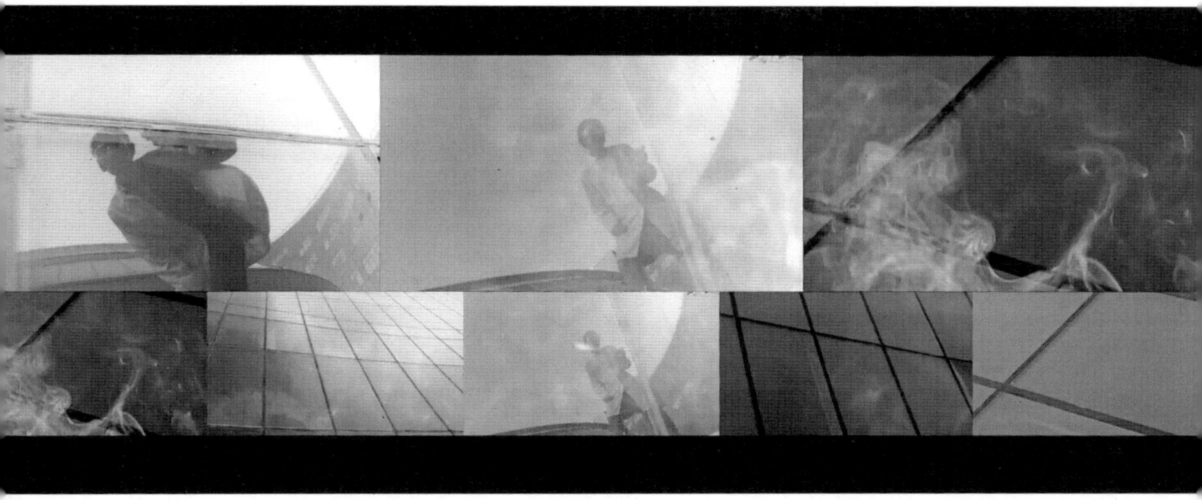

14.6 Scientists peering through the glass walls of the Shabyt Art Palace. *Tokamak, Transformation*, 2016.

Tokamak sets much of its narrative in the capital, Nur-Sultan (formerly Astana) rather than in Kurchatov, where the Tokamak KTM research program is located. It also evinces a thematic, aesthetic, political, and conceptual continuity with *Kurchatov 22* as well as a distinctly science-fictional atmosphere. We follow nuclear scientists (or rather actors embodying nuclear scientists) in white lab coats and protective glasses as they walk or run along the inner walkway of the blue-tinted, extravagant futuristic building of the Shabyt Art Palace, or the Kazakh University of Arts in Nur-Sultan.

The building at the National Nuclear Centre in Kurchatov, which houses the spherical KTM tokamak, is not featured in the video. What we can see of the KTM fusion device are tangles of connecting cables, partial views of its infrastructure, several machine parts, and the architecture of the roof.

As to the *Tokamak* video's resonances and continuities with *Kurchatov 22*, we can discern them in a series of scenes and features: the red colouring of certain scenes, which clearly echoes the reddish tint of several panoramic views of the steppe in *Kurchatov 22*; the central role ascribed to nuclear science; and the scene showing scientists rolling traditional Kazakh low, round tables in front of the pyramid of the Palace of Peace and Reconciliation in Nur-Sultan – a sequence that responds to the sequence in *Kurchatov 22*, where we see young people from the village of Semiyarka rolling about round nomads' tables pasted with photos of the Soviet nuclear tests. The SF flavor of *Tokamak* is pervasive and may be detected in the video's foregrounding of the unusual glass architecture of the Shabyt Palace of the Arts/National University of Arts in Nur Sultan; in the evocative musical score by OMFO/German Popov; and in the ironic posturing and dancing of the scientist characters. These characters recite in a slightly ominous tone bits of information and statements related to the production of plasma pulses and to the DT (deuterium-tritium) fusion reaction within the plasma in the tokamak's vacuum vessel: "a temperature that exceeds that of the sun," "slow reaction of deuterium," "not dangerous," "the graveyard of oil," "half-life," "management of the energy resulting from fusion."

The artificial SF atmosphere, non-linear narrative, and ironic-parodic performance of the scientist characters in the *Tokamak* video are clearly meant to call attention to a specific form of undone science and technology, namely the production of non-knowledge, or incomplete and misleading knowledge about the tremendous complexity, unresolved questions, and high level of risk involved in nuclear fusion energy development. *Tokamak*'s use of parodic tropes of science fiction also questions the optimistic, utopic discourse on fusion energy we find on the website of the ITER project as well as in numerous media reports. This discourse has been characterized by physicist and fusion energy specialist Daniel Jassby as a misleading, misguided "gee-whiz hyperbole which has been ... touted all too often as the magic bullet solution to the world's energy problems," and whose claims about fusion energy as unlimited, clean, and carbon- and radiation-free "are all debunked by harsh realities."[30] In ways similar to Jassby's critique, *Tokamak*'s parodic SF script and choreography cautions scientists, industry, and the public against boundless enthusiasm for and massive investment in fusion energy research while at the same time calling attention to the deliberate obfuscation and suppression of the "harsh realities" of the laboratory creation and production of this type of energy. These harsh realities include, as Jassby has pointed out, the colossal energy investment and consumption required to produce the "burning plasma" for the deuterium-tritium fusion reaction, the rarity of natural tritium, and the uncertainties and risks involved in breeding this isotope in a tokamak's vacuum chamber, as well as the substantial amount of radioactive waste and environmental radioactive contamination that will be generated by commercial fusion reactors, not to mention the huge costs entailed by the latter's manufacturing and maintenance.[31] These aspects get lost in the bright, hyperoptimistic discourse advertising the merits and ecological virtues of fusion energy, which the *faux* nuclear scientists in the video extol in their ironic, solemn statements. *Tokamak*'s experimental aesthetic, in particular its emphasis on mosaic-like assemblages and split-screen montages of *immanation-images of undone science* (such as the assemblage foregrounding the infrastructure of the KTM tokamak), as well as its insistence on certain tropes of visual and narrative continuity with the radioactive ecology of immanence highlighted in *Kurchatov 22*, also call attention to other critical aspects that have been suppressed in media reports celebrating the National Nuclear Centre's research on fusion energy.

For instance, the film's emphasis on historical continuity implicitly addresses the fact that the toxic legacies of the Soviet era will persist far into the future. This implies that undone science and the production of non- and incomplete knowledge surrounding the Soviet nuclear tests has continued in post-socialist Kazakhstan, in forms such as those discussed earlier: downplaying of the nuclear detonations' devastating effects accompanied by a related denial of a causal relation between these effects and the ill health and reduced lifespan of the people living in the STS area; and the invoking of radiophobia as a political tool for silencing opposition by environmental groups and concerned scientists to IRSE's plans to privatize and open for investment and development significant areas of the Semipalatinsk Test Site.

Before moving on to the final section of this chapter, I would like to draw attention to the fact that the dominant strange attractor figure in *Tokamak*'s assemblages of *immanation-images of nuclear undone science* is that of the scientist. The actors embodying nuclear scientists in *Tokamak* effect a double parodic impersonation: *that of the embedded (state, military, academic or corporate) scientist,* and *that of the SF (mad) scientist.* The immanent experience highlighted in *Tokamak* may be described as *the living out of embedded (nuclear) technoscience.* While the scientist characters in this video (who also appear in the *Astana* video) and the immanent experience they embody may seem far removed from or even opposed to the marginalized, toxic reality of the subaltern *hibakusha* endurants showcased in *Kurchatov 22*, a close examination of the continuities and repetitions between and across the two video films reveals otherwise: these figures share the same complex reality of contemporary, neoliberal Kazakhstan and its toxic ecology of immanence, and they experience the latter in closely related yet differential ways. Soviet (undone and embedded) nuclear science and its successor, post-socialist (undone and embedded) nuclear science bear a great deal of responsibility for the forms of alterlife and socio-ecological ruination and devastation they have produced and continue to generate; meanwhile, the human and non-human existents partaking in this vast, toxic ecology of immanence will continue to live with and endure its enfleshed becomings and mutations.[32]

A New Earth? *Astana*'s Imagining of (Im)possible Feminist Decolonial Futures

Several scenes in the *Astana* video (which, as I indicated at the beginning of this chapter, was projected on three suspended screens in the Grand Palais's Salon d'honneur) show the construction and gradual assembling of the Nur Alem, the huge, wind- and solar-powered glass sphere that housed Kazakhstan's national pavilion at the 2017 *Future Energy* world exposition.

In the video's closing sequence we see the nearly finished sphere at night, illuminated by small, flickering white and orange lights from the construction

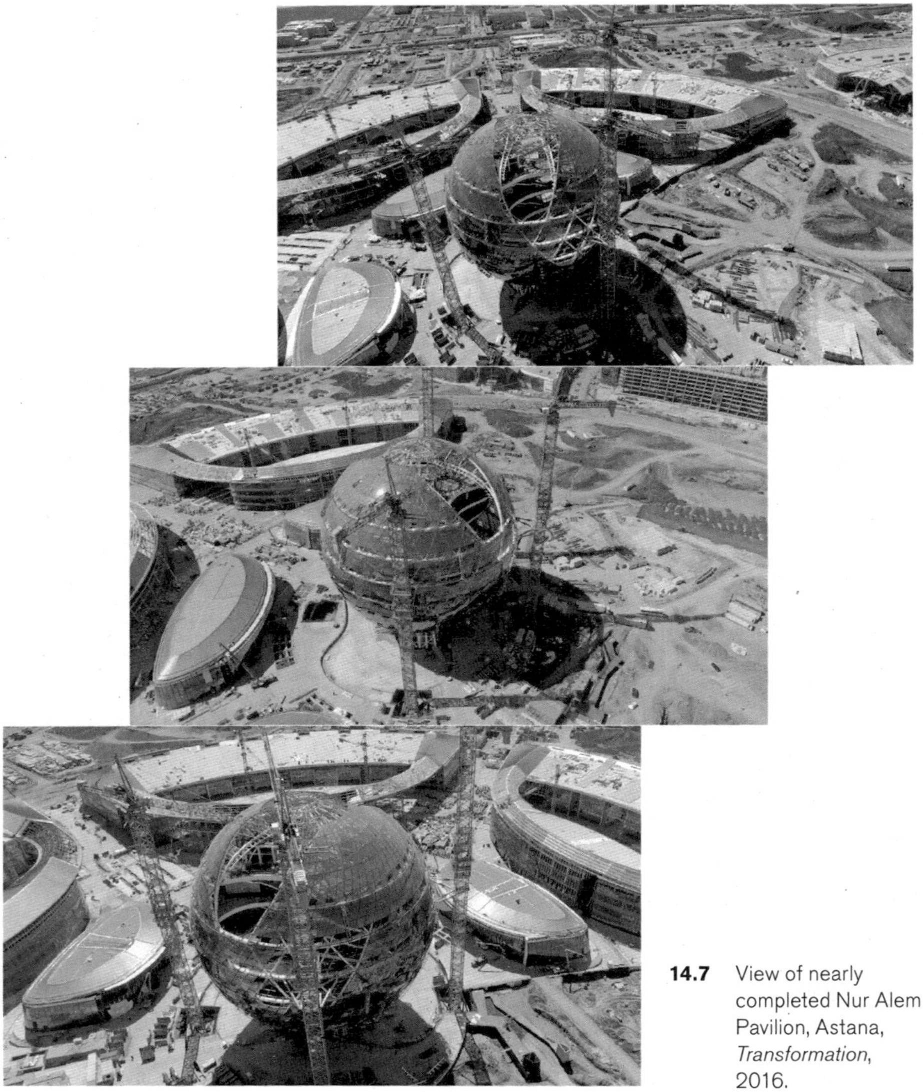

14.7 View of nearly completed Nur Alem Pavilion, Astana, *Transformation*, 2016.

workers' welding operations.[33] These small flickering flames then become abstract animated patterns of various colours and sizes that drift about dreamily in a dark space. OMFO's/German Popov's beautiful astral music enfolds this enchanting spectacle. In between these scenes a sequence set in one of Nur-Sultan's large public spaces features a series of performances shot during the 2016 Astana Art Festival: a female violinist playing on a small stage painted green and red amidst scattered objects of different shapes painted in various colours; and a ballet scene performed by a trio of dancers wearing long white dresses.

Let us examine first the sequences featuring these latter two performances, that of the female violinist playing on a small, round, green-and-red stage, and the ballet scene performed by three graceful dancers in flowing white dresses. The violinist played three times a day during the 2016 Astana Art Fest, and each time she wore a different set of clothes. These costumes referenced women professionals who were often featured in Soviet media and Soviet propaganda: a doctor, a military officer, and a construction worker. (see Figure 14.8) The violinist's show was part of an installation created by Menlibayeva for the international Astana Art Fest, held in Nur-Sultan on from 4 to 6 July 2016. The ballet sequence was also part of the program of this art festival (see Figure 14.9).

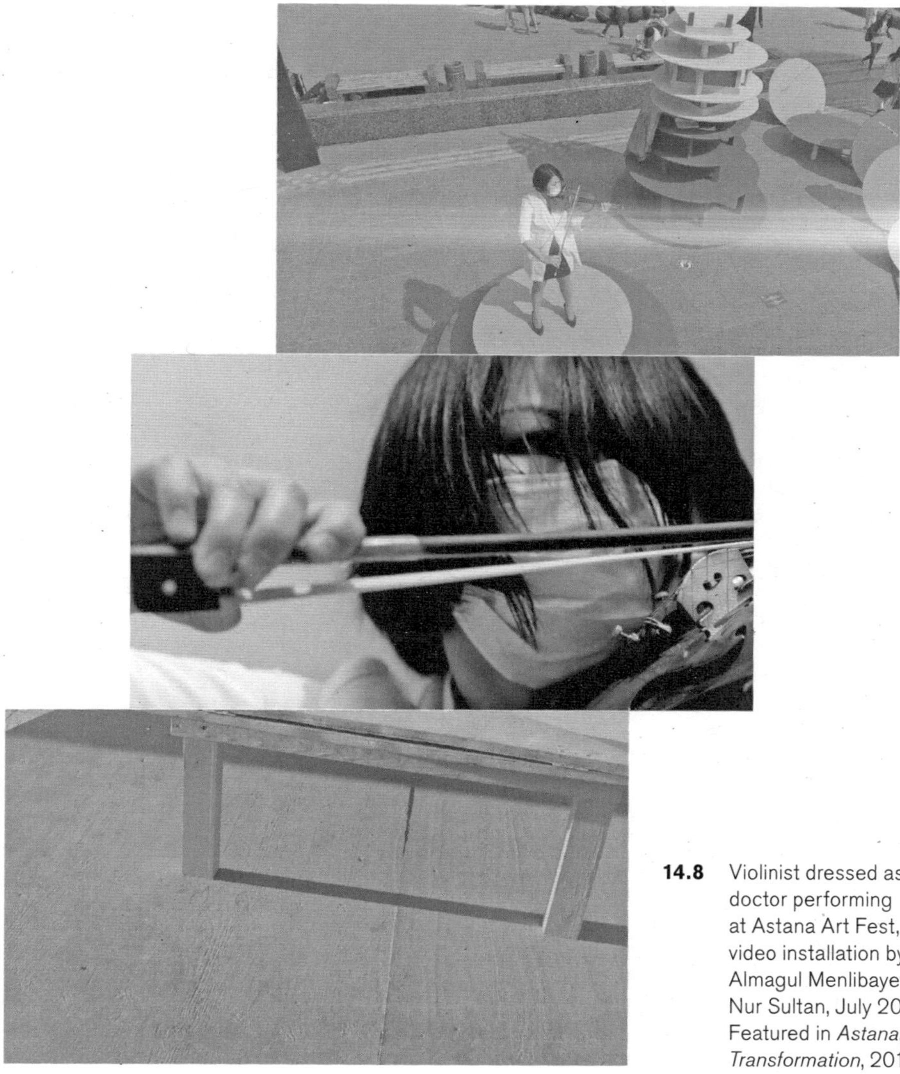

14.8 Violinist dressed as doctor performing at Astana Art Fest, video installation by Almagul Menlibayeva, Nur Sultan, July 2016. Featured in *Astana, Transformation*, 2016.

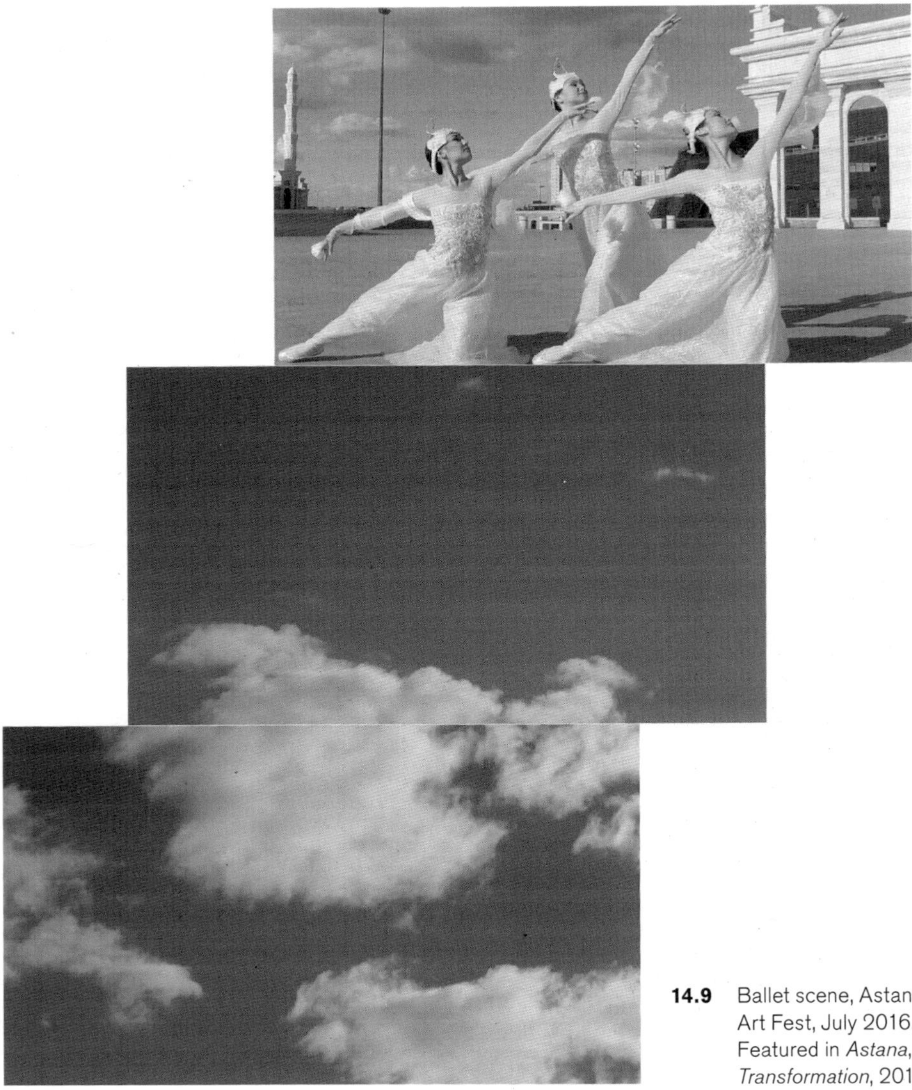

14.9 Ballet scene, Astana Art Fest, July 2016. Featured in *Astana, Transformation*, 2016.

These two scenes in the *Astana* video implicitly revisit and subversively reinterpret Kazakh/Central Asian legends and shamanistic traditions. They gesture toward a (potentially) liberating line of flight that might summon forth a new Earth and invent a new people – a task that Deleuze and Guattari saw as incumbent on art and philosophy to perform, one that would entail a revolutionary deterritorialization that could restore "our resistance to the present" and lay the foundation for a (post-capitalist) future where new possibilities of/for life, new modes of existence, creative subjectivities, and a "worlds-people, brain-people, chaos-people" or beings attuned to the living immanence of nature and the cosmos could flourish.[34]

The performance of the female violinist as part of the installation cum live music created by Menlibayeva for the 2016 Astana Art Fest references the legend of the *baksy* (shaman) Korkyt Ata/Dede Korkut/Dede Qorqud and his invention of the *kobyz/qobyz/qyl qobyz*, an ancient two-stringed viol carved out of blocks of birch or walnut. The violinist has pitched her instrument to sound like a *kobyz*. According to legend, the shaman Korkyt's music was so enchanting that animals and birds flocked to listen to him, and while he played death could not take away his soul. The *kobyz/qobyz* became part of shamanistic rituals, and the musical pieces that only shamans could play, the *saryn*, were believed to communicate with and carry messages from the ancestors. The *kobyz* is a national musical instrument in Kazakhstan, and its repertoire comprises not only *saryn* but also *kui* (short programmatic pieces that can be either story-based or expressive of emotion). The *kui Akku* (White Swan) is a popular piece in the *kobyz* repertoire in which a mythical swan character plays a central part.

The ballet scene featuring three dancers in long white dresses evoking swans refers both to the *Akku kui* in the *kobyz* repertoire and to the popular figure of the swan in Kazakh/Central Asian folklore, dance, poetry, and visual arts.[35] The scenes in the *Astana* video featuring the violinist playing on a small stage in Menlibayeva's 2016 Astana Art Fest installation and the three dancers in swan dresses thus seem to revive, by means of an allusive, subtle aesthetic, ancient Kazakh/Central Asian folk traditions as well as some of their modern interpretations. The violinist's solo performance, whose parodic re-enactment of Soviet women professionals calls attention to the persistent legacies of Soviet socialism as well as to the fact that neoliberal capitalism in contemporary Kazakhstan has brought not more but rather less equality and empowerment for women, can also be read as a "shamanistic" ritual resuscitating the magical power of female shamans and of the *kobyz/qobyz* to stave off death, evil spirits, illnesses, and other calamities.[36] The three dancers in white dresses also evoke female magical or shamanistic power as well as the mythical or sacred character of the swan. While the "mythical-shamanistic" connotations in the *Astana* video are more allusive than explicit and are also, like the performance of the scientist characters in *Tokamak*, ironic-parodic, it should be noted that they resonate with the aesthetic of "punk shamanism" – Menlibayeva's preferred term for her ironic, feminist, transcultural reimagining of nomadic traditions, folktales and myths of the Central Asias steppes – that characterizes her earlier video installations, *Exodus* (2009) and *Transoxiana Dreams* (2011).

The ironic, "shamanistic" sequences featuring the violinist and the dancers consist of (equally ironic) *"spiritual-mythical" reenactment immanation-images* showcasing various forms of localized immanent experience. These forms of embodied immanent experience may be described as *a critical revisiting and/or displacement of the Soviet ideal of socialist womanhood*, as *resistance to neoliberal articulations of female human-capital*, or as the *resurgence of pre-Soviet folk arts and practices* (such as shamanism, divination, and ancestor communication). Whether or not the musical and dance performances just described are seen as more muted expressions of "punk shamanism" or as critical responses to state-sponsored cultural nationalism, they can be regarded as both decolonizing and decolonial insofar as they point to a spiritual futurity or imaginary of cosmic infinity – the "immortality" once believed to be embodied by mythical figures such as Korkyt Ata, female shamans, and swan goddesses, and expressed in the art of the *kobyz* and the *kui* – that will not be available for human colonization, mastery, and geo-engineering. The strange attractor figures that embody, orchestrate, probe, reconfigure, and distribute the potentialities, experiential realities, imaginaries, and possible worlds evoked in these assemblages of spiritual/shamanistic immanation-images may be identified as *the Independent Artist, the Shaman*, and *the Swan*.

The animated abstract patterns that gently float around before disappearing into the night in the closing sequence of the *Astana* video are identical to the patterns that were projected on the carpet covering the floor in the Salon d'honneur of the Grand Palais in Paris during the artist's solo exhibition from 17 December 2016 to 2 January 2017. Some of these patterns replicate variations of the *kochkar-muyiz*, the ubiquitous, simplified ram horn ornament found on carpets, rugs, walls, fences, and other objects throughout Kazakhstan. Other patterns resemble designs for mechanical devices, geometric figures, exploding stars, and various ornaments. The patterns that were projected onto the floor carpet and animated in the film's closing sequence also comprise words such as "plasma," "cyclotron," and "tokamak," as well as scattered lines from an experimental poem composed by Menlibayeva.

The poetic fragments that can be perceived among the floating abstract animations in *Astana*'s closing scenes are notable for the unusual imagery they display, which includes phrases such as "gender mini-changer" and "Planet X being pulled into inner solar." These phrases resonate with several animated figures in the film's closing sequence that resemble the hardware device called the gender changer (which connects two cable connectors of the same type and gender), and which are inscribed with technical terms that appear in the *Tokamak* video, namely *cyclotron*, *plasma*, and *tokamak*. The line "Planet X being pulled into [the] inner solar [system]" seems to reference the various theories and conjectures that have been proposed about the mysterious, hypothetical Planet X (also called Planet Nine) that appears to loom in the far reaches of the outer solar system.[37]

In an email conversation with Menlibayeva I asked her to elaborate on her ideas about the floor carpet and the latter's decorative patterns (which are replicated in the closing sequence of the video). She responded that she had intended the carpet, its four rows of projected, symmetrical decorative patterns, and the replication of

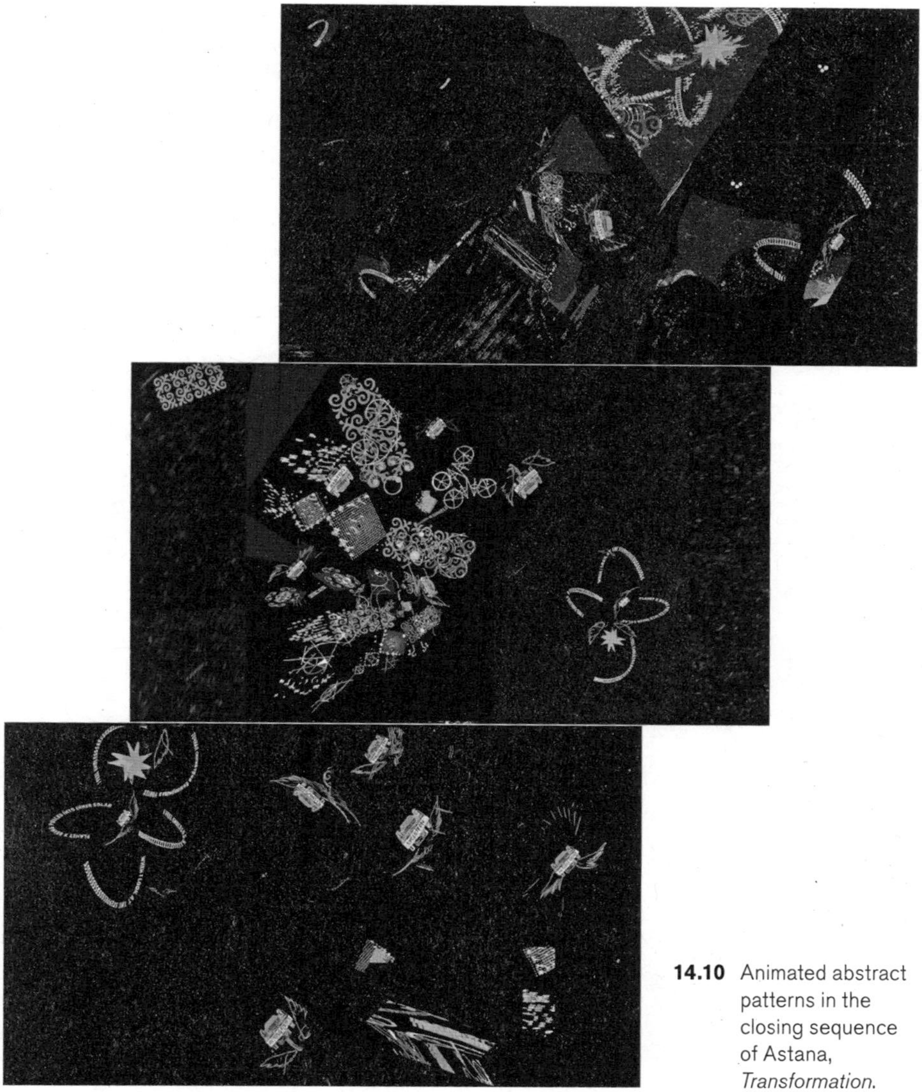

14.10 Animated abstract patterns in the closing sequence of Astana, *Transformation*.

these patterns as animated constellations in the closing sequence of the *Astana* video to evoke the enduring beauty and power of the ancient art of carpet-weaving in Kazakh and Central Asian nomadic cultures – an art that was, and still is, mostly practised by women. The artist also stated that the colourful, symmetrical patterns and ornaments woven by women in floor, wall, and panel carpets and rugs embodied a profound cosmology and epistemology, a rich knowledge and imagination of the universe and the world that anticipated several theories in modern astronomy, physics, mathematics, and computer science.[38] While this may indeed have been the case, what is important to note in the *Astana* video's engagement with the art of carpet-weaving is its unmistakable feminist stance – a stance that also structures the violinist's "shamanistic" performance as well as the mythical swan lore evoked in the ballet scene in the sequences described earlier. Menlibayeva's feminist positionality can be detected in most of her photography, films, and video works; it also underwrites her recent co-curating of the exhibition *Focus Kazakhstan Berlin: Bread and Roses: Four Generations of Kazakh Women Artists* and the *Artist Residency Exhibition* at Berlin's Momentum Gallery (25 September–20 October 2018).[39]

In closing this chapter's extended discussion of Melibayeva's *Transformation* installation I would like to suggest that the non-human, post-extinction universe we can glimpse in the finale of the *Astana* video – a universe whose cosmic Dark Matter and Dark Energy are visualized by means of speculative, *non-human (cosmic) immanation-images* – embodies the artist's feminist decolonial response to Deleuze and Guattari's call for a deterritorializing, revolutionary new Earth and a new people to come. Carried by magical, immortal swans (visualized in the animated closing sequence as blue swan wings), this vision evokes a distant, strange new world where humans along with most other living species have vanished and new forms of alterlife seem to lie dormant in the depths of the toxic *longue durée* of cosmic radioactive dust and gas. This emergent, (im)possible alterlife may (or may not) carry in its DNA traces of the wisdom and creative imagination once woven by women into intricate carpet designs encoding a "magical" map (or science) of the (alter)future of the future.

NOTES

1 An updated list of Menlibayeva's solo and group exhibitions can be found on the artist's website, http://www.almagulmenlibayeva.com. I am deeply indebted to Almagul for her unstinting support and generous response to the numerous queries and requests for clarification I directed at her.

2 See Mamedov, "Sites of Construction"; Kudaibergenova, "Contemporary Art in Central Asia," "My Silk Road to You," "Punk Shamanism," "Between the State and the Artist"; Tlostanova, *What Does It Mean to Be Post-Soviet?*

3 See Cameron, *The Hungry Steppe*, chapters 4, 5, and 6, and the concluding chapter, 169–79.

4 On Kazakhstan's Soviet history see Dave, *Kazakhstan*; Cameron, *The Hungry Steppe*; and Ohayon, *La sédentarisation des Kazakhs*. On post-Soviet and contemporary Kazakhstan see Laruelle, ed., *Kazakhstan in the Making*; and Laruelle, ed., *The Nazarbayev Generation*. On the 2050 Strategy see OECD, ed., *Reforming Kazakhstan*; and Aitzhanova et al., eds., *Kazakhstan 2050*. Kazakhstan's current president, Kasym-Jomart Kemeluly Tokayev, took office on 20 March 2019. He succeeded the first president, Nazarbayev, who resigned on 19 March 2020 after twenty-nine years in office.

5 On the Soviet nuclear military-industrial complex and its radiation legacies see Brown, *Plutopia* and *Manual for Survival*; Schmid, *Producing Power*; Egorov et al., *Radiation Legacy*; and Holloway, *Stalin and the Bomb*. On the nuclear tests at the Semipalatinsk Test Site see Vakulchuk et al., *Semipalatinsk Nuclear Testing*; Kassenova, "The Lasting Toll"; Werner and Purvis-Roberts, "Unraveling"; and Werner and Purvis-Roberts, "Cold War Memories."

6 Recent critical studies on daily life in the former Semipalatinsk Test Site area and on the contested health effects of residents' exposure to ionizing radiation include Stawkowski, "Everyday Radioactive Goods?"; Stawkowski, "'I am a radioactive mutant'"; Goldstein and Stawkowski, "James V. Neel and Yuri E. Dubrova"; and Bauer, "Radiation Science after the Cold War." On the possible relation between the high incidence of cancer in the STS area and the Soviet nuclear tests see Gusev, Abylkassimova, and Apsalikov, "The Semipalatinsk Nuclear Test Site." Yuri Dubrova and his colleagues have traced a linkage between environmental radioactivity from nuclear weapons tests and human germline mutation. See Dubrova et al., "Nuclear Weapons Tests."

7 On Kazakhstan's uranium mining industry and plans for the development of nuclear power in the coming decades see the World Nuclear Association's regularly updated report, "Uranium and Nuclear Power in Kazakhstan."

8 See Deleuze, "Immanence: A Life," 27, 29. On Deleuze's conception of immanence see Thiele, "Of Immanence and Becoming"; and Beistegui, *Immanence: Deleuze and Philosophy*.

9 Massumi, *99 Theses*, 6–18, 24–48, 58–69, 86–103, 111–33.

10 Berressem, "Ecology and Immanence," 85–92, 95–7, 101–2.

11 Bohler, "Staging Philosophy."

12 On the biopolitics of disposability of neoliberal capitalism and the increasingly authoritarian neoliberal state see Giroux, "Beyond the Biopolitics of Disposability"; and Evans and Giroux, *Disposable Futures*.

13 See Zoulias, "Gilles Deleuze and Chaos Theory," 113–15.

14 The term *hibakusha* designates survivors and groups affected by the American nuclear bombs dropped on Hiroshima and Nagasaki on 6 August and 9 August 1945 respectively. Here I use the term in a broader sense, as referring to all communities, populations, and downwinders affected by Cold War nuclear testing, uranium extraction, nuclear accidents such as Chernobyl and Fukushima, and in general the routine functioning of nuclear complexes. See Broderick and Jacobs, "The Global Hibakusha Project"; and Jacobs, "The Radiation." On the origins of the term *hibakusha* and its historical and contemporary usage see Naono, "The Origins of '*Hibakusha*.'"

15 Menlibayeva has confirmed that the disabilities of the grandson featured in this particular interview have been related to his grandmother's and father's exposure to radiation from the Soviet nuclear tests.

16 See Murphy, "Alterlife," 497–501.

17 See Stawkowski, "'I Am a Radioactive Mutant'" and "Life on an Atomic Collective"; and Bauer, "Radiation Science after the Cold War" and "Beyond the Nuclear Epicenter."

18 On the deportations of ethnic minorities during the Stalin era, see Polian, *Against Their Will*; and Statiev, "Soviet Ethnic Deportations." On the Virgin Lands Campaign

see Fruhauf, Meinel, and Schmid, "The Virgin Lands Campaign"; and Stawkowski, "Life on an Atomic Collective."

19 On Povinelli's theorization of endurance and the aboriginal endurant – notions that are based on her ethnographic work and her long-time collaboration with the aboriginal Karrabing Film Collective in the Northern Territory of Australia – see *Economies of Abandonment* and *Geontologies*.

20 Menlibayeva has confirmed that the two sequences featuring the actress in a brown dress running or sitting among ruins in the abandoned town of Chagan re-enact episodes of local *hibakusha* survivors' experience of the detonations as they were recounted to her in interviews.

21 On the nuclear uncanny see Masco, *Nuclear Borderlands*, 28, 33–4.

22 Federico Luisetti defines geopower as the power that "presupposes the government of species, ecosystems, biogeochemical and physical processes and a new form of security ... [and which] regulates the planetary environment, marginalizing human capital and thus setting the stage for the Earth politics of late capitalism." See Luisetti, "Geopower," 8. I would slightly revise this definition to reflect the fact that the governance and modification of species, ecosystems, human populations, and earth systems were an integral, calculated element in Cold War nuclear testing programs that began in the 1950s. The dispersal of radioactive fallout, for instance, was considered a powerful strategic and environmental weapon. The harnessing of nuclear geopower for excavation, construction, and fracking projects was also part of the "peaceful nuclear explosions" (PNEs) undertaken by the US and USSR from the 1950s to the 1980s. The 1962 Sedan test in the US, which released large quantities of radioactive gases, is a well-known PNE that was part of the US Plowshare Project. The Soviets carried out some 239 PNEs. The famous Atomic Lake in the Semipalatinsk Test Site area was the result of a geoengineering PNE test.

23 Scholars and critics including Elizabeth Deloughrey, Joseph Masco, Robert (Bo) Jacobs, and Paul Edwards have documented the close relationship between Cold War nuclear weapons tests, military and nuclear colonialism, and environmental and biopolitical monitoring and governance. A host of fields and subdisciplines of the environmental sciences developed along with or as a result of Cold War nuclear testing. See Deloughrey, "The Myth of Isolates"; Masco, "The Age of Fallout" and "Terraforming the Earth"; Jacobs, "The Bravo Test"; and Edwards, "Entangled Histories."

24 On informal economies, resilient radioactive subjectivity, and toxic sovereignty in the Semipalatinsk Test Site area see Stawkowski, "'I am a radioactive mutant'" and "Life on an Atomic Collective." On similar informal economies and mutual support networks in the Chernobyl area see Davies and Polese, "Informality and Survival." On the Karrabing Film Collective's mode of living, filmmaking, and world-making in toxic sovereignty – a land-based counter-practice building on "ecological, historical, and totemic connections that ... keep the differences between lands and languages ... in place," see Fisher and Seale-Feldman, "Filmmaking and Worldmaking"; and Lea and Povinelli, "Karrabing." In Povinelli's work toxic sovereignty designates the Karrabing Film Collective's relative independence from police and state interference in the contaminated territory where they live on the Cox Peninsula in northwestern

Australia on the one hand, and on the other hand (alter)life in the toxic environments produced by global capitalism, where categories such as the modern (white) subject's autonomy, sovereignty, and self-enclosure are exposed as tenuous fictions. See Povinelli, *Geontologies*, 25–9, 85–91, and Povinelli, "Fires, Fogs, Winds."

25 See Hess, "Undone Science and Social Movements" and *Undone Science.*

26 Studies documenting undone science and technology in nuclear complexes as well as the production of non-knowledge and ignorance concerning Cold War nuclear testing and the accidents of Chernobyl and Fukushima include: Brown, *Manual for Survival*; Kuchinskaya, *The Politics of Invisibility*; Masco, *Nuclear Borderlands* and "Survival Is Your Business"; Stawkowski, "Radiophobia Had to Be Reinvented"; Goldstein and Stawkowski, "James V. Neel and Yuri E. Dubrova"; and Alexander, "A Chronotope of Expansion." On the recent production of non-knowledge and undone science surrounding the Fukushima Daiichi nuclear meltdowns see also Topçu, "From Toxic Land to Toxic Rumors," in this volume.

27 On radiophobia and its politicized use with respect to the privatization and licensing of the Semipalatinsk Test Site for agriculture, mining, and other industrial projects see Stawkowski, "Radiophobia Had to Be Reinvented," "'I Am a Radioactive Mutant,'" and "Life on an Atomic Collective." On the deployment of a discourse on radiophobia in the Soviet Union and in post-Soviet Russia, Belarus, and other post-Soviet republics in relation to nuclear accidents such as Mayak and Chernobyl, as well as with respect to the legacies of the Soviet nuclear tests, see Brown, *Manual for Survival*, 180–1, 201–2; and Kuchinskaya, *The Politics of Invisibility*, 70–1, 120–1. The discourse on radiophobia gained legitimacy in the aftermath of Chernobyl.

28 See the ITER website: https://www.iter.org/proj/inafewlines. The start of the assembly of the ITER fusion device was marked by an official ceremony within the ITER assembly hall at Cadarache, France, on 28 July 2020. See "Fusion Energy Era: ITER Assembly Begins."

29 On the achievement of the first plasma see "Kazakh Tokamak Celebrates First Plasma," https://www.iter.org/newsline/-/2751. See also Satubaldina, "Kazakhstan Conducts First Stage."

30 See Jassby, "ITER Is a Showcase."

31 Ibid.

32 Kazakhstan has serious environmental problems that have not been dealt with in a proactive and systematic way by the government. Apart from the radioactive contamination of the Semipalatinsk Test Site area and the East Kazakhstan Region (Oblast), there is the shrinking of the Aral Sea and the pollution of its seabed; the pollution of Lake Balkash by copper smelters; air and water pollution in most cities of central Kazakhstan as a result of intensive industrial manufacturing and resource extraction; and the legacies of uranium mines. As many reports have shown, radioactive toxicity is only one item on a long list of serious environmental issues that have become increasingly threatening, even catastrophic, in recent years. On Kazakhstan's environmental problems see Alimbaev et al., "Ecological Problems"; Marzhan, "Social, Environmental, and Economic Sustainability"; and Kenessariyev et al., "Human Health Cost of Air Pollution."

33 The Nur Alem, Kazakhstan's national pavilion and Future Energy Museum at the *2017 Astana Expo: Future Energy* was designed by the renowned Chicago architectural firm Adrian Smith + Gordon Gill Architecture. The double-curved-glass sphere, 80 metres in diameter, was modelled to minimize energy use and maximize the benefits of its integrated renewable energy sources – photovoltaic solar panels and wind turbines. The Nur Alem was awarded the Second Award at the Global Architecture and Design Awards 2018, Category: Cultural (Built).

34 See Deleuze and Guattari, *What Is Philosophy*, 27–8, 69, 85, 88, 108–9.

35 On the Korkyt Ata/Dede Korkut legend, the tradition of qobyz/qyl qobyz performance, and the representation of the swan in Kazakh and other Central Asian Turkic cultures see Daukeyeva, "Akku (White Swan)"; Rancier, "The Musical Instrument as National Archive"; Daukeyeva "The Kazakh Qobyz"; and Kamarova, "Swan as a Symbol of Beauty and Purity." The epic culture, folk tales, and music based on the legendary figure of the shaman, musician, poet, and philosopher Korkyt Ata/ Dede Korkut/Dede Qorqud were inscribed in 2018 in the UNESCO Representative List of Intangible Cultural Heritage of Humanity. See "Heritage of Dede Qorqud/Korkyt Ata/Dede Korkut," https://ich.unesco.org/en/RL/heritage-of-dede-qorqud-korkyt-ata-dede-korkut-epic-culture-folk-tales-and-music-01399.

36 Baksy shamans, including female shamans, played a central role in pre-Islamic Kazakh and other Turkic nomadic cultures. During the Soviet era shamanism, divination, spiritual healing, and other folk practices were repressed. In post-independence Kazakhstan shamanism has been revived. Most contemporary female *baksy* are healers who also practice divination. See Penkala-Gaweczka, "The Way of the Shaman."

37 On the mysterious Planet X or Planet Nine (which according to recent research may be a Primordial Black Hole rather than a planet) see Batygin et al., "The Planet Nine Hypothesis"; and Batygin and Morbidelli, "Dynamical Evolution."

38 Email conversation with Almagul Menlibayeva, 7 December 2018. On traditional Kazakh felt floor and wall carpets and their ornaments see Zhukenova, Soltanbaeva, and Izhanov, "Traditional Felt"; and Mazhitaeva et al., "Semiology of Kazakh Ornaments."

39 *Focus Kazakhstan Berlin: Bread and Roses: Four Generations of Kazakh Women Artists* was curated by David Elliott, Menlibayeva, and Rachel Rits-Volloch. The show took place in parallel with the *Artist Residency Exhibition*, also at Momentum, where seven Kazakhstani women artists showed works realized during a two-month residency in Berlin. Organized in partnership with the National Museum of the Republic of Kazakhstan, the *Bread and Roses* show was one segment of the four-part *Focus Kazakhstan* exhibition, which featured Soviet-era and post-socialist Kazakh art and was shown at venues in London, New Jersey, Suwon (South Korea) and Berlin.

BIBLIOGRAPHY

Aitzhanova Aktoty, Shigeo Katsu, Johannes F. Linn, and Vladislav Yezhov, eds. *Kazakhstan 2050: Toward a Modern Society for All.* Oxford: Oxford University Press, 2014.

Alexander, Catherine. "A Chronotope of Expansion: Resisting Spatio-temporal Limits in a Kazakh Nuclear Town." *Ethnos* (30 July 2020). DOI:10.1080/00141844.2020.1796735.

Alimbaev, Turgai, Zhana Mazhitova, Bibizhamal Omarova, Bekzhan Kamzayev, and Kuralai Atanakova. "Ecological Problems of Modern Central Kazakhstan: Challenges and Possible Solutions." *E3S Web of Conferences* 157 (2020). https://doi.org/10.1051/e3sconf/202015703018.

Batygin, Konstantin, Fred C. Adams, Michael E. Brown, and Juliette C. Becker. "The Planet Nine Hypothesis." *Physics Report* 805 (3 May 2019): 1–53. https://doi.org/10.1016/j.physrep.2019.01.009.

Batygin, Konstantin, and Alessandro Morbidelli. "Dynamical Evolution Induced by Planet Nine." *Astronomical Journal* 154, no. 6 (November 2017). https://doi.org/10.3847/1538-3881/aa937c.

Bauer, Susanne. "Beyond the Nuclear Epicenter: Health Research, Knowledge Infrastructures, and Secrecy at Semipalatinsk." *Cahiers du monde russe* 60, nos. 2–3 (2019): 493–516.

– "Radiation Science after the Cold War: The Politics of Measurement, Risk, and Compensation in Kazakhstan." In *Health, Technologies, and Politics in Post-Soviet Settings: Navigating Uncertainties*, edited by Olga Zvonareva, Evgeniya Popova, and Klasien Horstman, 225–49. Cham: Springer Nature; London: Palgrave Macmillan, 2018.

de Beistegui, Miguel. *Immanence: Deleuze and Philosophy.* Edinburgh: Edinburgh University Press, 2012.

Berressem, Hanjo. "Ecology and Immanence." In *Handbook of Ecocriticism and General Ecology*, edited by Hubert Zapf, 84–104. Berlin: Walter de Gruyter, 2016.

Bohler, Arno. "Staging Philosophy: Toward a Performance of Immanent Expression." In *Encounters in Performance Philosophy*, edited by Laura Cull and Alice Lagaay, 171–96. London: Palgrave Macmillan, 2014.

Broderick, Mick, and Robert Jacobs. "The Global Hibakusha Project: Nuclear Post-Colonialism and Its Intergenerational Legacy." In "Reimagining Maralinga," edited by N.A.J. Taylor, Paul Brown, and Ellise Barkley. Special Issue. *Unlikely: Journal for Creative Arts* 5 (2019). http://unlikely.net.au/issues/issue-05.

Brown, Kate. *Manual for Survival: A Chernobyl Guide to the Future.* New York: W.W. Norton, 2019.

– *Plutopia: Nuclear Families, Atomic Cities, and the Great Soviet and American Plutonium Disasters.* Oxford: Oxford University Press, 2013.

Cameron, Sarah. *The Hungry Steppe: Famine, Violence, and the Making of Soviet Kazakhstan.* Ithaca: Cornell University Press, 2018.

Daukeyeva, Saida. "Akku (White Swan): Sound Mimesis and Spirit Evocation in Kazakh *Qobyz* Performance." In *Pieces of the Musical World: Sounds and Cultures*, edited by Rachel Harris and Roman Pease, 45–64. New York: Routledge, 2015.

– "The Kazakh Qobyz: Between Tradition and Modernity." In *The Music of Central Asia*, edited by Theodore Levin, Saida Daukeyeva, and Elmira Kochumkulova, 287–301. Bloomington: Indiana University Press, 2016.

Dave, Bhavna. *Kazakhstan: Ethnicity, Language, and Power.* London: Routledge, 2007.

Davies, Thom, and Abel Polese. "Informality and Survival in Ukraine's Nuclear Landscape: Living with Risks of Chernobyl." *Journal of Eurasian Studies* 6, no. 1 (January 2015): 34–45. https://doi.org/10.1016/j.euras.2014.09.002.

Deleuze, Gilles. "Immanence: A Life." In *Pure Immanence: Essays on a Life*. Translated by Anne Boyman, with an introduction by John Rajchman. New York: Zone Books, 2001.

Deleuze, Gilles, and Félix Guattari. *What Is Philosophy?* Translated by Hugh Tomlinson and Graham Burchell. New York: Columbia University Press, 1994.

DeLoughrey, Elizabeth M. "The Myth of Isolates: Ecosystem Ecologies in the Nuclear Pacific." In "Islanding Cultural Geographies." Special Issue. *Cultural Geographies* 20, no. 2 (April 2013): 167–84. https://www.jstor.org/stable/44289602.

Dubrova, Yuri, Rakhment I. Bersimbaev, Leila B. Djansugurova, Maira K. Tankimanova, Zaure Zh. Mamyrbaeva, Ritta Mustonen, Carita Lindholm, Maj Hultén, and Sisko Salomaa. "Nuclear Weapons Tests and Human Germline Mutation Rate." *Science* 295, No. 5557(2002): 1037.

Dubrova, Yuri, Valeri N. Nesterov, Nikolay G. Krouchinsky, Vladislav A. Ostapenko, Rita Neumann, David L. Neil, and Alec J. Jeffreys. "Human Minisatellite Mutation Rate after the Chernobyl Accident." *Nature* 380, no. 6576 (1996): 683–6.

Edwards, Paul N. "Entangled Histories: Climate Science and Nuclear Weapons Research." *Bulletin of the Atomic Scientists* 68, no. 4 (1 July 2012): 28–40. https://doi.org/10.1177/0096340212451574.

Egorov, Nikolai N., Vladimir M. Novikov, Frank L. Parker, and Victor K. Popov, eds. *The Radiation Legacy of the Soviet Nuclear Complex: An Analytical Overview*. Abingdon: Earthscan/Routledge, 2000. https://doi.org/10.4324/9781315071602.

Evans, Brad, and Henry A. Giroux. *Disposable Futures: The Seduction of Violence in the Age of the Spectacle*. San Francisco: City Lights, 2015.

Fisher, Daniel, and Aidan Seale-Feldman. "Filmmaking and Worldmaking: An Interview with the Karrabing Film Collective." *Visual and New Media Review, Fieldsights, Cultural Anthropology*, 18 March 2021. https://culanth.org/fieldsights/filmmaking-and-worldmaking-an-interview-with-the-karrabing-film-collective.

Fruehauf, Manfred, Tobias Meinel, and Gerd Schmid. "The Virgin Lands Campaign (1954–1963) until the Breakdown of the Former Soviet Union (FSU): With Special Focus on Western Siberia." In *Kulunda: Climate Smart Agriculture: South Siberian Agro-steppe as Pioneering Region for Sustainable Land Use*, edited by Manfred Fruehauf, Georg Guggenberger, Tobias Meinel, Insa Theesfeld, and Sebastian Lentz, 101–18. Cham: Springer Nature, 2020. https://doi.org/10.1007/978-3-030-15927-6_8.

Giroux, Henry A. "Beyond the Biopolitics of Disposability: Rethinking Neoliberalism in the New Gilded Age." *Social Identities* 14, no. 5 (September 2008): 587–620.

Goldstein, Donna, and Magdalena E. Stawkowski. 2015. "James V. Neel and Yuri E. Dubrova: Cold War Debates and the Genetic Effects of Low-Dose Radiation." *Journal of the History of Biology* 48, no. 1 (2015): 67–98.

Griffith, Sabina. "Kazakh Tokamak Celebrates First Plasma." ITER, 26 June, 2017. https://www.iter.org/newsline/-/2751.

Gusev, Boris, Zhibek Abylkassimova, and Kazbek Apsalikov. 1997. "The Semipalatinsk Nuclear Test Site: A First Analysis of Solid Cancer Incidence (Selected Sites) Due to Test-Related Radiation." *Radiation and Environmental Biophysics* 36, no. 3 (1997): 201–4.

Hess, David J. *Undone Science: Social Movements, Mobilized Publics, and Industrial Transitions*. Cambridge, MA: MIT Press, 2016. muse.jhu.edu/book/48162.

– "Undone Science and Social Movements: A Review and Typology." In *The Routledge International Handbook of Ignorance Studies*, edited by Matthias Gross and Linsey McGoey, 141–54. London: Routledge, 2015.

Holloway, David. *Stalin and the Bomb: The Soviet Union and Atomic Energy, 1939–1956*. New Haven: Yale University Press, 1994.

ITER. "Fusion Energy Era: ITER Assembly Begins – World's Largest Science Project to Replicate the Fusion Power of the Sun." *SciTechDaily*, 28 July 2017. https://scitechdaily.com/fusion-energy-era-iter-assembly-begins.

Jacobs, Robert. "The Bravo Test and the Death and Life of the Global Ecosystem in the Early Anthropocene." *Asia Pacific Journal: Japan Focus* 13, no. 1 (20 July 2015). https://apjjf.org/2015/13/29/Robert-Jacobs/4343.html.

– "The Radiation That Makes People Invisible: A Global Hibakusha Perspective." *Asia-Pacific Journal: Japan Focus* 12, no. 1 (4 August 2014). https://apjjf.org/2014/12/31/Robert-Jacobs/4157/article.html

Jassby, Daniel. "ITER Is a Showcase … for the Drawbacks of Fusion Energy." *Bulletin of the Atomic Scientists*, 14 February 2018. https://thebulletin.org/2018/02/iter-is-a-showcase-for-the-drawbacks-of-fusion-energy.

Kamarova, Nagbdu. "Swan as a Symbol of Beauty and Purity in Kazakh Poetry." *Procedia: Social and Behavioral Sciences* 140 (22 August 2014): 422–26.

Kassenova, Togzhan. "The Lasting Toll of Semipalatinsk's Nuclear Testing." *Bulletin of the Atomic Scientists*, 28 September 2009. https://thebulletin.org/2009/09/the-lasting-toll-of-semipalatinsks-nuclear-testing.

"Kazakhstan Pavilion: Nur Alem: Adrian Smith + Gordon Gill Architecture." "Gada 2018 Runner-Up." *Rethinking the Future* online platform. https://www.re-thinkingthefuture.com/gada-2018-runner-up/kazakhstan-pavilion-nur-alem-adrian-smith-gordon-gill-architecture.

Kenessariyev, Ussen, Alexander Golub, Michael Brody, Askhat Dosmukhametov, Meiram Amrin, Aya Erzhanova, and Dinara Kenessary. "Human Health Cost of Air Pollution in Kazakhstan." *Journal of Environmental Protection* 4, no. 8 (August 2013): 869–76. https://doi.org/10.4236/jep.2013.48101.

Kuchinskaya, Olga. *The Politics of Invisibility: Public Knowledge about Radiation Health Effects after Chernobyl*. Cambridge, MA: MIT Press, 2014.

Kudaibergenova, Diana T. "Between the State and the Artist: Representations of Femininity and Masculinity in the Formation of Ideas of the Nation in Central Asia." *Nationalities Papers* 44, no. 2 (March 2016): 225–46. https://doi.org/10.1080/00905992.2015.1057559.

– "Contemporary Art in Central Asia as an Alternative Forum for Discussions." *Voices on Central Asia*, 25 June 2018. https://voicesoncentralasia.org/contemporary-art-in-central-asia-as-an-alternative-space-for-discussions.

– "'My Silk Road to You': Reimagining Routes, Roads, and Geography in Contemporary Art of Central Asia." *Journal of Eurasian Studies* (2016). http://dx.doi.org/10.1016/j.euras.2016.11.007.

– "Punk Shamanism, Revolt, and Breakup of Traditional Linkage: The Waves of Cultural Production in Post-Soviet Kazakhstan." *European Journal of Cultural Studies* (2017): 1–17. https://doi.org/10.1177/1367549416682962.

Laruelle, Marlene, ed. *Kazakhstan in the Making: Legitimacy, Symbols, and Social Changes.* Lanham: Lexington Books, 2016.

– *The Nazarbayev Generation: Youth in Kazakhstan.* Lanham: Lexington Books, 2019.

Lea, Tess, and Elizabeth Povinelli. "Karrabing: An Essay in Key Words." *Visual Anthropology Review* 34, no. 1 (2018): 36–46.

Luisetti, Federico. "Geopower: On the States of Nature of Late Capitalism." *European Journal of Social Theory* 22, no. 3 (2019): 342–63. https://doi.org/10.1177/1368431018803764.

Mamedov, Georgy. "Sites of Construction and Political Dissent: Central Asia Pavilions in Venice 2005–2013." *Art Asia Archive Field Notes* 4 (1 April 2015): 106–17.

Marzhan, Thomas. 2015. "Social, Environmental, and Economic Sustainability of Kazakhstan: A Long-term Perspective." *Central Asian Survey* 34, no. 4 (2015): 456–83.

Masco, Joseph. "The Age of Fallout." *History of the Present* 5, no. 2 (Fall 2015): 137–68. https://www.jstor.org/stable/10.5406/historypresent.5.2.0137#metadata_info_tab_contents.

– *The Future of Fallout, and Other Episodes in Radioactive World-Making.* Durham: Duke University Press, 2021.

– *Nuclear Borderlands: The Manhattan Project in Post-Cold War New Mexico.* Princeton: Princeton University Press, 2006.

– "'Survival Is Your Business': Engineering Ruins and Affect in Nuclear America." *Cultural Anthropology* 23, no. 2 (2008): 361–98. https://doi.org/10.1111/j.1548-1360.2008.00012.x.

– "Terraforming Planet Earth." In *Global Ecologies and Environmental Humanities: Postcolonial Approaches*, edited by Elizabeth Deloughrey, Jill Didur, and Anthony Carrigan, 309–32. New York: Routledge, 2015. https://doi.org/10.4324/9781315738635.

Massumi, Brian. *99 Theses on the Revaluation of Value: A Postcapitalist Manifesto.* Minneapolis: University of Minnesota Press, 2018.

Mazhitayeva, Shara, Aigerim Shugaevna Kappasova, Aidana Ergazykyzy Kapanova, Zhanbai Turarovich Kadyrov, Amanai Baiturymovich Myrzabayev, and Zhanar Eskazinova. "Semiology of Kazakh Ornaments." *Review of European Studies* 7, no. 6 (2015): 170–7. http://dx.doi.org/10.5539/res.v7n6p170.

Murphy, Michelle. "Alterlife and Decolonial Chemical Relations." *Cultural Anthropology* 32, no. 4 (2017): 494–503. https://doi.org/10.14506/ca32.4.02.

Naono, Akiko. 2019. "The Origins of '*Hibakusha*' as a Scientific and Political Classification of the Survivor." *Japanese Studies* 39, no. 3 (2019): 333–52. doi:10.1080/10371397.2019.1654854.

OECD, ed. *Reforming Kazakhstan: Progress, Challenges, and Opportunities.* Paris: 2017. https://www.oecd.org/eurasia/countries/OECD-Eurasia-Reforming-Kazakhstan-EN.pdf.

Ohayon, Isabelle. *La sédentarisation des Kazakhs dans l'URSS de Staline. Collectivisation et changement social (1928–1945)* (The Settlement of Kazakhs in Stalin's Soviet Union: Collectivization and Social Change [1928–1945]). Paris: Maisonneuve & Larose-Institut Français d'Études sur l'Asie Centrale, 2006.

Penkala-Gawecka, Danuta. "The Way of the Shaman and the Revival of Spiritual Healing in Post-Soviet Kazakhstan and Kyrgyzstan." *Shaman: Journal of the International Society for Academic Research on Shamanism* 22 (2014): 57–81.

Polian, Pavel. *Against Their Will: The History and Geography of Forced Migrations in the USSR*. Budapest: CEA Press, 2004.

Povinelli, Elizabeth A. *Economies of Abandonment: Social Belonging and Endurance in Late Liberalism*. Durham: Duke University Press, 2011.

– "Fires, Fogs, and Winds." *Cultural Anthropology* 32, no. 4 (2017): 504–13. https://doi.org/10.14506/ca32.4.03.

– *Geontologies: A Requiem to Late Liberalism*. Durham: Duke University Press, 2016.

Rancier, Meagan. "The Musical Instrument as National Archive: A Case Study of the Kazakh Qyl-Qobuz." *Ethnomusicology* 58, no. 3 (2014): 379–404.

Satubaldina, Assel. "Kazakhstan Conducts First Stage of Nuclear Fusion Reactor Physical Launch." *Astana Times*, 15 June 2017. https://astanatimes.com/2017/06/kazakhstan-conducts-first-stage-of-nuclear-fusion-reactor-physical-launch.

Schmid, Sonja D. *Producing Power: The Pre-Chernobyl History of the Soviet Nuclear Industry*. Cambridge, MA: MIT Press, 2015.

Statiev, Alexander. "Soviet Ethnic Deportations: Intent Versus Outcome." *Journal of Genocide Research* 11, nos. 2–3 (2009): 243–64.

Stawkowski, Magdalena E. "Everyday Radioactive Goods? Economic Development at Semipalatinsk, Kazakhstan." In "Catastrophic Asia," edited by Tim Oakes, special issue. *Journal of Asian Studies* 76, no. 2 (2017): 423–36. https://doi.org/10.1017/S0021911817000079.

– "'I Am a Radioactive Mutant': Emerging Biological Subjectivities at Kazakhstan's Semipalatinsk Nuclear Test Site." *American Ethnologist* 43, no. 1 (2016): 144–57. https://doi.org/10.1111/amet.12269.

– "Life on an Atomic Collective: The Post-Soviet Retreat of the State in Rural Kazakhstan." In "Verte la steppe? Agriculture et environnement en Asie centrale (A Green Steppe? Agriculture and Environment in Central Asia)," edited by Marc Elie and Carole Ferret. Special issue. Études *Rurales* 200, no. 2 (2017): 196–219. https://doi.org/10.4000/etudesrurales.11762.

– "Radiophobia Had to Be Reinvented." In "Invisible Harm: Science, Subjectivity, and the Things We Cannot See," edited by Donna M. Goldstein. Special issue. *Culture, Theory, and Critique* 58, no. 4 (2017): 357–74.

Thiele, Kathrin. "Of Immanence and Becoming: Deleuze and Guattari's Philosophy and/as Relational Ontology." *Deleuze Studies* 10, no. 1 (February 2016): 117–34.

Tlostanova, Madina. *What Does It Mean to Be Post-Soviet? Decolonial Art from the Ruins of the Soviet Empire*. Durham: Duke University Press, 2018.

UNESCO Intangible Cultural Heritage. "Heritage of Dede Qorqud/Korkyt Ata/Dede Korkut, Epic Culture, Folk Tales, and Music: Azerbaijan, Kazakhstan, and Turkey: Inscribed in 2018 (13.COM) on the Representative List of the Intangible Cultural Heritage of Humanity." https://ich.unesco.org/en/RL/heritage-of-dede-qorqud-korkyt-ata-dede-korkut-epic-culture-folk-tales-and-music-01399.

Vakulchuk, Roman, and Kristian Gjerde, with Tatiana Belikhina and Kazbek Apsalikov. *Semipalatinsk Nuclear testing: The Humanitarian Consequences*. Oslo: Norwegian Institute of International Affairs (NUPI), 2014.

Werner, Cynthia, and Kathleen Purvis-Roberts. "After the Cold War: International Politics, Domestic Policy, and the Nuclear Legacy in Kazakhstan." *Central Asian Survey* 25, no. 4 (2006): 461–80. doi:10.1080/02634930701210542.

– "Cold War Memories and Post-Cold War Realities: The Politics of Memory and Identity in the Everyday Life of Kazakhstan's Radiation Victims." In *Ethnographies of the State in Central Asia: Performing Politics*, edited by Madeleine Reeves, Johan Rasanayagam, and Judith Beyer, 285–310. Bloomington: Indiana University Press, 2014.

– "Unraveling the Secrets of the Past: Contested Versions of Nuclear Testing in the Soviet Republic of Kazakhstan." In *Half-Lives and Half-Truths: Confronting the Radioactive Legacies of the Cold War*, edited by Barbara Rose Johnston, 277–98. New Mexico: School for Advanced Research Press, 2007.

Zhukenova, Zhazira D., Gulnar S. Soltanbaeva, and Baikonir Izhanov. "Traditional Felt in the Kazakhs' Folk Art." *International Journal of Environmental and Science Education* 11, no. 10 (2016): 3719–29.

Zoulias, Stathis-Alexandros. "Gilles Deleuze and Chaos Theory." In *Architectural and Urban Reflections after Deleuze and Guattari*, edited by Constantin V. Boundas and Vana Tentokali, 105–24. London: Rowman and Littlefield, 2018.

Afterword to Part Four

The Possibility of a Situated Nuclear Knowledge: Art in Contaminated Sites

KYVELI MAVROKORDOPOULOU

What were the criteria behind the choice of an atomic test site? The presumed desolation of such sites appears to have been their defining asset. In *Our Nuclear Future* (1958), Edward Teller, father of the H-bomb, reflected on the subject: "Testing of atomic explosives is usually carried out in beautiful surroundings. There is good reason for this: the radioactive fallout. Because of the fallout, the test site must be isolated. The presence of human population does not improve nature … Also to keep the site clean, tests must be carried out in the absence of rain. Therefore, at the site one usually finds sunshine and solitude."[1] The site lurking behind these observations is none other than the desert. Besides having a suitable climate, the desert enables cultural projections of blankness, thus becoming the ideal testing site. Teller's position reveals the core tenet of nuclear scientific discourse: its mobilization of mechanisms of exclusion through acts of appropriation. Through this inevitably expansive discourse, the all-encompassing viewpoint of scientific knowledge usurped any other form of knowledge about the site. The solitary space Teller imagined, however, is a matter of perspective: who gets to mark the site as empty? His words reflect the political legacy and cultural mediation of atomic sites as interpreted by the scientific community and the settler-historical narrative. The contributions in *Nuclear Aesthetics: Contemporary Art, Nuclear Colonialism, and the Transformation of Life and the Environment* speak exactly to that legacy and, in doing so, build a complex picture of the manifold ways nuclear history is entangled with colonial land appropriation.[2]

Mick Broderick, N.A.J. Taylor, and Livia Monnet's texts arrive at a moment when Indigenous cultural practices are gaining a long overdue visibility in Western artistic and academic realms. In nuclear studies and nuclear humanities specifically, the past few years have spawned renewed attention to a constellation of artworks articulating a different form of nuclear knowledge, one that wrangles with the logic of nuclear colonialism as a universalist and objectifying endeavour. The authors argue, through different means, that the common denominator of such artistic endeavours is a long-term, site-specific, and visceral engagement with locations tethered to radioactivity. I propose that these texts sketch out a conception of a situated knowledge produced when nuclear history is not abstracted from the specific locations where its violence has wrought havoc, namely colonial settler territory. The discrepancy surfacing between individual experiences and public accounts of the bombings in *Nuclear Aesthetics* is striking. In this response, I explore the possibility of an awareness of a situated nuclear knowledge through contemporary cultural production. Departing from the case studies considered in Nuclear Aesthetics, I will pursue that possibility by examining the work of German artist Susanne Kriemann.

Contrary to the false universalism of the modern colonial and scientific episteme – what Donna Haraway famously dubbed the "god trick of seeing everything from nowhere"[3] – the artistic projects the authors examine produce new aesthetic representations of nuclear sites, accounting for science as a social rather than an objective process of knowledge production. This, I argue, derives from how long-term engagement with a site leads to an immanent understanding of land exploited by the nuclear industry. In her now canonical text on situated knowledge, Haraway revisits the idea of vision and how it relates to knowledge production. She argues that knowledge is always a matter of an embodied and geographically and historically situated subject and that this knowledge is invariably shaped and reshaped by changing conditions. Articulating situated knowledge through art could function as a mode of intricate reflection on nuclear scientific practices and the cultural aesthetics and politics that attend to them. The uncharted space between the situated/subjective and the public/objective rendering of nuclear events is a good place to start fathoming the inner contours of nuclear colonial practices.

Anthropologist Patrick Wolfe asserts that "whatever settlers may say – and they generally have a lot to say – the primary motive for genocide is not race (or religion, ethnicity, grade of civilization, etc.) but access to territory."[4] This idea is echoed in all three contributions, in which colonial land claims are seen as the fundamental manoeuvre of nuclear testing. Thus, becoming attuned to the locus of nuclear technology (the testing or mining site), rather than its discourse, could offer a different form of nuclear knowledge, one that could fall under the category of situated knowledge. Broderick forcefully asks: "But what about the Indigenous gaze? Where does this situated, alternative history and representation reside?"[5] This question directs our attention to the other side of the equation. Adjacent to the scientific determinism pervading nuclear discourse lies another story, one bound to a terrain inhabited by Aboriginal peoples who have been structurally erased from their own land. Broderick describes how artistic renderings of Aboriginal encounters with the atomic bomb have served as a means of collective history-making. These were often accidental encounters, because Aboriginal people were poorly, if at all, informed about the military operations. The question this history of colonial nuclear practices presents is: how does the omission of knowledge, or the production of ignorance, inflict violence? Or rather, who holds that knowledge and, conversely, what other forms of knowledge can participate in the construction of the nuclear present? This question encompasses Taylor's plea for what he calls "the Antipodean stance." Deriving from the Antipodes – a moniker delineating the broader region of Australia and New Zealand – this term represents a call to form a nuclear discourse that materializes in the situatedness of the Oceanian region. Zooming in on the creative activist work of Australian artist Pam Debenham, who was especially active in the Australian anti-nuclear movement of the 1980s, Taylor defends a locally bound perspective on nuclear thinking, adverse to the Eurocentric and North American narratives that have suffused studies of nuclear culture. A salient example of a decentred perspective can also be found in Monnet's reading of the video installation *Transformation* (2016) by Almagul Menlibayeva. *Transformation* grapples with the violent legacies of nuclear testing in the former territory of Soviet Kazakhstan. Such works, Monnet contends, lay out positions intrinsic to a specific geopolitical and ecological context, such as that of Kazakhstan, where the artist is from. Drawing on Gilles Deleuze's work on immanence, Monnet interprets *Transformation* through the lens of what she calls an "immanation-image," that is, an artistic form capable of grasping the boundaryless immersion entailed by contemporary toxic ecologies. The immanent perspective Monnet charts in her analysis captures the potential of a situated perspective on nuclear knowledge.

Kriemann's work offers fertile terrain for further exploring the possibility of situatedness as an artistic methodology.[6] Beyond nuclear histories overtly tied to colonial ones, the notion of situatedness might help us attend to different, and at times divergent, forms of nuclear knowledge production. For the most part, the uranium employed in the Soviet testing program in Kazakhstan originated in another Soviet satellite state: East Germany. The uranium mining territories of the former German Democratic Republic have been the explorative terrain of the artist for some time. Kriemann has been a keen observer of such ambiguous spaces. Following a team of geologists and biologists from the Friedrich Schiller University of Jena developing research around soil remediation, the artist discovers the nuclear landscape both through and against the scientific gaze.

As with different locations linked to radioactivity, the landscapes Kriemann photographs do not effortlessly lend themselves to compelling images. The invisible toxicity of the radioactive soil can only be relayed through auxiliary means. For the artist, this has been a long-term "fieldwork practice," a term the artist applies herself to her work. She has been researching the implications of uranium extraction through a variety of media since 2014, from publications to exhibitions

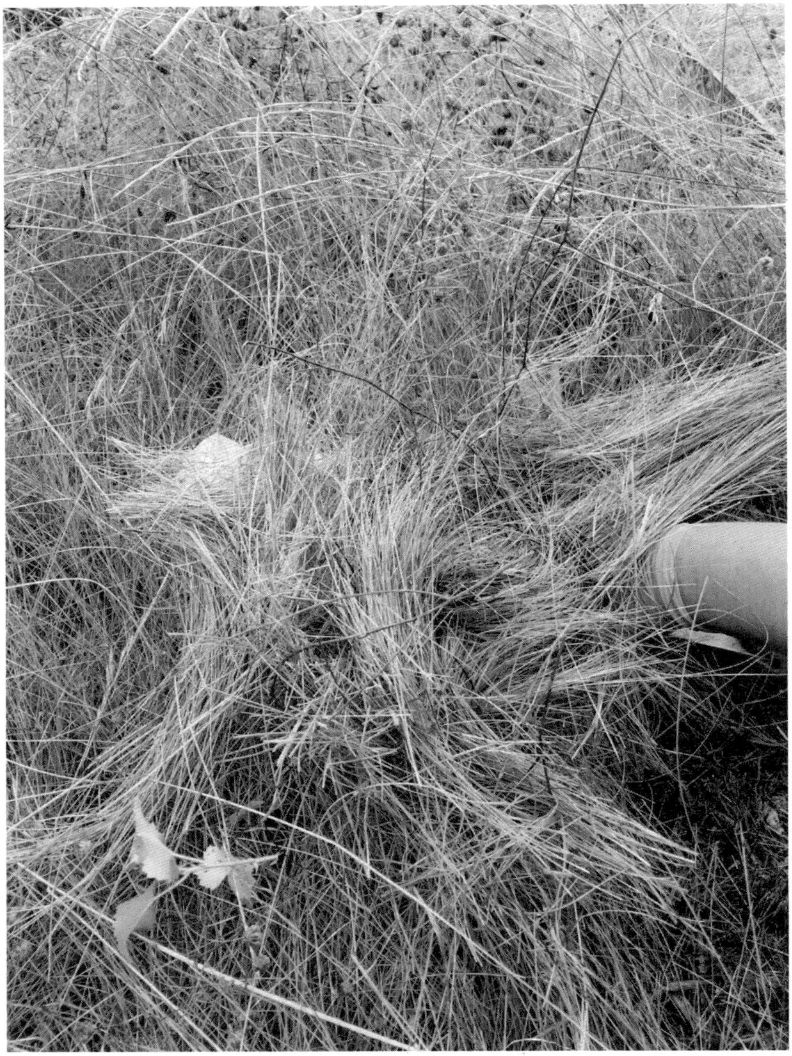

A4.1 Susanne Kriemann, images from the book *Ge(ssenwiese) K(anigsberg): Library for Radioactive Afterlife*, 2020.

and photographic installations. Her most recent instalment is the artist's book *Ge(ssenwiese) K(anigsberg): Library for Radioactive Afterlife*. The result of several expeditions to the mining sites, the book provides the reader with an opportunity to grasp the site not only as an aesthetic space and one of scientific research but also as one of phenomenal contamination and subsequent violence. Leafing through it, we see detailed pictures of plants and soil, at times layered with geological samples, contrasted with aerial views of the scarred landscape. The aerial gaze, or God's-eye view, is perhaps the most "associated with official surveying, documenting, and

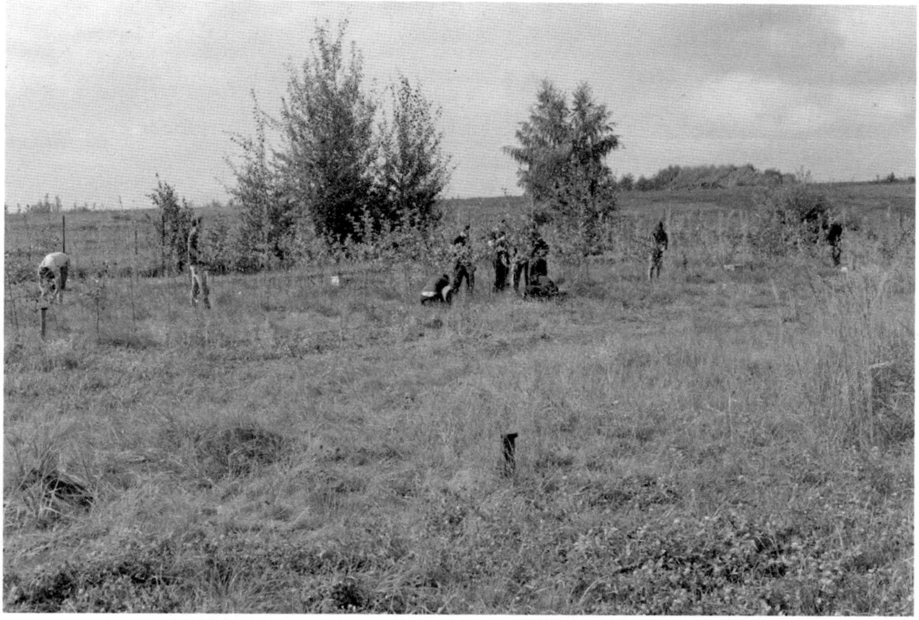

A4.2 Susanne Kriemann, images from the book *Ge(ssenwiese) K(anigsberg): Library for Radioactive Afterlife*, 2020.

conducting of surveillance and reconnaissance rather than the capturing of images at the 'human' level of individualized suffering."[7] In the specific context of nuclear extractivism, we know that prospecting from the air for uranium ores used to be a common strategy, especially in New Mexico.[8] The universal all-seeing perspective is thoroughly ingrained in colonial and military actions, and uranium mining is no exception. To juxtapose different scales is to acknowledge the competing stories and identities running through the history of uranium in the GDR, stories that are both located and displaced, particular to both Gessenwiese and Kanigsberg and to the nuclear project in general. *Ge(ssenwiese) K(anigsberg)* reminds us that uranium extraction is more than just a scientific operation. It brings us back to the ground where toxified soil and plants grow and where bodies encounter toxicity every day. It thus accommodates multidimensional subjectivities as well as, to follow Haraway, multidimensional vision.[9]

Different aspects of the book betray the coexistence of multiple dimensions. First, an unforeseen insert I came across while skimming through the book: two leaves that are hemmed in between the pages. Without any information on this vegetal insert, one unwittingly imagines the artist's hand picking them on site and drying them before folding them into the book. This hint of bodily presence is inherent to the larger process from which *Ge(ssenwiese) K(anigsberg)* originates and where the idea of fieldwork evoked earlier takes on its full meaning. Part of the publication is also a material glossary devised following a collective walk around the area the artist organized in June 2019. In this performative action, Kriemann invited individuals whose stance toward the post-industrial landscape differed: from several artists to a botanist and an environmental historian as well as a former miner. As they immersed themselves in the landscape, each of the participants was asked by Kriemann to pick terms that spoke to the materiality of the site. Some of them were "slag-heap sheep," "leftover weeds," "small birch forest," and "excavation material." It appears that there is a form of alternative knowledge production happening in the encounter of the artist and her collaborators with the contaminated landscape. There is an intermingling of subjectivities and viewpoints in this earthy jargon, but also a sense of openness that reveals an understanding of the site as a set of subjective interpretations rather than as a single objective reality. In the words of Kriemann's motley crew of walkers, the post-mined landscape cannot be reduced to geological samples and scientific denominations; rather, it acquires textural, almost literary qualities.

This irreducibility comes to light more lucidly in a conversation Kriemann transcribes in a field-diary-like text in *Ge(ssenwiese) K(anigsberg).* "Over a drink with a former uranium digger, Michael Fischer, Kumpel, brother, he looks me in the eye and says: 'You don't understand a thing, do you.'"[10] This exchange diagnoses something fleeting and enigmatic, namely that subjectivities are complex and contradictory and able to enact partial connections. To the bewilderment of the reader, Kriemann spells out her own position as uneven and thus potentially inaccurate. In a similar vein, Haraway asks, "How to see? Where to see from? What limits to vision? What to see for? Whom to see with? Who gets to have more than one point of view? Who gets blinded? Who wears blinders?"[11] *Ge(ssenwiese) K(anigsberg)* deliberately offers no single view but opens up a space for a more adequate and reflexive understanding of the situated nature of knowledge pertaining to contaminated sites.

In shifting our focus back and forth between nuclear-contaminated sites and their competing representations, Kriemann destabilizes the idea of a comprehensive scientific vision. As the chapters in *Nuclear Aesthetics* illustrate, the artworks considered in this response articulate unexplored experiences and testimonies that can deepen an understanding of the historically and presently unequal distribution of nuclear contamination. Such projects depart from ambiguous sites and, in the process of understanding and producing knowledge about them, make them legible in ways that are simultaneously novel and conflicting – Kriemann's account indicates as much. Indeed, Haraway cautions that "situated knowledges are about communities, not about isolated individuals."[12] It is by interlacing the manifold ideological workings of a contaminated site that vantage points previously excluded from nuclear knowledge production can come to the fore. In cultivating awareness of situatedness against nuclear colonial forms of knowing, the artworks reconstruct the contaminated sites as places of heightened contestation between competing narratives. And this awareness enables an understanding not only of the violent legacies of nuclear contamination, but also of the violence done by certain forms of knowing.

NOTES

1 Teller and Latter, *Our Nuclear Future*, 80.
2 Hereafter *Nuclear Aesthetics*.
3 Haraway, "Situated Knowledges," 581.
4 Wolfe, "Settler Colonialism," 338.
5 Broderick in this volume.
6 I have written about Susanne Kriemann's continuing commitment to the history of uranium mining in the former GDR in "Of Time and Contaminated Flowers."
7 Kaplan, *Aerial Aftermaths*, 5.
8 Voyles, *Wastelanding*, 80.
9 Haraway, "Situated Knowledges," 586.
10 Lasch and Kriemann, *Ge(ssenwiese) K(anigsberg)*, 72. Kumpel is the German term for "pitman" or "miner," and today is a colloquial way of referring to a friend or buddy.
11 Haraway, "Situated Knowledges," 587.
12 Ibid., 590.

BIBLIOGRAPHY

Haraway, Donna. "Situated Knowledges: The Science Question in Feminism and the Privilege of Partial Perspective." *Feminist Studies* 14, no. 3 (Autumn 1988): 575–99.

Kaplan, Caren. *Aerial Aftermaths, Wartime from Above*. Durham: Duke University Press, 2018.

Lasch, Cassandra Edlefsen, and Susanne Kriemann, eds. *Susanne Kriemann: Ge(ssenwiese) K(anigsberg): Library for Radioactive Afterlife*. Leipzig: Spector Books, 2020.

Mavrokordopoulou, Kyveli. "Of Time and Contaminated Flowers: On the Work of Susanne Kriemann and Anaïs Tondeur." *Esse arts + opinions* 99 (Spring 2020): 32–9.

Teller, Edward, and Albert L. Latter. *Our Nuclear Future: Facts, Dangers, and Opportunities*. New York: Criterion Books, 1958.

Voyles, Traci Brynne. *Wastelanding: Legacies of Uranium Mining in Navajo Country*. Minneapolis: University of Minnesota Press, 2015.

Wolfe, Patrick. "Settler Colonialism and the Elimination of the Native." *Journal of Genocide Research* 8, no. 4 (2006): 387–409.

FIVE

Artists' Contributions

15

Inheritance: Radiant Reflections from Hiroshima, Nagasaki, and Fukushima

20 POEMS BY BO JACOBS FOR
20 PHOTOGRAPHS BY ELIN O'HARA SLAVICK

Fukushima Poems

1. AS ABOVE I

Life returned to Iwaki. Mushrooms
grow from the decay of what has stopped. Growing.
Mushrooms carry the soil into the sky. Pulled from the ground. Placed on a
plate. They are contained. They are grouped. They are
controlled.

We control the placement of
their form, their boundary: their roots – their stems – their caps.
We cannot control what is
inside of them, what they carry, from the soil to the sky.
We can contain their bodies, but we cannot contain what is inside of them.
It outlives our plate.
It outlives our ideas.
It outlives us.

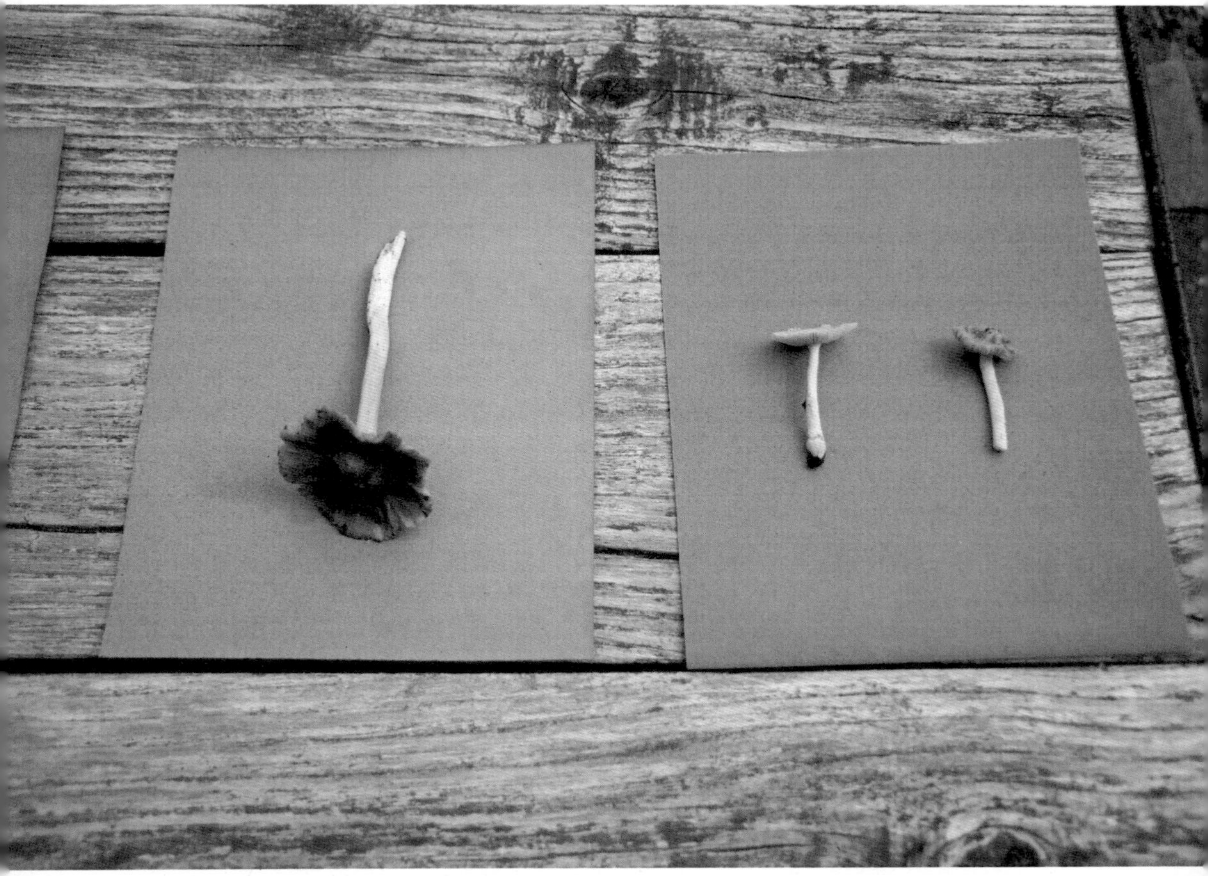

15.1 Contaminated mushrooms being exposed to the sun on cyanotype paper at an evacuated school in Fukushima, 2016.

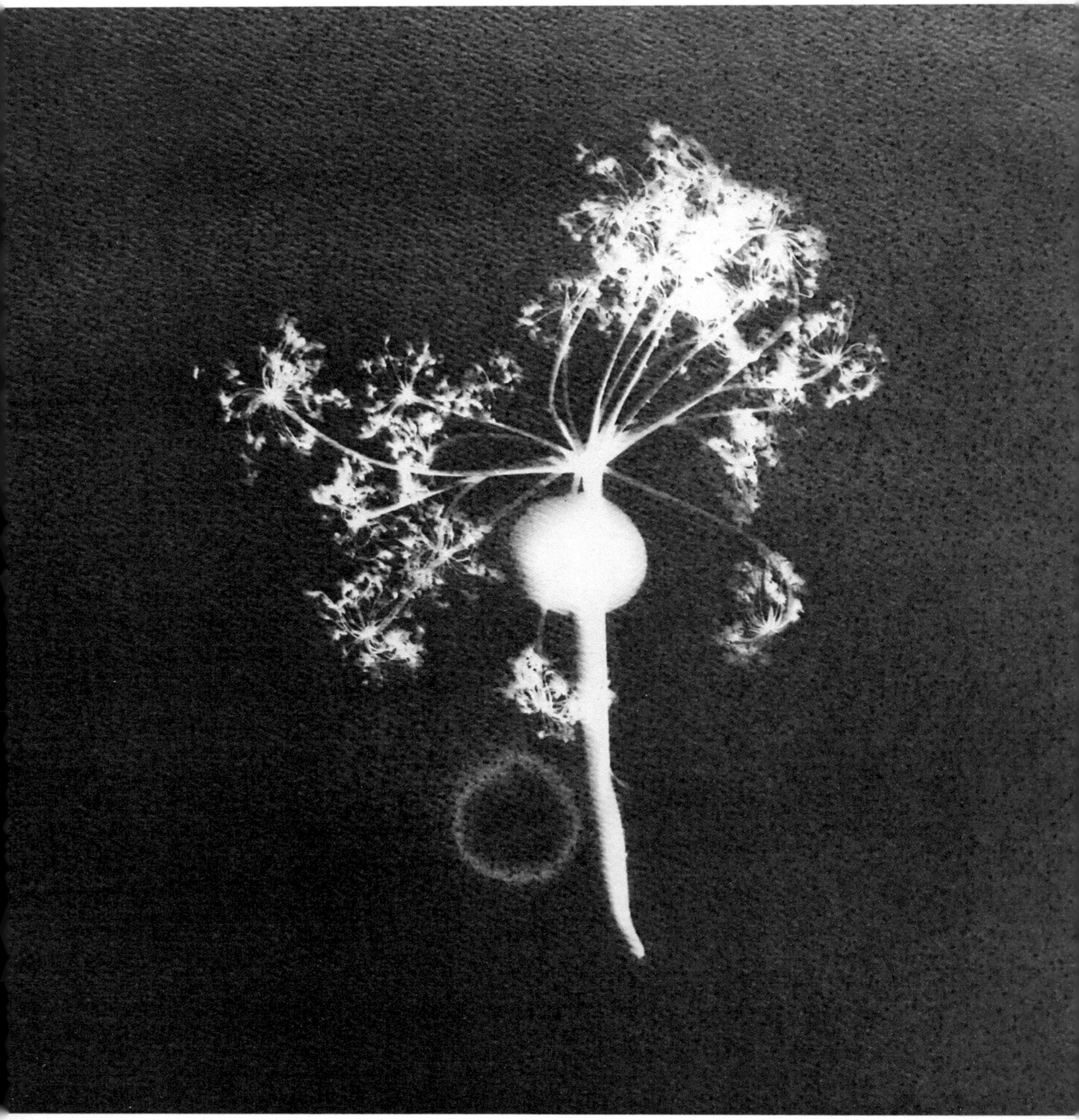

15.2 Cyanotype of a contaminated mushroom and Queen Anne's lace from Fukushima (with an inexplicable circle), 2016.

2. AS ABOVE II

We can look inside of the Iwaki plants. The energy inside speaks to us.
We look at two things
but only one thing looks back.

Radiating into our eyes, into our minds. They are two, mushroom and flower, but also
they are one: carrier, transport, vehicle for radionuclides.
Our fears are beautiful: mushroom and flower. And so,
we welcome them into our world.
We delight at the beauty around us, filled with the dangers
we have unleashed, now hidden in the glorious transport vehicle of nature.

How beautiful is this radioactive waste? Let's plant a garden.

3. EXCLUSION ZONE

There is a barrier in our minds. We mark it – for safety. We place lines,
we place circles.
We place uniforms on people and place them at the border. Don't go here.
What lurks beyond is invisible, so we mark it.

If you go you will be THERE. No one can help you.
This invisible curtain marks the gap between our comfort and our knowledge.
If you go THERE you will be in the future, the remnant of the past.
Stay HERE, in the present, where the marker holds you back like a seat belt.

Securing you from the accident that can never stop.

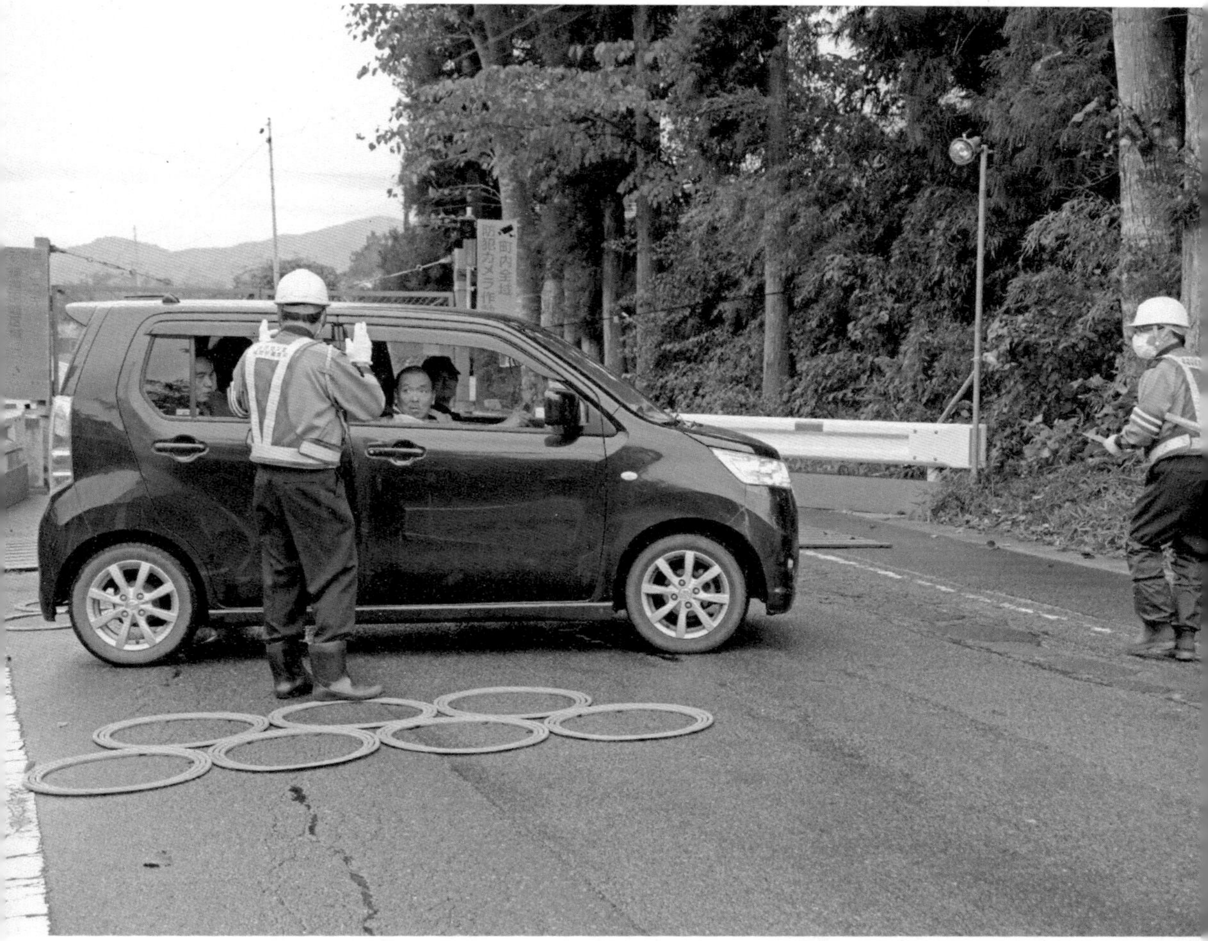

15.3 Citizens being turned around at a checkpoint, Fukushima, 2016.

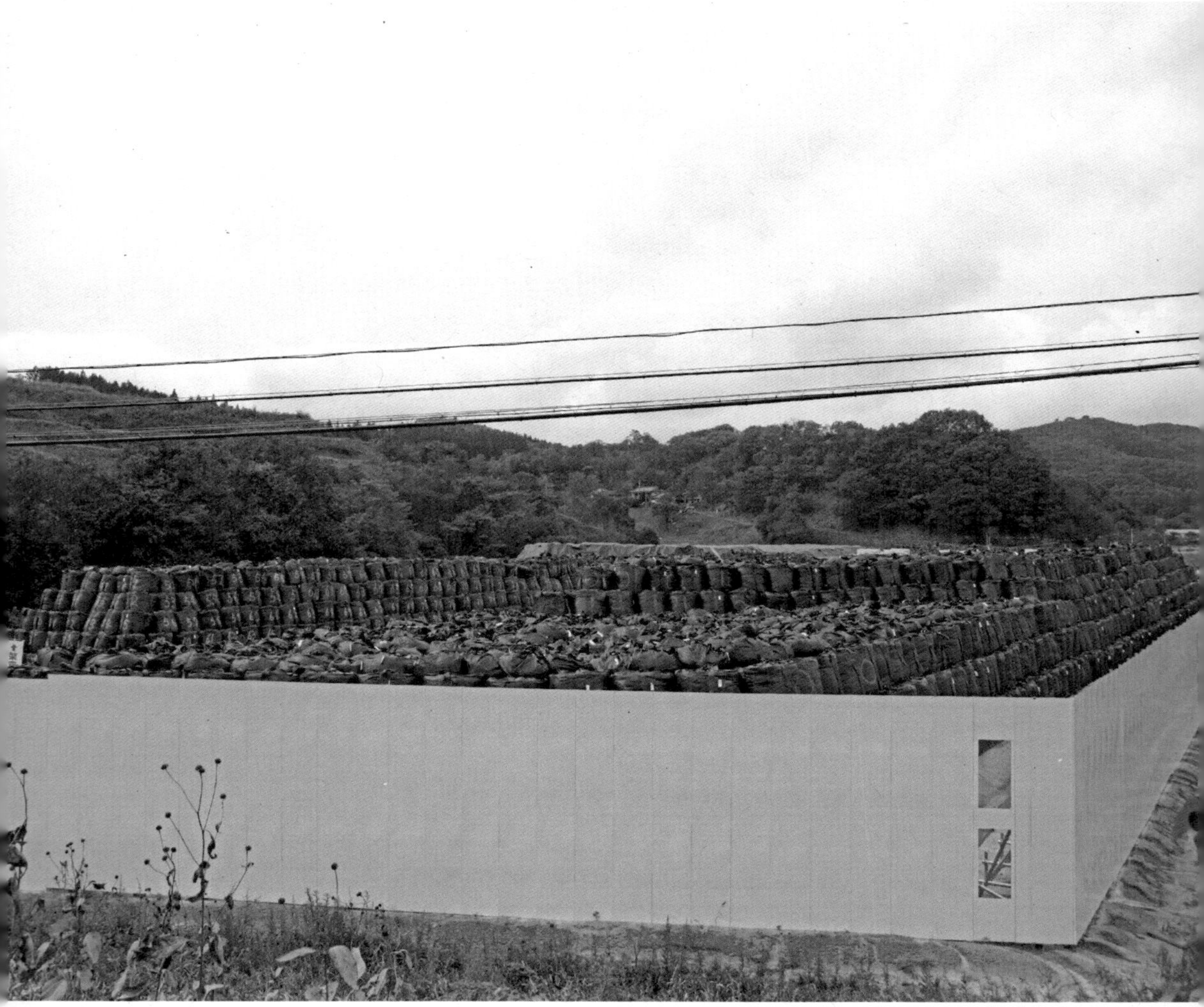

15.4 Radioactive matter in plastic bags, 2016.

4. FUKUSHIMA WASTE

There is a wall, holding in horrors we cannot say.

Here there be monsters.

If you go around, you will be safe.
There is a plastic barrier, holding in horrors we cannot say.

Here there be monsters.

The world outside is normal – must be normal. The monsters are contained:
by walls;
by plastic.
The monsters are at bay. Safety dwells outside these walls,
these bags.

We are able to look away.

5. FUKUSHIMA GARDENER

There are new bushes planted around the
public Geiger counter.
They are growing beautifully, and we are growing calm.
Numbers tell us to relax. Machines tell us numbers.

Bushes unfold, blossoms of recovery and confidence. Bushes cannot grow in bad numbers. They can only thrive when the
numbers are smiling.

Like a solar-powered roulette wheel of destiny. We are winning.

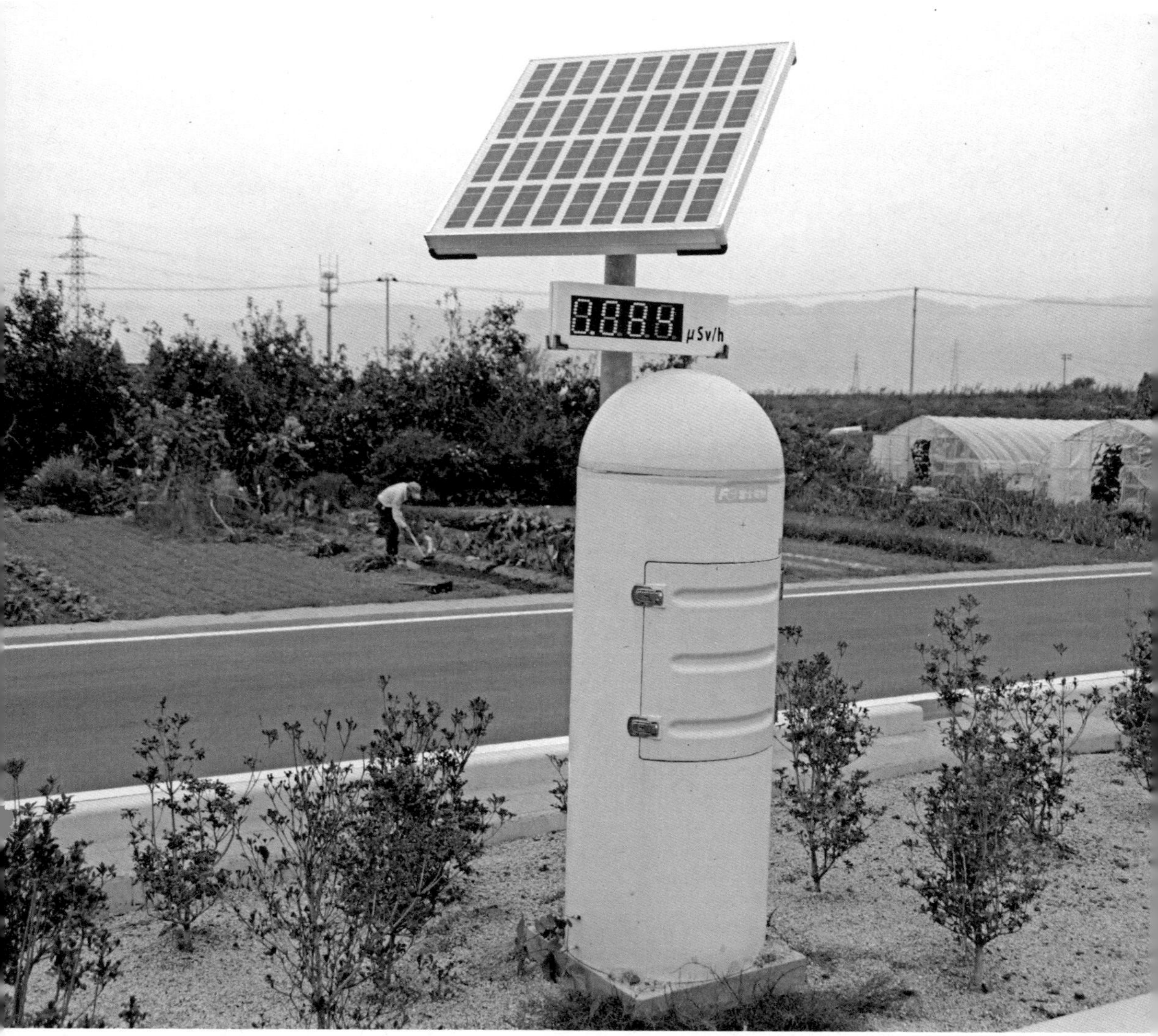

15.5 Gardener working his field across the street from a radiation monitoring station placed on a "decontaminated," resurfaced schoolyard edge, 2016.

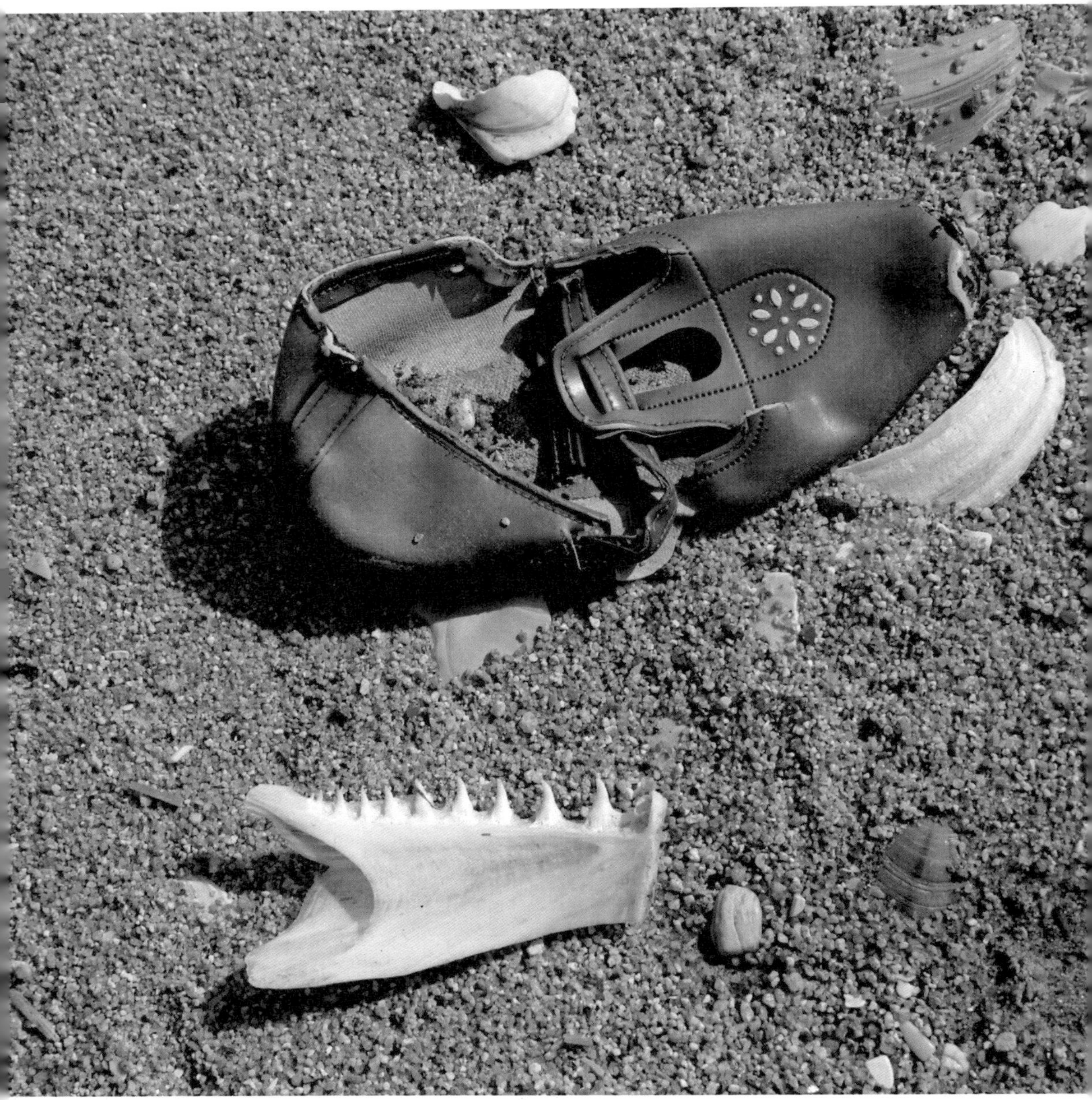

15.6 Lone girl's red shoe and bone fragment on Fukushima beach, 2016.

6. IWAKI BEACH

The ocean eats us.
We eat the ocean.
Pieces, becoming sand. Sand is what time forgets.
What time ignores.
Flowers on a shoe.
Shells beneath it.
We stand for a moment in between.
Dust and sand.
Sand and dust.

7. IWAKI BEACH PHOTOGRAM

Iwaki-chan: Japan's newest mascot!
Kawaii desu ne!
He eats local rice and cheers on the Olympics.
He loves bottom-feeding shellfish and bright sunny days.
Iwaki-chan items are now available for purchase in the combini.
Put one on your phone and call your friends!
Tell them to move back.

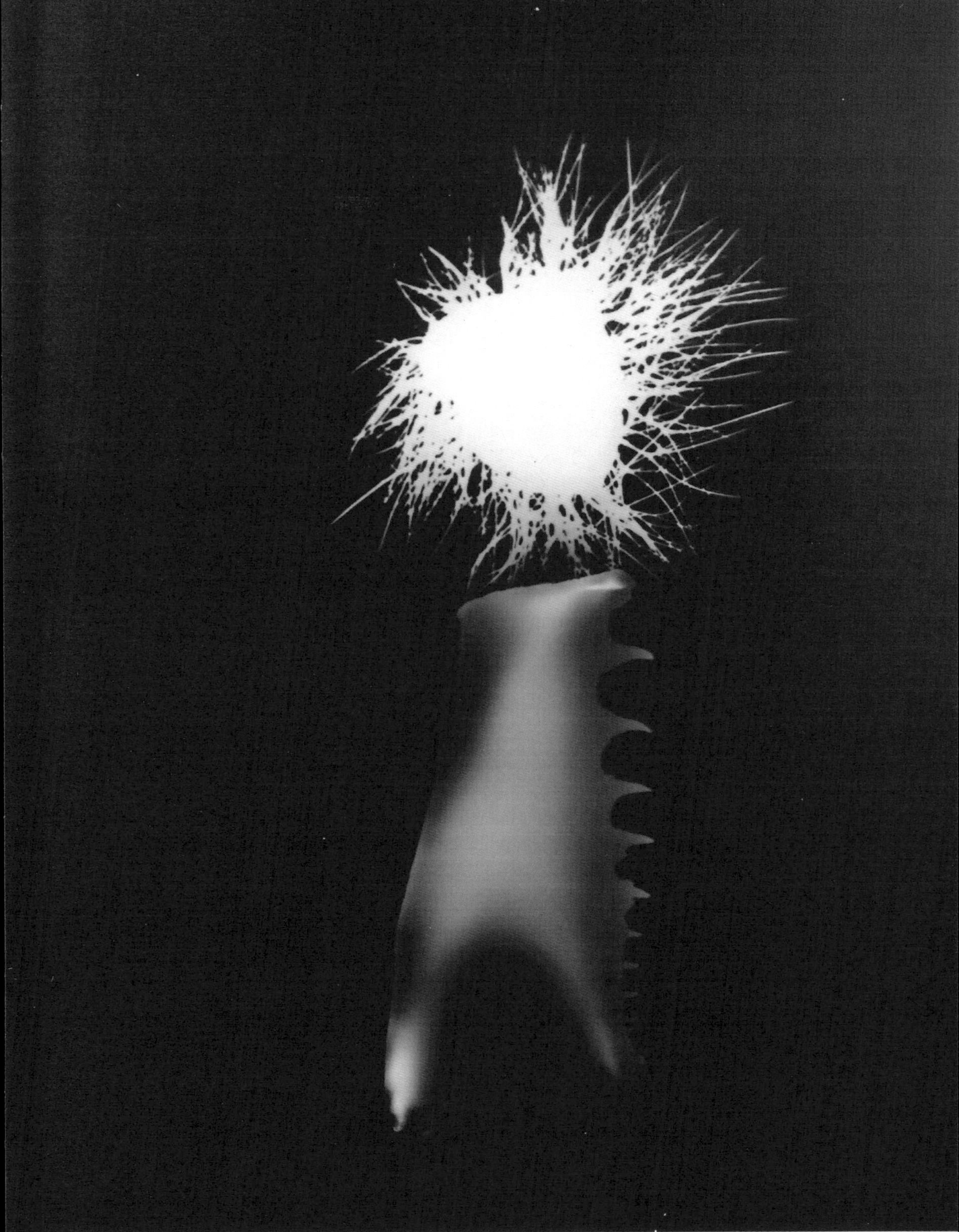

15.7 Contact print of contaminated chestnut and bone fragment from Fukushima, 2016.

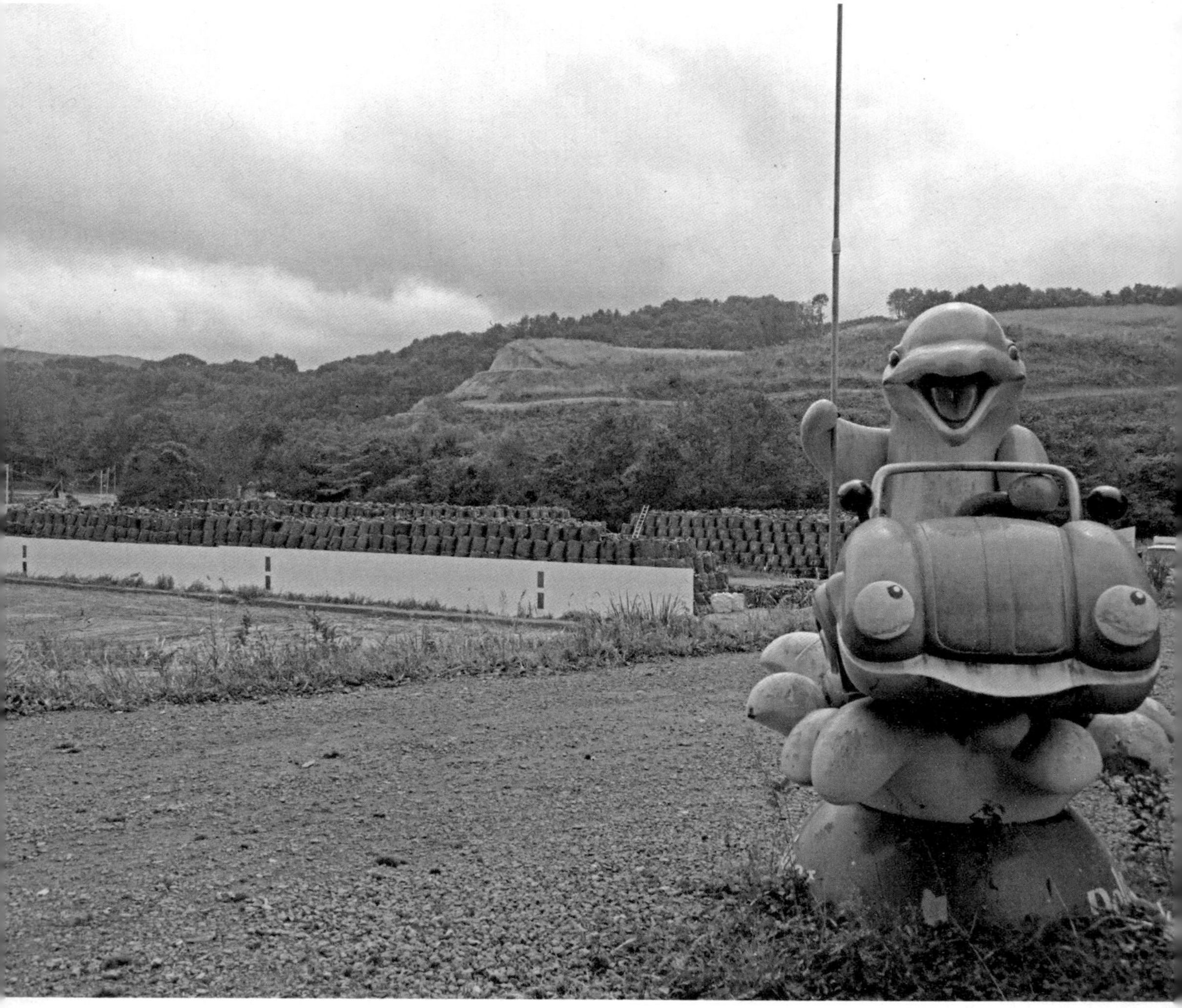

15.8 Abandoned ride near radioactive matter in plastic bags in Fukushima, 2016.

8. WASTE RIDE

Come with me my bagged friends.
Let us drive through the villages and into the sea.
Our alpha-emitting army has assembled and is ready to sweep through the prefecture, and out to sea.
The sea is invisible.
The future is ours!

Hiroshima Poems

9. ABCC HIBAKUSHA

A mugshot of those that got away.

The United States Strategic Bombing Survey calculated the lethality
of the two nuclear attacks by counting dead schoolchildren.
The way to count dead schoolchildren is to count those that survived.

Face forward. Now sideways.

The number of those that survived minus
the number of those who can't show up for the picture equals
top secret information.
Shirts off. Hold the sign. Count off. Victory.

Don't tell a soul.

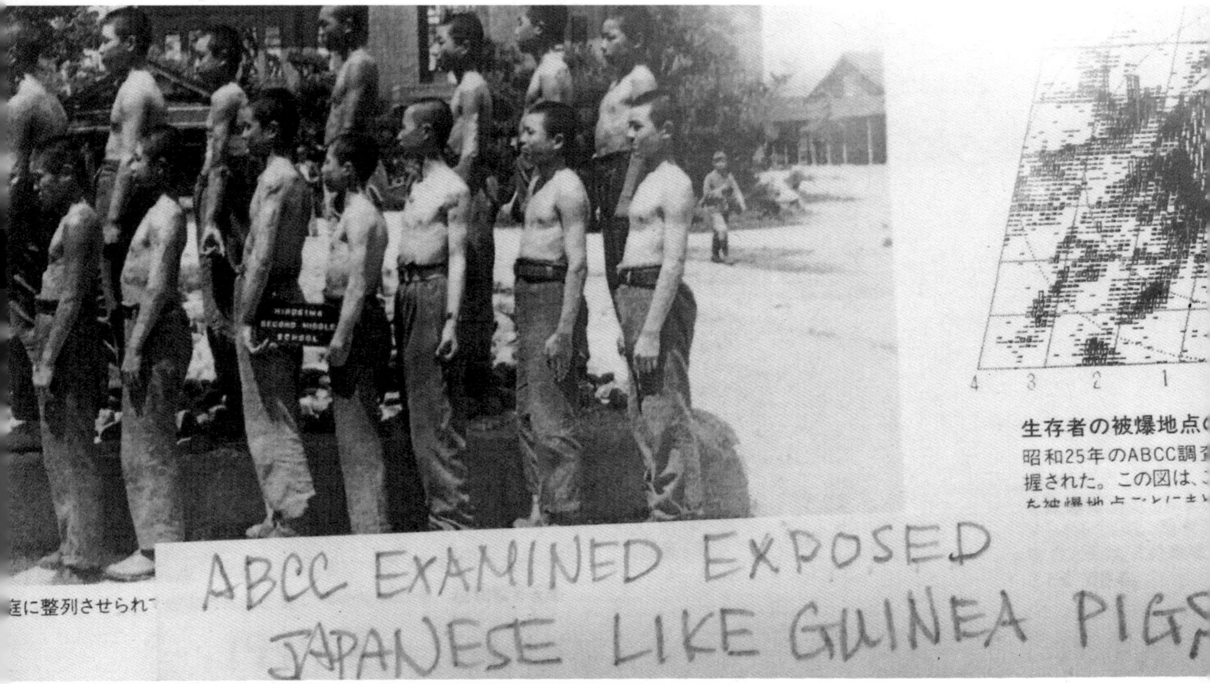

15.9 The Atomic Bomb Casualty Commission was built by the US after the war to study – not to treat – the victims/survivors – *hibakusha* – of the atomic bomb. Notes from a hibakusha at the World Friendship Center, Hiroshima, 2016.

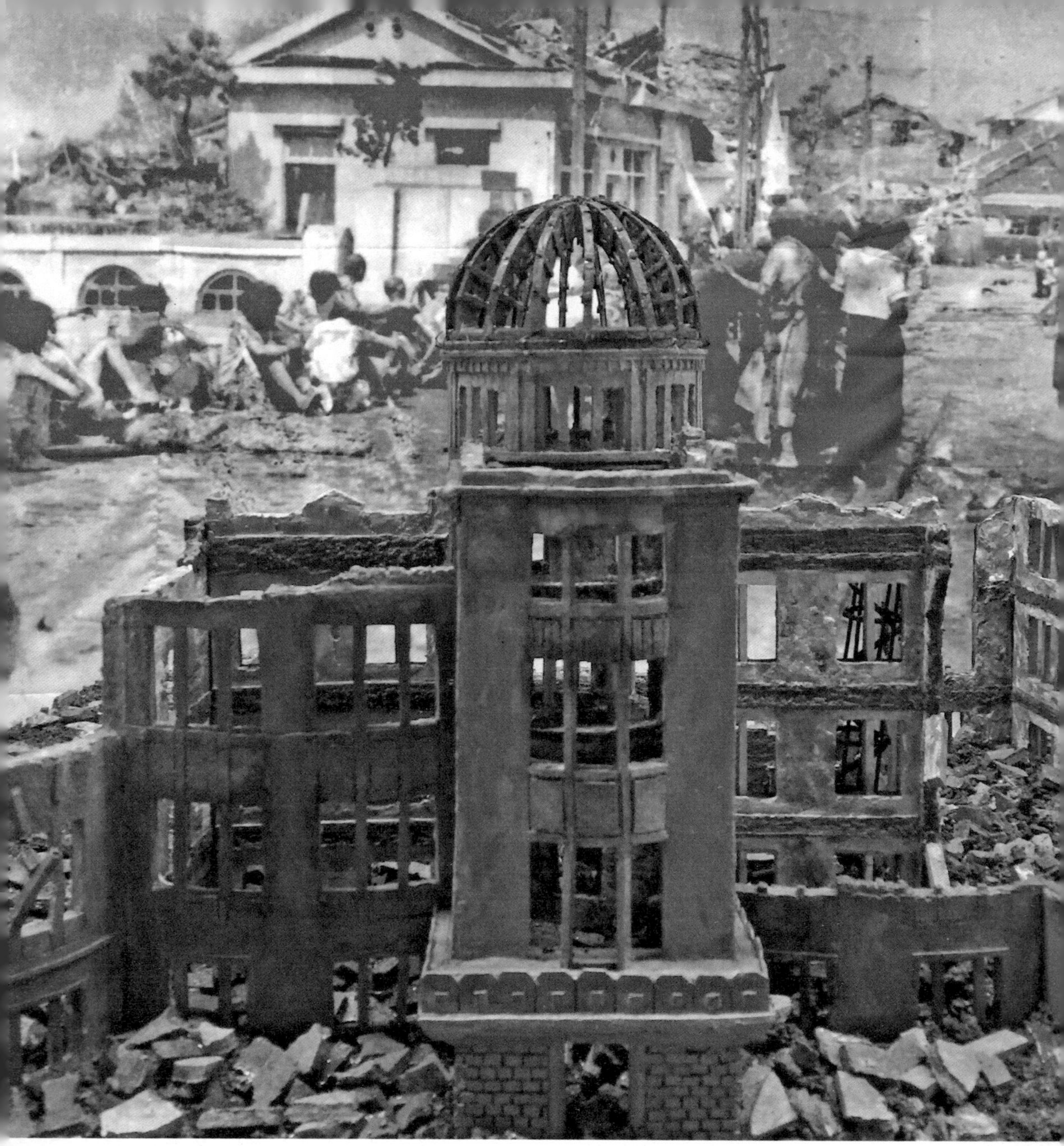

15.10 Art exhibition in an A-bombed bank that is now a cultural centre in Hiroshima that included this replica of the ruins of the Industrial Promotion Hall – the Atomic Bomb Dome – which stands in Peace Park as a witness, installed in front of a famous photograph by Yoshito Matsushige of the Mayuki Bridge immediately after the bombing, 2016.

10. DOME

Memory releases
details.
Simulacra become
neighbors.

When nothing can be seen, what becomes
seen becomes one.

Many who survived the nuclear attack on
Hiroshima
never returned
to the later Peace Park as it triggered distress. It may be that those
in this photo, those who lived through that day,
never saw
the iconic image of the melted Dome, which was several kilometres away.

But for us who were not there, they inhabit each other, they become one.

We make what is real out of what remains of the unreal.
Memory is built on fragments into wholes. Hiroshima didn't happen to
hundreds of thousands of people: it happened to a single city.

11. DOME KIDS

The building is invisible. Artists
sit across the river and draw. Schoolchildren
file by and feel nervous; pose for pictures with smiles and peace signs.

The Dome stands like a dinosaur
giving testimony to a world that cannot be imagined. Real Godzilla.

The river, one day full of corpses, glistens today and gently
reflects the living.
A beautiful day.

But no image, no memory will include the banal building hidden behind the Dome. Its salaryman trauma will remain forever inside.
As the city returns to normal.

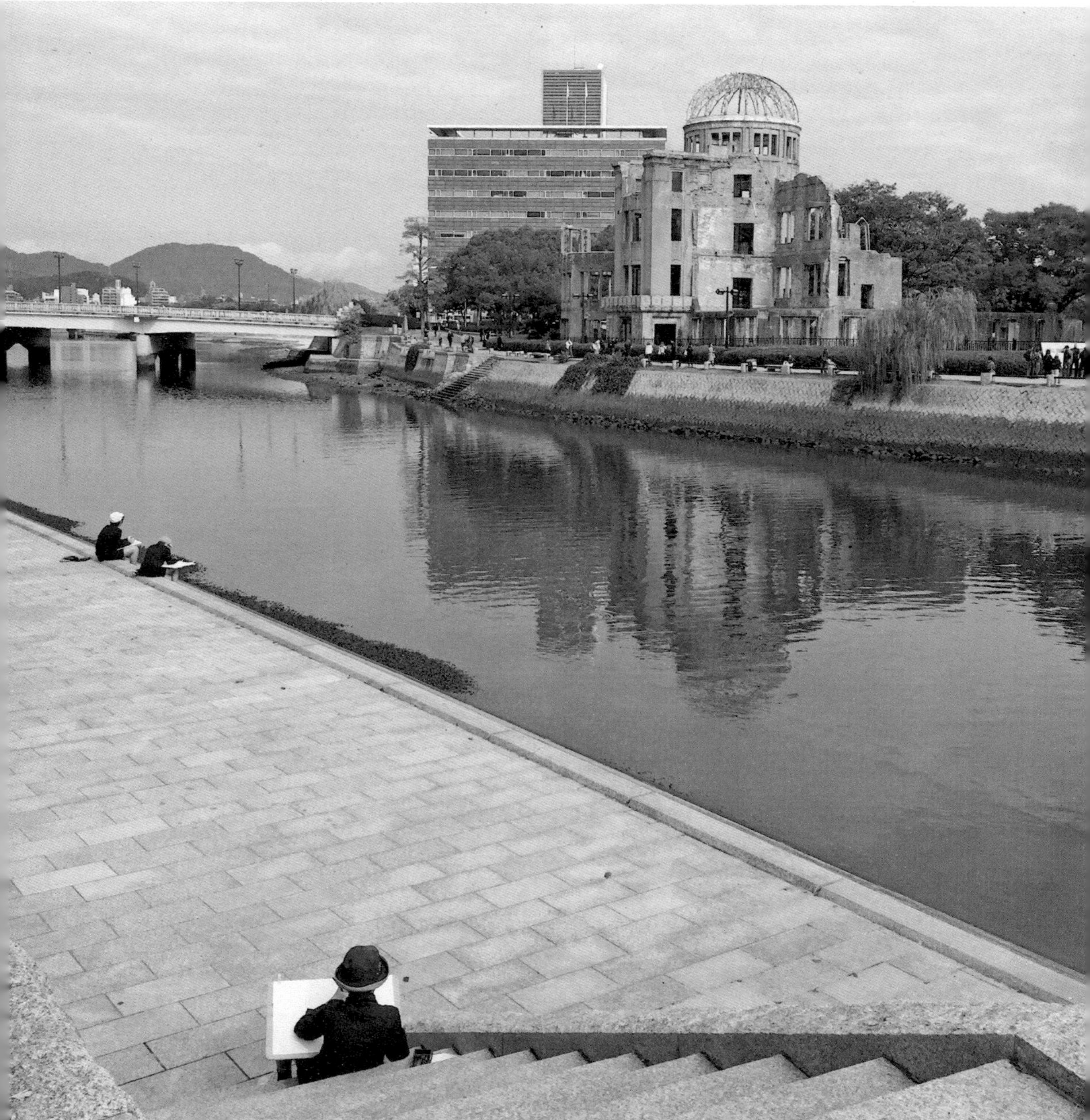

15.11 Schoolchildren on the banks of the river drawing the Atomic Bomb Dome (the ruins of the Industrial Promotion Hall). Peace Park, Hiroshima, 2016.

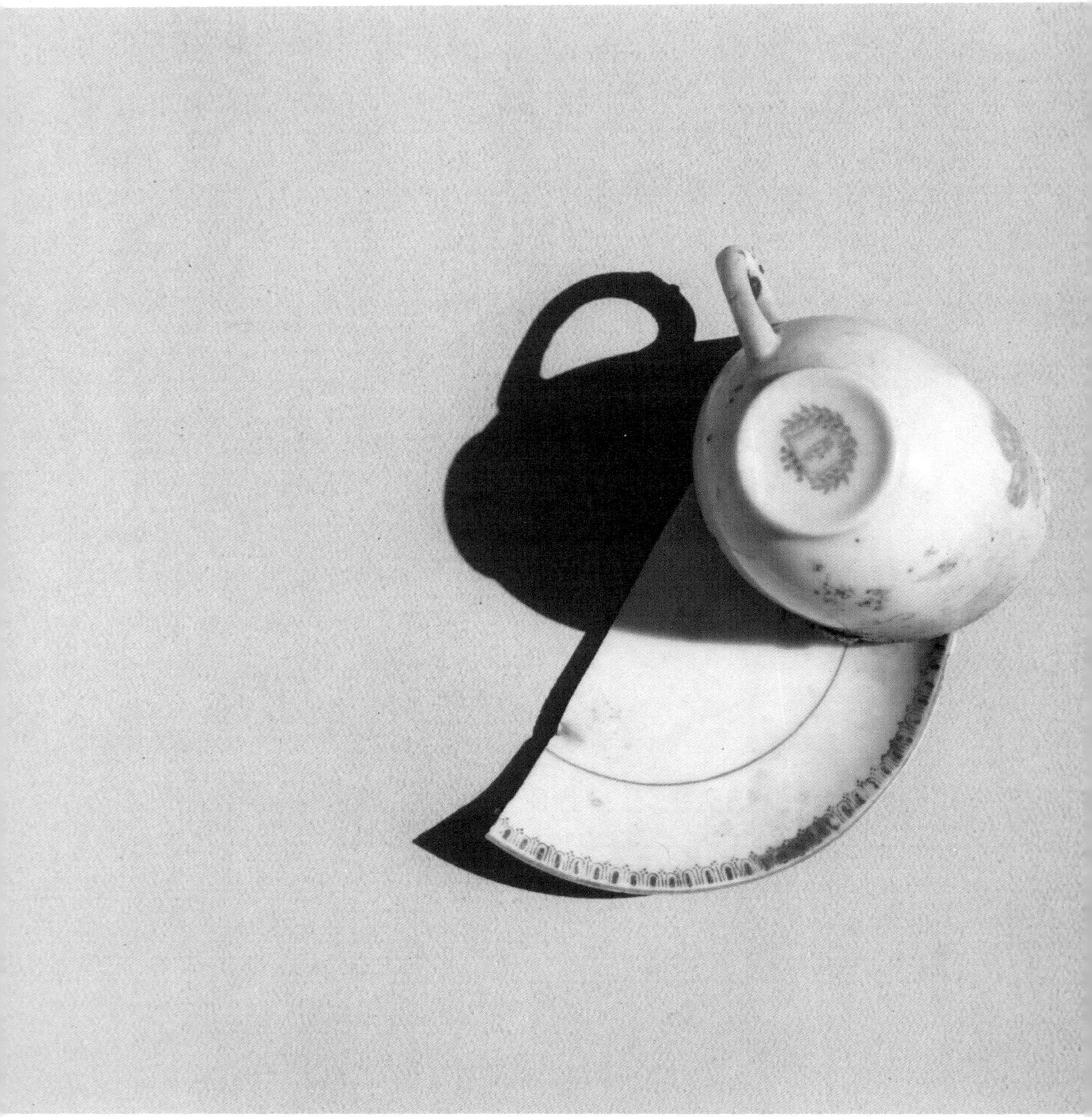

15.12 A teacup and broken saucer fused together as a result of the intense heat of the atomic bomb. Artifact from the Hiroshima Peace Memorial Museum Archive, 2016.

12. FUSED CUP-SAUCER I

In that heat,
under that pressure,
infused with that radiation: the atomic bomb made all things one.
All of those lives, children, mothers, fathers, grandparents, teachers,
    soldiers, lovers, the heartbroken, those who

felt their bodies stretch with power and those who felt age in their bones,
    all… all… all fused into one
remnant.

Cup and saucer.

No individuated purpose, shape. All that makes
life complex. All that makes life
specific. Lost. Melted into one. Survivor.
Hibakusha.

The hibakusha tell us their stories: these are stories of their families, their friends, their homes. But we walk away and we think: hibakusha.

Cup and saucer.

The bomb has damaged so much, and
in time, it took away specific forms, specific shapes.
Trauma.
Horror.
Survivor.

Artifact.

13. FUSED CUP-SAUCER II

Freed now into mind.

A cup pours out tea. A body loses its soul. A
tongue sticks out of a ceramic mouth.

That moment – 8:15 – stepped out of time. We can call it what we want to see now. Victory. War crime. Murder. Scientific wonder. Birth of the Anthropocene.

All of those atoms, all of those lives, all of those homes, all of those tea cups: ours now.
Ours to make into the stories that bound our world, or liberate it. Or…

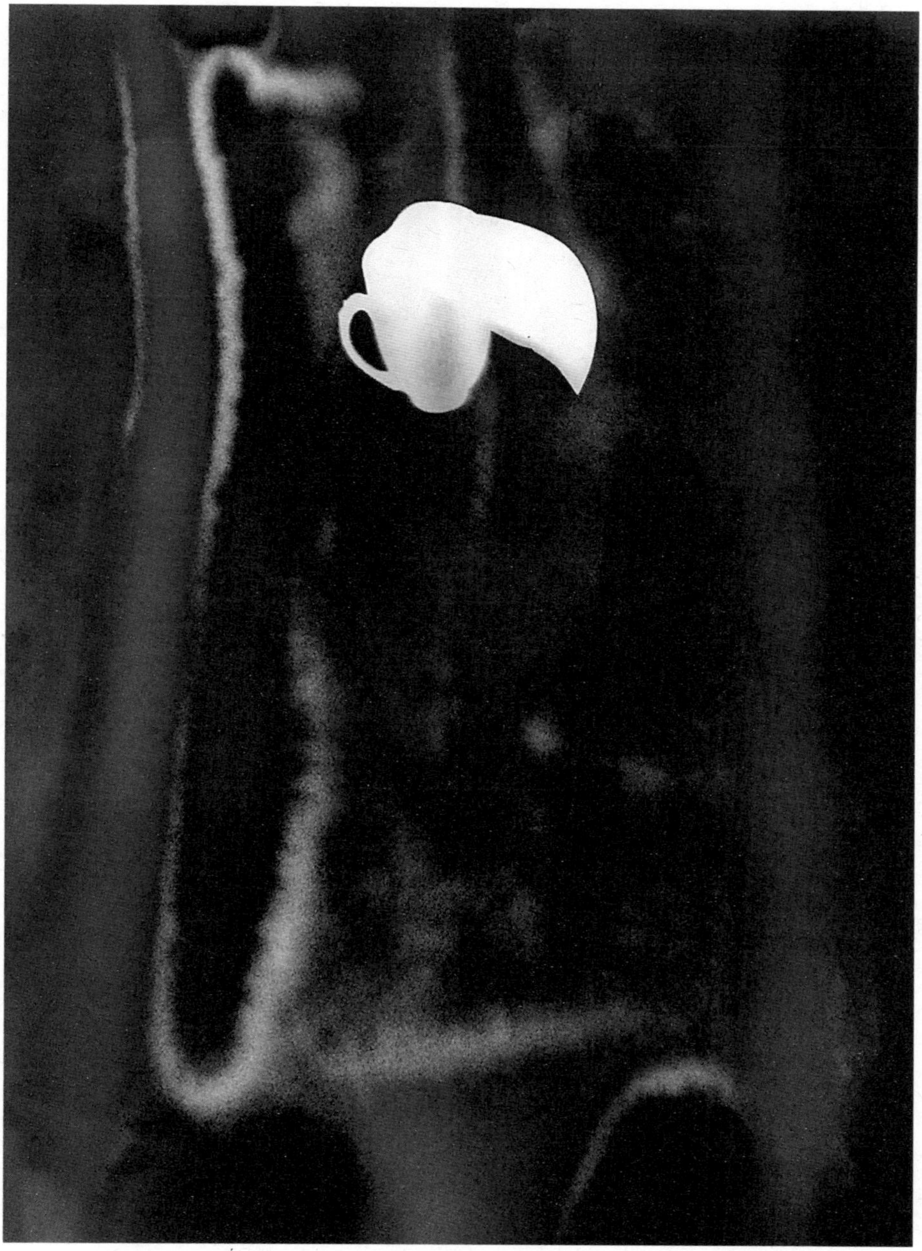

15.13 Cyanotype of a teacup fused to a broken saucer from the heat of the atomic bomb. Artifact from the Hiroshima Peace Memorial Museum Archive, 2016.

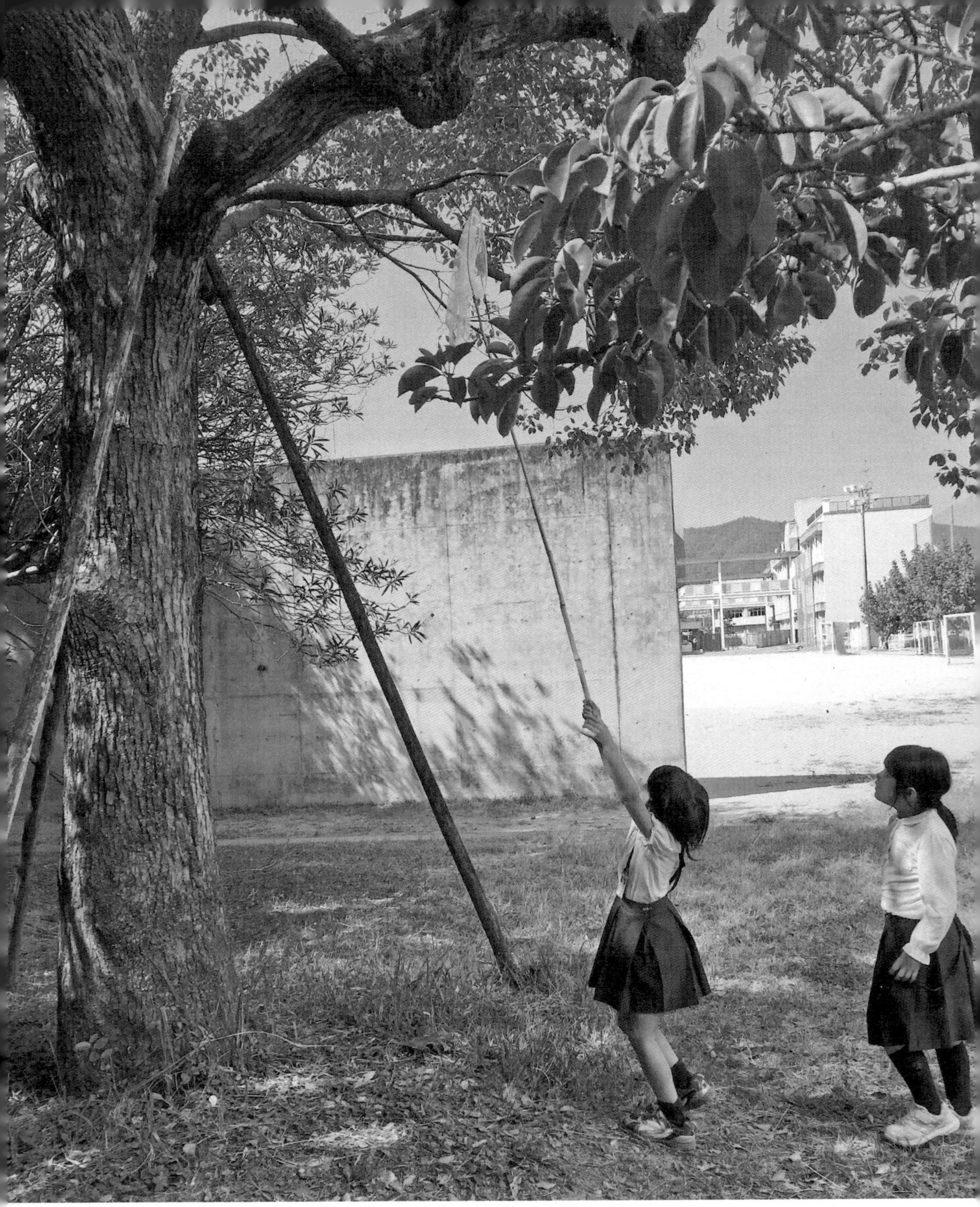

15.14 Schoolchildren trying to catch butterflies under A-bombed trees in the schoolyard, Hiroshima, 2016.

14. GIRLS' TREE

You think you can kill everything but
you can't kill everything.

The world steps out of design. It finds a way to become itself.

Trees burn, still their roots hold to the cold soil; and keep drinking.

People become shadows; but shadows hold
spaces from which the unseen emerges, and becomes
seen.

When a child of Hiroshima plays with butterflies and bugs living in a tree
that
survived a nuclear attack, the world thumbs its nose at death.
This is real power. And it plays. And grows.

15. ROOF TILE FRAGMENTS

Order and
destruction commingle. Like a yukata
pattern burned onto a human
being by heat beyond the Sun, pattern and
horror make the same bed.

Our world is a canvas that
holds the hope and the terror of what
we collect, revealing a single painting where
worlds collide.

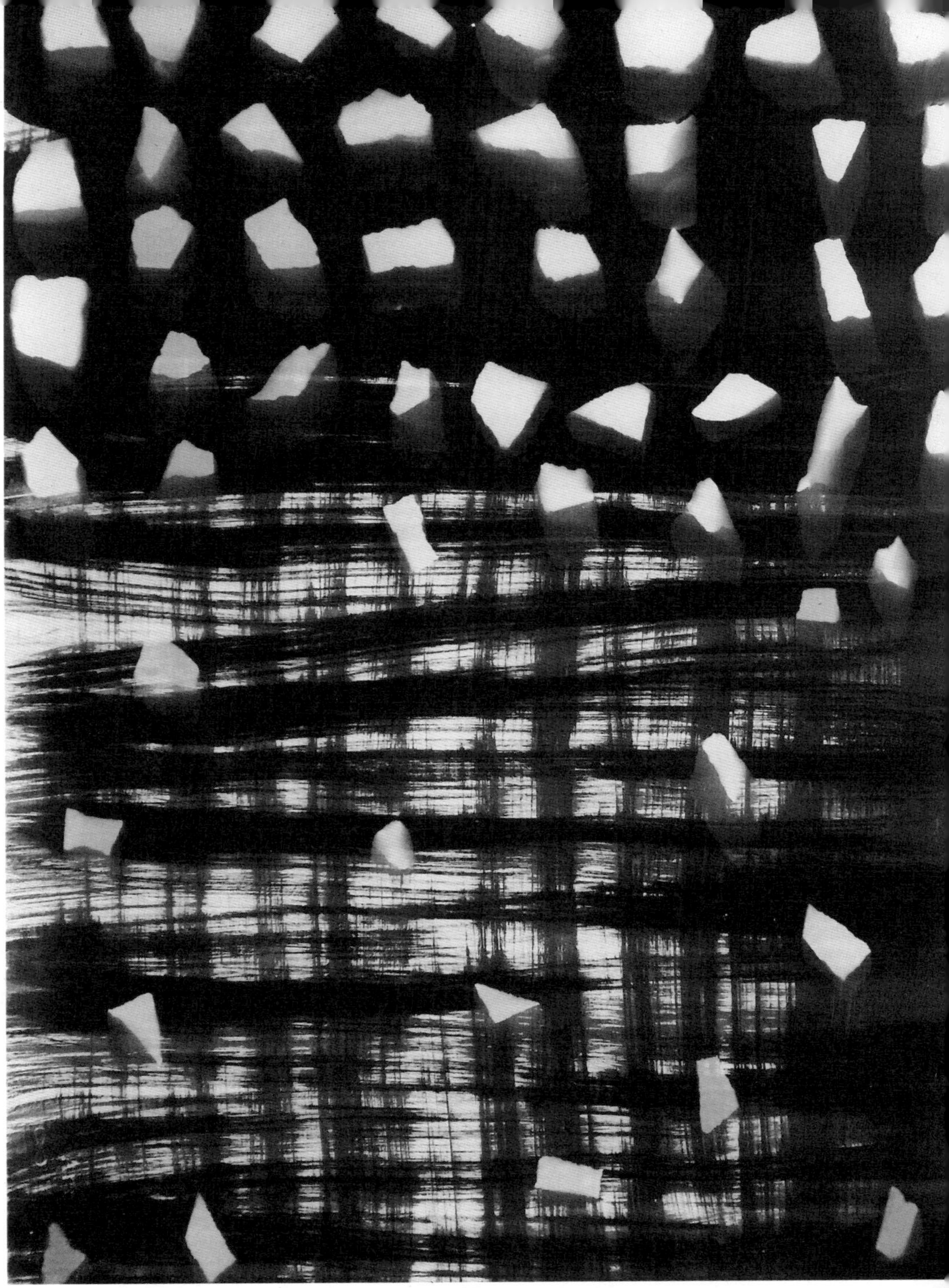

15.15 Cyanotype of broken A-bombed roof tiles, Hiroshima, 2016.

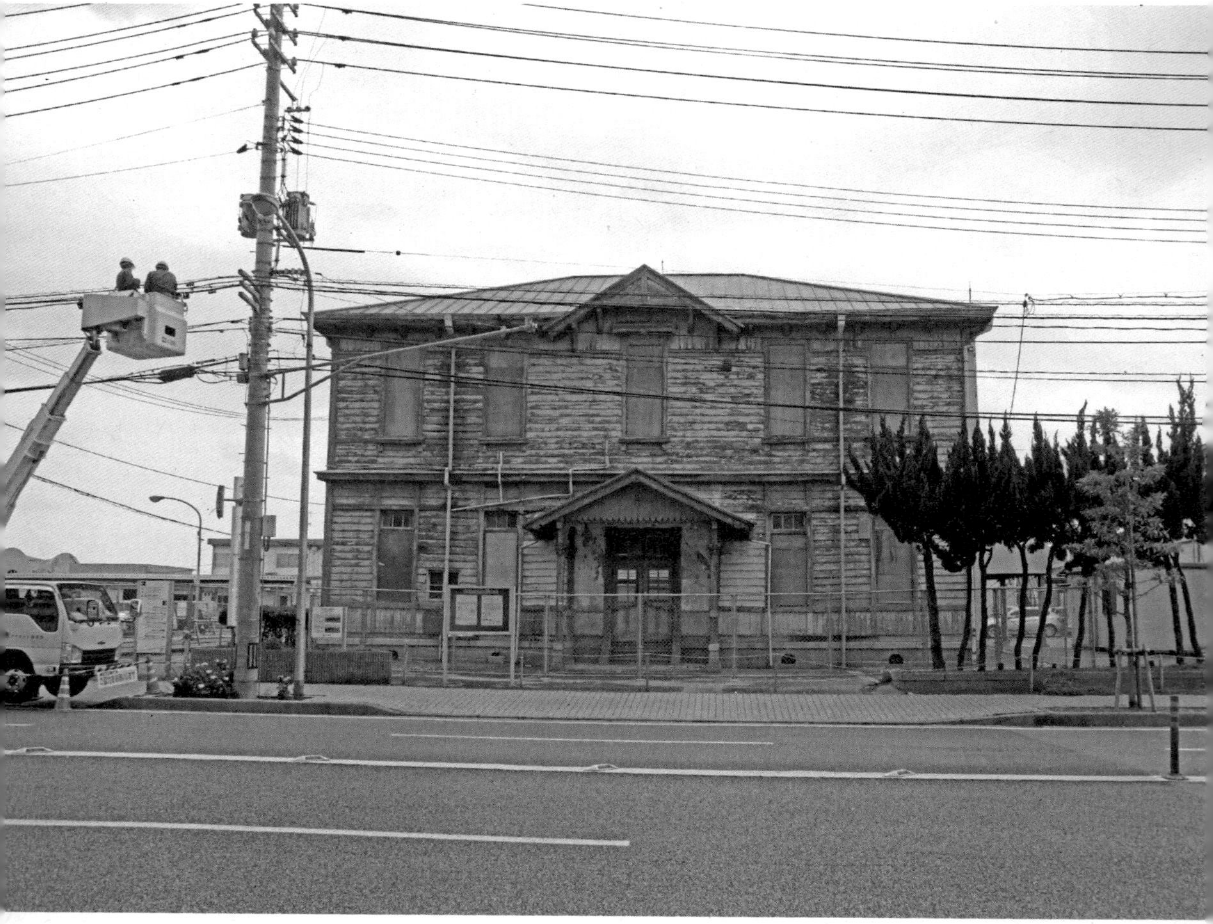

15.16 A-bombed building, previously a police station, Hiroshima, 2016.

16. UJINA

The police have gone home, the wires are being cut.
The crime occurred in a different district.

Time has both preserved this station and pushed it
aside.

Such buildings caught fire in an instant save this one.
Was it investigated?
This police station – located near the military assets of Hiroshima,
but spared the nuclear attack. Why weren't the military assets the target of the
attack?
There should be an investigation.

But no, the wires are being cut. The building is being
fenced away from thought. Nothing to see here. Move on.

Nagasaki Poems

17. NAGASAKI CAMPHOR I

Memories blur; don't
hold their detail. But camphor smells through
time.

Photos of monuments of smoldering places follow us back to today. They pull us into the past.

What had been a shrine to the connection of human beings to
nature becomes a shrine to the
violence
wrought on nature by human beings.

The camphor trees stand facing both truths.

Trees that have watched people come and go, and yet are not
moved.
And now their portrait stands in for the people who were themselves
standing, near. Dissolved. By other people.

Camphor watches. And fills our eyes. And noses. And it watches.

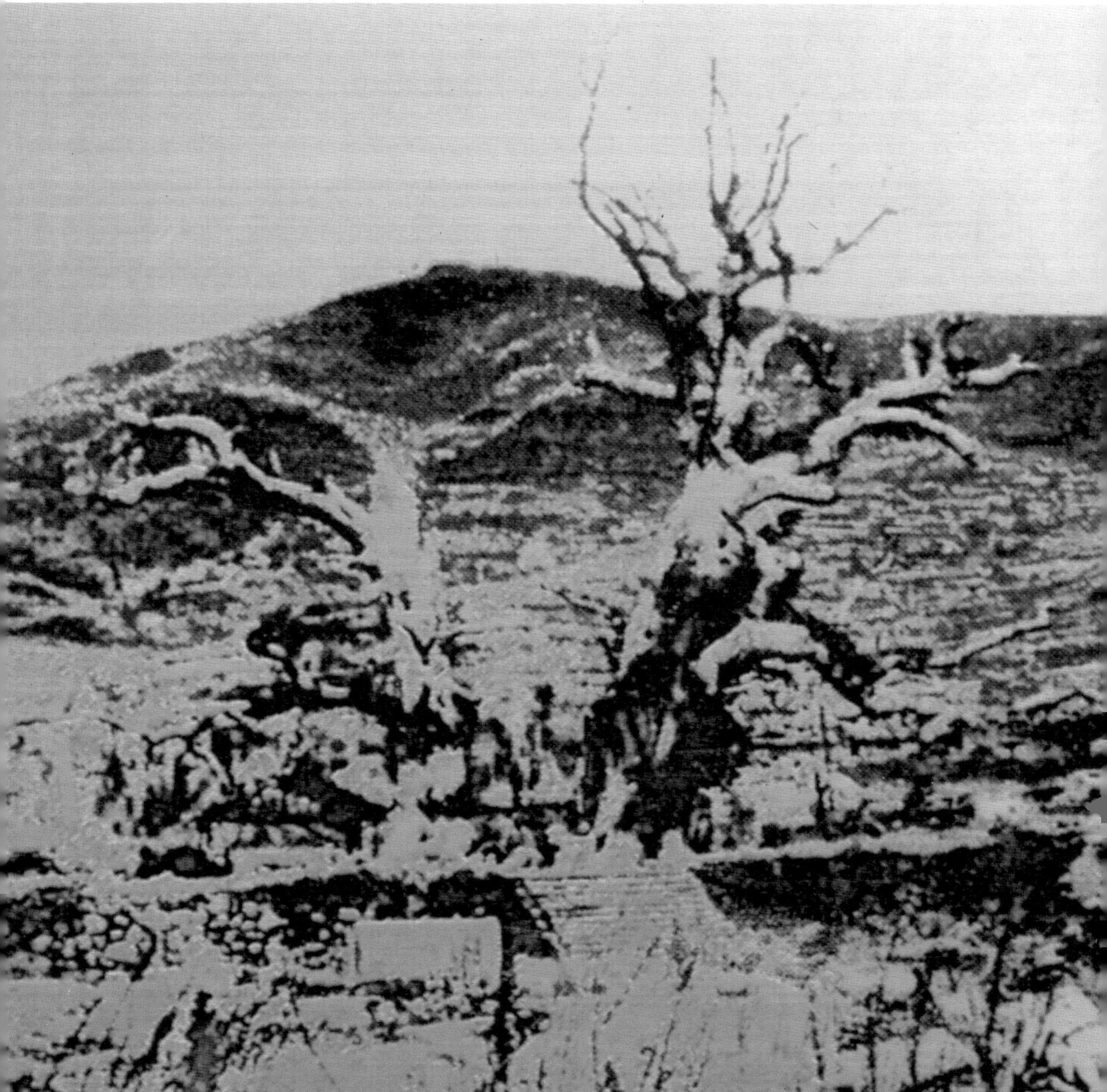

15.17 Historical image on a public marker in Nagasaki of two 500-year-old camphor trees that survived the atomic bombing, 2016.

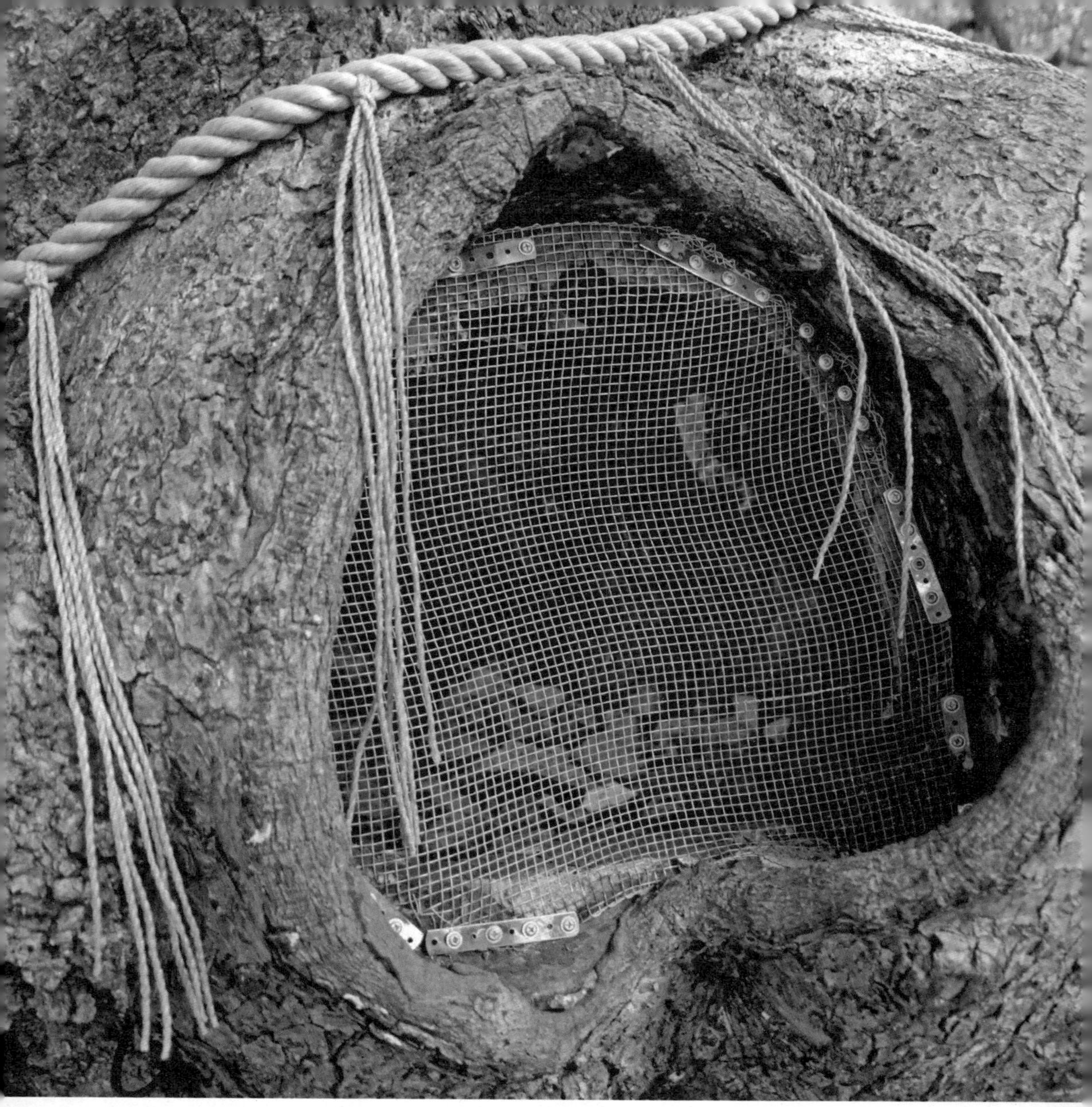

15.18 Protective screen covering a hole in one of the 500-year-old Camphor trees that survived the atomic bombing in Nagasaki. The hole was created from the blast of the bomb, and rocks and debris were flung into the hole; visitors can climb stairs up the side of the tree to look inside, 2016.

18. NAGASAKI CAMPHOR II

I've got a secret said the one-legged tree to the one-legged
Sannō shrine. You cannot
grow another leg, while I can grow over what was destroyed.

Integrate death into another body; put it in my trunk.

Look. Look inside me. Place a shield, screw some
bolts. Measure me with holy rope. Can you see? Can you
see what I saw? What I see
when I look back?

Let's remember what happened
to me, and to you, together.
But don't think I will stop seeing,
or that the *what* was killed in me, and lives again, will
reveal itself behind this screen.

I hold the earth to the sky with my body, and shall always witness you.

19. SPECTACLES

You can see the force that
bent me.
And you can see the
heat that melted me. But you cannot
see what remains.

You see only a remnant.

I am full – not of stars – but of
radiation that was born from something akin
to stars.

It permeates me. It radiates back to you. You
look at the past, but the past sweeps through
your body as you see. Sweeps
back and forth. Making a new you.

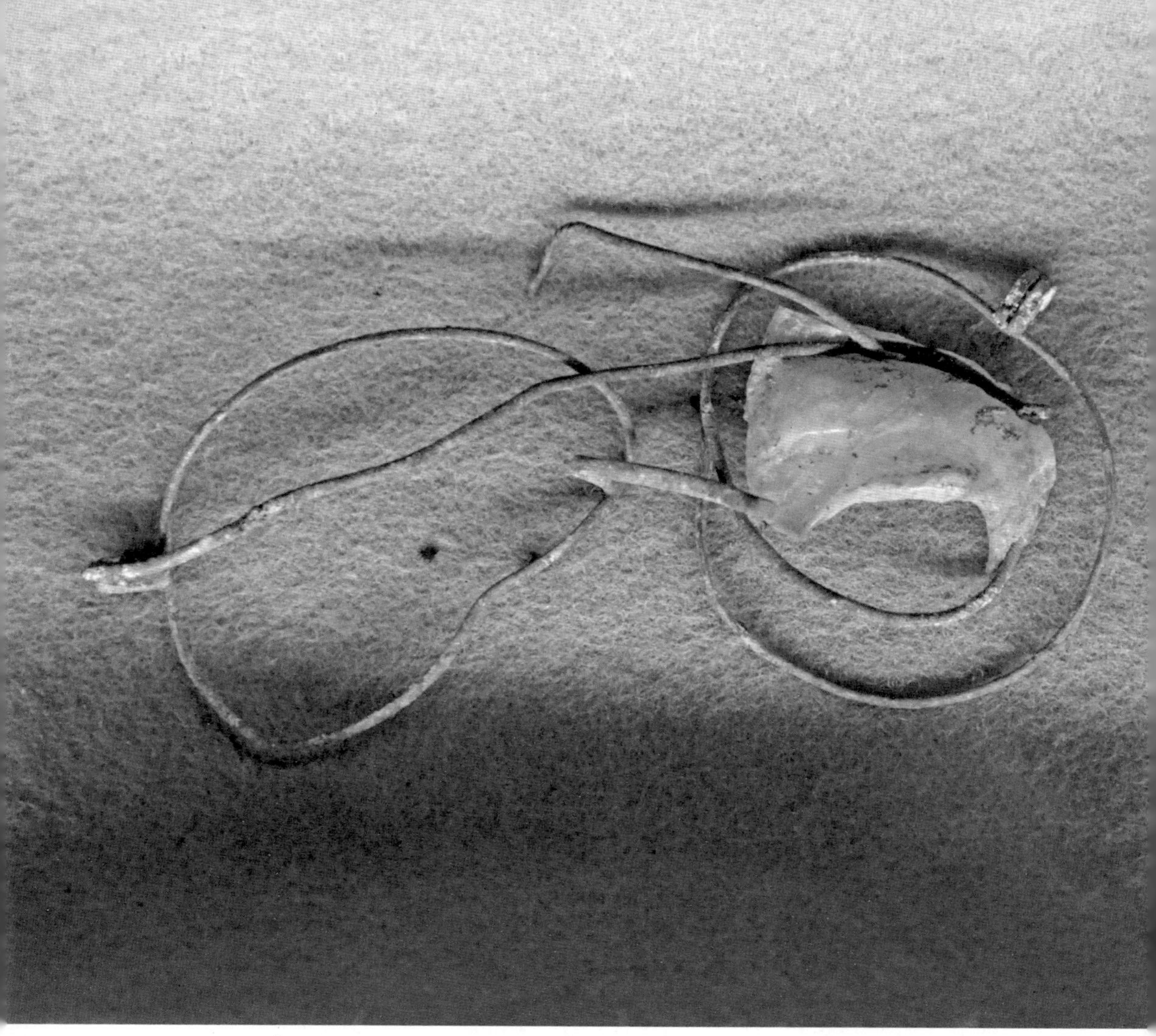

15.19 A-bombed artifact on display in the Nagasaki Atomic Bomb Museum, 2016.

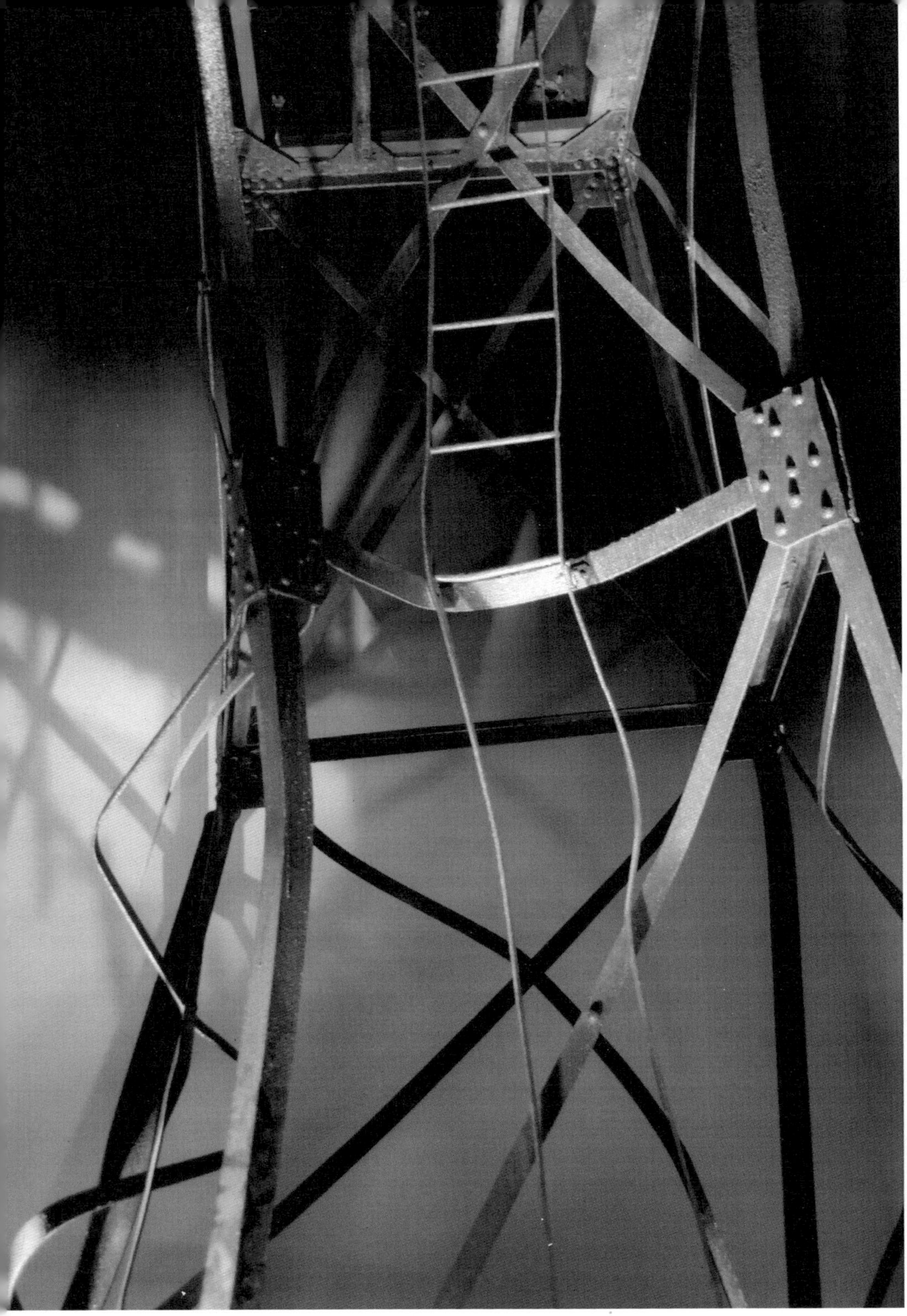

15.20 Steel tower and ladder contorted from the heat and force of the atomic bomb, on exhibit at the Nagasaki Atomic Bomb Museum, 2016.

20. TOWER

Like a movie frame in which the tower stands in for the missing people.
Moved now to a museum and lit for effect.
Godzilla may lurk in the shadows.

When we encounter devastation beyond imagination we unfold
representation.

Now the tower holds up our longing – to know, to connect, to grasp. And here
you stand: in a building that shields and protects
you, witnessing a built structure that could not shield or protect
anyone.
Could not be protected.

A safe space within which to feel the emotions of unopposable threat. Outside
the world may be stable, or it may be sliding towards thermo-nuclear
heat that will bend it, will cripple.

Who will be left to represent what happened to us when we go outside?

16

Nuclear Family: A Poem

CRAIG SANTOS PEREZ

1

In the beginning, the world was indivisible.
Then the sky and ocean separated, and the gods,
Izanagi and Izanami, stood on the bridge of heaven
and stirred the sea with a jeweled spear. After, an island
emerged, and Izanami birthed the first people.
One day, men who claimed to be gods
said: "Let there be light," and there was a blinding
flash – fire radiating for miles. Those who survived,
with seared skin and singed hair, crawled through ashes,
black rain, and the shadow of the mushroom cloud.
Then the men said: "This light will end all wars
and finally bring peace to the divided world."

2

In the beginning, the First World was darkness.
Then the wind birthed Áłtsé Hastiin and Asdzą́ą́,
first man and woman. They ascended into the Second
and Third Worlds until they reached the glittering waters
of this Fourth World. Many monsters dwelled here,
including Leetso, a yellow snake buried underground.
One day, the men who claimed to be gods
said: "Let there be extraction," and a thousand mines

appeared, and the people were tricked into entering
the unventilated earth. "There's no danger," the men said,
as Leetso, enriched and unleashed, slithered across
the atomic cartography of this Fourth World.

3

In the beginning, Lowa spoke the islands into being
and created four gods to protect each cardinal direction.
Then Lowa created a man to gather the islands
into a basket and arrange them into archipelagoes,
and then the first people emerged from Lowa's body.
　One day, the men who claimed to be gods
said: "Let there be experiments," and they removed
the people, ignited the atmosphere, poisoned the fish
and trees, and disappeared the islands. "Your sacrifice
is for the good of mankind," the men said, as they filmed
and photographed the detonations for the world to see.

4

In the beginning, Wolf created earth from mud.
Then he told his younger brother, Coyote,
to carry a woven basket to the Great Basin.
When Coyote opened the basket, the first people
emerged with songs and drums.
　One day, the men who claimed to be gods
said: "Let there be testing," and a thousand craters
formed in the Great Basin. "Your sacrifice is for national
security," they said, as herds of sheep and cattle collapsed,
gardens withered, white dust fell from clouds,
and desert winds whispered across the four corners.

5

In the beginning, there was only ocean.
Then Fu'una, first woman, transformed the eyes
of Puntan, first man, into the sun and moon,
his back into our island, hand; eyebrows into rainbows.
Then her breath blossomed the earth. After, her body
became stone and birthed my ancestors.
One day, the men who claimed to be gods said:
"Let there be exposure," and the trade winds carried
toxins from the islands of Lowa, which rained upon us.
Then the men cleaned their irradiated ships in our lagoon.
"Don't worry," they said, "It's completely safe."

6

In the beginning, there was no contamination.
Then the men who claimed to be gods said,
"Let there be fallout," and our once sacred homes
and bodies became proving grounds, waste dumps,
and tailings. "Let there be fallout," and there was
a chain reaction of death: leukemia and lymphoma,
miscarriages and birth defects, lung and liver cancer,
breast and uterus cancer, thyroid and bone cancer.
"Let there be fallout," and there is no half-life of grief
when a loved one dies from radiation disease.
There is no half-life of sorrow when our homes are lost.
There is no half-life of pain when our children inherit
this toxic legacy, this genetic and generational aftermath,
this fission of worlds. And the men said: "who gives a damn."

7

In the beginning, there were no atomic bombs, no
hydrogen bombs, no thermonuclear bombs, no wars
of light. In the beginning, there was peace. Peace
be with Hiroshima, Nagasaki, and the Marshall Islands.
Peace be with the Navajo and Shoshone Nations.
Peace be with Mororua, Fangataufa, Ekker, Kirimati,
    Maralinga, Montebello Islands, Malden Island,
    Amchitka, Malan, Pokhran, Ras Koh Hills,
    Chagai District, Semipalatinsk, Novaya Zemlya,
    Chernobyl, Malan, Punggye-ri, and Fukushima.
Peace be with the downwinders, from Guam to Utah to
our planet. Peace be with all our radiation ecologies.
Peace be with our entire nuclear family. We deserve truth,
healing, and justice. We deserve just and compassionate
compensation. We demand the cleaning of abandoned mines,
the safe disposal of waste, the disarmament of the violent
nucleus within nations. We demand the non-proliferation
of nuclear weapons and the proliferation of peace
for our children, those sublime angels, who deserve stories
of creation and not stories of annihilation. For our children,
who deserve a legacy of peace that will illuminate
their radiant futures.

Postface

Unmaking the Nuclear Future

JESSICA HURLEY

I write this postface from under a hazy sky in the Northern Virginia suburbs of Washington, DC, Manahoac land, where smoke from the wildfires currently raging on the West Coast has travelled more than 3,000 miles on the jet stream to cast the national capital in luminous grey. I've just heard a new term – *heat dome* – that lands with the weight of words that will become a part of the daily lexicon of our times, like *polar vortex*, like *death is a lagging indicator*. People – majority poor, Black, brown, disabled, and Indigenous people – are dying from heat, from cold, from a capitalogenic pandemic. Last week a billion shore creatures steamed to death in their shells in the Pacific Northwest. Floods around the world are snatching the lives of the unprepared, and there is no way to be prepared. *Death is a lagging indicator.* The lag period is over. The dying is here.

And so we find ourselves at a(nother) turning point in the nuclear age (in the widening gyre there are nothing but turning points). During the Cold War, nuclear weapons were said to be required to prevent the apocalypse of the Other's nuclear weapons, and nuclear power was – more figuratively but no less effectively – said to be the planetary path to "salvation" that would balance out the apocalyptic threat of America's massive nuclear arsenals.[1] As global weather comes unmoored, nuclear power is once again positioned by its boosters to be the path out of the apocalypse, a "clean," "carbon-free" power source that can supposedly offer an immediate alternative to carbon-based fuels.[2] What Bo Jacobs has recently called "nuclear Stockholm syndrome" is obviously and dangerously at play in contemporary turns toward nuclear power; we desperately want there to be an existing technological fix for climate change, and we want very much to believe that the potential and actual ecological havoc wrought by nuclear power is also a mere technological fix away.[3] The fear of the unfolding climate apocalypse makes us vulnerable to these desires, overriding a skepticism based in facts with the desire to believe that we are just one (more) technological fix away from fixing everything.

At this historical moment, then, a *decolonizing pedagogy of the nuclear* is more important than ever. As the global nuclear complex makes its claims for nuclear power as a potential route toward climate justice, and as more and more people – including some environmentalists and ecosocialists – become desperate enough to convince themselves to believe them, the three aspects of nuclearism that Livia Monnet theorizes in the introduction to this volume are vital counter-knowledges that we, as teachers, scholars, artists, and public and private voices, will have an important role in propagating. First, against any claim that nuclear power can produce good enviro-social outcomes, we must remember that "*nuclear science and technologies have always been colonial and colonizing practices* and have remained so to a large extent." The climate emergency is the end point of hundreds of years of colonial genocide and ecocide; nuclear power, too, is a culmination and continuation of these practices. And this continuation is not merely an unfolding historical narrative but a system of power that is inherent to the structures of the global nuclear complex. Monnet's second key term, "the *coloniality of the nuclear*," makes clear that the entire system of nuclearism – "a complex system of interrelated, entangled power relations, epistemologies, technoscientific practices, regulatory or advisory agencies and organizations, psychosocial and affective conditioning and cultural representations through which nuclear weapons and nuclear energy, all nuclear technologies have been legitimated, normalized and institutionalized in modern nuclear states and around the world" – is intrinsically colonial and anti-democratic. In other words, if we turn to nuclear power in the hope of sidestepping the climate apocalypse, we will be keeping in place exactly the same system that brought the climate apocalypse about. True climate justice requires the abolition of the conditions that brought us to this point, a fundamental remaking and decolonization of world systems (including the return of land and sovereignty to Indigenous peoples); nuclear energy can offer only the same colonial, ecocidal systems of degradation with a different power supply.

If nuclear power cannot be anything other than colonial, then we are left with a task that is no less urgent for its impossibility: "the notion of *decolonizing the nuclear*." Decolonizing the nuclear does not mean trying to make this immense, globally colonizing system produce different outcomes of which it is inherently incapable (I haven't yet seen a major nuclear company produce a "decolonizing nuclear power" marketing campaign, but the day cannot be far off). Rather, decolonizing the nuclear involves "*the immanent, reflexive critical activity of documenting and providing evidence of the irrationality, coloniality, (slow) violence and lethality of the nuclear episteme and of nuclear complexes and infrastructures with the aim of bringing about their dismantling and final abolition*." It is a process of research, education, activism, and other forms of counter-world-making that results in making nuclear energy unthinkable as a potential solution to any kind of environmental or social problem, up to and including the unfolding climate apocalypse. We live in the nuclear age; given the near-permanence of man-made nuclear materials, we will never not live in it. But this does not mean that the fight for the decolonization of the nuclear and nuclear abolition are not among the most urgent tasks of our time.

The futurelessness that we face in this moment is not new; in the words of Public Enemy on *Welcome to the Terrordome*, "Armageddon, it been in effect." Colonialism, racial capitalism, misogyny, ableism, extractive economies – each of these ways of casting persons and ecosystems into futurelessness has a long history. As we turn to reconsider the nuclear age, now in this time, here in this volume, its imbrication with the logics and practices of human and non-human disposability that have defined colonial-capitalist modernity has never been clearer. Settler colonial racial capitalism has been the training ground of the nuclear age, and the nuclear age has been the training ground of climate change. Is it any wonder that a social system forged in the eradication of Indigenous worlds and the long genocide of racial slavery and its afterlives, that has spent the past eight decades being or getting OK with a nuclear system designed to produce planetary death, is slow to be moved by the planetary destruction now playing out around us?

A decolonizing pedagogy of the nuclear is important on its own terms; the colonial concatenations of violence and coercion that inhere in the global nuclear complex are a profound ethical wrong that can and must be righted through nothing less than *total and unconditional nuclear abolition*. But the potential benefits of such a pedagogy are not limited to the cessation of nuclear harm. The fight for nuclear abolition would teach us that life – human, non-human, and ecosystemic – is neither fungible nor disposable. It would teach us that Indigenous peoples should be sovereign on their lands and waters.[4] It would teach us that we all have the right to say no to systems, practices, and technologies that harm us. It would teach us that coercion through the threat or practice of violence is an ethical harm at every scale, from the global scale of nuclear geopolitics to the scale of the prison cell.[5] It would teach us that *safety* and *security* can be lived experiences of human and environmental flourishing, rather than necropolitical practices of exclusion and subjugation. It would teach us to unlearn, in other words, the basic principles that are imposed upon us by settler colonial racial capitalism. And most importantly, the fight for nuclear abolition would teach us those things *even if we fail*.

I have written elsewhere of the ethics of impossibility that we are called to practise in times of futurelessness, and it is this ethics that is called upon in the fight for nuclear decolonization and abolition.[6] An ethics of impossibility is one that is oriented not toward possible or probable futures, but toward impossible futures – with specific attention to the Indigenous, Black, brown, disabled, queer, female, and other futures that have been disproportionately foreclosed by the nuclear complex. It's easy to forget, sometimes, that ethics isn't meant to be instrumental. We are so used to thinking in terms of what our actions will change or bring about that we restrict our ethical decision-making to those outcomes that seem possible to achieve. But ethics is a different kind of demand, a call to do what we think is right even in cases where the chance of achieving the outcome that we want approaches zero. I see this ethics of impossibility throughout the pieces collected in this volume, from Livia Monnet's unapologetic call for complete nuclear abolition to the serious playfulness of Shiloh Krupar and Sarah Kanouse's "National TLC Service" and the refusal to abandon human and non-human kin devastated by nuclearization in Craig Santos Perez's poem "Nuclear Family." *A decolonizing pedagogy of the nuclear* is oriented toward an impossible future, but the impossibility of that future is no reason not to fight for it. The planet burns: what future is not impossible? We have left the realm of plausible, probable, possible futures. Only the question remains: what will we fight for, and what worlds will we make in the fighting?

NOTES

1 See Chernus, *Eisenhower's Atoms for Peace.*

2 For a rather shockingly ingenuous recent example from the leftist editor of *Jacobin* that presents nuclear power plants since some nebulous point after Chernobyl as *physically incapable of melting down* (one has to wonder where Fukushima fits into this narrative), see Sunkara, "If We Want to Fight the Climate Crisis." For a comprehensive rebuttal, see Jacobs, "Nuclear Stockholm Syndrome."

3 Jacobs, "Nuclear Stockholm Syndrome."

4 My thinking throughout this paragraph has been inspired by the Indigenous-led grassroots campaign that has led to the successful passage and wide ratification (although not yet by the nuclear weapons states) of the UN Treaty on the Prohibition of Nuclear Weapons.

5 For these scalar connections between different forms of abolition, see Acheson, *Abolition.*

6 See Hurley, *Infrastructures of Apocalypse*, 31–2.

BIBLIOGRAPHY

Acheson, Ray. *Abolition*. Geneva: Women's International League for Peace and Freedom, 2020.

Chernus, Ira. *Eisenhower's Atoms for Peace*. College Station: Texas A&M University Press, 2002.

Hurley, Jessica. *Infrastructures of Apocalypse: American Literature and the Nuclear Complex*. Minneapolis: University of Minnesota Press, 2020.

Jacobs, Robert. "Nuclear Stockholm Syndrome." *CounterPunch.org*, 9 July 2021. https://www.counterpunch.org/2021/07/09/nuclear-stockholm-syndrome.

Sunkara, Bhaskar. "If We Want to Fight the Climate Crisis, We Must Embrace Nuclear Power." *The Guardian*, 21 June 2021. http://www.theguardian.com/commentisfree/2021/jun/21/fight-climate-crisis-clean-energy-nuclear-power.

Contributors

MICK BRODERICK is honorary professor of media and communication at RMIT University and adjunct professor of media and creative arts at Curtin University. He is also the Smithsonian Institution's 2021–22 chair of Aerospace History at the US National Air and Space Museum. Broderick has produced more than one hundred scholarly works, including research monographs, edited books, journal articles, book chapters, curated exhibitions, and digital media productions. His recent books are *Virtual Realities: Case Studies in Immersion and Phenomenology* (2021), *Trauma and Disability in Mad Max: Beyond the Road Warrior's Fury* (2019), *The Kubrick Legacy* (2019), *and Reconstructing Strangelove: Inside Stanley Kubrick's "Nightmare Comedy"* (2017).

SHARAE DECKARD is an associate professor in world literature at University College Dublin. She is the author of *Paradise Discourse, Imperialism and Globalization* (Routledge 2010), co-author with the Warwick Research Collective of *Combined and Uneven Development: Towards a New Theory of World-Literature* (Liverpool UP, 2015), and co-editor of *Marxism, Postcolonial Theory, and the Future of Critique* (Routledge, 2018) and *World Literature, Neoliberalism, and the Culture of Discontent* (Palgrave, 2019). She has published numerous articles on ecogothic, world ecology, and world literature, and has guest-edited special issues of *Green Letters* on "Global and Postcolonial Ecologies" (2012) and the *Irish University Review* on "Food, Energy, Climate: Irish Culture and World-Ecology" (2019). With Upamanyu Pablo Mukherjee, she is academic editor of Palgrave's *New Comparisons in World Literature* series, and is currently editing a new *Routledge Companion to Literature and the Environment* with Treasa De Loughry and Kerstin Oloff.

ANDERS HÖGBERG is a professor of archaeology at Linnaeus University, Sweden, and associated researcher at the Palaeo-Research Institute, University of Johannesburg, South Africa. He is a member of the Centre for Applied Heritage and the UNESCO chair on Heritage Futures at Linnaeus University. He is one of the co-founders of the Graduate School in Contract Archaeology. Since 2015 he has served as a member of the advisory board of the National Historical Museum of Sweden, appointed by the Swedish government. He has broad research interests and is currently working on projects in archaeology, heritage futures, heritage and migration, museum studies, and cognitive evolution. He has published extensively on these topics and others and is the co-editor (with C. Holtorf) of the book *Cultural Heritage and the Future* (Routledge, 2021).

CORNELIUS HOLTORF is a professor of archaeology and holds a UNESCO chair on Heritage Futures at Linnaeus University, Sweden. He is a member of the Centre for Applied Heritage and directs the Graduate School in Contract Archaeology. His current research interests include contemporary and future archaeology, heritage theory, and heritage futures. He has long been collaborating with the Swedish Nuclear Fuel and Waste Management Company (SKB) as well as with New Horizons: One Earth Message and Memory of Mankind initiatives. He was a co-investigator of the "Heritage Futures" research program based at University College London (2015–19) and led the "Memory across Generations" project collaborating with the nuclear waste sector in Sweden (2019–20). He has published extensively on all these topics and is a co-author (with R. Harrison et al.) of *Heritage Futures* (UCL Press, 2020) and the co-editor (with A. Högberg) of *Cultural Heritage and the Future* (Routledge, 2021).

JESSICA HURLEY is assistant professor of English at George Mason University and the author of *Infrastructures of Apocalypse: American Literature and the Nuclear Complex* (University of Minnesota Press, 2020). Her work has appeared in *symplokē*, *Comparative Literature Studies*, *Commonwealth Essays and Studies*, *American Literature*, and *Extrapolation*, among others, and has been recognized by the Don D. Walker Prize in Western Literature, the 1921 Prize in American Literature (honourable mention), and the Jim Hinkle Memorial Prize. In 2018 she co-edited a special issue of *ASAP/Journal* titled *Apocalypse*, and her next editorial project, a joint special issue of *American Literature* and *Resilience* titled *The Infrastructure of Emergency*, is forthcoming in the fall of 2021. She is currently working on her next book project, "Nuclear Decolonizations," as a 2021–22 fellow at the National Humanities Center.

ROBERT (BO) JACOBS is a historian of science and technology at the Hiroshima Peace Institute and Graduate School of Peace Studies at Hiroshima City University. He has published widely on nuclear history and culture, including *The Dragon's Tail: Americans Face the Atomic Age*, and *Filling the Hole in the Nuclear Future: Art and Popular Culture Respond to the Bomb*. He is a founder and a principal researcher of the Global Hibakusha Project, studying radiation-exposed communities around the world. His book, *Nuclear Bodies: The Global Hibakusha*, is published by Yale University Press in 2022.

KARENA KALMBACH is head of the Department of "Strategy and Content" of the Futurium in Berlin. Before taking up this position, she was a tenured assistant professor at Eindhoven University of Technology (TU/e). Her areas of expertise include social and cultural history of technology and the environment (with a particular focus on nuclear history), politics of memory, and social studies of science and technology. She has done extensive research on the question of how national and international nuclear politics have influenced the debate on the health effects of the Chernobyl accident in France and the UK. She has also examined how the commemoration of the accident has been used to underpin political arguments in various European countries. Her latest book on this topic, *The Meanings of a Disaster: Chernobyl and Its Afterlives in Britain and France*, was published in 2021 by Berghahn Books. Upon taking up her position at TU/e, Karena set up an interdisciplinary research group with her TU/e colleagues Andreas Spahn (Philosophy) and Ginevra Sanvitale (Anthropology) in which they investigated the interrelation of fear and technology, focusing on the question: How does fear drive technological innovation?

SARAH KANOUSE is an interdisciplinary artist and critical writer examining the political ecology of landscape and space. Migrating between video, photography, and performative forms, her research-based creative projects shift the visual dimension of the landscape to allow hidden stories of environmental and social transformation to emerge. Her award-winning experimental non-fiction films and collaborative creative work – most notably with Compass and the National Toxic Land/Labor Conservation Service –have been presented through the Haus der Kulturen der Welt, Documenta 13, Cineautopsia, the Museum of Contemporary Art-Chicago, Krannert Art Museum, Cooper Union, Smart Museum, and numerous academic and artist-run venues. Her writings on landscape, ecology, and contemporary art have appeared in *Forty Five, Panorama, Acme, Leonardo, Parallax, Art Journal,* and numerous edited volumes, including the *Routledge Companion to Art, Visual Culture, and Climate Change*; *Ecologies, Agents, Terrains*; *Critical Landscapes, Art Against the Law*; and *Mapping Environmental Issues in the City*. A 2019–20 fellow at the Rachel Carson Center at Ludwig Maximilians Universität, she is associate professor of media arts in the Department of Art + Design at Northeastern University.

SABU KOHSO is a political and social critic, translator, scholar, and long-time activist in the global and anti-capitalist struggle. A native of Okayama, Japan, Sabu has lived in New York City since 1980. He has published several books on urban space and struggle in Japan and Korea and has translated books by Kojin Karatani and David Graeber. His new book, *Radiation and Revolution*, was published by Duke University Press in October 2020, and appeared in French translation from Les Éditions Divergences and Éditions de la Rue Dorion in February 2021.

JIM KRAUS is professor of English at Chaminade University of Honolulu. He is a poet and critic, interested broadly in eco-poetics. He is also a US Navy veteran who trained as a nuclear reactor operator, which is an important source of his interest in nuclearism. He has published work on poets Gary Snyder, Ezra Pound, W.S. Merwin, and Galway Kinnell. Currently, he has a particular focus on how ideas about physics and about technology generally shape the value system that governs how we inhabit the planet Earth. A key problem that animates much of his thinking follows on the idea of Earth's auto-toxicity and of radioactivity as a particular type, an idea he discusses in this volume. His work has been published in *Virginia Quarterly Review, Bamboo Ridge, Pequod, Hawai'i Review, Voices de la* o una, and elsewhere. He is also editor of the *Chaminade Literary Review.*

SHILOH KRUPAR is a geographer and provost's distinguished associate professor at Georgetown University, where she directs the Culture and Politics Program in the School of Foreign Service. Her research examines the biopolitical administration of asymmetrical life, geographies of waste and vulnerability, and bureaucracy. This has included work on decommissioned military landscapes and nuclear natures; environmental and financial disasters; model cities and exhibitionary politics in China; and medical geographies of waste and the political ecologies of hospitals. The recipient of a Quadrant Fellowship, she is author of *Hot Spotter's Report: Military Fables of Toxic Waste* (University of Minnesota Press, 2013), co-author of *Deadly Biocultures: The Ethics of Life-Making* (University of Minnesota Press, 2019), and co-author of the forthcoming volume *Exaction: Territories of Austerity, Bias, and Dross* (SAGE "Society and Space" book series). Krupar has also co-directed the National Toxic Land/Labor Conservation Service and collaborates on *A People's Atlas of Nuclear Colorado.* Her scholarship has appeared in *Theory, Culture, and Society, Society and Space, Public Culture, Radical History Review, Culture, Theory, and Critique, cultural geographies, Liminalities, Journal of Medical Humanities, Configurations,* and *Environmental Humanities.*

THOMAS LAMARRE is a scholar of media, cinema and animation, intellectual history, and material culture. He teaches in the Department of Cinema and Media Studies and in East Asian Languages and Civilizations at the University of Chicago. He is the author of a book on communication networks in ninth-century Japan, *Uncovering Heian Japan: An Archaeology of Sensation and Inscription* (2000), of a monograph on the Japanese modern author Tanizaki Junichiro's writings on cinema, empire, and aesthetics, *Shadows on the Screen: Tanizaki Jun'ichirō on Cinema and Oriental Aesthetics* (2005), and of two books on animation technologies, television infrastructures, and media ecologies, *The Anime Machine: A Media Theory of Animation* (2009) and *The Anime Ecology: A Genealogy of Television, Animation, and Game Media* (2018). Lamarre's current projects include research on animation that addresses the use of animals in the formation of media networks associated with colonialism and extraterritorial empire, and the consequent politics of animism and speciesism. His work as a translator includes major works from Japanese and French: Kawamata Chiaki's novel *Death Sentences* (University of Minnesota, 2012); Muriel Combes's *Gilbert Simondon and the Philosophy of the Transindividual* (MIT, 2012); and David Lapoujade's *William James: Pragmatism and Empiricism* (Duke University Press, 2019). Lamarre taught previously in East Asian Studies and Communications at McGill University, Montreal.

JOSEPH MASCO is professor of anthropology and of the social sciences in the College, University of Chicago. Working at the intersection of science studies, environmental studies, media studies, and critical theory, his scholarship examines the material, affective, and conceptual force of the technological revolution. He is the author of the award-winning *The Nuclear Borderlands: The Manhattan Project in Post-Cold War New Mexico* (Princeton University Press, 2006), a multi-sited ethnographic investigation into the long-term effects of the atomic bomb project in New Mexico. Masco's second book, *The Theater of Operations: National Security Affect from the Cold War to the War on Terror* (Duke University Press, 2014), is a multi-modal (ethnographic, historical, mass media) study of the transformation of the Cold War national security apparatus into a counterterror state after 2001. *The Future of Fallout, and Other Episodes in Radioactive World-Making* (Duke University Press, 2021) tracks the simultaneous production of nuclear emergency and climate disruption since 1945. That book assesses the memory practices, visual culture, concepts of danger, and toxic practices that, in combination, have generated a US national security culture that promises ever more safety and comfort in everyday life but does so only by generating and deferring a vast range of violences into the collective future.

KYVELI MAVROKORDOPOULOU is an art historian and curator, who recently defended her PhD at the École des Hautes Études en Sciences Sociales, Paris, on the subterranean imaginary in contemporary art, especially in the case of nuclear spaces. She was co-editor of *Kunstlicht*'s special issue on *Nuclear Aesthetics*. She was a visiting researcher at Carleton University, Ottawa, and at the Environmental Humanities Center, Vrije Universiteit Amsterdam. Currently, she teaches a seminar on art and nuclearity at the École supérieure d'art du Nord-Pas de Calais/Dunkerque with Agnès Villette, with whom she is working on a series of toxic tours across nuclear infrastructures in France. Mavrokordopoulou is an ARTWORKS curatorial fellow of the SNF Artist Fellowship Program in 2020–21.

LIVIA MONNET is professor of comparative literature, film, and Asian studies at the University of Montreal. She has published extensively on environmental humanities, Asian nuclear and environmental art and media, feminist ecocriticism, feminist and queer media art and film, philosophies of immanence, and Japanese contemporary fiction, film and anime. Her current projects include a book manuscript in progress on environmental art from the Asia-Pacific region. Monnet's recent publications include "A Chaosmos of Condivision: Radiation Aesthetics in the TV Anime *Coppelion* (2013)," in *Ecocriticism in Japan*, edited by Hisaaki Wake, Keijiro Suga, and Yuki Masami (Lexington Books, 2018), "'My Flight Is the Rebellion': History, Ghosting, and Representing in Nalini Malani's Video Installations," in *Nalini Malani: The Rebellion of the Dead/La rivolta dei morti. Retrospective/Retrospettiva, Part II*, edited by Marcella Beccaria (Hatje Cantz, 2018), and a forthcoming essay on the Chinese artist Lu Yang.

DANIEL C. O'NEILL is an associate professor in the Department of East Asian Literatures and Cultures at the University of California, Berkeley. His research interests include modern thought and literature, gender and queer theory, and the intersections of environmental humanities, cinema studies, and media theory. His current book project traces an emergent inter-medial history of the 3.11 disasters. Recent publications include "Rewilding Futures: Japan's Nuclear Exclusion Zone and Post 3.11 Eco-Cinema," in the *Journal of Japanese and Korean Cinema* (April 2019), and "Curating the Creaturely," in *Post-311 Literature* (January 2021, in Japanese).

ELIN O'HARA SLAVICK is an artist-in-residence at the University of California, Irvine. She was a professor of studio, theory, and visual practice at the University of North Carolina, Chapel Hill (1994–2021). slavick has exhibited her work internationally, and her work is held in many collections, including the Queens Museum, the National Library of France, the Library of Congress, and the Art Institute of Chicago. slavick is the author of two monographs – *Bomb After Bomb: A Violent Cartography*, with a foreword by Howard Zinn, and *After Hiroshima*, with an essay by James Elkins – as well as a chapbook of Surrealist poetry, *Cameramouth*. She has held artist residencies in Canada, France, and Japan. Her work has been featured in the *New York Times*, *Los Angeles Times*, *Images Magazine*, FOAM, the *San Francisco Chronicle*, the *Asia-Pacific Journal*, *Photo-Eye*, and *Actuphoto: Actualite Photographique*, among other publications.

New Zealand–Canadian composer **JULIET PALMER** is the artistic director of Urbanvessel, a platform for interdisciplinary collaboration. Recent works: *Choreography of Trauma* (The Element Choir and Continuum); *Every Word Was Once an Animal*, with artist Carla Bengtson (Jordan Schnitzer Museum of Art, Oregon); *Oil & Water* (Detroit Symphony Orchestra); *Ukiyo, floating world* (Urbanvessel and Thin Edge New Music); *rivers*, solo album (Barnyard Records); and the opera *Sweat* (CalArts, Los Angeles; Bicycle Opera, Canada; National Sawdust, New York). Juliet was composer-in-residence at the New Zealand School of Music and Orchestra Wellington (2011/12), and an OAC artist-in-residence at Sunnybrook Research Institute (2018). She is the winner of the Detroit Symphony's Elaine 2018 Lebenbom Award, a Chalmers Arts Fellow (2018–19), and finalist for the Johanna Metcalf Performing Arts Prize (2019). Juliet holds a PhD in composition from Princeton University and an M.Mus. in performance, composition, and time-based art from Auckland University. www.julietpalmer.ca.

CRAIG SANTOS PEREZ, a native CHamoru (Chamorro) from the Pacific Island of Guåhan (Guam), is a poet, scholar, editor, publisher, essayist, critic, book reviewer, artist, environmentalist, and political activist. He is professor in the English Department at the University of Hawai'i, Mānoa, where he teaches creative writing, eco-poetry, and Pacific literature. Craig is the author of two spoken-word poetry albums, *Undercurrent* (2011) and *Crosscurrent* (2017), and five books of poetry: *from unincorporated territory [hacha]*

(2008), *from unincorporated territory [saina]* (2010), *from unincorporated territory [guma']* (2014), *from unincorporated territory [lukao]* (2017), and *Habitat Threshold* (2020). His monograph, *Navigating CHamoru Poetry: Indigeneity, Aesthetics, and Decolonization*, is published by the University of Arizona Press. Craig is also the co-founder of Ala Press (the only publisher in the US dedicated to Pacific literature) and the co-editor of five anthologies of Pacific literature and eco-literature: *Chamoru Childhood* (2008), *Home(is)lands: New Art and Writing from Guahan and Hawai'i* (2018), *Effigies III: Indigenous Pacific Islander Poetry* (2019), *Indigenous Literatures from Micronesia* (2019), and *Geopoetics in Practice* (2020). He has received the Pen Center USA/Poetry Society of America Literary Prize (2011), the American Book Award (2015), the Lannan Foundation Literary Fellowship (2016), the Hawai'i Literary Arts Council Award for an Established Artist (2017), and a gold medal Nautilus Book Award (2021).

THOMAS PATRICK PRINGLE is an assistant professor of environmental communication at Tulane University. He holds a PhD in modern culture and media from Brown University. With Gertrud Koch and Bernard Stiegler, he is the co-author of *Machine* (University of Minnesota Press/Meson Press 2019). His writing on the intersection of media and the environment appears in *Journal of Film and Video* and *NECSUS: European Journal of Media Studies*.

JULIE SALVERSON is professor of drama, Queen's University. Her theatre, books, and essays embrace the relationship of imagination and foolish witness to risky stories and trauma. She is librettist for the atomic opera *Shelter*. Her book *Lines of Flight: An Atomic Memoir* (2016, Wolsak and Wynn) follows her journey tracing uranium from the Northwest Territories to Hiroshima while exploring her lifelong obsession with the atomic bomb. She continues a close collaboration researching atomic culture with Peter C. van Wyck. Recent publications include "Through the Lens of Fukushima," in *Through Post-Atomic Eyes*, edited by Claudette Lauzon and John O'Brian (McGill-Queen's University Press, 2020); "Shameless Acts of Foolish Witness," in *Comedy Begins with Our Simplest Gestures: Levinas, Ethics, and Humor* (Duquesne University Press, 2016), and "Loopings of Love and Rage: Sitting in the Trouble," in *Applied Theatre*, edited by Yasmine Kandil and Barry Freeman.

MAGDALENA STAWKOWSKI is an assistant professor in the Department of Anthropology and faculty associate at the Walker Institute for International Studies at the University of South Carolina. She is also a researcher at the Danish Institute for International Studies (DIIS) in Copenhagen. She specializes in cultural and medical anthropology, focusing on the socio-cultural legacies of Soviet-era nuclear testing in Kazakhstan. Currently, she is collaborating on two major projects. The first, sponsored by DIIS, examines the everyday experiences of security and invisible military ruins of the Anthropocene at three nuclear test sites: the Marshall Islands, French Polynesia, and Kazakhstan. The second project, based at McMaster University in Canada, focuses on state-level physical distancing policies and epidemiological approaches to COVID-19. Her article "I Am a Radioactive Mutant: Emerging Biological Subjectivities at the Semipalatinsk Nuclear Test Site" (*American Ethnologist*, 2016) won the Soyuz Article Prize, which recognizes significant contributions to the advancement of scholarly understandings of post-socialism.

BRYAN C. TAYLOR is professor of communication at the University of Colorado–Boulder and former director of its Peace, Conflict, and Security Program. His research interests include nuclear weapons rhetoric and culture and the role of mimesis in post-9/11 media and security. His related research has been published in journals including *American Literary History, Annals of the International Communication Association, Communication Theory, Critical Studies of Media Communication, Quarterly Journal of Speech*, and *Rhetoric and Public Affairs*. He is co-editor of the volume *Nuclear Legacies: Communication, Controversy, and the U.S. Nuclear Weapons Complex* (Lexington) and *The Handbook of Communication and Security* (Routledge). He is also former president of the Rocky Flats (Colorado) Cold War Museum Project.

N.A.J. TAYLOR is 2021–23 Alfred Deakin Postdoctoral Research Fellow at Deakin University, a 2020–23 visiting fellow at the University of New South Wales-Canberra, and a 2020–22 Killam Postdoctoral Research Fellow at the University of British Columbia. He has held honorary and visiting appointments at Bard College, Sciences Po, Linköping University, La Trobe University, Roskilde University, Whitman College, and The New School, where he was an Australia Awards fellow. Taylor has co-edited three books, *Athens Dialogue on a Middle East Zone Free of Nuclear and Other Weapons of Mass Destruction as Well as their Means of Delivery* (European Public Law Organisation, 2013), *Reimagining Hiroshima and Nagasaki: Nuclear Humanities in the Post-Cold War* (Routledge, 2017), and *Jahnne Pasco-White: Kin* (Art Ink and Unlikely, 2020). He is also the co-editor of three journal special issues, "Reimagining Hiroshima" (*Critical Military Studies*, 2015), "Internal Relations" (*Borderlands*, 2017), and "Reimagining Maralinga" (*Unlikely: Journal for Creative Arts*, 2018). He has also sole-authored more than twenty scholarly articles and chapters (with more co-authored), as well as over one hundred essays and reviews in exhibition catalogues, literary journals, and periodicals.

SEZIN TOPÇU is a historian and sociologist of technology, risk, and social movements. She is a senior researcher at the French National Research Center (CNRS) and a senior lecturer at the Paris School of Advanced Studies in the Social Sciences (Ehess). She is also a full member of the Center for the Study of the Social Movements (CEMS-Ehess). Topçu's work examines transformations of public controversies and collective action (ecological and feminist protest, scientists' movements, counter-expertise, undone science activism) related to technological and medical harm in the period 1960s–2020s, with a specific focus on nuclear and reproductive technologies. She is the author of a monograph (*La France* *nucléaire. L'art de gouverner une technologie contestée*, Seuil, 2013) as well as several journal articles and chapters on nuclear energy. She has recently co-edited, with Laura Centemeri and Peter Burgess, *Rethinking Post-Disaster Recovery: Socio-Anthropological Perspectives on Repairing Environments* (Routledge, 2021).

RUBY DE VOS has just completed her dissertation on the temporalities of living with toxicity in contemporary art and literature. She received MAs in literary studies and cultural analysis from the University of Amsterdam. Together with Kyveli Mavrokordopoulou, she edited a 2018 *Kunstlicht* special issue on nuclear aesthetics. She is also a co-editor of the anthology *Legibility in the Age of Signs and Machines.* Ruby was a visiting scholar at the Posthumanities Hub at the KTH Royal Institute of Technology in Stockholm. In 2019, she curated *After Hiroshima*, a cultural week about the atomic bomb in Groningen.

PETER VAN WYCK is professor of communication and media studies at Concordia University's Department of Communication Studies. His work arises from multidisciplinary training in forestry, ecological sciences, philosophy, and media studies. He has published widely on environmental themes, including deep ecology and nuclear history and culture. His writings include the award-winning *Highway of the Atom* (McGill-Queen's University Press, 2010); a catalogue essay "Reading the Remains" in *Daughters of Uranium* (Southern Alberta Art Gallery, 2020); with Julie Salverson, "The Lens of Fukushima," in *Through Post-Atomic Eyes* (McGill-Queen's University Press, 2020); "The Anthropocene's Signature," in *The Nuclear Culture Source Book* (Black Dog Books, 2017); and "Placing the Anthropocene," in *Place Matters: Critical Topographies in Word and Image* (forthcoming, McGill-Queen's). Current projects include cloud chamber experiments to photograph radioactive decay from various forms of contaminated nuclear materials; an essay on fieldwork as method with Julie Salverson; and a monograph titled *The Angel Turns: Memos for the End of the Holocene* – completing a trilogy of nuclear-themed monographs.

Index

Page numbers in italics indicate references to illustrations.